Linear Algebra

Linear Algebra

Sterling K. Berberian
Prof. Emer., Mathematics
The University of Texas at Austin

Dover Publications, Inc.
Mineola, New York

Bibliographical Note

This Dover edition, first published in 2014, is an unabridged republication of the work originally published in 1992 by Oxford University Press, New York. The author has added a new Errata and Comments section specially for this edition.

Library of Congress Cataloging-in-Publication Data

Berberian, Sterling K., 1926–
Linear algebra / Sterling K. Berberian.—Dover edition.
p. cm.
Originally published: Oxford : Oxford University Press, 1992.
Includes index.
ISBN-13: 978-0-486-78055-9
ISBN-10: 0-486-78055-4
1. Algebras, Linear. I. Title.

QA184.B47 2014
512'.5—dc23

2014010527

Manufactured in the United States by LSC Communications
78055404 2018
www.doverpublications.com

For Jim and Susan

Preface

This book grew out of my experiences in teaching two one-semester courses in linear algebra, the first at the immediate post-calculus level, the second at the upper-undergraduate level. The latter course takes up general determinants and standard forms for matrices, and thus requires some familiarity with permutation groups and the factorization of polynomials into irreducible polynomials; this material is normally covered in a one-semester abstract algebra course taken between the two linear algebra courses.

Part 1 of the book (Chapters 1–6) mirrors the first of the above-mentioned linear algebra courses, Part 2 the second (Chapters 7–13); the information on factorization needed from the transitional abstract algebra course is thoroughly reviewed in an Appendix.

The underlying plan of Part 1 is simple: first things first. For the benefit of the reader with little or no experience in formal mathematics (axioms, theorem-proving), the proofs in Part 1 are especially detailed, with frequent comments on the logical strategy of the proof. The more experienced reader can simply skip over superfluous explanations, but the theorems themselves are cast in the form needed for the more advanced chapters of the book; apart from Chapter 6 (an elementary treatment of 2×2 and 3×3 determinants), nothing in Part 1 has to be redone for Part 2.

Part 2 goes deeper, addressing topics that are more demanding, even difficult: general determinant theory, similarity and canonical forms for matrices, spectral theory in real and complex inner product spaces, tensor products. {Not to worry: in mathematics, 'difficult' means only that it takes more time to make it easy.}

Courses with different emphases can be based on various combinations of chapters:

(A) An introductory course that serves also as an initiation into formal mathematics: Chapter 1–6.

(B) An advanced course, for students who have met matrices and linear mappings before (on an informal, relatively proofless level), and have had a theorem-proving type course in elementary abstract algebra (groups and rings): a rapid tour of Chapters 1–5, followed by a selection of chapters from Part 2 tailored to the needs of the class. The selection can be tilted towards algebra or towards analysis/geometry, as indicated in the flow chart following this preface.

It is generally agreed that a course in linear algebra should begin with a discussion of examples; I concur wholeheartedly (Chapter 1, §§1 and 2). Now comes the hard decision: which to take up first, (i) linear equations and matrices, or (ii) vector spaces and linear mappings? I have chosen the latter course, partly on grounds of efficiency, partly because matrices are addictive and linear mappings are not. My experience in introductory courses is that once a class has tasted the joy of matrix computation, it is hard to get anyone to focus on something so austere as a linear mapping on a vector space. Eventually (from Chapter 4 onward) the reader will, I believe, perceive the true relation between linear mappings and matrices to be symbiotic: each is indispensable for an understanding of the other. It is equally a joy to see computational aspects (matrices, determinants) fall out almost effortlessly when the proper conceptual foundation has been laid (linear mappings).

A word about the role of Appendix A ('Foundations'). The book starts right off with vectors (I usually cover Section 1 of Chapter 1 on Day 1), but before taking up Section 2 it is a good idea to go over briefly Appendix A.2 on set notations. Similarly, before undertaking Chapter 2 (linear mappings), a brief discussion of Appendix A.3 on functions is advisable. Appendix A.1 (an informal discussion of the logical organization of proofs) is cited in the text wherever it can heighten our understanding of what's going on in a proof. In short, Appendix A is mainly for reference; what it contains is important and needs to be talked about, often and in little bits, but not for a whole hour at a stretch.

The wide appeal of linear algebra lies in its importance for other branches of mathematics and its adaptability to concrete problems in mathematical sciences. Explicit applications are not abundant in the book but they are not entirely neglected, and applicability is an ever-present consideration in the choice of topics. For example, Hilbert space operator theory and the representation of groups by matrices (the applications with which I am most familiar) are not taken up explicitly in the text, but the chapters on inner product spaces consciously prepare the way for Hilbert space and the reader who studies group representations will find the chapters on similarity and tensor products helpful. Systems of linear equations and the reduction of matrices to standard forms are applications that belong to everyone; they are treated thoroughly. For the class that has the leisure to take them up, the applications to analytic geometry in Chapter 6 are instructive and rewarding.

In brief, linear algebra is a feast for all tastes. *Bon appétit!*

Sterling Berberian

Austin, Texas
August 1990

Flow chart of chapters

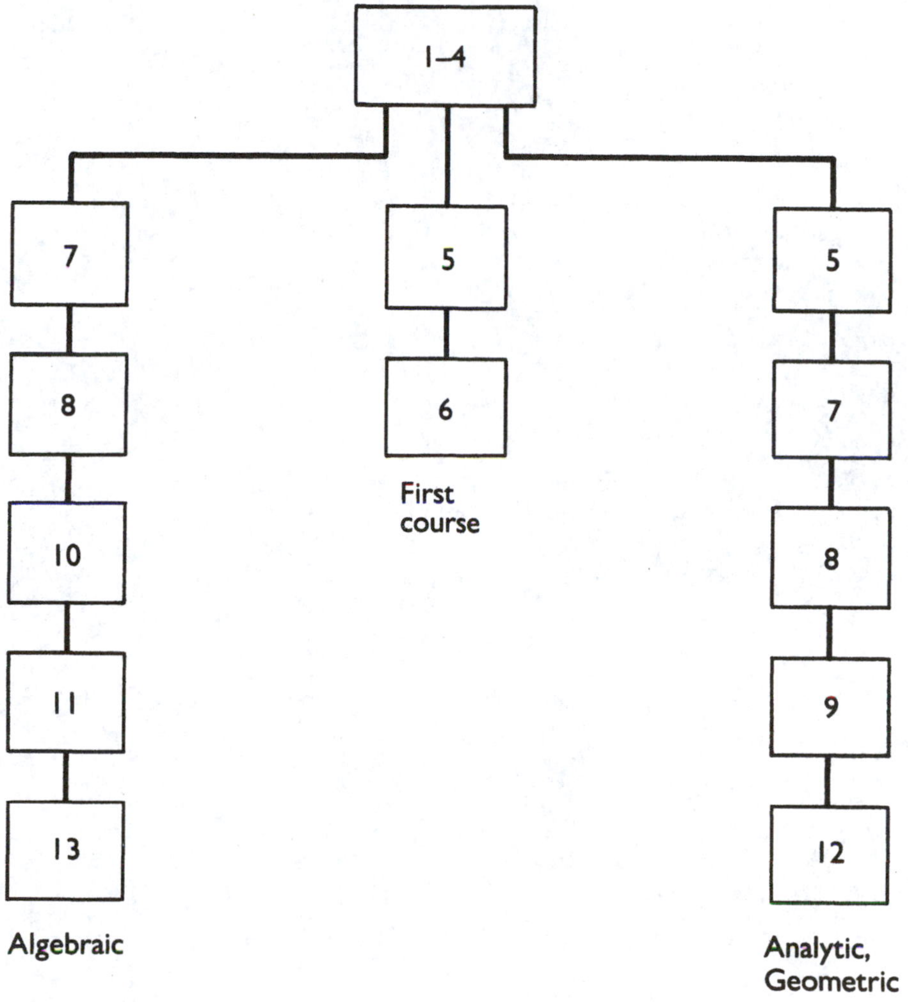

Contents

Part one

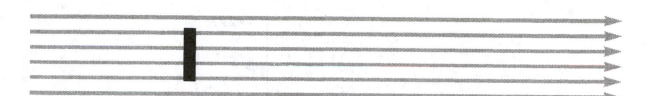

1

Vector spaces

Fig. 1

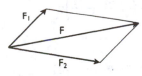

Fig. 2

An arrow has direction and magnitude (its length) (Fig. 1). Quantities with magnitude and direction are familiar from physics (for example, force and velocity); such quantities are called *vectorial*, and it's often useful to represent them by arrows—for instance, when determining the net effect F of two forces F_1 and F_2 acting on a point (Fig. 2). (Viewing F_1 and F_2 as adjacent sides of a parallelogram, F is the diagonal they include.)

That's an intuitive glimpse of vectors. The way mathematicians look at them (as a part of 'abstract algebra') leads to the idea of a 'vector space'.[1] In between this austere view and the intuitive idea, the example of 'geometric' vectors in 'ordinary 3-space' is fun and instructive; let's look at it.

1.1 Motivation (vectors in 3-space)

We assume the points of 3-space to have coordinates in the usual cartesian way: pick an origin, choose three mutually perpendicular axes, and assign coordinates to a point by measuring signed distances along the chosen axes (Fig. 3). If P_1 and P_2 are points of 3-space, Fig. 4 shows an arrow with initial point P_1 and final point P_2. The progress from P_1 to P_2 is indicated by measuring the change in each of the coordinates; thus, if $P_1(x_1, y_1, z_1)$ and

[1] Single quotes around an expression mean that no formal definition of the expression is proposed (at least, not yet). Just take it informally (don't insist on understanding it) and read on . . . .

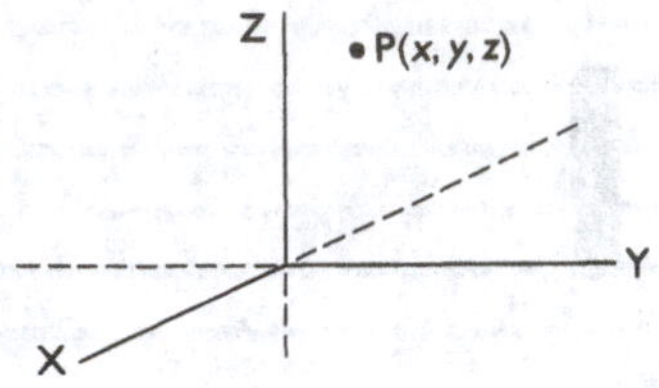

Fig. 3

Fig. 4

$P_2(x_2, y_2, z_2)$ are the coordinates, then the components of change in the coordinate directions are the numbers

$$x_2 - x_1, y_2 - y_1, z_2 - z_1.$$

One organizes this data (an ordered triple of real numbers) into the symbol

$$\overrightarrow{P_1P_2} = [x_2 - x_1, y_2 - y_1, z_2 - z_1].$$

(Note the order of subtraction: 'the twos minus the ones'.) For example,

$$P_1(2, -3, 0), P_2(1, 2, -5), \overrightarrow{P_1P_2} = [-1, 5, -5].$$

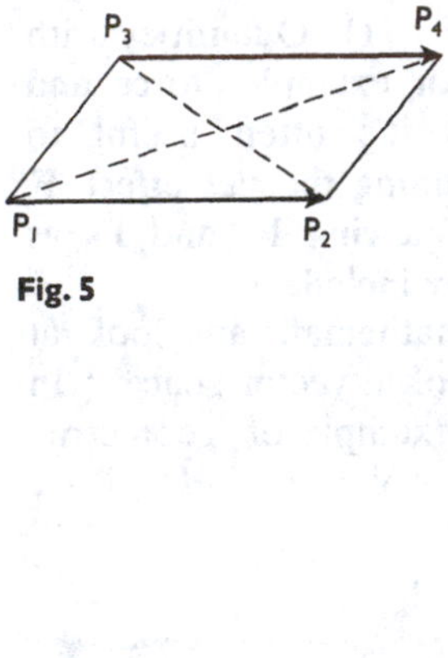

Fig. 5

For physical reasons, two arrows are thought of as representing 'the same vector' if they point in the same direction and have the same length (think of a vector as a force eligible to be applied at *any* point of the space). Thus (Fig. 5) the arrow from P_1 to P_2 represents the same vector as the arrow from P_3 to P_4 if and only if the figure $P_1P_2P_4P_3$ is a parallelogram (in particular, lies in a plane!). This will happen if and only if the midpoint of the segment P_1P_4 coincides with the midpoint of the segment P_2P_3, that is,

$$\left(\frac{x_1 + x_4}{2}, \frac{y_1 + y_4}{2}, \frac{z_1 + z_4}{2}\right) = \left(\frac{x_2 + x_3}{2}, \frac{y_2 + y_3}{2}, \frac{z_2 + z_3}{2}\right);$$

comparing coordinates, we see that this means

$$x_2 - x_1 = x_4 - x_3, \quad y_2 - y_1 = y_4 - y_3, \quad z_2 - z_1 = z_4 - z_3;$$

in other words,

$$[x_2 - x_1, y_2 - y_1, z_2 - z_1] = [x_4 - x_3, y_4 - y_3, z_4 - z_3].$$

To summarize, the arrow from P_1 to P_2 represents the same vector as the arrow from P_3 to P_4 if and only if

$$\overrightarrow{P_1P_2} = \overrightarrow{P_3P_4}.$$

It is thus appropriate to call $\overrightarrow{P_1P_2}$ *the vector determined by the arrow from* P_1 *to* P_2. The numbers

$$x_2 - x_1, y_2 - y_1, z_2 - z_1$$

are called the *components* of the vector.

Every ordered triple $u = [a, b, c]$ can be written in the form $\overrightarrow{P_1P_2}$ (in many ways!); for instance, if $P_1 = (2, 3, 5)$ and $P_2 = (2 + a, 3 + b, 5 + c)$ then $u = \overrightarrow{P_1P_2}$. Accordingly, we propose to call such triples *vectors*.

Real numbers will also be called *scalars*, to dramatize their difference from vectors (ordered triples of real numbers). {In physics, quantities having only magnitude are called 'scalar' quantities.}

We remarked earlier that the net effect (or 'resultant') of two forces can be calculated by a 'parallelogram rule'. This suggests the following question: Given a parallelogram $P_1P_2P_4P_3$ in 3-space, how can the components of $\overrightarrow{P_1P_4}$ be calculated from those of $\overrightarrow{P_1P_3}$ and $\overrightarrow{P_1P_2}$? Assuming coordinates $P_1(x_1, y_1, z_1)$, etc., by definition

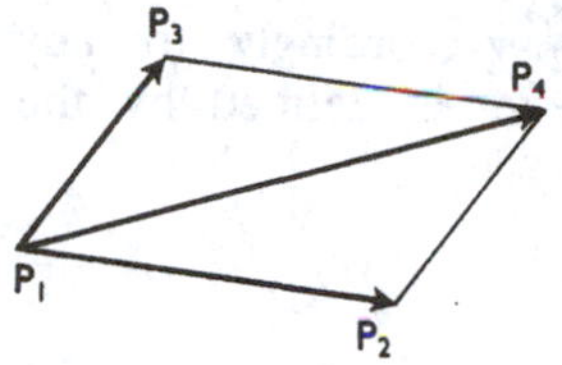

Fig. 6

$$\overrightarrow{P_1P_2} = [x_2 - x_1, y_2 - y_1, z_2 - z_1],$$
$$\overrightarrow{P_1P_3} = [x_3 - x_1, y_3 - y_1, z_3 - z_1],$$
$$\overrightarrow{P_1P_4} = [x_4 - x_1, y_4 - y_1, z_4 - z_1].$$

Since $P_1P_2P_4P_3$ is a parallelogram, we have $\overrightarrow{P_1P_2} = \overrightarrow{P_3P_4}$, which means

$$x_2 - x_1 = x_4 - x_3, \quad y_2 - y_1 = y_4 - y_3, \quad z_2 - z_1 = z_4 - z_3,$$

in other words,

$$x_4 = x_2 + x_3 - x_1, \quad y_4 = y_2 + y_3 - y_1, \quad z_4 = z_2 + z_3 - z_1,$$

that is,

$$x_4 - x_1 = x_2 + x_3 - 2x_1 = (x_2 - x_1) + (x_3 - x_1),$$
$$y_4 - y_1 = y_2 + y_3 - 2y_1 = (y_2 - y_1) + (y_3 - y_1),$$
$$z_4 - z_1 = z_2 + z_3 - 2z_1 = (z_2 - z_1) + (z_3 - z_1).$$

This shows that *each component of* $\overrightarrow{P_1P_4}$ *is the sum of the corresponding components of* $\overrightarrow{P_1P_2}$ *and* $\overrightarrow{P_1P_3}$. For this reason, the resultant vector $\overrightarrow{P_1P_4}$ is also called the 'sum' of the vectors $\overrightarrow{P_1P_2}$ and $\overrightarrow{P_1P_3}$.

In general, the **sum** of two vectors $u = [a, b, c]$, $u' = [a', b', c']$ is defined by the formula

$$u + u' = [a + a', b + b', c + c'];$$

so to speak, the vectors are added 'componentwise'. The message of the preceding paragraph is that

$$\overrightarrow{P_1P_4} = \overrightarrow{P_1P_2} + \overrightarrow{P_1P_3}$$

(provided that $P_1P_2P_4P_3$ is a parallelogram!).

The magnitude of the vector $\overrightarrow{P_1P_2}$ is represented by its length, which is the distance $|P_1P_2|$ from P_1 to P_2:

$$|P_1P_2| = [(x_2 - x_1)^2 + (y_2 - y_1)^2 + (z_2 - z_1)^2]^{\frac{1}{2}}.$$

This suggests defining length for any vector $u = [a, b, c]$; it is usually called the *norm* of u, written $\|u\|$, and it is defined by the formula

$$\|u\| = (a^2 + b^2 + c^2)^{\frac{1}{2}}.$$

In particular,

$$\|\overrightarrow{P_1P_2}\| = |P_1P_2|.$$

The vector $\overrightarrow{P_2P_1}$ has the same magnitude as $\overrightarrow{P_1P_2}$ but points in the opposite direction; since

$$\begin{aligned}\overrightarrow{P_2P_1} &= [x_1 - x_2,\ y_1 - y_2,\ z_1 - z_2] \\ &= [-(x_2 - x_1),\ -(y_2 - y_1),\ -(z_2 - z_1)],\end{aligned}$$

it is appropriate to call $\overrightarrow{P_2P_1}$ the 'negative' of $\overrightarrow{P_1P_2}$. Accordingly, for any vector $u = [a, b, c]$, the *negative* of u, written $-u$, is defined by the formula

$$-u = [-a, -b, -c].$$

In particular,

$$\overrightarrow{P_2P_1} = -\overrightarrow{P_1P_2}.$$

When $P_1 = P_2$, the components of $\overrightarrow{P_1P_2}$ are all 0. The vector $[0, 0, 0]$ is called the *zero vector* and will be denoted θ (to distinguish it from the number 0). Thus

$$\theta = [0, 0, 0] = \overrightarrow{PP}$$

for every point P. Also,

$$u + (-u) = \theta$$

for every vector u; in particular,

$$\overrightarrow{P_1P_2} + \overrightarrow{P_2P_1} = \theta$$

for every pair of points P_1, P_2 of 3-space.

The negative $-u$ of a vector u is obtained by 'multiplying' the vector u by -1 (in the sense that each component of u is multiplied by -1); the effect is to reverse the direction of u, while conserving its length. More generally, any vector $u = [a, b, c]$ can be multiplied by any scalar r: one defines

$$ru = [ra, rb, rc],$$

called the *scalar multiple* of u by r. In particular, $-u = (-1)u$. Since

$$\begin{aligned}||ru||^2 &= (ra)^2 + (rb)^2 + (rc)^2 \\ &= r^2(a^2 + b^2 + c^2) = |r|^2||u||^2,\end{aligned}$$

we have

$$||ru|| = |r| \cdot ||u||.$$

In passing from u to ru, length is multiplied by a factor of $|r|$. If $r = 0$ then $ru = \theta$; if $r < 0$ then (guided by the discussion of vectors $\overrightarrow{PQ}$) we say that ru has direction opposite to that of u, and if $r > 0$ we say that ru has the same direction as u. (However, if $u = \theta$ then $ru = \theta$ and statements about 'direction' are without content.)

To summarize, our study of arrows in 3-space leads us to consider the set of all ordered[2] triples $u = [a, b, c]$, $u' = [a', b', c'], \ldots$ of real numbers; we call such triples *vectors* (a, b, c being the *components* of u). There are three natural operations that we perform on vectors:

$$u + u' = [a + a', b + b', c + c'],$$
$$ru = [ra, rb, rc],$$
$$\|u\| = (a^2 + b^2 + c^2)^{\frac{1}{2}},$$

which we call the *sum* (of the vectors u and u'), *scalar multiple* (of the vector u by the real number r) and *norm* (of the vector u). Each of these operations has a natural geometric (or physical) interpretation when the vectors are thought of as arising from arrows (or, say, forces).

In this book, we undertake a systematic study of sets, called 'vector spaces', in which operations $u + u'$ and ru are defined, satisfying appropriate laws (for instance, $u + u' = u' + u$). We shall see in the following two sections that there is an immense variety of examples of such spaces, thus an immense economy in developing their common properties simultaneously. The appropriate strategy is to set down a list of axioms[3] for a 'vector space', meaningful and verifiable for the examples we want to encompass, and to study the logical consequences of the axioms; what we learn will then be applicable to *all* of the relevant examples.

When a norm $\|u\|$ for vectors u is also defined, one speaks of a 'normed vector space'. In Chapter 5 we study a special type of normed vector space, called *Euclidean space*, whose 'geometry' accords with our intuition in '3-dimensional space' but whose 'dimension' can be any positive integer n (goodbye intuition!).

Before setting down the axioms for an abstract vector space (in §1.3), let's enlarge our list (presently of length 1) of concrete examples; this is the purpose of the next section.

▶ *Exercises*

1. Given the point P(2, −1, 3) and the vector $u = [-3, 4, 5]$, find the point Q such that $\overrightarrow{PQ} = u$.

2. Given the points P(2, −1, 3), Q(3, 4, 1), R(4, −3, 4), S(5, 2, 2). True or false (explain): PQSR is a parallelogram.

3. Let $u = [2, 2, 1]$, $v = [-2, 1, 2]$, $w = [1, -2, 2]$. Show that

$$\|u + v\|^2 = \|u\|^2 + \|v\|^2,$$
$$\|(u + v) + w\|^2 = \|u + v\|^2 + \|w\|^2$$

and interpret geometrically.

4. Show that for every triple of points P, Q, R,

[2] The term 'ordered' refers to the concept of equality: triples $[a, b\, c]$ and $[a', b', c']$ are regarded as **equal** if $a = a'$, $b = b'$ and $c = c'$. In particular, $[2, 1, 5]$ is *not* equal to $[1, 2, 5]$ (order counts!). In contrast, the **set** with elements a, b, c is written $\{a, b, c\}$, or $\{b, a, c\}$, etc.; changing the order in which the elements of the set are listed doesn't change the set.

[3] An axiom (or postulate) is a statement *accepted* as true, for the purpose of exploring its logical consequences (or demonstrating consequences one suspects to be true). A familiar example: Euclid's axioms for geometry (including the famous 'parallel postulate').

$$\overrightarrow{PQ} + \overrightarrow{QR} = \overrightarrow{PR}.$$

{Method 1: Calculate components. Method 2: Draw a picture.}

5. Given the points $P_1(x_1, y_1, z_1)$ and $P_2(x_2, y_2, z_2)$, let M be the point such that

$$\overrightarrow{OM} = \tfrac{1}{2} \cdot \overrightarrow{OP_1} + \tfrac{1}{2} \cdot \overrightarrow{OP_2},$$

where $O(0, 0, 0)$ is the origin. Find the coordinates of M and interpret geometrically.

1.2 $\mathbf{R}^n$ and $\mathbf{C}^n$

$\mathbf{R}^n$ and $\mathbf{C}^n$ are the most important examples of 'vector spaces' (defined officially in the next section). The essential idea is that what was done in the preceding section for ordered triples $[a, b, c]$ can be done equally well for ordered n-ples $[x_1, x_2, \ldots, x_n]$.

1.2.1 Definition Fix a positive integer n and let $\mathbf{R}^n$ be the set of all ordered n-ples $[x_1, x_2, \ldots, x_n]$ of real numbers. {For example, $[3, -1/2, \sqrt{2}, \pi]$ is an element of $\mathbf{R}^4$.}

The elements of $\mathbf{R}^n$ will be called **vectors** and the elements of $\mathbf{R}$ (that is, real numbers) will be called **scalars**.

Vectors $[x_1, x_2, \ldots, x_n]$ and $[y_1, y_2, \ldots, y_n]$ are said to be **equal** if $x_i = y_i$ for $i = 1, \ldots, n$; this is expressed by writing $[x_1, x_2, \ldots, x_n] = [y_1, y_2, \ldots, y_n]$. If $x = [x_1, x_2, \ldots, x_n]$ then x_i is called the ith **component** of the vector x. {Thus, two vectors are equal if and only if they are 'componentwise equal'.}

If $x = [x_1, x_2, \ldots, x_n]$ and $y = [y_1, y_2, \ldots, y_n]$, the **sum** of x and y, denoted $x + y$, is the vector defined by the formula

$$x + y = [x_1 + y_1, \ldots, x_n + y_n];$$

and if c is a scalar, then the **multiple** of x by c, denoted cx, is the vector defined by the formula

$$cx = [cx_1, \ldots, cx_n].$$

{Briefly, sums and scalar multiples of vectors are defined 'componentwise'.}

Similarly, $\mathbf{C}^n$ denotes the set of all ordered n-ples of complex numbers, with $x + y$ and cx defined by the above formulas; in this context, the 'scalars' are the elements of $\mathbf{C}$ (that is, complex numbers).

In the arguments of this section, it doesn't matter whether the set of scalars is $\mathbf{R}$ or $\mathbf{C}$. To cover both cases simultaneously, we let F stand for either $\mathbf{R}$ or $\mathbf{C}$ and we write F^n for either $\mathbf{R}^n$ or $\mathbf{C}^n$. (More generally, F can be any **field**.[1])

[1] The term 'field' is explained in Appendix A.4; we use it mainly as a reminder that the set of scalars is a well-organized algebraic system (in particular, every nonzero scalar has a reciprocal). Chapters 1–6, 9 and 12 stress the fields $\mathbf{R}$ and/or $\mathbf{C}$, but whenever F is used to denote the field of scalars, it can be assumed that arguments involving it are valid with F *any* field (occasionally with extra restrictions, as spelled out in the context).

1.2.2 Definition The element of F^n all of whose components are zero is denoted θ and is called the **zero vector**; thus $\theta = [0, \ldots, 0]$. If $x = [x_1, x_2, \ldots, x_n]$ is any vector, the vector $[-x_1, \ldots, -x_n]$ is called the **negative** of x and is denoted $-x$.

The key properties of sums and scalar multiples are collected in the following theorem:

1.2.3 Theorem *Let* F *be a field (for example,* **R** *or* **C**), *let n be a positive integer, and let* F^n *be defined as in* 1.2.1.

(1) *If* $x, y \in F^n$ *then* $x + y \in F^n$. (F^n is said to be 'closed under addition'.)

(2) *If* $c \in F$ *and* $x \in F^n$ *then* $cx \in F^n$. (F^n is said to be 'closed under multiplication by scalars'.)

(3) $x + y = y + x$ *for all* $x, y \in F^n$. (The addition of vectors is 'commutative'.)

(4) $(x + y) + z = x + (y + z)$ *for all* x, y, z *in* F^n. (The addition of vectors is 'associative'.)

(5) $x + \theta = x = \theta + x$ *for every vector* x. (The zero vector is 'neutral' for addition.)

(6) $x + (-x) = \theta = (-x) + x$ *for every vector* x.

(7) $c(x + y) = cx + cy$ *and* $(c + d)x = cx + dx$ *for all vectors* x, y *and all scalars* c, d. (Multiplication by scalars is 'distributive'.)

(8) $(cd)x = c(dx)$ *for all scalars* c, d *and all vectors* x. (Multiplication by scalars is 'associative'.)

(9) $1x = x$ *for every vector* x. ('Neutrality' of 1 for scalar multiplication.)

Proof. Sums and scalar multiples are *defined* to be elements of F^n, so (1) and (2) are obvious.

(3) Say $x = [x_1, x_2, \ldots, x_n]$, $y = [y_1, y_2, \ldots, y_n]$. To show that the vectors $x + y$ and $y + x$ are equal, let's apply the criterion of 1.2.1: we have to show that for each index i $(i = 1, \ldots, n)$, the ith component of $x + y$ is equal to the ith component of $y + x$. These components are, respectively, $x_i + y_i$ and $y_i + x_i$ (by the definition of sum for vectors), and $x_i + y_i = y_i + x_i$ by the commutativity of addition in F.

(4)–(9) are proved using the same strategy: the equality of two vectors is established by showing that their corresponding components are equal (by a known property of the scalar field F).

For example, let's prove the first equality in (7). The ith components of cx and cy are cx_i and cy_i, so the ith component of $cx + cy$ is $cx_i + cy_i$; but this is equal to $c(x_i + y_i)$ (distributive law in F), which is the ith component of $c(x + y)$. Conclusion: $cx + cy = c(x + y)$.

The remaining assertions are now 'obvious', but it is good practice to write out the details anyway (understanding grows out of the end of a pencil). ■[1]

The elements of F^n are also denoted $x = (x_1, x_2, \ldots, x_n)$ (parentheses instead of brackets). For instance, $\mathbf{R}^3$ denotes 3-space coordinatized in the

[1] This blob marks the end of a proof.

usual way, consisting of points (x, y, z); it also denotes the set of vectors $[a, b, c]$ obtained from pairs of points (as in the preceding section). Thus, the symbol $\mathbf{R}^3$ is used equivocally, for ordered triples as points ('geometric' objects) and for ordered triples as vectors ('algebraic' objects). The use of parentheses or brackets is a way of distinguishing between these two conceptions of $\mathbf{R}^3$.

In practice, the distinction is frequently blurred: one also writes $(x_1, x_2, \ldots, x_n)$ for vectors and one speaks of x_i as the ith coordinate of $[x_1, x_2, \ldots, x_n]$. Confusion is easily avoided by a liberal use of the words 'point' or 'vector' as appropriate.

▶ ***Exercises***

1. For the vectors $u = (1, -2, 7)$, $v = (2, -1, 3)$, $w = (4, 1, -5)$ in $\mathbf{R}^3$, compute $2u - 3v + w$.
2. Consider the vectors $u = (2, 1)$, $v = (-5, 3)$, $w = (3, 4)$ in $\mathbf{R}^2$. Do there exist real numbers a, b such that $au + bv = w$? What if $v = (6, 3)$?
3. Give a detailed proof of (8) of Theorem 1.2.3 (just checking!). {Write out all steps of the proof and give a reason for each step.}
4. In Definition 1.2.1, $n = 1$ is not ruled out. What do the elements of $\mathbf{R}^1$ look like? Describe sums and scalar multiples in $\mathbf{R}^1$.

1.3 Vector spaces: the axioms, some examples

The properties of F^n listed in Theorem 1.2.3 were not chosen at random; they are just what it takes to be a 'vector space' in the sense of the following definition:

1.3.1 Definition

Let F be a field (for example, $\mathbf{R}$ or $\mathbf{C}$). A **vector space over** F is a set V admitting two 'operations', called *addition* and *multiplication by scalars*, subject to the set of rules given below. The elements of V are called *vectors* and the elements of F are called *scalars*. For each pair of vectors x, y, there is determined a vector, denoted $x + y$, called the *sum* of x and y. For each vector x and each scalar c, there is determined a vector, denoted cx, called the *scalar multiple* of x by c. Thus we have two ways of 'combining' vectors and scalars:

(1) if $x \in \mathrm{V}$ and $y \in \mathrm{V}$ then $x + y \in \mathrm{V}$ (V is *closed under addition*);

(2) if $x \in \mathrm{V}$ and $c \in \mathrm{F}$ then $cx \in \mathrm{V}$ (V is *closed under multiplication by scalars*).

The following rules are assumed for sums and scalar multiples (they are also called the *axioms*, or *postulates*, or *laws* of a vector space):

(3) $x + y = y + x$ for all x, y in V (*commutative law* for addition);

(4) $(x + y) + z = x + (y + z)$ for all x, y, z in V (*associative law* for addition);

(5) there exists a vector θ in V such that $x + \theta = x = \theta + x$ for all x in V (existence of a *zero vector*);

(6) for each x in V there exists a vector $-x$ in V such that $x + (-x) = \theta = (-x) + x$ (existence of *negatives*);

(7) $c(x + y) = cx + cy$, $(c + d)x = cx + dx$ for all vectors x, y and all scalars c, d (*distributive laws*);

(8) $(cd)x = c(dx)$ for all vectors x and all scalars c, d (*associative law* for scalar multiplication);

(9) $1x = x$ for all vectors x (*unity law* for scalar multiplication).

{The reader who doesn't like the looks of the next couple of paragraphs is forgiven for jumping ahead to 1.3.2.}

The above definition bristles with the terms 'for all', 'there exists', 'such that', 'if. . . then', terms for which there are efficient symbolic abbreviations (Appendix A.1.5, A.2.11): $\forall$, $\exists$, $\ni$, $\Rightarrow$. When these logical symbols are used, the statements become both compact and unreadable; writing out the symbolic form is still good practice in grasping the logical structure of the statements. That's not the way to write a book, but it's useful for taking notes and for recording thought-experiments (for example, when struggling with the exercises); with a pledge not to make an obsession of it, let's practice a little by rewriting the axioms (1)–(9) for a vector space in symbolic form:

(1) $x, y \in \mathrm{V} \;\Rightarrow\; x + y \in \mathrm{V}$;

(2) $x \in \mathrm{V},\, c \in \mathrm{F} \;\Rightarrow\; cx \in \mathrm{V}$;

(3) $x + y = y + x \quad (\forall x, y \in \mathrm{V})$;

(4) $(x + y) + z = x + (y + z) \quad (\forall x, y, z \in \mathrm{V})$;

(5) $\exists\ \theta \in \mathrm{V} \;\ni\; x + \theta = x = \theta + x \quad (\forall x \in \mathrm{V})$;

(6) $(\forall\ x \in \mathrm{V})\ \exists\ -x \in \mathrm{V} \;\ni\; x + (-x) = \theta = (-x) + x$;

(7) $c(x + y) = cx + cy,\ (c + d)x = cx + dx \quad (\forall\ x, y \in \mathrm{V}$ and $\forall\ c, d \in \mathrm{F})$;

(8) $(cd)x = c(dx) \quad (\forall\ c, d \in \mathrm{F}$ and $\forall\ x \in \mathrm{V})$;

(9) $1x = x \quad (\forall\ x \in \mathrm{V})$.

The 'compact symbolic form' is not unique; for instance, (1) is expressed equally well by

$$x + y \in \mathrm{V} \quad (\forall\ x, y \in \mathrm{V}),$$

and (3) by

$$x, y \in \mathrm{V} \;\Rightarrow\; x + y = y + x.$$

Exercise. Cover up the verbal definitions in 1.3.1 and try to re-create them by decoding the above symbolic forms.

Back to business:

1.3.2 Definition When $\mathrm{F} = \mathbf{R}$ in Definition 1.3.1, V is called a **real vector space**; when $\mathrm{F} = \mathbf{C}$, V is called a **complex vector space**.

When several vector spaces appear in the same context, it is assumed (unless signalled otherwise) that they are vector spaces over the same field (for example, all real or all complex).

Now let's explore the wealth of examples of vector spaces.

1.3.3 Example For each positive integer n, $\mathbf{R}^n$ is a real vector space and $\mathbf{C}^n$ is a complex vector space for the operations defined in the preceding section (Theorem 1.2.3).

1.3.4 Example Let F be a field, let T be a nonempty set, and let V be the set of all functions[1] $x:\mathrm{T}\to\mathrm{F}$. For x, y in V, $x = y$ means that $x(t) = y(t)$ for all $t \in \mathrm{T}$. If $x, y \in \mathrm{V}$ and $c \in \mathrm{F}$, define functions $x + y$ and cx by the formulas

$$(x + y)(t) = x(t) + y(t), \quad (cx)(t) = cx(t)$$

for all $t \in \mathrm{T}$. (So to speak, sums and scalar multiples in V are defined 'pointwise'.) Let θ be the function defined by $\theta(t) = 0$ for all $t \in \mathrm{T}$, and, for $x \in \mathrm{V}$, let $-x$ be the function defined by $(-x)(t) = -x(t)$ for all $t \in \mathrm{T}$. It is straightforward to check that V is a vector space over F; it is denoted[2] $\mathrm{V} = \mathcal{F}(\mathrm{T}, \mathrm{F})$ and is called the space of F-valued functions on T.

Exercise. Write out a detailed verification of the properties (3)–(9) for $\mathrm{V} = \mathcal{F}(\mathrm{T}, \mathrm{F})$. In the case that $\mathrm{T} = \{1, 2, \ldots, n\}$ for some positive integer n, do you see any similarity between V and F^n?

1.3.5 Example Let $\mathrm{F} = \mathbf{R}$ or $\mathbf{C}$. If p is a polynomial with coefficients in F, regard p as a function on F in the natural way: if

$$p = a_0 + a_1 t + a_2 t^2 + \ldots + a_n t^n,$$

where the coefficients $a_0, \ldots, a_n$ are in F and t is an 'indeterminate'[3], define the function $p:\mathrm{F}\to\mathrm{F}$ by the formula

$$p(c) = a_0 + a_1 c + a_2 c^2 + \ldots + a_n c^n$$

for all $c \in \mathrm{F}$ (i.e., by substituting c for t). {For example, if $p = t^2 + 5$ then $p(-2) = 9$.} Write $\mathcal{P}$ for the set of all such polynomial functions. If, in the vector space $\mathrm{V} = \mathcal{F}(\mathrm{T}, \mathrm{F})$ of Example 1.3.4, one takes $\mathrm{T} = \mathrm{F}$, then $\mathcal{P} \subset \mathrm{V}$; since the sum of two polynomials is a polynomial, and a scalar multiple of a polynomial is a polynomial, one sees that $\mathcal{P}$ is also a vector space over F.

1.3.6 Example If m is a fixed positive integer and if, in the preceding example, one limits attention to the polynomials $p = a_0 + a_1 t + a_2 t^2 + \ldots + a_m t^m$ (that is, either $p = 0$ or p is a nonzero polynomial of degree $\leq m$)[4], then the resulting set

[1] The terminology of functions is summarized in Appendix A.3.

[2] In general, $\mathcal{F}(\mathrm{X}, \mathrm{Y})$ denotes the set of all functions $f:\mathrm{X}\to\mathrm{Y}$. When $\mathrm{X} = \mathrm{Y}$ this is abbreviated $\mathcal{F}(\mathrm{X})$.

[3] A working definition of 'indeterminate': a symbol that is manipulated as if it were a number (*any* number).

[4] The degree of p is the largest integer k such that $a_k \neq 0$; the degree of the zero polynomial is not defined.

$\mathcal{P}_m$ of functions is also a vector space (the set of polynomials in question is closed under addition and under multiplication by scalars).

1.3.7 Example Let F be a field and let V be the set of all 'infinite sequences' $x = (a_1, a_2, a_3, \dots)$ of elements of F. If also $y = (b_1, b_2, b_3, \dots)$, $x = y$ means that $a_i = b_i$ for all i. Define sums and scalar multiples in V 'term-by-term', that is, by the formulas

$$(a_1, a_2, a_3, \dots) + (b_1, b_2, b_3, \dots) = (a_1 + b_1, a_2 + b_2, a_3 + b_3, \dots),$$
$$c(a_1, a_2, a_3, \dots) = (ca_1, ca_2, ca_3, \dots).$$

Then V is a vector space over F. Do you see any similarity between V and the space $\mathcal{F}(\mathbf{P}, \mathrm{F})$ obtained by setting $\mathrm{T} = \mathbf{P}$ in Example 1.3.4? (**P** is the set of positive integers.[5])

1.3.8 Example A minor variation on the preceding example: consider sequences indexed by the set $\mathbf{N} = \{0, 1, 2, 3, \dots\}$ of nonnegative integers, that is, sequences $x = (a_0, a_1, a_2, \dots)$ with $a_i \in \mathrm{F}$ for all $i \in \mathbf{N}$. {This notation is better adapted to Example 1.3.5.}

1.3.9 Example Let V be the space described in the preceding example (1.3.8). Consider the set W of all $x \in \mathrm{V}$ such that, from some index onward, the a_i are all 0. {So to speak, W is the set of all 'finitely nonzero' sequences, or sequences that are 'ultimately zero'. For example, $x = (0, -3, 1, 5, 0, 0, 0, \dots)$.} Since $\mathrm{W} \subset \mathrm{V}$ and since W is closed under sums and scalar multiples, it is easy to see that W is also a vector space over F. Assuming $\mathrm{F} = \mathbf{R}$ or $\mathbf{C}$, do you see any similarity between W and the space $\mathcal{P}$ of polynomial functions (1.3.5)?

1.3.10 Example Let $\mathrm{V}_1, \dots, \mathrm{V}_n$ be vector spaces over F and let $\mathrm{V} = \mathrm{V}_1 \times \dots \times \mathrm{V}_n$ be their cartesian product[6], that is, the set of all n-ples $(x_1, \dots, x_n)$ with $x_i \in \mathrm{V}_i$ for $i = 1, \dots, n$. Write $(x_1, \dots, x_n) = (y_1, \dots, y_n)$ if $x_i = y_i$ for all i. For $x = (x_1, \dots, x_n)$, $y = (y_1, \dots, y_n)$ in V and for $c \in \mathrm{F}$, define

$$x + y = (x_1 + y_1, \dots, x_n + y_n),\ cx = (cx_1, \dots, cx_n).$$

Arguments formally the same as for F^n (1.2.3) show that V is a vector space over F for these operations; the only difference is that, instead of citing algebraic properties of the scalar field F, we cite the algebraic laws in the coordinate spaces V_i. This example is important enough to merit an official definition:

1.3.11 Definition With notations as in Example 1.3.10, V is called the **product vector space** (or 'direct sum') of the vector spaces $\mathrm{V}_1, \dots, \mathrm{V}_n$. For $n = 2$, we write simply $\mathrm{V} = \mathrm{V}_1 \times \mathrm{V}_2$; for $n = 3$, $\mathrm{V} = \mathrm{V}_1 \times \mathrm{V}_2 \times \mathrm{V}_3$.

[5] Appendix A.2.9.

[6] Appendix, A.2.8.

These examples[7] give us an idea of the mathematical objects that can be put *into* the context of the axioms for a vector space. In the next section we turn to the question of what can be gotten *out* of the axioms, that is, what properties can be *inferred* from the properties (1)–(9) postulated in the definition of a vector space.

▶ Exercises

1. Let V be a vector space over a field F. Let T be a nonempty set and let $W = \mathcal{F}(T, V)$ be the set of all functions $x:T \to V$. Show that W can be made into a vector space over F in a natural way. {Hint: Use the definitions in Example 1.3.4 as a guide.}

2. The definition of a vector space (1.3.1) can be formulated as follows. A vector space over F is a nonempty set V together with a pair of mappings

$$\sigma:V \times V \to V, \quad \mu:F \times V \to V$$

(σ suggests 'sum' and μ suggests 'multiple') having the following properties: $\sigma(x, y) = \sigma(y, x)$ for all x, y in V; $\sigma(\sigma(x, y), z) = \sigma(x, \sigma(y, z))$ for all x, y, z in V; etc. The exercise: Write out the 'etc.' in detail.

3. Let V be a complex vector space (1.3.2). Show that V is also a real vector space (sums as usual, scalar multiplication restricted to real scalars). These two ways of looking at V may be indicated by writing V_C and V_R.

4. Every real vector space V can be 'embedded' in a complex vector space W in the following way: let $W = V \times V$ be the real vector space constructed as in Example 1.3.10 and define multiplication by complex scalars by the formula

$$(a + bi)(x, y) = (ax - by, bx + ay)$$

for $a, b \in \mathbf{R}$ and $(x, y) \in W$. {In particular, $i(x, y) = (-y, x)$. Think of (x, y) as '$x + iy$'.}

Show that W satisfies the axioms for a complex vector space. (W is called the *complexification* of V.)

1.4 Vector spaces: first consequences of the axioms

Our first deductions from the axioms (1.3.1) have to do with the *uniqueness* of certain vectors mentioned in the axioms:

1.4.1 Theorem *Let* V *be a vector space over a field* F, *with notations as in 1.3.1.*

(i) *If θ' is a vector such that $\theta' + x = x$ for all $x \in V$, then $\theta' = \theta$; thus, the vector θ whose existence is postulated in 1.3.1 is unique.*

[7] CAUTION: The Minister of Education has determined that skipping over the examples may be hazardous to your understanding of the subject. (It's not too late to go back over them!)

(ii) *If $x + y = \theta$, necessarily $y = -x$; thus, the vector $-x$ whose existence is postulated in 1.3.1 is uniquely determined by x.*

(iii) *$\theta + \theta = \theta$; and if z is a vector such that $z + z = z$, necessarily $z = \theta$.*

Proof.

(i) We have $\theta' = \theta' + \theta = \theta$, where the first equality holds by the property of θ postulated in (5) of 1.3.1, and the second equality holds by the assumption on θ'. (See also Exercise 5.)

(ii) Suppose $x + y = \theta$. Adding $-x$ (the vector provided by (6) of 1.3.1) to both sides of the equation, we have, successively, $-x + (x + y) = -x + \theta$, $(-x + x) + y = -x$, $\theta + y = -x$, $y = -x$.

(iii) If $z + z = z$, then $\theta = z + (-z) = (z + z) + (-z) = z + (z + (-z)) = z + \theta = z$. On the other hand, $\theta + \theta = \theta$ by (5) of Definition 1.3.1. ■

1.4.2 Corollary *For every vector x, $-(-x) = x$.*

Proof. Since $-x + x = \theta$ by (6) of Definition 1.3.1, we have $x = -(-x)$ by (ii) of the theorem. ■

1.4.3 Corollary *For every vector x, $0x = \theta$; for every scalar c, $c\theta = \theta$.*

Proof. Let $z = 0x$. Using one of the distributive laws (1.3.1) at the appropriate step, we have

$$z + z = 0x + 0x = (0 + 0)x = 0x = z,$$

therefore $z = \theta$ by (iii) of the theorem. The equality $c\theta = \theta$ is proved similarly, using the other distributive law. {Can you write out the details?} ■

1.4.4 Corollary *For every vector x and every scalar c, $c(-x) = -(cx) = (-c)x$.*

Proof. $\theta = c\theta = c(x + (-x)) = cx + c(-x)$, therefore $c(-x) = -(cx)$ by (ii) of the theorem. Similarly, $\theta = 0x = (c + (-c))x = cx + (-c)x$, therefore $(-c)x = -(cx)$. ■

The preceding compact proof looks nice in print but shows little of the reasoning behind the proof. Here's a format for writing out the proof of $c(-x) = -(cx)$ that shows the reason for each step:

$$\begin{aligned} \theta &= c\theta && \text{[by 1.4.3]} \\ &= c(x + (-x)) && \text{[by (6) of 1.3.1]} \\ &= cx + c(-x) && \text{[by (7) of 1.3.1]} \end{aligned}$$

therefore

$$c(-x) = -(cx) \qquad \text{[by (ii) of 1.4.1].}$$

Once one is comfortably settled in vector spaces, the abbreviated argument given earlier is clearly preferable, but at the beginning it is helpful to look at things 'in slow motion'. For practice, rewrite the proof of $(-c)x = -(cx)$ in the fuller form.

1.4.5 Corollary *For every vector* x, $(-1)x = -x$.

Proof. $(-1)x = -(1x) = -x$. ■

Corollary 1.4.3 says that if one of the 'factors' of a scalar multiple cx is zero, then $cx = \theta$; the following theorem says that that's the *only* way a scalar multiple can be θ:

1.4.6 Theorem (No divisors of θ) *Let* V *be a vector space,* c *a scalar,* $x \in$ V. *Then* $cx = \theta$ *if and only if* $c = 0$ *or* $x = \theta$.

Proof. We are to show that

$$cx = \theta \Leftrightarrow c = 0 \text{ or } x = \theta.$$

{The implication $\Rightarrow$ is the *only if* assertion, and $\Leftarrow$ is the *if* assertion.}

Only if: Suppose $cx = \theta$. If $c = 0$, fine. Assuming $c \neq 0$, we have to show that $x = \theta$. Let $d = c^{-1}$ be the reciprocal of c. Then $d(cx) = d\theta = \theta$ by 1.4.3, that is, $(dc)x = \theta$; but $dc = 1$ and $1x = x$, so $x = \theta$.

If: This is the assertion of 1.4.3. ■

1.4.7 Corollary (Cancellation laws) *Let* x, y *be vectors and* c, d *scalars.*

(i) *If* $cx = cy$ *and* $c \neq 0$, *then* $x = y$.

(ii) *If* $cx = dx$ *and* $x \neq \theta$, *then* $c = d$.

Proof. (i) Since

$$c(-x + y) = c(-x) + cy = -(cx) + cy = -(cx) + cx = \theta,$$

we have $-x + y = \theta$ by 1.4.6, therefore $y = -(-x) = x$.

(ii) Similarly,

$$(c - d)x = (c + (-d))x = cx + (-d)x = dx + (-(dx)) = \theta,$$

therefore $c - d = 0$ by 1.4.6. ■

Negatives are just a step away from subtraction (which is nothing more than a handy notation):

1.4.8 Definition For vectors x and y, the vector $x + (-y)$ is denoted $x - y$.

For example, if $x + y = z$ then (adding $-y$ to both sides) $x = z - y$.

The theorems in this section are 'plausible' and their proofs are 'obvious'. The essence of these results is that we don't have to be afraid of the computational reflexes developed in elementary algebra; they are not likely to lead us astray. We can now relax, 'do what comes naturally', and turn to more interesting matters. As we progress, the language becomes richer (more concepts), the assertions of the theorems subtler and their proofs more challenging. The game doesn't really begin until the first 'Hey, I don't see that!' and the fun doesn't begin until the subsequent 'Ah, I see it now!'. And all the better if it is followed by 'I wonder if . . .'.

Exercises

1. In a vector space, $x - y = \theta$ if and only if $x = y$; $x + y = z$ if and only if $x = z - y$.[1]
2. Let V be a vector space over a field F. If x is a fixed nonzero vector, then the mapping $f:\mathrm{F} \to \mathrm{V}$ defined by $f(c) = cx$ is injective. If c is a fixed nonzero scalar, then the mapping $g:\mathrm{V} \to \mathrm{V}$ defined by $g(x) = cx$ is bijective.
3. If V is a vector space and y is a fixed vector, then the mapping $\tau:\mathrm{V} \to \mathrm{V}$ defined by $\tau(x) = x + y$ is bijective (it is called *translation* by the vector y).
4. If a is a nonzero scalar and b is a vector in the space V, then the equation $ax + b = \theta$ has a unique solution x in V.
5. If, in a vector space, θ' is a vector such that $\theta' + x = x$ for even a single vector x, then $\theta' = \theta$. {Hint: Add $-x$ to both sides of the equation.}

1.5 Linear combinations of vectors

1.5.1 Definition Let V be a vector space (1.3.1). A vector $x \in \mathrm{V}$ is said to be a **linear combination** of vectors $x_1, \ldots, x_n$ in V if there exist scalars $c_1, \ldots, c_n$ such that

$$x = c_1x_1 + \ldots + c_nx_n.$$

The scalars c_i are called the **coefficients** of the x_i in the linear combination.

What is meant by such 'finite sums'? For $n = 2$, $c_1x_1 + c_2x_2$ means what we think it means—the sum of c_1x_1 and c_2x_2. In principle, vectors can only be added two at a time, so there are two reasonable interpretations of $c_1x_1 + c_2x_2 + c_3x_3$:

$$(c_1x_1 + c_2x_2) + c_3x_3 \quad \text{and} \quad c_1x_1 + (c_2x_2 + c_3x_3);$$

by the associative law, it does not matter which interpretation is made—the resulting vectors are the same. Now that the expression $c_1x_1 + c_2x_2 + c_3x_3$ is under control, one can define

$$c_1x_1 + c_2x_2 + c_3x_3 + c_4x_4$$

to be $(c_1x_1 + c_2x_2 + c_3x_3) + c_4x_4$, and so on. What does 'and so on' mean? It means that sums of greater length are defined 'recursively' (or 'inductively'): once the sum of $c_1x_1, \ldots, c_kx_k$ is under control, we define

$$c_1x_1 + \ldots + c_{k+1}x_{k+1}$$

to be $(c_1x_1 + \ldots + c_kx_k) + c_{k+1}x_{k+1}$. Given any finite list of vectors $c_1x_1, \ldots, c_nx_n$, we reach the sum $c_1x_1 + \ldots + c_nx_n$ of 'length n' in a finite number of steps.

[1] An exercise in the form of a simple statement (without coercive verbs such as 'Prove', 'Show', . . .) is an invitation to prove the statement.

To make the foregoing discussion 'more rigorous' would take us down a path where the returns diminish quickly. Let's be quite informal: in forming finite sums $x_1 + \ldots + x_n$ (or linear combinations $c_1x_1 + \ldots + c_nx_n$), the 'subsums' can be performed in any order and the terms may be permuted at will; for instance,

$$\begin{aligned} x_1 + x_2 + x_3 + x_4 &= (x_1 + x_2) + (x_3 + x_4) \\ &= [x_1 + (x_2 + x_3)] + x_4 \\ &= [x_4 + (x_3 + x_1)] + x_2. \end{aligned}$$

(Justification: the associative and commutative laws for vector addition.)

1.5.2 Example Every polynomial

$$a_0 + a_1t + a_2t^2 + \ldots + a_nt^n$$

is a sum of monomials a_it^i, thus is a linear combination of the powers $t^0 = 1$, $t^1 = t$, t^2, $t^3, \ldots, t^n$.

1.5.3 Example In F^n, any vector $x = (a_1, \ldots, a_n)$ can be written as a linear combination

$$\begin{aligned} x &= (a_1, 0, \ldots, 0) + (0, a_2, 0, \ldots, 0) + \ldots + (0, \ldots, 0, a_n) \\ &= a_1(1, 0, \ldots, 0) + a_2(0, 1, 0, \ldots, 0) + \ldots + a_n(0, \ldots, 0, 1); \end{aligned}$$

writing $e_1 = (1, 0, \ldots, 0)$, $e_2 = (0, 1, 0, \ldots, 0), \ldots, e_n = (0, \ldots, 0, 1)$, we have $x = a_1e_1 + \ldots + a_ne_n$.

1.5.4 Example The function $f(t) = \sin(t + \pi/6)$ is a linear combination of the functions $\sin t$ and $\cos t$, namely, $f = (\sqrt{3}/2)\sin + (1/2)\cos$. (Why?)

1.5.5 Example The hyperbolic sine function $\sinh t$ is a linear combination of the exponential functions e^t and e^{-t}. (What are the coefficients?)

The summation notation used in algebra and calculus can be carried over to vector sums: the sum $x_1 + \ldots + x_n$ is also written $\sum_{i=1}^{n} x_i$. In particular, a linear combination $c_1x_1 + \ldots + c_nx_n$ can be written $\sum_{i=1}^{n} c_ix_i$.

The usual rules for manipulating such expressions are valid. For example, given vectors x_1, x_2, y_1, y_2, we have

$$\begin{aligned} \sum_{i=1}^{2} (x_i + y_i) &= (x_1 + y_1) + (x_2 + y_2) \\ &= (x_1 + x_2) + (y_1 + y_2) \\ &= \sum_{i=1}^{2} x_i + \sum_{i=1}^{2} y_i \end{aligned}$$

(notice the intervention of the commutative and associative laws for the addition of vectors).

▶ ***Exercises***

1. Express the function $f(t) = \cos(t - \pi/3)$ as a linear combination of the functions $\sin t$ and $\cos t$.

2. Express the function e^t as a linear combination of the hyperbolic functions $\cosh t$ and $\sinh t$.

3. Convince yourself of the correctness of the following formulas pertaining to linear combinations in a vector space:

(i) $$c\sum_{i=1}^{n} x_i = \sum_{i=1}^{n} cx_i.$$

(ii) $$\sum_{i=1}^{n} a_i x_i + \sum_{i=1}^{n} b_i x_i = \sum_{i=1}^{n} (a_i + b_i)x_i$$

(iii) $$\sum_{i=1}^{m} a_i x_i + \sum_{j=1}^{n} b_j y_j = \sum_{k=1}^{m+n} c_k z_k,$$

where $c_k = a_k$ and $z_k = x_k$ for $k = 1, 2, \ldots, m$, while $c_k = b_{k-m}$ and $z_k = y_{k-m}$ for $k = m+1, m+2, \ldots, m+n$.

(iv) $$\left(\sum_{i=1}^{m} c_i\right)\left(\sum_{j=1}^{n} x_j\right) = \sum_{i=1}^{m}\left(\sum_{j=1}^{n} c_i x_j\right) = \sum_{j=1}^{n}\left(\sum_{i=1}^{m} c_i x_j\right) = \sum_{i,j} c_i x_j,$$

the last expression signifying the sum, with mn terms, of the vectors $c_i x_j$ for all possible combinations of i and j.

4. In $\mathbf{R}^3$, express the vector $(2, -3, 0)$ as a linear combination of e_1, e_2, e_3 (cf. Example 1.5.3).

5. Show that in a vector space, $x - y$ is a linear combination of x and y.

6. True or false (explain): In $\mathbf{R}^3$,

(i) $(1, 2, 3)$ is a linear combination of $(2, 1, 0)$ and $(-3, 2, 0)$;

(ii) $(1, 2, 0)$ is a linear combination of $(2, 1, 0)$ and $(-3, 2, 0)$.

7. In $\mathbf{R}^3$, let $x_1 = (1, 2, 2)$, $x_2 = (2, -2, 1)$. True or false (explain):

(i) the vector $x = (0, 6, 3)$ is a linear combination of x_1, x_2.

(ii) the vector $x = (2, 1, -2)$ is a linear combination of x_1, x_2.

8. In the vector space $\mathcal{P}$ of real polynomial functions (Example 1.3.5), let $p(t) = 2t^3 - 5t^2 + 6t - 4$, $q(t) = t^3 + 6t^2 + 3t + 5$, $r(t) = 4t^2 - 3t + 7$. True or false (explain): r is a linear combination of p and q.

1.6 Linear subspaces

Let's begin with an example. Let P be a plane through the origin in 3-space, for example the plane determined by the equation

$$(*) \qquad 2x - 3y + z = 0.$$

We can think of P as the set of all vectors $u = [x, y, z]$ in $\mathbf{R}^3$ determined by arrows from the origin $(0, 0, 0)$ to points (x, y, z) whose coordinates satisfy the equation $(*)$. From the discussion in §1.1 we see that if u, v are in P and c is a real number, then $u + v$ and cu are also in P (Fig. 7). {This is clear either geometrically or from substituting the components of $u + v$ and cu into the equation $(*)$.} So to speak P, regarded as a subset of the vector space $\mathbf{R}^3$, is 'closed' under addition and under multiplication by scalars.

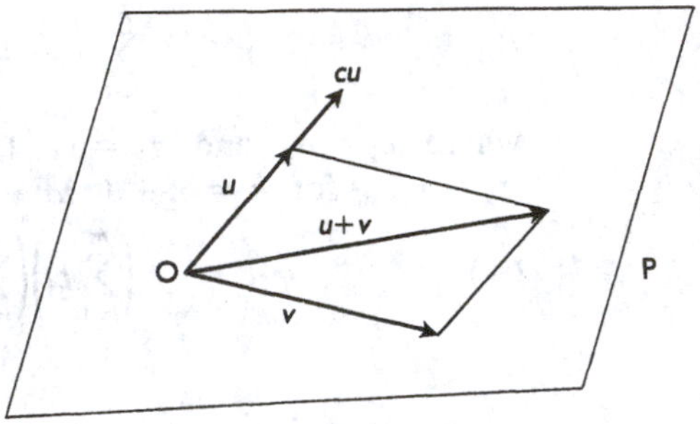

Fig. 7

1.6.1 Definition Let V be a vector space. A subset M of V is called a **linear subspace** of V if it has the following three properties: (i) $\theta \in M$; (ii) if $x, y \in M$ then $x + y \in M$; (iii) if $x \in M$ and c is a scalar, then $cx \in M$. {Briefly, M contains the zero vector and is closed under the linear operations.}

It follows that M is also closed under linear combinations: if $x_1, \ldots, x_n$ are in M and $c_1, \ldots, c_n$ are scalars, then $c_1x_1 + \ldots + c_nx_n$ is in M.

1.6.2 Theorem *Every linear subspace of a vector space is itself a vector space.*

Proof. With notations as in 1.6.1, let us check that M has all of the properties required of a vector space (1.3.1). Since M is closed under addition and scalar multiplication, conditions (1) and (2) of 1.3.1 are fulfilled. Properties (3), (4) and (7)–(9) are inherited from V. (For example, $x + y = y + x$ is true for all x, y in V, so it is true in particular when x, y are in M.) We know that $\theta \in M$, so M satisfies condition (5). Finally, if $x \in M$ then $-x = (-1)x \in M$, so M satisfies condition (6). ■

1.6.3 Examples Every vector space V has at least the linear subspaces $M = \{\theta\}$ and $M = V$ (called the *trivial* linear subspaces). If V is a line through the origin in 3-space, then V has only the trivial linear subspaces. If V is a plane through the origin, then each line in V through the origin is a linear subspace of V. In $\mathbf{R}^3$, each line through the origin and each plane through the origin is a linear subspace.

1.6.4 Example Consider a system of m linear equations in n unknowns $x_1, \ldots, x_n$:

$$(*)\quad \begin{cases} a_{11}x_1 + a_{12}x_2 + \ldots + a_{1n}x_n = 0 \\ a_{21}x_1 + a_{22}x_2 + \ldots + a_{2n}x_n = 0 \\ \ldots \\ a_{m1}x_1 + a_{m2}x_2 + \ldots + a_{mn}x_n = 0 \end{cases}$$

the a_{ij} being fixed scalars, that is, elements of a field F. (The system is called *homogeneous* because all of the right sides are 0.) A *solution* of the system is an n-ple $x = (x_1, \ldots, x_n)$ of scalars that satisfy all of the equations (*). The zero vector $\theta = (0, \ldots, 0)$ is always a solution, called the *trivial* solution (it can happen that there are no others—consider the 'system' $1x_1 = 0$). The set M of all solution-vectors $x = (x_1, \ldots, x_n)$ is a linear subspace of F^n. (Proof?) We'll be returning to this very instructive example on many occasions.

1.6.5 Example Let $\mathcal{F} = \mathcal{F}(\mathbf{R})$ be the vector space of all functions $x:\mathbf{R} \to \mathbf{R}$ (the 'real-valued functions of a real variable'), with the pointwise linear operations (1.3.4). Let $\mathcal{C}$ be the set of all $x \in \mathcal{F}$ that are continuous at every point of $\mathbf{R}$. Then $\mathcal{C}$ is a linear subspace of $\mathcal{F}$. (What are the statements from calculus that justify this assertion?)

1.6.6 Example With the preceding notations, let $\mathcal{D}$ be the set of all $x \in \mathcal{F}$ that are differentiable at every point of $\mathbf{R}$. Then $\mathcal{D}$ is a linear subspace of $\mathcal{F}$. (Why?) In fact, $\mathcal{D}$ is a linear subspace of $\mathcal{C}$. (Why?)

1.6.7 Example Let $\mathcal{P}$ be the set of all polynomial functions on $\mathbf{R}$ with real coefficients (1.3.5), that is, all functions $p:\mathbf{R} \to \mathbf{R}$ of the form

$$p(t) = a_0 + a_1 t + \ldots + a_n t^n \quad (t \in \mathbf{R}),$$

where n is any nonnegative integer and the coefficients a_i are real numbers. Then $\mathcal{P}$ is a linear subspace of the vector space $\mathcal{D}$ of differentiable functions (1.6.6).

1.6.8 Example With $\mathcal{P}$ the vector space of real polynomial functions (1.6.7), fix a nonnegative integer m and let $\mathcal{P}_m$ be the set of all $p \in \mathcal{P}$ such that $p = 0$ or p has degree $\leq m$ (1.3.6). Then $\mathcal{P}_m$ is a linear subspace of $\mathcal{P}$. In particular, $\mathcal{P}_0$ is the space of constant functions, $\mathcal{P}_1$ the space of all linear functions, $\mathcal{P}_2$ the space of all linear or quadratic functions, etc.

1.6.9 Definition If M and N are subsets of a vector space V, one writes

$$\mathrm{M} + \mathrm{N} = \{x + y:\ x \in \mathrm{M},\ y \in \mathrm{N}\}$$
$$\mathrm{M} \cap \mathrm{N} = \{x \in \mathrm{V}:\ x \in \mathrm{M}\ \&\ x \in \mathrm{N}\}$$

(called the *sum* and *intersection*, respectively, of M and N).

When applied to linear subspaces, these formulas produce linear subspaces (the theme is 'new subspaces from old'):

1.6.10 Theorem *If* M *and* N *are linear subspaces of a vector space* V, *then so are* $\mathrm{M} + \mathrm{N}$ *and* $\mathrm{M} \cap \mathrm{N}$.

Proof. Since M and N both contain the zero vector, $\theta = \theta + \theta$ shows that M + N does, too. If $x, x' \in$ M and $y, y' \in$ N, and if c is a scalar, then

$$(x + y) + (x' + y') = (x + x') + (y + y') \in \mathrm{M} + \mathrm{N}$$

(because M and N are closed under addition) and

$$c(x + y) = cx + cy \in \mathrm{M} + \mathrm{N}$$

(because M and N are closed under multiplication by scalars), thus M + N is closed under the linear operations. The three criteria of Definition 1.6.1 are met: M + N is a linear subspace of V.

The proof for M ∩ N is even easier. (Isn't it?). ■

1.6.11 Corollary *If* M *and* N *are linear subspaces of a vector space* V, *then* M + N *is the smallest linear subspace of* V *containing both* M *and* N.

Proof. If $x \in$ M then $x = x + \theta$ shows that $x \in$ M + N, thus M ⊂ M + N. Similarly N ⊂ M + N. If P is any linear subspace of V that contains both M and N, then M + N ⊂ P (because P is closed under addition). Thus M + N is a linear subspace of V containing both M and N, and it is contained in every other such subspace. ■

1.6.12 Corollary *If* M *and* N *are linear subspaces of a vector space* V, *then* M ∩ N *is the largest linear subspace of* V *contained in both* M *and* N.

Proof. M ∩ N is a linear subspace of V contained in both M and N; and if P is a linear subspace of V contained in both M and N, then P ⊂ M ∩ N. ■

More generally, if $\mathscr{M}$ is any set of linear subspaces M of V, the *intersection* of $\mathscr{M}$, denoted $\cap\mathscr{M}$, is defined to be the set

$$\{x \in \mathrm{V}: \quad x \in \mathrm{M} \text{ for every } \mathrm{M} \in \mathscr{M}\}.$$

Let $\mathrm{N} = \cap\mathscr{M}$. Since $\theta \in$ M for every $\mathrm{M} \in \mathscr{M}$, we have $\theta \in$ N. If $x, y \in$ N and c is a scalar, then $x, y \in$ M for every $\mathrm{M} \in \mathscr{M}$, therefore $x + y$ and cx belong to M for every $\mathrm{M} \in \mathscr{M}$, consequently $x + y$ and cx belong to N. Thus $\cap\mathscr{M}$ is a linear subspace of V, clearly the largest that is contained in every $\mathrm{M} \in \mathscr{M}$.

One can also show that there is a *smallest* linear subspace of V *containing* every $\mathrm{M} \in \mathscr{M}$ (Exercise 7).

▶ *Exercises*

1. Let M and N be the subsets of $\mathbf{R}^2$ defined as follows:

$$\mathrm{M} = \{(a, 0): \quad a \in \mathbf{R}\}, \quad \mathrm{N} = \{(0, b): \quad b \in \mathbf{R}\}.$$

Show that M and N are linear subspaces of $\mathbf{R}^2$ and describe the linear subspaces M ∩ N and M + N.

2. Let $\mathscr{F} = \mathscr{F}(\mathbf{R})$ as in Example 1.6.5, fix a nonempty subset T of $\mathbf{R}$, and let

$$\mathrm{M} = \{x \in \mathscr{F}: \quad x(t) = 0 \text{ for all } t \in \mathrm{T}\}.$$

Show that M is a linear subspace of $\mathcal{F}$.

3. With $\mathcal{F} = \mathcal{F}(\mathbf{R})$ as in 1.6.5, let

$$M = \{x \in \mathcal{F}: \quad x = 0 \text{ on } (-\infty, 0]\},$$
$$N = \{x \in \mathcal{F}: \quad x = 0 \text{ on } (0, +\infty)\}.$$

Prove that $M + N = \mathcal{F}$ and $M \cap N = \{\theta\}$.

4. Let M and N be linear subspaces of a vector space V, and let

$$M \cup N = \{x \in V: \quad x \in M \text{ or } x \in N\}$$

be the union of M and N.

True or false (explain): $M \cup N$ is a linear subspace of V.

{If true, give a proof. If false, give a counterexample (an example of V, M, N for which $M \cup N$ is not a linear subspace of V).}

5. Let M, N, P be linear subspaces of a vector space V.

(i) Show that if $P \supset M$, then $P \cap (M + N) = M + (P \cap N)$. (This is called the *modular law* for linear subspaces.)

(ii) In general, $P \cap (M + N) \neq (P \cap M) + (P \cap N)$. {Hint: Let $V = \mathbf{R}^2$ and let P, M, N be three distinct lines through the origin.}

6. Let V be a vector space and let A be any subset of V. There exist linear subspaces of V that contain A (for instance, V itself is such a subspace). Let $\mathcal{N} = \{N: N$ is a linear subspace of V containing A$\}$. Show that $\mathcal{N}$ contains a smallest element. {Try $\cap\mathcal{N}$.} Thus, there exists a smallest linear subspace of V containing A.

7. Let V be a vector space and let $\mathcal{M}$ be any set of linear subspaces of V. Let A be the *union* of the subspaces in $\mathcal{M}$, that is,

$$A = \cup\mathcal{M} = \{x \in V: \quad x \in M \text{ for some } M \in \mathcal{M}\}.$$

Apply Exercise 6 to conclude that there exists a smallest linear subspace of V that contains every $M \in \mathcal{M}$.

8. A linear subspace M of a vector space V is closed under subtraction: if $x, y \in M$ then $x - y \in M$.

9. Let V be a vector space and let $\mathcal{M}$ be a set of linear subspaces of V, having the following property: if $M, N \in \mathcal{M}$ then there exists $P \in \mathcal{M}$ such that $M \subset P$ and $N \subset P$. (That is, any two elements of $\mathcal{M}$ are contained in a third.) Prove that $\cup\mathcal{M}$ is a linear subspace of V.

10. The vectors $(1, 2, 2)$, $(2, -2, 1)$ in $\mathbf{R}^3$ determine a plane P through the origin.

True or false (explain): (i) the vector $(3, 0, 3)$ lies in P; (ii) the vector $(2, 1, -2)$ lies in P.

11. Let M and N be linear subspaces of V. Prove that the following conditions are equivalent[1]:

[1] This means that any of the conditions (a), (b), (c) implies any other. One strategy is to prove that (a) ⇒ (b), (b) ⇒ (c), and (c) ⇒ (a) (i.e., prove the implications in 'cyclic order').

(a) $M \cap N = \{\theta\}$;

(b) if $y \in M$, $z \in N$ and $y + z = \theta$, then $y = z = \theta$;

(c) if $y + z = y' + z'$, where $y, y' \in M$ and $z, z' \in N$, then $y = y'$ and $z = z'$.

12. Let M and N be linear subspaces of V. If $M + N = V$ and $M \cap N = \{\theta\}$ (cf. Exercise 11), V is said to be the *direct sum* of M and N, written $V = M \oplus N$. Prove that the following conditions are equivalent:

(a) $V = M \oplus N$;

(b) for each $x \in V$, there exist unique elements $y \in M$ and $z \in N$ such that $x = y + z$.

13. Let V be the real vector space of all functions $x:\mathbf{R} \to \mathbf{R}$ (1.3.4). Call $y \in V$ *even* if $y(-t) = y(t)$ for all $t \in \mathbf{R}$, and call $z \in V$ *odd* if $z(-t) = -z(t)$ for all $t \in \mathbf{R}$. Let

$$M = \{y \in V:\ \ y \text{ is even}\},\quad N = \{z \in V:\ \ z \text{ is odd}\}.$$

(i) Prove that M and N are linear subspaces of V and that $V = M \oplus N$ in the sense of Exercise 12. {Hint: If $x \in V$ consider the functions y and z defined by $y(t) = \frac{1}{2}[x(t) + x(-t)]$, $z(t) = \frac{1}{2}[x(t) - x(-t)]$.}

(ii) What does (i) say for $x(t) = e^t$?

(iii) What does (i) say for a polynomial function x?

14. Let V be a vector space, and let A, B, M, N be linear subspaces of V such that $A \cap B = M \cap N$. Prove that $A = (A + (B \cap M)) \cap (A + (B \cap N))$. {Hint: Exercise 5.}

15. Suppose $V = M \oplus N$ in the sense of Exercise 12, and let A be a linear subspace of V such that $A \supset M$. Prove that $A = M \oplus (A \cap N)$. {Hint: Exercise 5.}

16. Let V and W be vector spaces over the same field F and let $V \times W$ be the product vector space (1.3.11). If M is a linear subspace of V, and N is a linear subspace of W, show that $M \times N$ is a linear subspace of $V \times W$.

17. If $V = V_1 \times V_2$ (1.3.11), $M_1 = \{(x_1, \theta):\ x_1 \in V_1\}$ and $M_2 = \{(\theta, x_2):\ x_2 \in V_2\}$, then $V = M_1 + M_2$ and $M_1 \cap M_2 = \{\theta\}$ (where θ stands for the zero vector of all vector spaces in sight). In the terminology of Exercise 12, V is the direct sum of its subspaces M_1 and M_2, that is, $V = M_1 \oplus M_2$. (Generalization?)

18. Let M and N be linear subspaces of V whose union $M \cup N$ is also a linear subspace. Prove that either $M \subset N$ or $N \subset M$.

{Hint: Assume to the contrary that neither of M, N is contained in the other; choose vectors $y \in M$, $z \in N$ with $y \notin N$, $z \notin M$ and try to find a home for $y + z$.}

2

Linear mappings

A vector space is more than a set: it is a set equipped with linear operations (sums and scalar multiples). A linear subspace is more than a subset: it is a nonempty subset closed under the linear operations. We thus expect that the most interesting mappings[1] between vector spaces are those that relate in a natural way to the linear operations in the vector spaces. This chapter is about such mappings, and once they get onstage it's hard to shove them out of the spotlight.

2.1 Linear mappings

The most important mappings between vector spaces are those that 'preserve' sums and scalar multiples:

2.1.1 Definition Let V and W be vector spaces over the same scalar field F. A mapping $T:\mathrm{V} \to \mathrm{W}$ is said to be **linear** if it is 'additive' and 'homogeneous', that is, if

$$T(x + y) = Tx + Ty \quad \text{and} \quad T(cx) = c(Tx)$$

for all x, y in V and all c in F. (Such a mapping is also called a *linear transformation*.)

It follows from linearity that general linear combinations are also preserved:

[1] 'Mapping' is a synonym for 'function' (Appendix A.3).

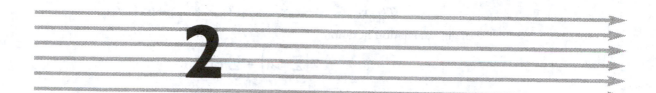

$$T(c_1x_1 + \ldots + c_nx_n) = c_1(Tx_1) + \ldots + c_n(Tx_n),$$

or, in summation notation,

$$T\left(\sum_{i=1}^{n} c_ix_i\right) = \sum_{i=1}^{n} c_i(Tx_i).$$

A linear mapping also preserves zero and negatives:

2.1.2 Theorem *If $T:\mathrm{V}\rightarrow\mathrm{W}$ is a linear mapping, then $T\theta = \theta$ and $T(-x) = -(Tx)$ for all x in V.*

Proof. We write θ for the zero vector of either V or W. If $z = T\theta$, then

$$z + z = T\theta + T\theta = T(\theta + \theta) = T\theta = z,$$

therefore $z = \theta$ by 1.4.1. For all $x \in \mathrm{V}$,

$$Tx + T(-x) = T(x + (-x)) = T\theta = \theta,$$

therefore $T(-x) = -(Tx)$ by 1.4.1. ■

The following example of a linear mapping is important enough to be promoted to a theorem:

2.1.3 Theorem *Let V be a vector space over F and let $x_1, \ldots, x_n$ be a finite list of (not necessarily different) vectors in V. Then the mapping $T:\mathrm{F}^n \rightarrow \mathrm{V}$ defined by the formula*

$$T(a_1, \ldots, a_n) = a_1x_1 + \ldots + a_nx_n$$

is linear; it is the unique linear mapping such that $Te_i = x_i$ for all i, where $e_1, \ldots, e_n$ are the vectors described in 1.5.3.

Proof. The argument combines the definitions of the operations in F^n with the vector space axioms satisfied by V. If $a = (a_1, \ldots, a_n)$ and $b = (b_1, \ldots, b_n)$ are vectors in F^n and if $k \in \mathrm{F}$ then

$$\begin{aligned}
T(a + b) &= T(a_1 + b_1, \ldots, a_n + b_n) \\
&= (a_1 + b_1)x_1 + \ldots + (a_n + b_n)x_n \\
&= (a_1x_1 + \ldots + a_nx_n) + (b_1x_1 + \ldots + b_nx_n) \\
&= Ta + Tb, \\
T(ka) &= T(ka_1, \ldots, ka_n) \\
&= (ka_1)x_1 + \ldots + (ka_n)x_n \\
&= k(a_1x_1) + \ldots + k(a_nx_n) \\
&= k(a_1x_1 + \ldots + a_nx_n) \\
&= k(Ta),
\end{aligned}$$

thus T is linear.

Recall that e_i is the element of F^n with 1 in the ith coordinate and 0s elsewhere (Example 1.5.3). The formula $Te_i = x_i$ follows from the relations $1x_i = x_i$ and $0x_j = \theta$ for $j \neq i$. If S is another linear mapping such that

$Se_i = x_i$ for all i then, by linearity, $Sa = Ta$ for every linear combination a of $e_1, \ldots, e_n$, in other words for every $a \in \mathrm{F}^n$ (1.5.3). ■

A feeling for the importance of linearity can be gained by working carefully through the following list of examples.

2.1.4 Example Let $\mathcal{P}$ be the vector space of polynomial functions $p:\mathbf{R} \to \mathbf{R}$ (1.3.5) and let $T:\mathcal{P} \to \mathcal{P}$ be the mapping defined by $Tp = p'$ (the derivative of p). The additivity and homogeneity of T are familiar properties of differentiation.

2.1.5 Example With $\mathcal{P}$ as in the preceding example, define $S:\mathcal{P} \to \mathcal{P}$ as follows: for $p \in \mathcal{P}$, Sp is the prederivative of p with constant term 0, that is, Sp is the polynomial function such that $(Sp)' = p$ and $(Sp)(0) = 0$. {For example, if $p(t) = t + 5$ then $(Sp)(t) = \frac{1}{2}t^2 + 5t$.} The mapping S is linear (for example, by the familiar properties of 'indefinite integrals').

2.1.6 Example Each of the following formulas defines a linear mapping $T:\mathbf{R}^3 \to \mathbf{R}^3$:

$$T(x_1, x_2, x_3) = (x_2, x_1, x_3)$$
$$T(x_1, x_2, x_3) = (2x_1 + x_2 - x_3, x_2, x_3)$$
$$T(x_1, x_2, x_3) = (0, x_2, x_3)$$
$$T(x_1, x_2, x_3) = (5x_1, 5x_2, 5x_3).$$

2.1.7 Example The mappings $T:\mathbf{R}^3 \to \mathbf{R}^2$ and $S:\mathbf{R}^2 \to \mathbf{R}^3$ defined by $T(x_1, x_2, x_3) = (x_2, x_3)$ and $S(x_1, x_2) = (0, x_1, x_2)$ are linear.

2.1.8 Example If V is any vector space and a is a fixed scalar, then the mapping $T:\mathrm{V} \to \mathrm{V}$ defined by $Tx = ax$ is linear. {The commutativity of multiplication in the field of scalars plays a crucial role here: $T(cx) = a(cx) = (ac)x = (ca)x = c(ax) = c(Tx)$.}

2.1.9 Example Consider a system of m linear equations in n unknowns $x_1, \ldots, x_n$:

$$(*) \quad \begin{cases} a_{11}x_1 + a_{12}x_2 + \ldots + a_{1n}x_n = c_1 \\ a_{21}x_1 + a_{22}x_2 + \ldots + a_{2n}x_n = c_2 \\ \ldots \\ a_{m1}x_1 + a_{m2}x_2 + \ldots + a_{mn}x_n = c_m \end{cases}$$

where the a_{ij} and c_i are given scalars (say in $\mathbf{R}$). The problem is to find an n-ple $(x_1, \ldots, x_n)$ of scalars such that when these scalars are substituted into $(*)$, all of the equations are satisfied. This is not always possible (consider the 'system' $0x_1 = 1$). We'll see later (in Chapter 4) how to decide whether there are any solutions and, if so, how to compute them; for the moment, let's just see how the problem can be formulated in vector space terms. For any n-ple $x = (x_1, \ldots, x_n)$, whether a solution or not, one can compute the left sides of the m equations $(*)$; write Tx for the resulting m-ple of scalars:

$$Tx = (a_{11}x_1 + \ldots + a_{1n}x_n, \ldots, a_{m1}x_1 + \ldots + a_{mn}x_n).$$

This defines a mapping $T:\mathbf{R}^n \to \mathbf{R}^m$. The vector $c = (c_1, \ldots, c_m)$ belongs to $\mathbf{R}^m$; the problem is to find (if possible) a vector x in $\mathbf{R}^n$ such that $Tx = c$. The vectors c for which such a solution is possible are precisely the vectors in the range of T. The mapping T is linear (proof?). {Suggested notation: Consider vectors $x = (x_1, \ldots, x_n)$, $y = (y_1, \ldots, y_n)$ and a scalar r.}

2.1.10 Example The mapping $T:\mathbf{R}^3 \to \mathbf{R}$ defined by $T(x_1, x_2, x_3) = 2x_1 + x_2 - x_3$ is linear. {Here $\mathbf{R}$ is regarded as a vector space over itself, sums and scalar multiples being the usual sum and product in $\mathbf{R}$. In other words, we make no distinction between $\mathbf{R}^1$ and $\mathbf{R}$.}

2.1.11 Definition If V is a vector space over F, a **linear form** on V is a linear mapping $f:\mathrm{V} \to \mathrm{F}$. {Here F is regarded as a vector space over itself with its usual sum and product (cf. 2.1.10).}

In other words, a linear form is a linear mapping whose values are scalars.

▶ ***Exercises***

1. Fix a vector $y \in \mathbf{R}^3$ and define $T:\mathbf{R}^3 \to \mathbf{R}^3$ by the formula $Tx = x \times y$ (the *cross product* of x and y). Then T is linear. {Review: If $x = (x_1, x_2, x_3)$ and $y = (y_1, y_2, y_3)$, then $x \times y = (d_1, d_2, d_3)$, where the d_i are the 2×2 determinants

$$d_1 = \begin{vmatrix} x_2 & x_3 \\ y_2 & y_3 \end{vmatrix}, \quad d_2 = \begin{vmatrix} x_3 & x_1 \\ y_3 & y_1 \end{vmatrix}, \quad d_3 = \begin{vmatrix} x_1 & x_2 \\ y_1 & y_2 \end{vmatrix},$$

defined by $d_1 = x_2y_3 - y_2x_3$, etc.}

2. Let T be a set, $\mathrm{V} = \mathcal{F}(\mathrm{T}, \mathbf{R})$ the vector space of all functions $x:\mathrm{T} \to \mathbf{R}$ (Example 1.3.4). Fix a point $t \in \mathrm{T}$ and define $f:\mathrm{V} \to \mathbf{R}$ by the formula $f(x) = x(t)$. Then f is a linear form on V.

3. If $T:\mathrm{V} \to \mathrm{W}$ is a linear mapping, then $T(x - y) = Tx - Ty$ for all x, y in V.

4. If V is a vector space, prove that the mapping $T:\mathrm{V} \times \mathrm{V} \to \mathrm{V}$ defined by $T(x, y) = x - y$ is linear. (It is understood that $\mathrm{V} \times \mathrm{V}$ has the product vector space structure described in Example 1.3.10.)

5. With notations as in Theorem 2.1.3, an element $(a_1, \ldots, a_n)$ of F^n such that $T(a_1, \ldots, a_n) = \theta$ is called a (linear) *relation* among the vectors $x_1, \ldots, x_n$. For example, suppose that $\mathrm{V} = \mathrm{F}^2$, $n = 3$ and $x_1 = (2, -3)$, $x_2 = (4, 1)$, $x_3 = (8, 9)$.

(i) Find a formula for the linear mapping $T:\mathrm{F}^3 \to \mathrm{F}^2$ defined in Theorem 2.1.3.

(ii) Show that $(-2, 3, -1)$ is a relation among x_1, x_2, x_3.

6. The proof of Theorem 2.1.2 uses only the additivity of the mapping T. Give a proof using only its homogeneity. {Hint: Corollaries 1.4.3, 1.4.5.}

7. If $\mathcal{P}$ is the vector space of real polynomial functions (Example 1.3.5) and $f:\mathcal{P} \to \mathbf{R}$ is the mapping defined by $f(p) = p'(1)$ (the value of the derivative of p at 1), then f is a linear form on $\mathcal{P}$. What is the geometric meaning of $f(p) = 0$?

8. Let $S:\mathcal{P} \to \mathcal{P}$ be the linear mapping of Example 2.1.5. Define $R:\mathcal{P} \to \mathcal{P}$ by $Rp = Sp + p(5)1$, where $p(5)1$ is the constant function defined by the real number $p(5)$. Then R is linear and $(Rp)' = p$. {So to speak, the constant of integration can be tailor-made for p in a linear fashion.} More generally, look at the mapping $Rp = Sp + f(p)1$, where f is any linear form on $\mathcal{P}$.

2.2 Linear mappings and linear subspaces: kernel and range

Linear subspaces and linear mappings both relate to the vector space structure, so it's not surprising that they relate to each other:

2.2.1 Theorem *Let* V *and* W *be vector spaces,* T:V $\to$ W *a linear mapping.*

(i) *If* M *is a linear subspace of* V, *then* T(M) *is a linear subspace of* W.

(ii) *If* N *is a linear subspace of* W, *then* T^{-1}(N) *is a linear subspace of* V.

Proof. When we say that T:V $\to$ W is a linear mapping, it is understood (2.1.1) that V and W are vector spaces over the same field of scalars (for example, both real or both complex).

(i) By T(M) we mean the image[1] under T of the vectors in M, that is,

$$T(\mathrm{M}) = \{Tx: \ x \in \mathrm{M}\}.$$

We have to verify that T(M) satisfies the three conditions of 1.6.1. Since $\theta \in$ M and $T\theta = \theta$, T(M) contains the zero vector. Assuming y, y' in T(M) and c a scalar, we must show that $y + y'$ and cy are in T(M). Say $y = Tx$, $y' = Tx'$, with x, $x' \in$ M. Then $x + x' \in$ M and $T(x + x') = Tx + Tx' = y + y'$, thus $y + y' \in T$(M). Also, $cx \in$ M and $T(cx) = c(Tx) = cy$, therefore $cy \in T$(M).

(ii) By T^{-1}(N) we mean the inverse image[1] of N under T, that is,

$$T^{-1}(\mathrm{N}) = \{x \in \mathrm{V}: \ Tx \in \mathrm{N}\}.$$

Since $T\theta = \theta \in$ N, we have $\theta \in T^{-1}$(N). Assuming x, x' in T^{-1}(N) and c a scalar, we must show that $x + x'$ and cx are in T^{-1}(N). We know that Tx and Tx' are in N, therefore

$$T(x + x') = Tx + Tx' \in \mathrm{N} \quad \text{and} \quad T(cx) = c(Tx) \in \mathrm{N},$$

thus $x + x'$ and cx are in T^{-1}(N). ■

There are two special cases of great importance:

[1] Appendix A.3.

2.2.2 Corollary *If $T: \mathrm{V} \to \mathrm{W}$ is a linear mapping, then its range $T(\mathrm{V})$ is a linear subspace of* W, *and $T^{-1}(\{\theta\}) = \{x \in \mathrm{V}: \ Tx = \theta\}$ is a linear subspace of* V.

Proof. Let $\mathrm{M} = \mathrm{V}$ and $\mathrm{N} = \{\theta\}$ in the theorem. ■

2.2.3 Definition With the preceding notations, $T^{-1}(\{\theta\})$ is called the **kernel** (or 'null space') of T and is denoted Ker T; thus,

$$\operatorname{Ker} T = \{x \in \mathrm{V}: \ Tx = \theta\}.$$

The range of T is also called the **image** of T and is denoted Im T; thus,

$$\operatorname{Im} T = T(\mathrm{V}) = \{Tx: \ x \in \mathrm{V}\}.$$

Finding the kernel and range is often the first step in the study of a linear mapping.

2.2.4 Example Given a system of m homogeneous linear equations in n unknowns, say with real coefficients a_{ij},

$$(*) \begin{cases} a_{11}x_1 + a_{12}x_2 + \ldots + a_{1n}x_n = 0 \\ a_{21}x_1 + a_{22}x_2 + \ldots + a_{2n}x_n = 0 \\ \ldots \\ a_{m1}x_1 + a_{m2}x_2 + \ldots + a_{mn}x_n = 0 \end{cases}$$

let $T: \mathbf{R}^n \to \mathbf{R}^m$ be the linear mapping described in Example 2.1.9. The kernel of T is the set of solution-vectors of the system $(*)$. The range of T is the set of vectors $c = (c_1, \ldots, c_m)$ in $\mathbf{R}^m$ for which the equation $Tx = c$ has a solution x in $\mathbf{R}^n$, that is, for which there exists a vector $x = (x_1, \ldots, x_n)$ satisfying the system of linear equations

$$(**) \begin{cases} a_{11}x_1 + a_{12}x_2 + \ldots + a_{1n}x_n = c_1 \\ a_{21}x_1 + a_{22}x_2 + \ldots + a_{2n}x_n = c_2 \\ \ldots \\ a_{m1}x_1 + a_{m2}x_2 + \ldots + a_{mn}x_n = c_m. \end{cases}$$

It is not a trivial matter to determine the kernel and range of T. (This question is addressed in Chapter 4; cf. §4.9, Exercise 3.)

2.2.5 Example Let $\mathcal{P}$ be the vector space of real polynomial functions and let $T: \mathcal{P} \to \mathcal{P}$ be the linear mapping defined by differentiation: $Tp = p'$ for all $p \in \mathcal{P}$ (2.1.4). The kernel of T consists of the constant functions. The range of T is $\mathcal{P}$. (Why?)

2.2.6 Example Let $f: \mathbf{R}^3 \to \mathbf{R}$ be the linear form (2.1.11) defined by $f(x) = 2x_1 - 3x_2 + x_3$ for all $x = (x_1, x_2, x_3) \in \mathbf{R}^3$. The kernel of f is a plane through the origin (example at the beginning of §1.6). The range of f is $\mathbf{R}$. (Why?)

2.2.7 Example Let $T:\mathbf{R}^3 \to \mathbf{R}^2$ be the linear mapping defined by $T(x_1, x_2, x_3) = (x_2, x_3)$. The kernel of T is a line through the origin. (Which line?) The range of T is $\mathbf{R}^2$.

2.2.8 Example Let $S:\mathbf{R}^2 \to \mathbf{R}^3$ be the linear mapping defined by $S(x_1, x_2) = (0, x_1, x_2)$. The kernel of S is $\{\theta\}$. The range of S is the 'y,z-plane' in $\mathbf{R}^3$.

The kernel provides a simple test for injectivity of a linear mapping:

2.2.9 Theorem *Let $T:\mathrm{V} \to \mathrm{W}$ be a linear mapping.*

(i) *For x, x' in* V, *$Tx = Tx'$ if and only if $x - x' \in \operatorname{Ker} T$.*
(ii) *T is injective if and only* $\operatorname{Ker} T = \{\theta\}$.

Proof.
(i) By the linearity of T, $T(x - x') = Tx - Tx'$, therefore $T(x - x') = \theta$ if and only if $Tx = Tx'$.
(ii) If $\operatorname{Ker} T = \{\theta\}$, we see from (i) that $Tx = Tx' \Rightarrow x = x'$, thus T is injective. Conversely, suppose T is injective; if $x \in \operatorname{Ker} T$ then $Tx = \theta = T\theta$, therefore $x = \theta$ by injectivity. ■

► ***Exercises***

1. With notations as in Examples 2.2.7 and 2.2.8, compute $T(Sy)$ for $y = (y_1, y_2) \in \mathbf{R}^2$. Compute $S(Tx)$ for $x = (5, -3, 4)$; also for $x = (0, -3, 4)$.

2. Let V be a vector space over F, $f:\mathrm{V} \to \mathrm{F}$ a linear form on V (Definition 2.1.11). Assume that f is not identically zero and choose a vector y such that $f(y) \neq 0$. Let N be the kernel of f. Prove that every vector x in V may be written $x = z + cy$ with $z \in \mathrm{N}$ and $c \in \mathrm{F}$, and that z and c are uniquely determined by x. {Hint: If $x \in \mathrm{V}$, compute the value of f at the vector $x - [f(x)/f(y)]y$.}

3. Let $S:\mathrm{V} \to \mathrm{W}$ and $T:\mathrm{V} \to \mathrm{W}$ be linear mappings and let $\mathrm{M} = \{x \in \mathrm{V}:\ Sx \in T(\mathrm{V})\}$. Prove that M is a linear subspace of V.

4. If $T:\mathrm{V} \times \mathrm{V} \to \mathrm{V}$ is the linear mapping $T(x, y) = x - y$ (§2.1, Exercise 4), determine the kernel and range of T.

5. Let V be a real vector space, W its complexification (§1.3, Exercise 4); write $\mathrm{W}_{\mathbf{R}}$ for W regarded as a real vector space (§1.3, Exercise 3). For $(x, y) \in \mathrm{W}$, we have

$$(x, y) = (x, \theta) + (\theta, y) = (x, \theta) + i(y, \theta).$$

Prove that $x \mapsto (x, \theta)$ is an injective mapping $\mathrm{V} \to \mathrm{W}_{\mathbf{R}}$. {'Identifying' $x \in \mathrm{V}$ with $(x, \theta) \in \mathrm{W}$, one can suggestively write $\mathrm{W} = \mathrm{V} + i\mathrm{V}$; so to speak, V is the 'real part' of its complexification.}

6. Let $\mathcal{P}$ be the vector space of real polynomial functions (1.3.5) and let $T:\mathcal{P} \to \mathcal{P}$ be the linear mapping defined by $Tp = p - p'$, where p' is the derivative of p. Prove that T is injective.

2.3 Spaces of linear mappings: $\mathscr{L}(\mathrm{V}, \mathrm{W})$ and $\mathscr{L}(\mathrm{V})$

2.3.1 Definition Let V and W be vector spaces over the same scalar field F. The set of *all* linear mappings $T{:}\mathrm{V} \to \mathrm{W}$ is denoted $\mathscr{L}(\mathrm{V}, \mathrm{W})$. When $\mathrm{V} = \mathrm{W}$, $\mathscr{L}(\mathrm{V}, \mathrm{V})$ is abbreviated $\mathscr{L}(\mathrm{V})$.

What 'all' means will be clear when we know more about the structure of vector spaces (Theorem 3.8.1); for the time being, our supply of examples of linear mappings (for general vector spaces V and W) is relatively meager.

2.3.2 Example (Zero mapping) The mapping $T{:}\mathrm{V} \to \mathrm{W}$ defined by setting $Tx = \theta$ for all $x \in \mathrm{V}$ is linear; it is called the **zero linear mapping** from V to W and is denoted 0. Thus $0 \in \mathscr{L}(\mathrm{V}, \mathrm{W})$, $0x = \theta$ for all $x \in \mathrm{V}$.

2.3.3 Example (Identity mapping) The mapping $T{:}\mathrm{V} \to \mathrm{V}$ defined by $Tx = x$ for all $x \in \mathrm{V}$ is linear; it is called the **identity linear mapping** on V and is denoted I (or I_{V}, if it is necessary to keep track of the space on which it acts). Thus $I \in \mathscr{L}(\mathrm{V})$, $Ix = x$ for all $x \in \mathrm{V}$.

2.3.4 Example If f is a linear form on V (2.1.11) and y is a fixed vector in W, the mapping $T{:}\mathrm{V} \to \mathrm{W}$ defined by $Tx = f(x)y$ for all $x \in \mathrm{V}$ is linear. {Proof: $T(x + x') = f(x + x')y = [f(x) + f(x')]y = f(x)y + f(x')y = Tx + Tx'$ and $T(cx) = f(cx)y = [cf(x)]y = c[f(x)y] = c(Tx)$.} The dependence of T on f and y is expressed by writing $T = f \otimes y$ (which suggests a kind of 'product' of f and y). Thus $f \otimes y \in \mathscr{L}(\mathrm{V}, \mathrm{W})$, $(f \otimes y)(x) = f(x)y$ for all $x \in \mathrm{V}$.

2.3.5 Example If a is a fixed scalar, then the mapping $T{:}\mathrm{V} \to \mathrm{V}$ defined by $Tx = ax$ is linear (2.1.8), thus $T \in \mathscr{L}(\mathrm{V})$; T is called a *scalar* linear mapping. (In a sense made precise in Definition 2.3.9, T is a scalar multiple of the identity, $T = aI$.)

2.3.6 Example If x is a fixed vector in V, then the mapping $T{:}\mathrm{F} \to \mathrm{V}$ defined by $Tc = cx$ is linear (proof?), thus $T \in \mathscr{L}(\mathrm{F}, \mathrm{V})$.

2.3.7 Example The elements of $\mathscr{L}(\mathrm{V}, \mathrm{F})$ are the linear forms on V (2.1.11).

Given a pair of vector spaces V and W over F, we have a set $\mathscr{L}(\mathrm{V}, \mathrm{W})$. Can it be made into a vector space? Why should we ask the question? At any rate, the problem would be to define sums and scalar multiples of linear mappings, and the motivation should come from 'real life'. That's what the next example is about.

2.3.8 Example Consider (as one might in Calculus) the differential equation $3y' + 2y = 0$, where the prime means derivative with respect to the variable x. A *solution* of the equation is a function $y = f(x)$ such that $3f'(x) + 2f(x) = 0$ for all x in the domain of f (if possible, for all $x \in \mathbf{R}$; if not, at least for x in some interval). For this equation, there is the trivial solution: the function identically zero, that is, $f(x) = 0$ for all $x \in \mathbf{R}$; what is wanted is an *interesting* solution (and, if possible, a list of *all* solutions). If f is a solution, then $f' = (-2/3)f$ shows that f is not only differentiable, but its derivative is also differentiable; and $f'' = (-2/3)f' = (4/9)f$, so f'' is also differentiable, $f''' = (-8/27)f$, and so on. In brief, a solution f must have derivatives of all orders. {You will long ago have thought of the solution $f(x) = e^{-2x/3}$ and are wondering what this has to do with linear mappings! Patience.} Let V be the set of all functions $f:\mathbf{R} \to \mathbf{R}$ that have derivatives of all orders. If f and g are such functions, then so are $f + g$ and cf $(c \in \mathbf{R})$, thus V is a vector space for the pointwise linear operations. {In other words, V is a linear subspace of the vector space $\mathcal{F}(\mathbf{R}, \mathbf{R})$ described in 1.3.4.} The formula $Df = f'$ defines a linear mapping in V, that is, $D \in \mathcal{L}(V)$; we are looking for functions $f \in V$ such that $3Df + 2f = 0$, or $3Df + 2If = 0$ (where I is the identity linear mapping). It is irresistibly tempting to 'factor' the expression on the left side, whether it means anything or not:

$$(3D + 2I)f = 0.$$

For the moment, *finding* such functions f is not the point of interest; what is interesting is that we seem to be looking at a 'linear combination' of linear mappings, formally $3D + 2I$.

Back to general vector spaces:

2.3.9 Definition For $S, T \in \mathcal{L}(V, W)$ and for a scalar a, define mappings $S + T:V \to W$ and $aT:V \to W$ by the formulas

$$(S + T)x = Sx + Tx, \quad (aT)x = a(Tx)$$

for all $x \in V$. So to speak, $S + T$ is the *pointwise sum* of S and T, and aT is the *pointwise scalar multiple* of T by a. (If the definition leaves you cold, try reading over the preceding example again.)

2.3.10 Lemma *If $S, T \in \mathcal{L}(V, W)$ and a is a scalar, then $S + T, aT \in \mathcal{L}(V, W)$.*

Proof. The problem is to show that the mappings $S + T:V \to W$ and $aT:V \to W$ are linear. For all x, x' in V and all scalars c,

$$\begin{aligned}(S + T)(x + x') &= S(x + x') + T(x + x')\\ &= (Sx + Sx') + (Tx + Tx')\\ &= (Sx + Tx) + (Sx' + Tx')\\ &= (S + T)x + (S + T)x'\end{aligned}$$

(the first equality by the definition of $S + T$, the second by the linearity of S and T, etc.), and

$$\begin{aligned}(S + T)(cx) &= S(cx) + T(cx)\\ &= c(Sx) + c(Tx)\\ &= c(Sx + Tx)\\ &= c((S + T)x),\end{aligned}$$

thus $S + T$ is linear. The proof that aT is linear is equally straightforward. (Isn't it?) ■

2.3.11 Theorem *If* V *and* W *are vector spaces over* F, *then* $\mathcal{L}(V, W)$ *is also a vector space over* F *for the operations defined in* 2.3.9.

Proof. By the lemma, $\mathcal{L}(V, W)$ is 'closed' for the proposed operations; it remains to verify the remaining axioms (3)–(9) of Definition 1.3.1.

(3) The equality $S + T = T + S$ follows from the corresponding equality in W: for all $x \in V$,

$$(S + T)x = Sx + Tx = Tx + Sx = (T + S)x.$$

(4) Associativity of addition is proved similarly.

(5) The zero linear mapping 0 satisfies $T + 0 = T$ for all $T \in \mathcal{L}(V, W)$.

(6) For $T \in \mathcal{L}(V, W)$ the mapping $-T = (-1)T$ is linear and does what is expected of it: $T + (-T) = 0$.

The verification of the remaining axioms (7)–(9) is equally straightforward. ■

Besides $S + T$ and aT, there is another important way of combining linear mappings:

2.3.12 Definition If U, V, W are vector spaces over F, and if $T: U \to V$ and $S: V \to W$ are mappings, we write $ST: U \to W$ for the **composite** mapping[1], defined by $(ST)x = S(Tx)$ for all $x \in U$. {Thus, ST is another notation for $S \circ T$ (Fig. 8).}

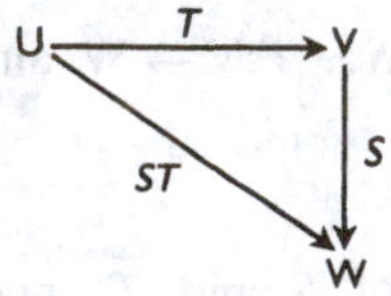

Fig. 8

The composite of two *linear* mappings is linear:

2.3.13 Theorem *If* $T \in \mathcal{L}(U, V)$ *and* $S \in \mathcal{L}(V, W)$, *then* $ST \in \mathcal{L}(U, W)$.

Proof. For x, x' in U and c scalar,

$$\begin{aligned}(ST)(x + x') &= S(T(x + x')) = S(Tx + Tx') \\ &= S(Tx) + S(Tx') = (ST)x + (ST)x',\end{aligned}$$

$$(ST)(cx) = S(T(cx)) = S(c(Tx)) = c(S(Tx)) = c((ST)x). \ ■$$

Composition of functions is associative, so if $T: U \to V$, $S: V \to W$ and $R: W \to X$, then $(RS)T = R(ST)$; one writes simply RST.

Theorem 2.3.13 yields something extra when $U = V = W$: if $S, T \in \mathcal{L}(V)$ then also $ST \in \mathcal{L}(V)$. Thus $\mathcal{L}(V)$ is closed under *three* operations: sums, scalar multiples, and products (i.e. composition). Some interesting interrelations between these operations are given in the exercises.

For $T \in \mathcal{L}(V)$, the **powers** of T are defined as in elementary algebra: $T^1 = T$, $T^2 = TT$, $T^3 = T^2T$, $T^4 = T^3T$, etc., and, to round things out, $T^0 = I$ (the identity linear mapping).

[1] Appendix A.3.

► ***Exercises***

1. Complete the details in the proof of Theorem 2.3.11.

2. Let V be a vector space. If $R, S, T \in \mathcal{L}(\mathrm{V})$ and c is a scalar, the following equalities are true:
   (i) $(RS)T = R(ST)$;
   (ii) $(R + S)T = RT + ST$;
   (iii) $R(S + T) = RS + RT$;
   (iv) $(cS)T = c(ST) = S(cT)$;
   (v) $TI = T = IT$ (I the identity mapping).

3. If $S, T \in \mathcal{L}(\mathbf{R}^3)$ are defined by the formulas
$$S(x, y, z) = (2x - y + 5z, 2y - z, 3z),$$
$$T(x, y, z) = (-x + y, 3x - y, 2x + z),$$
find the formulas for $S + T$, $3T$, $S - T$, and ST.

4. If $T:\mathrm{V} \to \mathrm{W}$ is a linear mapping and g is a linear form on W, show that the formula $f(x) = g(Tx)$ defines a linear form f on V. {Shortcut: $f = gT = g \circ T$.} The formula $T'g = f$ defines a mapping $T':\mathcal{L}(\mathrm{W}, \mathrm{F}) \to \mathcal{L}(\mathrm{V}, \mathrm{F})$, where F is the field of scalars. Show that T' is linear. {T' is called the *transpose* (or 'adjoint') of T.}

5. If V is a vector space over F, the vector space $\mathcal{L}(\mathrm{V}, \mathrm{F})$ of linear forms on V is called the *dual space* of V and is denoted V′. Thus, the correspondence $T \mapsto T'$ described in Exercise 4 defines a mapping $\mathcal{L}(\mathrm{V}, \mathrm{W}) \to \mathcal{L}(\mathrm{W}', \mathrm{V}')$. Show that this mapping is linear, that is, $(S + T)' = S' + T'$ and $(cT)' = cT'$ for all S, T in $\mathcal{L}(\mathrm{V}, \mathrm{W})$ and all scalars c.

6. With notations as in Exercise 4, prove that if T is surjective then T' is injective; in other words, show that if $T(\mathrm{V}) = \mathrm{W}$ then $T'g = 0 \Rightarrow g = 0$.

7. As in Example 2.1.4, let $\mathrm{V} = \mathcal{P}$ be the real vector space of all polynomial functions $p:\mathbf{R} \to \mathbf{R}$, $D:\mathrm{V} \to \mathrm{V}$ the linear mapping defined by $Dp = p'$ (the derivative of p); let $D':\mathrm{V}' \to \mathrm{V}'$ be the transpose of D as defined in Exercise 4. (Caution: The prime is being used with three different meanings.)

   For each $t \in \mathbf{R}$ let φ_t be the linear form on V defined by $\varphi_t(p) = p(t)$. Let $[a, b]$ be a closed interval in $\mathbf{R}$ and let φ be the linear form on V defined by
$$\varphi(p) = \int_a^b p(t)dt.$$
   Prove: $D'\varphi = \varphi_b - \varphi_a$. {Hint: Fundamental Theorem of Calculus.}

8. For each real number α, define $T_\alpha \in \mathcal{L}(\mathbf{R}^2)$ by the formula
$$T_\alpha(x, y) = (x \cos\alpha - y \sin\alpha, x \sin\alpha + y \cos\alpha).$$
   (i) Find $T_\alpha(1, 0)$ and $T_\alpha(0, 1)$,
   (ii) Show that $T_\alpha T_\beta = T_{\alpha+\beta}$.

(iii) Geometric interpretation?

(iv) Describe T_0.

(v) Prove that T_α is bijective. {Hint: Contemplate $\beta = -\alpha$ in (ii).}

9. Let U, V, W be vector spaces over F. Prove:

(i) For fixed $T \in \mathcal{L}(\mathrm{U}, \mathrm{V})$, $S \mapsto ST$ is a linear mapping $\mathcal{L}(\mathrm{V}, \mathrm{W}) \to \mathcal{L}(\mathrm{U}, \mathrm{W})$ (see Figure 8 in 2.3.12).

(ii) For fixed $S \in \mathcal{L}(\mathrm{V}, \mathrm{W})$, $T \mapsto ST$ is a linear mapping $\mathcal{L}(\mathrm{U}, \mathrm{V}) \to \mathcal{L}(\mathrm{U}, \mathrm{W})$.

10. If $T{:}\mathrm{U} \to \mathrm{V}$ and $S{:}\mathrm{V} \to \mathrm{W}$ are linear mappings, prove that $\operatorname{Im}(ST) \subset \operatorname{Im} S$ and $\operatorname{Ker}(ST) \supset \operatorname{Ker} T$.

11. Let $T{:}\mathrm{V} \to \mathrm{V}$ be a linear mapping such that $\operatorname{Im} T \subset \operatorname{Ker}(T - I)$, where I is the identity mapping (2.3.3). Prove that $T^2 = T$. {Recall that $T^2 = TT$.}

12. Suppose $\mathrm{V} = \mathrm{M} \oplus \mathrm{N}$ in the sense of §1.6, Exercise 12. For each $x \in \mathrm{V}$ let $x = y + z$ be its unique decomposition with $y \in \mathrm{M}$, $z \in \mathrm{N}$ and define $Px = y$, $Qx = z$. Prove that $P, Q \in \mathcal{L}(\mathrm{V})$, $P^2 = P$, $Q^2 = Q$, $P + Q = I$ and $PQ = QP = 0$.

13. Let $T{:}\mathrm{V} \to \mathrm{V}$ be a linear mapping such that $T^2 = T$, and let

$$\mathrm{M} = \operatorname{Im} T, \qquad \mathrm{N} = \operatorname{Ker} T.$$

Prove that $\mathrm{M} = \{x \in \mathrm{V}\colon \ Tx = x\} = \operatorname{Ker}(T - I)$ and that $\mathrm{V} = \mathrm{M} \oplus \mathrm{N}$ in the sense of §1.6, Exercise 12.

14. Find $S, T \in \mathcal{L}(\mathbf{R}^2)$ such that $ST \neq TS$.

15. Let $T{:}\mathrm{U} \to \mathrm{V}$ and $S{:}\mathrm{V} \to \mathrm{W}$ be linear mappings. Prove:

(i) If S is injective, then $\operatorname{Ker}(ST) = \operatorname{Ker} T$.

(ii) If T is surjective, then $\operatorname{Im}(ST) = \operatorname{Im} S$.

16. Let V be a real or complex vector space and let $T \in \mathcal{L}(\mathrm{V})$ be such that $T^2 = I$. Define

$$\mathrm{M} = \{x \in \mathrm{V}\colon \ Tx = x\}, \qquad \mathrm{N} = \{x \in \mathrm{V}\colon \ Tx = -x\}.$$

Prove that M and N are linear subspaces of V such that $\mathrm{V} = \mathrm{M} \oplus \mathrm{N}$. {Hint: For every vector x, $x = \frac{1}{2}(x + Tx) + \frac{1}{2}(x - Tx)$.}

17. Let $\mathrm{V} = \mathcal{F}(\mathbf{R})$ be the real vector space of all functions $x{:}\mathbf{R} \to \mathbf{R}$ (1.3.4). For $x \in \mathrm{V}$ define $Tx \in \mathrm{V}$ by the formula $(Tx)(t) = x(-t)$. Analyze T in the light of Exercise 16. {Remember §1.6, Exercise 13?}

18. If $T \in \mathcal{L}(\mathrm{U}, \mathrm{V})$ and $S \in \mathcal{L}(\mathrm{V}, \mathrm{W})$, prove that $(ST)' = T'S'$ (cf. Exercise 4).

19. Let $S, T \in \mathcal{L}(\mathrm{V}, \mathrm{W})$ and let $\mathrm{M} = \{x \in \mathrm{V}\colon \ Sx = Tx\}$. Prove that M is a linear subspace of V. {Hint: Consider $\operatorname{Ker}(S - T)$.}

20. If $T:\mathrm{V} \to \mathrm{W}$, $S:\mathrm{W} \to \mathrm{V}$ are linear mappings such that $ST = I$, prove that $\operatorname{Ker} T = \{\theta\}$ and $\operatorname{Im} S = \mathrm{V}$.

21. If $\mathrm{V} = \{\theta\}$ or $\mathrm{W} = \{\theta\}$ then $\mathcal{L}(\mathrm{V}, \mathrm{W}) = \{0\}$. If $\mathrm{V} \neq \{\theta\}$ then $I_{\mathrm{V}} \neq 0$.

22. A lightning proof of Theorem 2.3.11 can be based on an earlier exercise: Since $0 \in \mathcal{L}(\mathrm{V}, \mathrm{W})$, Lemma 2.3.10 shows that $\mathcal{L}(\mathrm{V}, \mathrm{W})$ is a linear subspace of the vector space $\mathcal{F}((\mathrm{V}, \mathrm{W})$ defined as in §1.3, Exercise 1, therefore $\mathcal{L}(\mathrm{V}, \mathrm{W})$ is also a vector space (1.6.2).

23. If $T:\mathrm{U} \to \mathrm{V}$ and $S:\mathrm{V} \to \mathrm{W}$ are linear mappings, prove that $\operatorname{Ker}(ST) = T^{-1}(\operatorname{Ker} S)$.

24. Let $\mathrm{V} = \mathcal{P}$ be the vector space of real polynomial functions, $D:\mathrm{V} \to \mathrm{V}$ the differentiation mapping $Dp = p'$ (2.1.4). Let u be the monomial $u(t) = t$ and define another linear mapping $M:\mathrm{V} \to \mathrm{V}$ by the formula $Mp = up$. (So to speak, M is multiplication by t.) Prove that $DM - MD = I$ (the identity mapping).

{Hint: The claim is that $(up)' - up' = p$ for all p. Remember the 'product rule'?}

25. Let $D:\mathcal{P} \to \mathcal{P}$ be the differentiation mapping of Exercise 24 and let $S:\mathcal{P} \to \mathcal{P}$ be the linear mapping such that $(Sp)' = p$ and $(Sp)(0) = 0$ (2.1.5). Prove:

(i) If $R:\mathcal{P} \to \mathcal{P}$ is any linear mapping such that $(Rp)' = p$ for all $p \in \mathcal{P}$, then $DR = I$.

(ii) If $T:\mathcal{P} \to \mathcal{P}$ is a linear mapping such that $DT = 0$, then there exists a linear form f on $\mathcal{P}$ such that (in the notation of Example 2.3.4) $T = f \otimes 1$, where 1 is the constant function 1. {Hint: $\operatorname{Im} T \subset \operatorname{Ker} D$.}

(iii) With R as in (i), $R = S + f \otimes 1$ for a suitable linear form f on $\mathcal{P}$. {Hint: Consider $T = R - S$.} Remember §2.1, Exercise 8?

26. Let $S, T \in \mathcal{L}(\mathrm{V})$. If $ST - I$ is injective then so is $TS - I$. {Hint: $S(TS - I) = (ST - I)S$.}

2.4 Isomorphic vector spaces

In $\mathbf{R}^3$, a plane through the origin is a lot like $\mathbf{R}^2$. Vague as it is, this statement is surely true for, say, the xy-plane $\{(x, y, 0):\ x, y \in \mathbf{R}\}$, whose elements are the 'same' as those of $\mathbf{R}^2$ except for the gratuitous 0; the feeling of 'sameness' is based on the perception that the point (x, y) of $\mathbf{R}^2$ corresponds to the point $(x, y, 0)$ of the xy-plane.

A polynomial is known by the coefficients it keeps. For example, the cubic polynomial $5 + 4t - 2t^3$ is completely determined by the list of coefficients $(5, 4, 0, -2)$. The vector space of polynomials of degree ≤ 3 (Example 1.3.6) is in some sense the 'same' as $\mathbf{R}^4$, with the point (a_0, a_1, a_2, a_3) of $\mathbf{R}^4$ corresponding to the polynomial $a_0 + a_1 t + a_2 t^2 + a_3 t^3$.

On the other hand, $\mathbf{R}^2$ and $\mathbf{R}^3$ are decidedly 'different'. {Intuitively, there are 'copies' of $\mathbf{R}^2$ inside $\mathbf{R}^3$, but not the other way around.}

The point of these examples is that the 'same' vector space can present itself in different disguises. Our understanding of vector spaces is heightened by our ability to decide when they are essentially the same or essentially different. The mathematical tool for this is the concept of isomorphism[1].

2.4.1 Definition A vector space V is said to be **isomorphic** to a vector space W if there exists a bijective[2] linear mapping $T:\mathrm{V}\rightarrow\mathrm{W}$. Such a mapping is called an **isomorphism** of V onto W. We write $\mathrm{V}\cong\mathrm{W}$ to indicate that V is isomorphic to W.

2.4.2 Example Let $\mathrm{V}=\mathbf{R}^2$ and let W be the linear subspace of $\mathbf{R}^3$ consisting of all vectors $(x_1, x_2, 0)$. Then $\mathrm{V}\cong\mathrm{W}$; a specific isomorphism $T:\mathrm{V}\rightarrow\mathrm{W}$ is $T(x_1, x_2) = (x_1, x_2, 0)$. {Another is $T(x_1, x_2) = (x_2, x_1, 0)$.} It is straightforward to check that the proposed mapping is linear and bijective.

2.4.3 Example Let n be a fixed positive integer, $\mathrm{V}=\mathbf{R}^{n+1}$ and $\mathrm{W}=\mathcal{P}_n$ the vector space of real polynomials of degree $\leq n$ (1.3.6). The mapping $\mathrm{V}\rightarrow\mathrm{W}$ defined by $(a_0, a_1, \ldots, a_n)\mapsto a_0 + a_1t + \ldots + a_nt^n$ is linear and bijective, therefore $\mathrm{V}\cong\mathrm{W}$. {Linearity and surjectivity are obvious. If $a_0 + a_1t + \ldots + a_nt^n$ is the zero polynomial then all of its coefficients must be zero; thus the kernel of the mapping is zero, whence injectivity (2.2.9).}

2.4.4 Example Let $\mathrm{T}=\{1, 2, \ldots, n\}$ and let $\mathrm{V}=\mathcal{F}(\mathrm{T}, \mathbf{R})$ be the vector space of all functions $x:\mathrm{T}\rightarrow\mathbf{R}$, with the pointwise linear operations (1.3.4). Then $\mathrm{V}\cong\mathbf{R}^n$ via the linear bijection $x\mapsto(x(1), x(2), \ldots, x(n))$.

2.4.5 Example The field $\mathbf{C}$ of complex numbers can be thought of as a vector space over $\mathbf{R}$: addition in $\mathbf{C}$ as usual, and multiplication only by real scalars (cf. §1.3, Exercise 3). Then $\mathbf{R}^2\cong\mathbf{C}$ (as real vector spaces), via the linear bijection $(a, b)\mapsto a + bi$.

The inverse of a bijective linear mapping is automatically linear:

2.4.6 Theorem *Let* V *and* W *be vector spaces over* F, $T:\mathrm{V}\rightarrow\mathrm{W}$ *a bijective linear mapping. Then the inverse mapping* $T^{-1}:\mathrm{W}\rightarrow\mathrm{V}$ *is also linear.*

Proof. Let $y,\ y'\in\mathrm{W}$ and $c\in\mathrm{F}$; the problem is to show that $T^{-1}(y + y') = T^{-1}y + T^{-1}y'$ and $T^{-1}(cy) = c(T^{-1}y)$. Thought: To show that two vectors u, v of V are equal, it suffices to show that $Tu = Tv$. {Reason: T is injective.} Let's apply this idea to the vectors $u = T^{-1}(y + y')$ and $v = T^{-1}y + T^{-1}y'$:

$$Tu = TT^{-1}(y + y') = y + y',$$

[1] The term 'isomorphism', which is meant to convey the idea of 'same structure', is a combination of the Greek words ísos (equal) and morphé (form).

[2] A mapping is said to be *bijective* (Appendix A.3) if it is one-one and onto (that is, both injective and surjective).

whereas

$$Tv = T(T^{-1}y + T^{-1}y') = T(T^{-1}y) + T(T^{-1}y') = y + y'.$$

The argument uses the additivity of T, and proves that T^{-1} is additive.

The same technique works for the pair $u = T^{-1}(cy)$ and $v = c(T^{-1}y)$, using the homogeneity of T. ■

The message of Theorem 2.4.6 is that the relation of isomorphism is symmetric: if $\mathrm{V} \cong \mathrm{W}$ then also $\mathrm{W} \cong \mathrm{V}$. Two other properties of isomorphism should be mentioned:

2.4.7 Theorem

Let U, V, W *be vector spaces over the same field of scalars. Then:*

(i) $\mathrm{V} \cong \mathrm{V}$;

(ii) *if* $\mathrm{V} \cong \mathrm{W}$ *then* $\mathrm{W} \cong \mathrm{V}$;

(iii) *if* $\mathrm{U} \cong \mathrm{V}$ *and* $\mathrm{V} \cong \mathrm{W}$, *then* $\mathrm{U} \cong \mathrm{W}$.

Proof.

(i) The identity mapping $I:\mathrm{V} \to \mathrm{V}$ is bijective and linear.

(ii) This is 2.4.6.

(iii) If $T:\mathrm{U} \to \mathrm{V}$ and $S:\mathrm{V} \to \mathrm{W}$ are linear bijections, then the composite mapping $ST:\mathrm{U} \to \mathrm{W}$ is also linear and bijective. ■

These three properties of isomorphism are called 'reflexivity', 'symmetry', and 'transitivity'. In view of symmetry, the relation $\mathrm{V} \cong \mathrm{W}$ can be expressed by saying that V and W are isomorphic (the roles of V and W are symmetric).

2.4.8 Terminology (optional)

When vector spaces are fitted into the context of general algebraic structures, a linear mapping is also called a *homomorphism* (of vector spaces), an injective linear mapping is called a *monomorphism*, and a surjective linear mapping is called an *epimorphism*; a linear mapping of a vector space V into itself (i.e., an element of $\mathcal{L}(\mathrm{V})$) is called an *endomorphism* of V. In general algebraic structures, a bijective homomorphism is called an *isomorphism*, and an isomorphism of an algebraic structure onto itself is called an *automorphism* (self-isomorphism) of the structure; thus, an automorphism of a vector space V is a bijective element of $\mathcal{L}(\mathrm{V})$.

▶ ***Exercises***

1. Write out complete proofs of the assertions in Examples 2.4.2–2.4.5.

2. Regard $\mathbf{C}^n$ as a real vector space in the natural way (sums as usual, multiplication only by real scalars); then $\mathbf{C}^n \cong \mathbf{R}^{2n}$. Prove this by exhibiting a bijection $\mathbf{C}^n \to \mathbf{R}^{2n}$ that is linear for the real vector space structures.

3. Let X be a set, V a vector space, $f:\mathrm{X} \to \mathrm{V}$ a bijection. There exists a unique vector space structure on X for which f is a linear mapping (hence a vector space isomorphism).

{Hint: Define sums and scalar multiples in X by the formulas

$$x + y = f^{-1}(f(x) + f(y)), \qquad cx = f^{-1}(cf(x)).$$

This trick is called 'transport of structure'.}

4. Let V be the set of all n-ples $x = (x_1, \dots, x_n)$ of real numbers >0, that is,

$$V = \{(x_1, \dots, x_n): \quad x_i \in (0, +\infty) \text{ for } i = 1, \dots, n\}.$$

For $x = (x_1, \dots, x_n)$, $y = (y_1, \dots, y_n)$ in V and $c \in \mathbf{R}$, define

$$x \oplus y = (x_1y_1, \dots, x_ny_n), \qquad c \cdot x = (x_1^c, \dots, x_n^c)$$

(here x_iy_i and x_i^c are the usual product and power). Define a mapping $T{:}\mathrm{V} \to \mathbf{R}^n$ by the formula $T(x_1, \dots, x_n) = (\log x_1, \dots, \log x_n)$.

(1) Show that $T(x \oplus y) = Tx + Ty$ and $T(c \cdot x) = c(Tx)$ for all x, y in V and all $c \in \mathbf{R}$.

(2) True or false (explain): V is a real vector space for the operations $\oplus$ and $\cdot$.

5. Let V be a vector space, $T \in \mathcal{L}(\mathrm{V})$ bijective, and c a nonzero scalar. Prove that cT is bijective and that $(cT)^{-1} = c^{-1}T^{-1}$.

6. Let $T{:}\mathrm{V} \to \mathrm{W}$ be a bijective linear mapping. For each $S \in \mathcal{L}(\mathrm{V})$, define $\varphi(S) = TST^{-1}$ (note that the product is defined). Prove that φ: $\mathcal{L}(\mathrm{V}) \to \mathcal{L}(\mathrm{W})$ is a bijective linear mapping such that $\varphi(RS) = \varphi(R)\varphi(S)$ for all R, S in $\mathcal{L}(\mathrm{V})$.

7. Let V be a vector space, $T \in \mathcal{L}(\mathrm{V})$. Prove:

(i) If $T^2 = 0$ then $I - T$ is bijective.

(ii) If $T^n = 0$ for some positive integer n, then $I - T$ is bijective.

{Hint: (i) In polynomial algebra, $(1 - t)(1 + t) = 1 - t^2$.}

8. Let U, V, W be vector spaces over the same field. Form the product vector spaces U × V, V × W (Definition 1.3.11), then the product vector spaces (U × V) × W and U × (V × W). Prove that $(\mathrm{U} \times \mathrm{V}) \times \mathrm{W} \cong \mathrm{U} \times (\mathrm{V} \times \mathrm{W})$.

9. Prove that $\mathbf{R}^2 \times \mathbf{R}^3 \cong \mathbf{R}^5$.

10. With notations as in Exercise 8, prove that $\mathrm{V} \times \mathrm{W} \cong \mathrm{W} \times \mathrm{V}$.

11. With notations as in Example 2.4.2, prove that $(x_1, x_2) \mapsto (x_1, x_1 + x_2, 0)$ is an isomorphism $\mathbf{R}^2 \to \mathrm{W}$.

2.5 Equivalence relations and quotient sets

Where are we? The fundamental concepts introduced so far are *vector space*, *linear subspace* and *linear mapping*. Brushing past a host of examples, two constructions ('new spaces from old') stand out: for a pair of vector spaces V and W, the *vector space* $\mathcal{L}(\mathrm{V}, \mathrm{W})$ of *linear mappings* of V into W (Definition 2.3.1) and the *product vector space* V × W (Definition 1.3.11).

One more construction of this stature remains to be taken up (in the next section): *quotient vector spaces*. The present section, mostly about sets and functions, prepares the way for quotient vector spaces.

2.5.1 Theorem *Let $f:\mathrm{A}\to\mathrm{B}$ be any function. For x, y in A, write $x \sim y$ if $f(x) = f(y)$. Then:*

(i) $x \sim x$ *for all* $x \in \mathrm{A}$;
(ii) *if* $x \sim y$ *then* $y \sim x$;
(iii) *if* $x \sim y$ *and* $y \sim z$, *then* $x \sim z$.

Proof. This is immediate from the properties of equality in B. For example, if $f(x) = f(y)$ and $f(y) = f(z)$, then $f(x) = f(z)$, which proves (iii). ■

This seems thin soup indeed, but there are many interesting instances of 'relations' $x \sim y$ having the properties (i)–(iii). Here's an example from arithmetic:

2.5.2 Example Fix a positive integer n. For integers $x, y \in \mathbf{Z}$, declare $x \sim y$ in case $x - y$ is an integral multiple of n. {For example, if $n = 6$ then $5 \sim 23$ because $5 - 23 = -18 = (-3)6$.}. Thus, $x \sim y$ means that $x - y = kn$ for some $k \in \mathbf{Z}$; writing $\mathbf{Z}n = \{kn: k \in \mathbf{Z}\}$ for the set of all integral multiples of n, we have

$$x \sim y \Leftrightarrow x - y \in \mathbf{Z}n.$$

Properties (i)–(iii) are easily verified; for instance, (iii) is proved by noting that if $x - y = kn$ and $y - z = jn$, then addition of these equations yields $x - z = (k + j)n$. In this example, the traditional notation for $x \sim y$ is

$$x \equiv y \pmod{n},$$

and x is said to be *congruent to* y *modulo* n.

Here's the general idea that underlies such situations:

2.5.3 Definition Let X be a nonempty set. Suppose that for each ordered pair (x, y) of elements of X, we are given a statement $\mathrm{S}(x, y)$ about x and y. Write $x \sim y$ if the statement $\mathrm{S}(x, y)$ is true. We say that $\sim$ is an **equivalence relation** in X if the following three conditions hold:

(i) $x \sim x$ for all $x \in \mathrm{X}$ (*reflexivity*);
(ii) if $x \sim y$ then $y \sim x$ (*symmetry*);
(iii) if $x \sim y$ and $y \sim z$ then $x \sim z$ (*transitivity*).

When $x \sim y$ we say that x is **equivalent** to y (for the relation $\sim$).

The specific example that leads to 'quotient vector spaces' is the following:

2.5.4 Theorem *Let* V *be a vector space*, M *a linear subspace of* V. *For x, y in* V, *write $x \sim y$ if $x - y \in \mathrm{M}$. Then $\sim$ is an equivalence relation in* V.

Proof.
(i) For all $x \in \mathrm{V}$, $x - x = \theta \in \mathrm{M}$, therefore $x \sim x$.

(ii) If $x - y \in \mathrm{M}$ then $y - x = -(x - y) \in \mathrm{M}$.
(iii) If $x - y \in \mathrm{M}$ and $y - z \in \mathrm{M}$, then $x - z = (x - y) + (y - z) \in \mathrm{M}$. ■

In a set with an equivalence relation, it is of interest to gather together all of the elements equivalent to a given element:

2.5.5 Definition If $\sim$ is an equivalence relation in the set X and if $x \in \mathrm{X}$, the set of all elements of X that are equivalent to x is denoted $[x]$ and is called the **equivalence class** of x; thus,

$$[x] = \{y \in \mathrm{X}: \ y \sim x\}.$$

2.5.6 Example With $f{:}\mathrm{A} \to \mathrm{B}$ and $\sim$ as in 2.5.1, $[x] = f^{-1}(\{f(x)\})$; thus, $[x]$ is the set of all points of A where f has the same value that it has at x.

{Proof: If $y \in \mathrm{A}$ then $y \in [x] \Leftrightarrow y \sim x \Leftrightarrow f(y) = f(x) \Leftrightarrow f(y) \in \{f(x)\} \Leftrightarrow y \in f^{-1}(\{f(x)\})$.}

2.5.7 Example With V, M and $\sim$ as in 2.5.4, $[x] = \{x + z: \ z \in \mathrm{M}\}$.

{Proof. For $y \in \mathrm{V}$, $y \sim x \Leftrightarrow y - x \in \mathrm{M} \Leftrightarrow y - x = z$ for some $z \in \mathrm{M}$.}

2.5.8 Definition With notation as in 2.5.7, the set $\{x + z: \ z \in \mathrm{M}\}$ is written $x + \mathrm{M}$ and is called the **coset** of x **modulo** M.

One can think of $x + \mathrm{M}$ as the result of 'translating' M by adding the fixed vector x to each of its elements.

2.5.9 Example As in 2.2.6, let $f{:}\mathbf{R}^3 \to \mathbf{R}$ be the linear form defined by

$$f(x) = 2x_1 - 3x_2 + x_3$$

for all $x = (x_1, x_2, x_3) \in \mathbf{R}^3$. Let M be the kernel of f,

$$\mathrm{M} = f^{-1}(\{0\}) = \{x \in \mathbf{R}^3: \ 2x_1 - 3x_2 + x_3 = 0\};$$

thus, M is a plane through the origin. The range of f is $\mathbf{R}$: $\operatorname{Im} f = f(\mathbf{R}^3) = \mathbf{R}$. For every $c \in \mathbf{R}$,

$$f^{-1}(\{c\}) = \{x \in \mathbf{R}^3: \ 2x_1 - 3x_2 + x_3 = c\};$$

this is a plane, either equal to M (if $c = 0$) or parallel to M (if $c \neq 0$).

For x, y in $\mathbf{R}^3$, $f(x) = f(y) \Leftrightarrow x - y \in \mathrm{M}$, that is, the equivalence relation defined as in 2.5.1 coincides with the one defined in 2.5.4. If $u \in \mathbf{R}^3$, the equivalence class of u is given by

$$[u] = u + \mathrm{M} = f^{-1}(\{f(u)\}) = \{x \in \mathbf{R}^3: \ 2x_1 - 3x_2 + x_3 = f(u)\},$$

thus the set of equivalence classes is the set of planes consisting of M and the planes parallel to it; in particular, for each $u \in \mathbf{R}^3$, $[u] = u + \mathrm{M}$ is the plane through u parallel to M. {Except that if $u \in \mathrm{M}$, then $[u] = f^{-1}(\{f(u)\}) = f^{-1}(\{0\}) = \mathrm{M}$.}

For example, if $u = (-5, 1, 2)$ then $f(u) = -11$ and $[u]$ is the plane with equation $2x_1 - 3x_2 + x_3 = -11$.

2.5.10 Remarks In the preceding example, the set of equivalence classes $x + \mathrm{M}$ $(x \in \mathbf{R}^3)$ effects a 'decomposition' of $\mathbf{R}^3$ into a set of parallel planes; each point of $\mathbf{R}^3$ lies on precisely one of these planes.

In Example 2.5.6, the equivalence classes $[x] = f^{-1}(\{f(x)\})$ decompose the domain A of f into the 'sets of constancy' of f; for each $y \in f(\mathrm{A})$, say $y = f(x)$, the set on which f takes the value y is $f^{-1}(\{y\}) = f^{-1}(\{f(x)\}) = [x]$.

The technical term for what is going on in these examples is 'partition':

2.5.11 Definition Let X be a nonempty set. A **partition** of X is a set $\mathcal{A}$ of subsets of X such that:

(1) every $\mathrm{A} \in \mathcal{A}$ is nonempty;

(2) if $\mathrm{A}, \mathrm{B} \in \mathcal{A}$ and $\mathrm{A} \neq \mathrm{B}$, then $\mathrm{A} \cap \mathrm{B} = \varnothing$;

(3) for each $x \in \mathrm{X}$, there exists $\mathrm{A} \in \mathcal{A}$ with $x \in \mathrm{A}$.

Condition (2) says (in contrapositive form[1]) that if two sets in $\mathcal{A}$ have an element in common, then they are identical; thus, an element of X can belong to *at most* one set in $\mathcal{A}$. Condition (3) says that each element of X belongs to *at least* one set in $\mathcal{A}$. Conditions (2) and (3) together say that each element of X belongs to *exactly* one set in $\mathcal{A}$.

Figure 9 represents a partition of a set X into five pairwise disjoint subsets. {A little reflection about points on boundaries (where are they?) shows that the representation still leaves something to the imagination.}

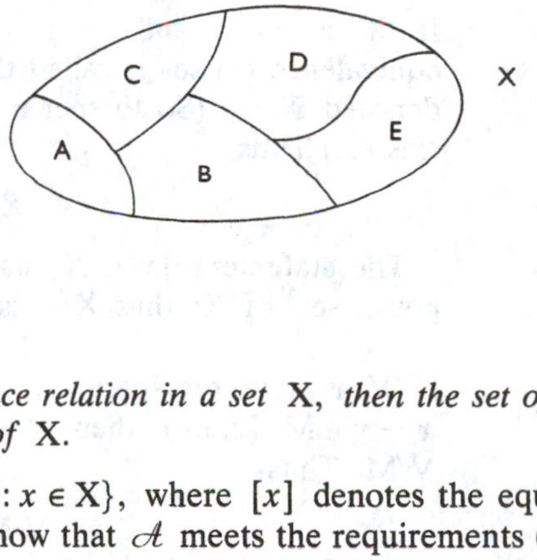

Fig. 9

2.5.12 Theorem *If* $\sim$ *is an equivalence relation in a set* X, *then the set of equivalence classes for* $\sim$ *is a partition of* X.

Proof. Let $\mathcal{A} = \{[x] : x \in \mathrm{X}\}$, where $[x]$ denotes the equivalence class of x (2.5.5); we have to show that $\mathcal{A}$ meets the requirements (1)–(3) of 2.5.11.

First, reflexivity ($x \sim x$) shows that $x \in [x] \in \mathcal{A}$; that takes care of (1) and (3).

To prove (2), suppose $z \in [x] \cap [y]$; we have to show that $[x] = [y]$. The strategy is to show that $[x] \subset [y]$ and $[y] \subset [x]$ (a 'double-inclusion argument'). At any rate, $z \in [x]$ and $z \in [y]$, so $z \sim x$ and $z \sim y$; thus (symmetry) $x \sim z$ and $z \sim y$, therefore $x \sim y$ (transitivity).

Let's show that $[x] \subset [y]$. Assuming $t \in [x]$, we have to show that $t \in [y]$.

[1] Appendix A.1.8.

We know that $t \sim x$ (because $t \in [x]$) and $x \sim y$, therefore $t \sim y$; thus $t \in [y]$, as desired.

Since $[x] \cap [y] = [y] \cap [x]$, x and y are on an equal footing, so $[y] \subset [x]$ follows from the preceding paragraph by interchanging the roles of x and y. ■

2.5.13 Corollary *With notations as in the theorem,*

(1) $x \sim y \Leftrightarrow [x] = [y]$;

(2) *for any* x, y *in* X, *either* $[x] = [y]$ *or* $[x] \cap [y] = \varnothing$.

Proof.

(2) If $[x] \cap [y] \neq \varnothing$ then $[x] = [y]$ by the theorem.

(1) If $x \sim y$ then x belong to both $[y]$ and $[x]$, therefore $[x] = [y]$ by (2). Conversely, if $[x] = [y]$ then $x \in [x] = [y]$, so $x \sim y$. ■

2.5.14 Corollary *If* V *is a vector space and* M *is a linear subspace of* V, *then*

(1) $x - y \in \mathrm{M} \Leftrightarrow x + \mathrm{M} = y + \mathrm{M}$;

(2) *for any* x, y *in* V, *either* $x + \mathrm{M} = y + \mathrm{M}$ *or* $(x + \mathrm{M}) \cap (y + \mathrm{M}) = \varnothing$.

Proof. This is Corollary 2.5.13 applied to the equivalence relation of Example 2.5.7. ■

Let's give a name to the set whose elements are the equivalence classes for an equivalence relation:

2.5.15 Definition If X is a set and $\sim$ is an equivalence relation in X, the set of all equivalence classes is called the **quotient set** of X for the relation $\sim$ and is denoted X/$\sim$. (So to speak, the points of X are 'divided' into equivalence classes.) Thus,

$$\mathrm{X}/\sim \;=\; \{[x]\colon x \in \mathrm{X}\}.$$

The statement $[x] \subset \mathrm{X}$ may be written $[x] \in \mathcal{P}(\mathrm{X})$, where $\mathcal{P}(\mathrm{X})$ is the power set[2] of X; thus X/$\sim$ is a subset of $\mathcal{P}(\mathrm{X})$, that is, $\mathrm{X}/\sim\; \subset \mathcal{P}(\mathrm{X})$.

2.5.16 Example If V is a vector space, M is a linear subspace of V, and $x \sim y$ means $x - y \in \mathrm{M}$ (2.5.4), then V/$\sim$ is replaced by the more suggestive notation V/M. Thus

$$\mathrm{V/M} = \{x + \mathrm{M}\colon x \in \mathrm{V}\}$$

is the set of all cosets of V modulo M (2.5.8).

One more technical matter needs to be discussed:

2.5.17 Definition If $\sim$ is an equivalence relation in a set X, there is a natural mapping $q\colon \mathrm{X} \to \mathrm{X}/\sim$, namely, the mapping that assigns to each point x in X its equivalence class $[x]$:

$$q(x) = [x] \qquad \text{for all } x \in \mathrm{X}.$$

[2] Appendix A.2.7.

Since $X/\sim$ is the set of all classes $[x]$, it is obvious that q is surjective; it is called the **quotient mapping** of X onto $X/\sim$.

2.5.18 Remarks With notations as in the preceding definition, $q(x) = q(y) \Leftrightarrow [x] = [y] \Leftrightarrow x \sim y$ (2.5.13); thus *every* equivalence relation may be derived from a function by the method of 2.5.1.

Finally, suppose $\mathcal{A}$ is *any* partition of a set X (2.5.11). Each point of X belongs to a unique set $A \in \mathcal{A}$; for $x, y \in X$, *declare* $x \sim y$ if x and y belong to the *same* set A of $\mathcal{A}$. A moment's thought while looking at Figure 9 shows that $\sim$ is an equivalence relation, for which the equivalence classes are precisely the sets in $\mathcal{A}$.

The bottom line: equivalence relations, partitions, and sets of constancy of a function are three ways of looking at the same thing.

▶ *Exercises*

1. Define an *arrow* to be an ordered pair (P_1, P_2) of points in cartesian 3-space (§1.1). Let X be the set of all arrows. Define $(P_1, P_2) \sim (P_3, P_4)$ to mean that $P_1P_2P_4P_3$ is a parallelogram; this is an equivalence relation in X. What are the equivalence classes?

2. Let X be the set of all *nonzero* vectors in $\mathbf{R}^n$, that is, $X = \mathbf{R}^n - \{\theta\}$. For x, y in X, write $x \sim y$ if x is proportional to y, that is, if there exists a scalar c (necessarily nonzero) such that $x = cy$; this is an equivalence relation in X.

3. For x, y in $\mathbf{R}$, write $x \sim y$ if $x - y$ is a rational number (that is, $x - y \in \mathbf{Q}$). This is an equivalence relation in $\mathbf{R}$. The quotient set is denoted $\mathbf{R}/\mathbf{Q}$.

4. If $\mathcal{S}$ is a nonempty set of vector spaces over the field F, then the relation of isomorphism (2.4.1) is an equivalence relation in $\mathcal{S}$.

5. Let $f:\mathbf{R}^3 \to \mathbf{R}^2$ be the linear mapping defined by $f(x, y, z) = (x - 2y, y + 3z)$, and let $\sim$ be the equivalence relation derived from f as in Theorem 2.5.1.

(i) Show that $(6, 3, -1) \sim (0, 0, 0)$.

(ii) True or false (explain): $(-2, 0, 4) \sim (4, 3, 3)$.

6. Let X be a nonempty set. Let R be a subset of the cartesian product[3] $X \times X$, such that (i) R contains the 'diagonal' of $X \times X$, that is, $(x, x) \in R$ for all $x \in X$; (ii) R is 'symmetric in the diagonal', that is, if $(x, y) \in R$ then $(y, x) \in R$; and (iii) if $(x, y) \in R$ and $(y, z) \in R$ then $(x, z) \in R$. Does this suggest a way of defining an equivalence relation $x \sim y$ in X?

7. Let $\mathcal{L}$ be the set of all lines L in the cartesian plane that are not vertical (i.e. not parallel to the y-axis). Define a function $m:\mathcal{L} \to \mathbf{R}$ by $m(L) = $ slope of L. Let $\sim$ be the equivalence relation on $\mathcal{L}$ derived from m (2.5.1), that is, $L \sim L'$ means that L and L′ have the same slope.

[3] Appendix A.2.8.

(i) Describe the equivalence class [L] of L.

(ii) True or false (explain): There exists a bijection $\mathscr{L}/\sim \to \mathbf{R}$.

8. Let P be a fixed point in the cartesian plane and let $\mathscr{L}_P$ be the set of all nonvertical lines that pass through the point P (cf. Exercise 7). True or false (explain): There exists a bijection $\mathscr{L}_P \to \mathbf{R}$.

9. Let $f:X \to Y$ be any mapping, $x \sim x'$ the equivalence relation defined by $f(x) = f(x')$ (2.5.1), $q:X \to X/\sim$ the quotient mapping (2.5.17).

Define a mapping $g:X/\sim \to f(X)$ as follows: if $u \in X/\sim$ define $g(u) = f(x)$, where x is any element of X such that $u = q(x)$. {If also $u = q(x')$ then $x' \sim x$, so $f(x') = f(x)$; thus $g(u)$ depends only on u, not on the particular x chosen to represent the class u.} Prove (cf. Fig. 10):

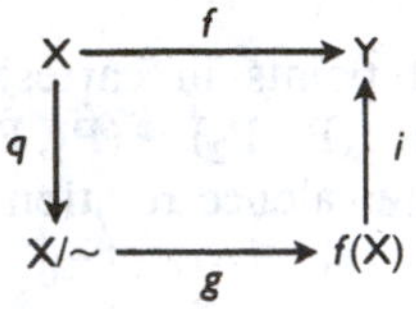

Fig. 10

(i) g is bijective;

(ii) $f = i \circ g \circ q$, where $i:f(X) \to Y$ is the insertion mapping (Appendix A.3.8).

In particular, if $f:X \to Y$ is surjective, then $g:X/\sim \to Y$ is bijective.

10. For x, y in $\mathbf{R}$, write $x \sim y$ if $x - y$ is an integer (that is, $x - y \in \mathbf{Z}$); this is an equivalence relation in $\mathbf{R}$. The quotient set is denoted $\mathbf{R}/\mathbf{Z}$; the equivalence class $[x]$ of $x \in \mathbf{R}$ is the set $x + \mathbf{Z} = \{x + n: n \in \mathbf{Z}\}$.

(i) There is a connection between functions defined on $\mathbf{R}/\mathbf{Z}$ and functions defined on $\mathbf{R}$ that are periodic[4] of period 1; can you make it precise?

(ii) There exists a bijection $g:\mathbf{R}/\mathbf{Z} \to \mathbf{U}$, where $\mathbf{U} = \{(a, b) \in \mathbf{R}^2: a^2 + b^2 = 1\}$ is the 'unit circle' in $\mathbf{R}^2$. {Hint: Apply Exercise 9 to the function $f:\mathbf{R} \to \mathbf{U}$ defined by $f(t) = (\cos 2\pi t, \sin 2\pi t)$.}

11. For x, y in $\mathbf{R}$, write $x \sim y$ if $x - y$ is an integral multiple of 2π (that is, $x - y \in 2\pi\mathbf{Z}$). With an eye on Exercise 10, discuss periodic functions of period 2π and describe a bijection of $\mathbf{R}/2\pi\mathbf{Z}$ onto the unit circle $\mathbf{U}$.

2.6 Quotient vector spaces

If $T:V \to W$ is a linear mapping, then the kernel of T is a linear subspace of V (2.2.2). Conversely, every linear subspace of V is the kernel of a suitable linear mapping on V:

2.6.1 Theorem *Let* V *be a vector space,* M *a linear subspace of* V,

$$V/M = \{x + M: x \in V\}$$

[4] A function $f:\mathbf{R} \to Y$ is said to be *periodic of period* p if $f(x + p) = f(x)$ for all $x \in \mathbf{R}$ (Y can be any set).

the set of all cosets of V *modulo* M (2.5.16), *and* $Q{:}\mathrm{V} \to \mathrm{V/M}$ *the quotient mapping* (2.5.17),

$$Qx = x + \mathrm{M} \qquad (x \in \mathrm{V}).$$

Then V/M *can be made into a vector space in such a way that* Q *becomes a linear mapping with kernel* M; *explicitly, sums and scalar multiples in* V/M *are defined by the formulas*

(i) $$(x + \mathrm{M}) + (y + \mathrm{M}) = x + y + \mathrm{M},$$
(ii) $$c(x + \mathrm{M}) = cx + \mathrm{M}.$$

Proof. Let $u, v \in \mathrm{V/M}$ and let c be a scalar; the problem is to define elements $u + v$ and cu of V/M. Say $u = x + \mathrm{M}$, $v = y + \mathrm{M}$. We would like to define $u + v = x + y + \mathrm{M}$ (this would be consistent with the desire to make Q additive, since it would say that $Qx + Qy = Q(x + y)$). The proposed sum $x + y + \mathrm{M}$ appears to depend on specific vectors x and y chosen to represent the cosets u and v, but in fact it does not: if also $u = x' + \mathrm{M}$ and $v = y' + \mathrm{M}$, then $x - x' \in \mathrm{M}$ and $y - y' \in \mathrm{M}$ (2.5.14), therefore

$$(x + y) - (x' + y') = (x - x') + (y - y') \in \mathrm{M},$$

so $x + y + \mathrm{M} = x' + y' + \mathrm{M}$. This shows that sums in V/M are unambiguously defined by the formula (i).

Similarly, if $x - x' \in \mathrm{M}$ then $cx - cx' = c(x - x') \in \mathrm{M}$, so scalar multiples in V/M are unambiguously defined by the formula (ii).

We have thus defined sums $u + v$ and scalar multiples cu in V/M in such a way that the formulas (i) and (ii) hold, in other words,

$$(*) \qquad Q(x + y) = Qx + Qy, \qquad Q(cx) = c(Qx)$$

for all x, y in V and all scalars c. To see that V/M is a vector space, we have to check the axioms (3)–(9) of 1.3.1. Let $u, v, w \in \mathrm{V/M}$ and let c, d be scalars. Say $u = Qx$, $v = Qy$, $w = Qz$.

(3) Since $x + y = y + x$, applying Q to both sides and citing $(*)$, we have $u + v = v + u$.

(4) Application of Q to $(x + y) + z = x + (y + z)$ produces $(u + v) + w = u + (v + w)$.

(7)–(9) are proved similarly; for example, $(c + d)u = cu + du$ results on applying Q to $(c + d)x = cx + dx$ and citing both of the properties in $(*)$.

(5) Since $x + \theta = x$, application of Q yields $u + Q\theta = u$, thus $Q\theta = \mathrm{M}$ serves as a 'zero vector' for V/M.

(6) Applying Q to $x + (-x) = \theta$, we see that $u + Q(-x) = Q\theta$, so $Q(-x) = -x + \mathrm{M}$ serves as the 'negative' of $u = x + \mathrm{M}$.

Now that we know that V/M is a vector space, $(*)$ says that Q is a linear mapping. Finally, $Qx = Q\theta \Leftrightarrow x - \theta \in \mathrm{M}$; in other words, $x \in \operatorname{Ker} Q \Leftrightarrow x \in \mathrm{M}$. ■

2.6.2 Definition With notations as in 2.6.1, V/M is called the **quotient vector space** of V modulo the linear subspace M.

The most important application of quotient vector spaces is the subject of the next section.

2.6.3 Remarks (Optional) Was it *really* necessary to make the 'independence of representatives' argument in the proof of 2.6.1? Let's omit it and see how we run into trouble. Given $u, v \in \mathrm{V/M}$, choose $x \in u$, $y \in v$ and declare $u + v = x + y + \mathrm{M}$. Do this for every pair u, v in V/M. We now have an addition defined in V/M. Does it make Q additive? That is, given *any* x, $y \in \mathrm{V}$, is $Qx + Qy = Q(x + y)$? Let $u = Qx$, $v = Qy$. The question is, is $u + v = x + y + \mathrm{M}$? The answer is not obvious: there is no reason to suppose that x and y were the representatives chosen in the course of defining $u + v$. The 'independence of representatives' argument settles such matters once and for all.

▶ ***Exercises***

1. Prove that if $\mathrm{V} = \mathrm{M} \oplus \mathrm{N}$ (§1.6, Exercise 12) then $\mathrm{V/M} \cong \mathrm{N}$. {Hint: Restrict the quotient mapping $\mathrm{V} \to \mathrm{V/M}$ to N (Appendix A.3.8), and calculate the kernel and range of the restricted mapping.}

2. Let M and N be linear subspaces of the vector spaces V and W, respectively. Form the product vector space $\mathrm{V} \times \mathrm{W}$ (Definition 1.3.11). Then $\mathrm{M} \times \mathrm{N}$ is a linear subspace of $\mathrm{V} \times \mathrm{W}$ and

$$(\mathrm{V} \times \mathrm{W})/(\mathrm{M} \times \mathrm{N}) \cong (\mathrm{V/M}) \times (\mathrm{W/N}).$$

Prove this by showing that

$$(x, y) + \mathrm{M} \times \mathrm{N} \mapsto (x + \mathrm{M}, y + \mathrm{N})$$

defines a function $(\mathrm{V} \times \mathrm{W})/(\mathrm{M} \times \mathrm{N}) \to (\mathrm{V/M}) \times (\mathrm{W/N})$ and that this function is linear and bijective. {The essential first step is to show that if

$$u = (x, y) + \mathrm{M} \times \mathrm{N} = (x', y') + \mathrm{M} \times \mathrm{N},$$

then $(x + \mathrm{M}, y + \mathrm{N}) = (x' + \mathrm{M}, y' + \mathrm{N})$, that is, the proposed functional value at u depends only on u and not on the particular ordered pair $(x, y) \in u$ chosen to represent it (cf. the proof of Theorem 2.6.1).}

3. Let $f{:}\mathbf{R}^3 \to \mathbf{R}$ be the linear form $f(x) = 2x_1 - 3x_2 + x_3$ and let M be its kernel. As in Example 2.5.9, think of V/M as the set of planes consisting of M and the planes parallel to it. If u and v are two of these planes, their sum $u + v$ in V/M can be visualized as follows: choose a point P on the plane u and a point Q on the plane v; add the arrows OP and OQ using the parallelogram rule, obtaining an arrow OR; the plane through R parallel to M is $u + v$.

4. Let M be a linear subspace of V, $u = x + \mathrm{M}$ and $v = y + \mathrm{M}$ two elements of V/M.

(i) Let $\mathrm{A} = \{s + t\colon\ s \in u, t \in v\}$ be the sum of the sets $x + \mathrm{M}$ and $y + \mathrm{M}$ in the sense of Definition 1.6.9. Then A is also a coset, namely $\mathrm{A} = x + y + \mathrm{M}$. {This offers a definition of $u + v$ that is obviously independent of the vectors x and y chosen to represent u and v.}

(ii) If c is a nonzero scalar, then the set $B = \{cs : s \in u\}$ is a coset, namely $B = cx + M$. What if $c = 0$?

5. Let V be a vector space, M and N linear subspaces of V, and x, y vectors in V. Prove:

(i) $x + M \subset y + N$ if and only if $M \subset N$ and $x - y \in N$.

(ii) $(x + M) \cap (y + N) \neq \varnothing$ if and only if $x - y \in M + N$.

(iii) If $z \in (x + M) \cap (y + N)$ then $(x + M) \cap (y + N) = z + M \cap N$.

6. Suppose $V = M \oplus N$ (§1.6, Exercise 12) and let $x, y \in V$. Prove that $(x + M) \cap (y + N)$ contains exactly one point. {Hint: Exercise 5.} Interpret geometrically if $V = \mathbf{R}^2$; if $V = \mathbf{R}^3$.

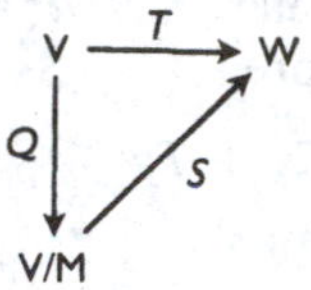

Fig. 11

7. Let $T: V \to W$ be a linear mapping, M a linear subspace of V such that $M \subset \operatorname{Ker} T$, and $Q: V \to V/M$ the quotient mapping. Prove:

(i) There exists a (unique) linear mapping $S: V/M \to W$ such that $T = SQ$ (Fig. 11).

(ii) S is injective if and only if $M = \operatorname{Ker} T$.

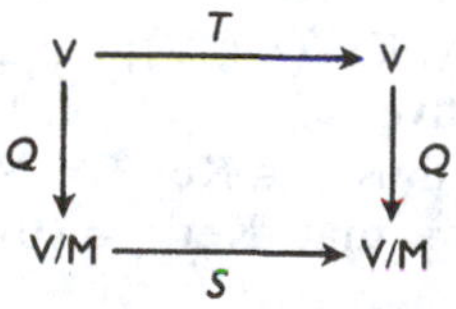

Fig. 12

8. Let M be a linear subspace of V, $Q: V \to V/M$ the quotient mapping, $T: V \to V$ a linear mapping such that $T(M) \subset M$. Prove that there exists a (unique) linear mapping: $S: V/M \to V/M$ such that $SQ = QT$ (Fig. 12). {Hint: Apply Exercise 7 to the linear mapping $QT: V \to V/M$.}

2.7 The first isomorphism theorem

At the core of the 'First isomorphism theorem' is an equivalence relation:

2.7.1 Lemma *If $T: V \to W$ is a linear mapping and N is the kernel of T, then*

$$T^{-1}(\{Tx\}) = x + N$$

for all $x \in V$.

Proof. Let $x \sim y$ be the equivalence relation in V defined by $Tx = Ty$ (2.5.1); the equivalence class of x is $T^{-1}(\{Tx\})$ (2.5.6). On the other hand, $x \sim y \Leftrightarrow x - y \in N$ (2.2.9), so the equivalence class of x is $x + N$ (2.5.7). ■

2.7.2 Theorem (First Isomorphism Theorem) *If $T: V \to W$ is a linear mapping and N is the kernel of T, then*

(1) N *is a linear subspace of* V;

(2) $T(V)$ *is a linear subspace of* W;

(3) $\mathrm{V/N} \cong T(\mathrm{V})$, *via the mapping* $x + \mathrm{N} \mapsto Tx$.

Concisely,

$$\mathrm{V}/\mathrm{Ker}\, T \cong \mathrm{Im}\, T$$

for every linear mapping T *with domain* V.

Proof. Statements (1) and (2) are a review of Corollary 2.2.2.

(3) The problem is to define a bijective linear mapping $S{:}\mathrm{V/N} \to T(\mathrm{V})$ (2.4.1).

If $u \in \mathrm{V/N}$, say $u = x + \mathrm{N}$, then T is constant on the subset u of V (2.7.1); it follows that a mapping $S{:}\mathrm{V/N} \to T(\mathrm{V})$ is defined unambiguously by the formula $S(x + \mathrm{N}) = Tx$.

Let $Q{:}\mathrm{V} \to \mathrm{V/N}$ be the quotient mapping (2.6.1), which is surjective and linear. From the definition of S, $S(Qx) = Tx$ for all $x \in \mathrm{V}$.

Linearity of S: Let u, $v \in \mathrm{V/N}$ and let c be a scalar. Say $u = Qx$, $v = Qy$. Then $u + v = Qx + Qy = Q(x + y)$, therefore

$$\begin{aligned} S(u + v) &= S(Q(x + y)) = T(x + y) = Tx + Ty \\ &= S(Qx) + S(Qy) = Su + Sv; \end{aligned}$$

also, $cu = c(Qx) = Q(cx)$, so

$$S(cu) = S(Q(cx)) = T(cx) = c(Tx) = c(Su).$$

Bijectivity of S: $T(\mathrm{V})$ is the set of all vectors Tx ($x \in \mathrm{V}$); so is $S(\mathrm{V/N})$; thus $S(\mathrm{V/N}) = T(\mathrm{V})$, that is, S is surjective.

If $u \in \mathrm{Ker}\, S$, say $u = Qx$, then $\theta = Su = Tx$; thus $x \in \mathrm{Ker}\, T = \mathrm{N}$, so $u = \mathrm{N}$ is the zero vector of V/N. This shows that $\mathrm{Ker}\, S = \{0\}$, consequently S is injective (2.2.9). ■

► ***Exercises***

1. Where there's a first, there's a second. The following result is called the *Second isomorphism theorem*. Let V be a vector space, M and N linear subspaces of V, $\mathrm{M} \cap \mathrm{N}$ their intersection and $\mathrm{M} + \mathrm{N}$ their sum (1.6.9). Then $\mathrm{M} \cap \mathrm{N}$ is a linear subspace of M, N is a linear subspace of $\mathrm{M} + \mathrm{N}$, and

$$\mathrm{M}/(\mathrm{M} \cap \mathrm{N}) \cong (\mathrm{M} + \mathrm{N})/\mathrm{N}.$$

{Hint: Let $Q{:}\mathrm{V} \to \mathrm{V/N}$ be the quotient mapping and let $T = Q|\mathrm{M}$ be the restriction of Q to M (Appendix A.3.8), that is, $T{:}\mathrm{M} \to \mathrm{V/N}$ and $Tx = x + \mathrm{N}$ for all $x \in \mathrm{M}$. Show that T has range $(\mathrm{M} + \mathrm{N})/\mathrm{N}$ and kernel $\mathrm{M} \cap \mathrm{N}$.}

2. Let M, N be linear subspaces of the vector spaces V, W and let

$$T{:}\mathrm{V} \times \mathrm{W} \to (\mathrm{V/M}) \times (\mathrm{W/N})$$

be the linear mapping defined by $T(x, y) = (x + \mathrm{M}, y + \mathrm{N})$. Determine the kernel of T and deduce another proof of §2.6, Exercise 2.

3. Let V be a vector space over F, $f{:}\mathrm{V} \to \mathrm{F}$ a linear form on V (Definition 2.1.11), N the kernel of f. Prove that if f is not identically zero, then

$V/N \cong F$. {Hint: If $f(z) \neq 0$ and c is any scalar, evaluate f at the vector $(c/f(z))z$.}

4. Let V be a vector space, $V \times V$ the product vector space (1.3.11), and $\Delta = \{(x, x): x \in V\}$ (called the *diagonal* of $V \times V$). Prove that Δ is a linear subspace of $V \times V$ and that $(V \times V)/\Delta \cong V$. {Hint: §2.2, Exercise 4.}

5. Let N be the set of vectors in $\mathbf{R}^5$ whose last two coordinates are zero:

$$N = \{(x_1, x_2, x_3, 0, 0): \ x_1, x_2, x_3 \in \mathbf{R}\}.$$

Prove that N is a linear subspace of $\mathbf{R}^5$, $\mathbf{R}^3 \cong N$, and $\mathbf{R}^5/N \cong \mathbf{R}^2$. {Taking some liberties with the notation, $\mathbf{R}^5/\mathbf{R}^3 = \mathbf{R}^2$.}

6. Let T be a set, A a subset of T, $V = \mathcal{F}(T, \mathbf{R})$ and $W = \mathcal{F}(A, \mathbf{R})$ the vector spaces of functions constructed in 1.3.4. Let N be the set of all functions $x \in V$ such that $x(t) = 0$ for all $t \in A$. Then N is a linear subspace of V and $V/N \cong W$.

{Hint: Consider the mapping $x \mapsto x|A$ $(x \in V)$, where $x|A$ is the restriction of the function x to A (Appendix A.3.8).}

7. Another approach to the First Isomorphism Theorem (bypassing §2.6) can be based on two earlier exercises. The correspondence $x + N \mapsto Tx$ defines a bijection $S{:}V/N \to T(V)$ (Lemma 2.7.1 and §2.5, Exercise 9). The operations on V/N defined by the formulas

$$u + v = S^{-1}(Su + Sv), \quad cu = S^{-1}(c(Su))$$

make V/N a vector space and S an isomorphism (§2.4, Exercise 3). {These operations on V/N coincide with those defined in §2.6; the virtue of the construction in §2.6 is that we don't have to know in advance that N is the kernel of some linear mapping on V.}

3

Structure of vector spaces

The vector spaces most amenable to study are those that are 'finitely generated' in the following sense: there is a finite list of vectors $x_1, \ldots, x_r$ such that *every* vector in the space is a linear combination of the x_i. We see in this chapter that every finitely generated vector space V over a field F looks like one of the 'standard' spaces F^n (1.2.1); more precisely, $V \cong F^n$ for a unique n. This is the vector space analogue of assigning coordinates to the points of space; an isomorphism $V \cong F^n$ assigns numerical 'coordinates' to each vector, thereby making it possible to keep track of vectorial operations and mappings by means of numbers. It is in this vague sense of rendering concrete and computable that the term 'structure' in the chapter heading is intended. The central concept of the chapter is that of *basis*, which is the vector space analogue of a 'coordinate system'; choosing a basis is like choosing coordinate axes.

If $V \cong F^n$ then n is called the *dimension* of V. If T is a linear mapping defined on the vector space V, the dimension of its kernel is called the *nullity* of T, and the dimension of its range is called the *rank* of T. The First Isomorphism Theorem proved in §2.7 leads to a valuable formula connecting rank and nullity (§3.6).

3.1 Linear subspace generated by a subset

3.1.1 Definition Let V be a vector space, A a nonempty set of vectors in V. The **linear span** of A is the set of all vectors that can be expressed as a linear combination of vectors in A. The linear span of A is denoted [A].

Thus, $x \in [A]$ if and only if $x = c_1x_1 + \ldots + c_nx_n$ for suitable vectors $x_1, \ldots, x_n$ in A and suitable scalars $c_1, \ldots, c_n$.

For example, if V is the vector space $\mathcal{P}$ of polynomial functions (1.3.5) and $A = \{1, t, t^2\}$, then [A] is the set of all polynomial functions of degree at most 2. If $V = F^n$ and $A = \{e_1, \ldots, e_n\}$ (see 1.5.3), then $[A] = V$. In both examples, [A] is a linear subspace of V; this is true in general:

3.1.2 Theorem *Let* V *be a vector space,* A *a nonempty subset of* V, [A] *the linear span of* A. *Then:*

(1) [A] *is a linear subspace of* V;
(2) $A \subset [A]$;
(3) *if* M *is a linear subspace of* V *such that* $A \subset M$, *then* $[A] \subset M$.

Thus, [A] *is the smallest linear subspace of* V *that contains* A.

Proof.
(2) If $x \in A$ then $x = 1x \in [A]$.
(1) Choose $z \in A$ (by assumption, A, is nonempty); then $\theta = 0z \in [A]$.
Assuming x, y in [A] and c a scalar, we have to show that $x + y$ and cx are in [A]. Say

$$x = a_1x_1 + \ldots + a_mx_m, \qquad y = b_1y_1 + \ldots + b_ny_n,$$

where the a_i, b_j are scalars and the x_i, y_j are elements of A. Then $x + y$ is a linear combination of the elements $x_1, \ldots, x_m, y_1, \ldots, y_n$ of A with coefficients $a_1, \ldots, a_m, b_1, \ldots, b_n$, therefore $x + y \in [A]$. Also, $cx = (ca_1)x_1 + \ldots + (ca_m)x_m \in [A]$.
(3) Let M be a linear subspace of V containing A. Since M is itself a vector space (1.6.2), it is immediate from $A \subset M$ that $[A] \subset M$. ■

3.1.3 Remark With the convention that $[\varnothing] = \{\theta\}$, Theorem 3.1.2 remains true with the word 'nonempty' omitted from its statement.

3.1.4 Terminology In view of 3.1.2, [A] is also called the linear subspace of V **generated** by A, and A is said to be a **generating set** (or 'system of generators') for [A]. If $[A] = V$, we say that A is **generating** (for V).

3.1.5 Corollary *If* $x_1, \ldots, x_n$ *is a finite list of vectors in* V *(not necessarily distinct) and if* $A = \{x_1, \ldots, x_n\}$, *then* [A] *is the set of all linear combinations of* $x_1, \ldots, x_n$.

Proof. Let M be the set of linear combinations in question:

$$M = \{a_1x_1 + \ldots + a_nx_n\colon\ a_1, \ldots, a_n \text{ scalars}\}.$$

The inclusion $M \subset [A]$ is immediate from the definition of [A].

To prove the reverse inclusion $[A] \subset M$, it suffices by 3.1.2 to show that M is a linear subspace of V containing A. If F is the field of scalars and $T\colon F^n \to V$ is the mapping defined by $T(a_1, \ldots, a_n) = a_1x_1 + \ldots + a_nx_n$, then M is the range of T; since T is linear (2.1.3), M is a linear subspace of V (2.2.2). If $e_i \in F^n$ is the vector with 1 in the ith coordinate and 0's elsewhere, then $Te_i = x_i$; this shows that $A \subset M$. ■

3.1.6 Theorem *If $T\colon V \to W$ is a linear mapping and A is a subset of V, then* $[T(A)] = T([A])$.

Proof. {What does the theorem say? Start with a subset A of V. We can first map A into $T(A)$, then form the linear subspace $[T(A)]$ generated by the image; or, we can first form the linear subspace [A] generated by A, then map, obtaining $T([A])$. The theorem says that both procedures lead to the same linear subspace of W.}

Since [A] is a linear subspace of V containing A, $T([A])$ is a linear subspace of W (2.2.1) containing $T(A)$, therefore $[T(A)] \subset T([A])$ (3.1.2). On the other hand, $T^{-1}([T(A)])$ is a linear of V (2.2.1), containing A (because $T(A) \subset [T(A)]$), therefore $[A] \subset T^{-1}([T(A)])$, so $T([A]) \subset [T(A)]$.

Alternate proof. If $x_1, \ldots, x_n \in A$ and $c_1, \ldots, c_n$ are scalars, then

$$T(c_1x_1 + \ldots + c_nx_n) = c_1(Tx_1) + \ldots + c_n(Tx_n);$$

the left sides of these equations fill out $T([A])$, and the right sides, $[T(A)]$. ■

3.1.7 Corollary *If $T\colon V \to W$ is a* ***surjective*** *linear mapping and A is a generating set for V, then $T(A)$ is generating for W.*

Proof.

$$\begin{aligned} W &= T(V) && (T \text{ is surjective}) \\ &= T([A]) && (\text{A is generating for V}) \\ &= [T(A)] && (\text{by 3.1.6}), \end{aligned}$$

thus W is generated by $T(A)$. ■

▶ *Exercises*

1. If $T\colon V \to W$ is linear and A is a subset of V such that $T(A)$ is generating for W, then T is surjective.

2. If M and N are linear subspaces of V such that $V = M + N$ (1.6.9) and $T\colon V \to V/N$ is the quotient mapping, then $T(M) = V/N$ but M need not equal V; thus we cannot conclude in Exercise 1 that A is generating for V. Can you exhibit an example of such V, M and N?

3. (*Substitution principle*) Suppose $x, x_1, \ldots, x_n$ are vectors such that x is a linear combination of $x_1, \ldots, x_n$ but not of $x_1, \ldots, x_{n-1}$. Prove:

$$[\{x_1, \ldots, x_n\}] = [\{x_1, \ldots, x_{n-1}, x\}].$$

{Hint: When x is represented as a linear combination of $x_1, \ldots, x_n$, the coefficient of x_n must be nonzero.}

4. Let $S: V \to W$ and $T: V \to W$ be linear mappings, and let A be a subset of V such that $V = [A]$. Prove that if $Sx = Tx$ for all $x \in A$, then $S = T$. (Informally, a linear mapping is determined by its values on a generating set.)

5. Let V be a vector space, A a nonempty subset of V. For each finite subset $\{x_1, \ldots, x_n\}$ of A, let

$$M(x_1, \ldots, x_n) = \{a_1x_1 + \ldots + a_nx_n\colon\ a_1, \ldots, a_n \text{ scalars}\};$$

as noted in the proof of Corollary 3.1.5, this is a linear subspace of V. Let $\mathscr{M}$ be the set of all linear subspaces of V obtained in this way (by considering all possible finite subsets of A); by definition, $[A] = \bigcup \mathscr{M}$ (the union of the subspaces belonging to $\mathscr{M}$). Prove that [A] is a linear subspace of V by verifying the criterion of §1.6, Exercise 9.

6. Let V and W be vector spaces, $V \times W$ the product vector space (1.3.11); let $A \subset V$ and $B \subset W$. Prove:

(i) $[A \times B] \subset [A] \times [B]$.

(ii) In general, $[A \times B] \neq [A] \times [B]$. {Hint: Consider $A = \varnothing$, $B = W$; or $V = W = \mathbf{R}$ and $A = B = \{1\}$.}

(iii) $[(A \times \{\theta\}) \cup (\{\theta\} \times B)] = [A] \times [B]$. {Hint:

$$\left(\sum_{i=1}^{m} c_i x_i, \sum_{j=1}^{n} d_j y_j\right) = \sum_{i=1}^{m} c_i(x_i, \theta) + \sum_{j=1}^{n} d_j(\theta, y_j).\}$$

3.2 Linear dependence

Let V be a vector space and let $x_1, \ldots, x_n$ be a finite list of vectors in V. Do there exist scalars $c_1, \ldots, c_n$ such that $c_1x_1 + \ldots + c_nx_n = \theta$? The answer is obviously 'yes': the choice $c_1 = \ldots = c_n = 0$ does the job.

Let's revise the question: Do there exist scalars $c_1, \ldots, c_n$, *not all zero*, such that $c_1x_1 + \ldots + c_nx_n = \theta$? Now the answer is, 'It depends on the list of vectors.' For example, consider a list x_1 of length 1: if $x_1 = \theta$ then the answer is 'yes' ($c_1 = 1$ works); if $x_1 \neq \theta$ then the answer is 'no', since $c_1x_1 = \theta$ forces $c_1 = 0$ (1.4.6).

The second question thus divides finite lists into two types; in this section we concentrate on those for which the answer is 'yes' (in the next, on those for which the answer is 'no').

3.2.1 Definition Let V be a vector space. A finite list of vectors $x_1, \ldots, x_n$ in V is said to be **linearly dependent** (briefly, 'dependent') if there exist scalars $c_1, \ldots, c_n$ such that $c_1x_1 + \ldots + c_nx_n = \theta$ and such that at least one of the coefficients c_i is nonzero; such an equation (in which at least one of the c_i is nonzero) is called a **dependence relation** among the x_i. The concept is also expressed by saying that 'the vectors $x_1, \ldots, x_n$ *are* linearly dependent'.

Note that rearrangement of the vectors in a list has no bearing on the question of linear dependence. For example, if the vectors x_1, x_2, x_3 are linearly dependent then so are the vectors x_2, x_3, x_1. {Reason: The commutative law for the addition of vectors.}

3.2.2 Example The list x_1 is linearly dependent if and only if $x_1 = \theta$. (The crux of the matter is 1.4.6.)

3.2.3 Example The list x_1, x_2 is linearly dependent if and only if one of the vectors is a multiple of the other. {*Proof*. If, for example, $x_2 = cx_1$, then $cx_1 + (-1)x_2 = \theta$ shows that x_1, x_2 are linearly dependent. Conversely, if $c_1x_1 + c_2x_2 = \theta$ and if, for example, c_1 is nonzero, then $x_1 = (-c_2/c_1)x_2$.}

3.2.4 Theorem *Let* V *be a vector space over* F, *let* $x_1, \ldots, x_n$ *be a finite list of vectors in* V, *and let* $T:\mathrm{F}^n \to \mathrm{V}$ *be the linear mapping defined by* $T(a_1, \ldots, a_n) = a_1x_1 + \ldots + a_nx_n$ (2.1.3). *The following conditions are equivalent*:

(a) $x_1, \ldots, x_n$ *are linearly dependent*;

(b) T *is not injective*.

Proof. Condition (b) means that the kernel of T is not $\{\theta\}$ (2.2.9), that is, there exists a nonzero vector $(c_1, \ldots, c_n)$ in F^n such that $c_1x_1 + \ldots + c_nx_n = \theta$; this is also the meaning of condition (a), thus (a) ⇔ (b). ■

The next theorem stretches the idea of Example 3.2.3 to lists of arbitrary length:

3.2.5 Theorem *Let* V *be a vector space and let* $x_1, \ldots, x_n$ *be a finite list of vectors in* V *with* $n \geq 2$. *The following conditions are equivalent*:

(a) $x_1, \ldots, x_n$ *are linearly dependent*;

(b) *some* x_j *is a linear combination of the* x_i *with* $i \neq j$.

Proof.

(b) ⇒ (a): Suppose $x_j = \sum_{i \neq j} c_ix_i$. Let $c_j = -1$; then $\sum_{i=1}^{n} c_ix_i = \theta$ with $c_j \neq 0$.

(a) ⇒ (b): Suppose $c_1x_1 + \ldots + c_nx_n = \theta$ is a dependence relation, say with $c_j \neq 0$. Then $c_jx_j = -\sum_{i \neq j} c_ix_i$, therefore $x_j = \sum_{i \neq j} (-c_i/c_j)x_i$. ■

3.2.6 Corollary *If the vectors* $x_1, \ldots, x_n$ *are not distinct, then they are linearly dependent*.

Proof. Suppose $x_j = x_k$ for a pair of indices j, k with $j \neq k$. Then $x_j = \sum_{i \neq j} c_ix_i$, where $c_k = 1$ and $c_i = 0$ for all indices i not equal to either j or k, thus the list $x_1, \ldots, x_n$ is linearly dependent by criterion (b) of Theorem 3.2.5. ■

The following variation on condition (b) of 3.2.5 focuses the search for dependence relations:

3.2.7 Theorem *The following conditions on a finite list of vectors $x_1, \dots, x_n$ are equivalent:*

(a) *$x_1, \dots, x_n$ are linearly dependent;*

(b) *either $x_1 = \theta$ or there exists an index $j > 1$ such that x_j is a linear combination of $x_1, \dots, x_{j-1}$.*

Proof.

(a) ⇒ (b): Let $c_1, \dots, c_n$ be scalars, not all 0, such that $c_1x_1 + \dots + c_nx_n = \theta$. Let j be the largest index for which $c_j \neq 0$; then $c_1x_1 + \dots + c_jx_j = \theta$, since the terms for indices larger than j (in case $j < n$) are θ. If $j = 1$ then $x_1 = \theta$ (Example 3.2.2); if $j > 1$, then $x_j = \sum_{i=1}^{j-1} (-c_i/c_j)x_i$.

(b) ⇒ (a): Suppose that $j > 1$ and that x_j is a linear combination of $x_1, \dots, x_{j-1}$. If $j = n$, then (a) holds by criterion (b) of 3.2.5; if $j < n$ then x_j is a linear combination of $x_1, \dots, x_{j-1}, x_{j+1}, \dots, x_n$ (use coefficients 0 for the indices $j+1, \dots, n$), so 3.2.5 is again applicable. On the other hand, if $x_1 = \theta$ then the relation $1x_1 + 0x_2 + \dots + 0x_n = \theta$ demonstrates the linear dependence of $x_1, \dots, x_n$. ■

Note that if *any* vector in the list $x_1, \dots, x_n$ is zero, then the list is linearly dependent.

▶ ***Exercises***

1. Let $V = \mathcal{F}(\mathbf{R})$ be the vector space of all real-valued functions of a real variable (Example 1.3.4). Define $f: \mathbf{R} \to \mathbf{R}$ by $f(t) = \sin(t + \pi/6)$. Prove that the functions f, sin, cos are linearly dependent in V.

2. In the vector space of real polynomial functions (Example 1.3.5), show that the functions p_1, p_2, p_3, p_4 defined by $p_1(t) = t^2 + t + 1$, $p_2(t) = t^2 + 5$, $p_3(t) = 2t + 1$, $p_4(t) = 3t^2 + 5t$ are linearly dependent.

3. Show that in $\mathbf{R}^3$ the vectors $e_1, e_2, e_3, (3, -2, 4)$ are linearly dependent (cf. 1.5.3).

4. If $x_1, \dots, x_n$ are linearly dependent, then so are $x_1, \dots, x_n, x_{n+1}$.

5. If $T: V \to W$ is a linear mapping and $x_1, \dots, x_n$ are linearly dependent vectors in V, then the vectors $Tx_1, \dots, Tx_n$ of W are linearly dependent.

6. A finite list of vectors $x_1, \dots, x_n$ is linearly dependent if and only if there exists an index j such that $x_j \in [\{x_i : i \neq j\}]$ (cf. 3.1.5).

7. If $T: V \to W$ is an injective linear mapping and if $x_1, \dots, x_n$ are vectors in V such that $Tx_1, \dots, Tx_n$ are linearly dependent in W, then $x_1, \dots, x_n$ are linearly dependent in V.

8. Let V be a vector space. A subset A of V is said to be *linearly dependent* if there exists a finite list of **distinct** vectors $x_1, \dots, x_n$ in A that is linearly dependent in the sense of Definition 3.2.1. Prove:

   (i) A subset A of V is linearly dependent if and only if some vector in A is in the linear span of the remaining vectors of A, that is, there exists $x \in A$ such that $x \in [A - \{x\}]$.

   (ii) If $A \subset B \subset V$ and A is linearly dependent, then B is also linearly dependent.

3.3 Linear independence

3.3.1 Definition A finite list of vectors $x_1, \ldots, x_n$ is said to be **linearly independent** (briefly, 'independent') if it is *not* linearly dependent in the sense of 3.2.1; in other words, the only scalars $c_1, \ldots, c_n$ for which $c_1x_1 + \ldots + c_nx_n = \theta$ are $c_1 = \ldots = c_n = 0$. The concept is also expressed by saying that 'the vectors $x_1, \ldots, x_n$ *are* linearly independent'.

In particular, x_1 is linearly independent if and only if $x_1 \neq \theta$ (3.2.2); x_1, x_2 are linearly independent if and only if neither is a multiple of the other (3.2.3). If $x_1, \ldots, x_n$ are linearly independent, then they are distinct (3.2.6).

The results of this section are mostly restatements of those in the preceding section; the repetition is worthwhile because the terms 'dependent' and 'independent' are both widely used and it's essential to be comfortable with them both.

3.3.2 Theorem *Let* V *be a vector space over the field* F, *let* $x_1, \ldots, x_n$ *be a finite list of vectors in* V, *and let* $T:\mathrm{F}^n \to \mathrm{V}$ *be the linear mapping defined by* $T(a_1, \ldots, a_n) = a_1x_1 + \ldots + a_nx_n$ (2.1.3). *The following conditions are equivalent*:

(a) $x_1, \ldots, x_n$ *are linearly independent*;

(b) T *is injective*;

(c) $c_1x_1 + \ldots + c_nx_n = \theta \Rightarrow c_1 = \ldots = c_n = 0$.

Proof. Condition (c) says that the kernel of T is zero, so (b) and (c) are equivalent by 2.2.9. Conditions (a) and (b) are equivalent because their negations are equivalent by 3.2.4. ■

3.3.3 Example The vectors $e_1, \ldots, e_n$ of F^n described in 1.5.3 are linearly independent, since $c_1e_1 + \ldots + c_ne_n = (c_1, \ldots, c_n)$.

3.3.4 Theorem *Let* V *be a vector space and let* $x_1, \ldots, x_n$ *be a finite list of vectors in* V *with* $n \geq 2$. *The following conditions are equivalent*:

(a) $x_1, \ldots, x_n$ *are linearly independent*;

(b) *no* x_j *is a linear combination of the* x_i *with* $i \neq j$.

Proof. The negations of (a) and (b) are equivalent by 3.2.5. ■

3.3.5 Theorem *The following conditions on a finite list of vectors* $x_1, \ldots, x_n$ *are equivalent*:

(a) $x_1, \ldots, x_n$ *are linearly independent*;

(b) $x_1 \neq \theta$ *and, for every index* $j > 1$, x_j *is not a linear combination of* $x_1, \ldots, x_{j-1}$.

Proof. The negations of (a) and (b) are equivalent by 3.2.7. ■

3.3.6 Corollary *If $x_1, \ldots, x_n$ are linearly independent, then they are nonzero.*

Proof. This is clear from condition (b) of 3.3.5, since every permutation of the x_i is also linearly independent. ■

Every linear mapping preserves linear dependence (§3.2, Exercise 5). For the preservation of independence, more is needed:

3.3.7 Theorem *If $T:\mathrm{V} \to \mathrm{W}$ is an **injective** linear mapping and if $x_1, \ldots, x_n$ are linearly independent vectors in* V*, then $Tx_1, \ldots, Tx_n$ are linearly independent in* W.

Proof. Assuming $c_1(Tx_1) + \ldots + c_n(Tx_n) = \theta$, we have to show that the c_i are all 0. By linearity, $T(c_1x_1 + \ldots + c_nx_n) = \theta$; since T is injective, $c_1x_1 + \ldots c_nx_n = \theta$ (2.2.9), therefore $c_1 = \ldots = c_n = 0$ by the independence of the x_i. ■

The following theorem is for application in a later section (3.5.8):

3.3.8 Theorem *Let* V *be a vector space and $x_1, \ldots, x_n$ a finite list of vectors in* V*. Let $1 \le r < n$, let $\mathrm{M} = [\{x_{r+1}, \ldots, x_n\}]$ be the linear subspace generated by $\{x_{r+1}, \ldots, x_n\}$ (3.1.4), and let $Q:\mathrm{V} \to \mathrm{V/M}$ be the quotient mapping (2.6.1). The following conditions are equivalent:*

(a) *$x_1, \ldots, x_n$ are independent in* V;

(b) *$Qx_1, \ldots, Qx_r$ are independent in* V/M *and $x_{r+1}, \ldots, x_n$ are independent in* V.

Proof.

(a) ⇒ (b): Obviously $x_{r+1}, \ldots, x_n$ are independent. Assuming $c_1(Qx_1) + \ldots + c_r(Qx_r) = Q\theta$, we have to show that $c_1 = \ldots = c_r = 0$. By the linearity of Q, we have $Q(c_1x_1 + \ldots + c_rx_r) = Q\theta$, thus $c_1x_1 + \ldots + c_rx_r \in \operatorname{Ker} Q = \mathrm{M}$. By the definition of M,

$$c_1x_1 + \ldots + c_rx_r = c_{r+1}x_{r+1} + \ldots + c_nx_n$$

for suitable scalars $c_{r+1}, \ldots, c_n$ (3.1.5); then

$$c_1x_1 + \ldots + c_rx_r + (-c_{r+1})x_{r+1} + \ldots + (-c_n)x_n = \theta,$$

so *all* of the coefficients are 0 by the independence of the x_i.

(b) ⇒ (a): Suppose $c_1x_1 + \ldots + c_nx_n = \theta$. Write $z = c_{r+1}x_{r+1} + \ldots + c_nx_n$; since $z \in \mathrm{M} = \operatorname{Ker} Q$, $Q\theta = c_1(Qx_1) + \ldots + c_r(Qx_r) + Qz = c_1(Qx_1) + \ldots + c_r(Qx_r)$, therefore $c_1 = \ldots = c_r = 0$; but then the original relation simplifies to $c_{r+1}x_{r+1} + \ldots + c_nx_n = \theta$, therefore also $c_{r+1} = \ldots = c_n = 0$. Thus $c_i = 0$ for $i = 1, \ldots, n$. ■

▶ ***Exercises***

1. Show that the vectors $(2, -1, 1)$, $(-1, 2, 1)$, $(1, 1, 1)$ in $\mathbf{R}^3$ are linearly independent.

2. If $x_1, \ldots, x_n, x_{n+1}$ are linearly independent, then so are $x_1, \ldots, x_n$.

3. If $x_1, \ldots, x_n$ are linearly independent, then the following implication is

true: (*) $a_1x_1 + \dots + a_nx_n = b_1x_1 + \dots + b_nx_n \Rightarrow a_i = b_i$ for all i. Conversely, the validity of (*) implies that $x_1, \dots, x_n$ are linearly independent.

4. Prove that if the vectors x_1, x_2, x_3 are linearly independent, then so are the vectors x_1, $x_1 + x_2$, $x_1 + x_2 + x_3$. Then state and prove a generalization (i.e., a theorem that contains this as a special case).

5. Let a, b, c be distinct real numbers. Prove that the vectors $(1, 1, 1)$, (a, b, c), (a^2, b^2, c^2) in $\mathbf{R}^3$ are linearly independent. (Generalization?)

6. If $T{:}\mathrm{V} \rightarrow \mathrm{W}$ is a linear mapping and $x_1, \dots, x_n$ are vectors in V such that $Tx_1, \dots, Tx_n$ are linearly independent, then $x_1, \dots, x_n$ are linearly independent.

7. (i) Let V be a real or complex vector space. Prove that if x, y, z are linearly independent vectors in V, then so are $x + y$, $y + z$, $z + x$.

(ii) What if V is a vector space over 'the littlest field' $\{0, 1\}$ (Appendix A.4.2)? {Hint: $2x = \theta$ for every vector x.}

8. Let $x_1, \dots, x_n, x_{n+1}$ be linearly independent vectors, $a_1, \dots, a_n$ scalars. For $i = 1, \dots, n$ define $y_i = x_i + a_ix_{n+1}$. Prove that $y_1, \dots, y_n$ are linearly independent.

9. In the vector space $\mathcal{F}(\mathbf{R})$ of all functions $\mathbf{R} \rightarrow \mathbf{R}$ (Example 1.3.4), sin and cos are linearly independent; so are the functions sin, cos, p, where p is any nonzero polynomial function (Hint: Differentiate.); so are sin, cos, p, exp (the base e exponential function).

10. Prove that the functions sin, sinh are linearly independent.

11. Prove that the functions $\sin t$, $\sin 2t$, $\sin 3t$ are linearly independent.

12. Let $\mathrm{V} = \mathbf{C}^2$ and regard V also as a real vector space by restriction of scalars (§1.3, Exercise 3). Let $x_1 = (1, i)$, $x_2 = (i, -1)$. Show that x_1, x_2 are linearly independent in $\mathrm{V}_{\mathbf{R}}$ but linearly dependent in $\mathrm{V}_{\mathbf{C}}$.

13. Let $\mathcal{P}$ be the vector space of real polynomial functions (Example 1.3.5), let n be a positive integer, and suppose $p \in \mathcal{P}$ has degree n. For $k = 0, \dots, n$ let $p^{(k)}$ be the k'th derivative of p. (In particular, $p^{(0)} = p$, $p^{(1)} = p'$, $p^{(2)} = p''$, etc.) Prove that the polynomials $p^{(n)}$, $p^{(n-1)}, \dots, p''$, p', p are linearly independent. {Hint: Determine the degree of $p^{(k)}$, then look at Theorem 3.3.5.}

14. With $\mathrm{V} = \mathcal{P}$ as in Exercise 13, let $D \in \mathcal{L}(\mathrm{V})$ be the linear mapping defined by $Dp = p'$ (the derivative of p). Prove that for every positive integer n, the elements I, D, $D^2, \dots, D^n$ of $\mathcal{L}(\mathrm{V})$ are linearly independent.

15. A *set* A of vectors is said to be *linearly independent* if every finite list $x_1, \dots, x_n$ of distinct vectors in A is linearly independent (in other words, A is not linearly dependent in the sense of §3.2, Exercise 8). Prove:

(1) A is linearly independent if and only if no vector of A is a linear combination of the remaining vectors of A, that is, for every vector $x \in A$, $x \notin [A - \{x\}]$.

(2) The empty set $\varnothing$ is linearly independent 'by default'. {Hint: Independence of A means $x \in A \Rightarrow x \notin [A - \{x\}]$. Remember 'vacuous implication' (Appendix A.1.5)?}

(3) If A is linearly independent and $B \subset A$, then B is linearly independent.

(4) If $T{:}V \to W$ is an injective linear mapping and A is a linearly independent subset of V, then $T(A)$ is a linearly independent subset of W.

(5) Let $f{:}\mathbf{R}^2 \to \mathbf{R}$ be the linear form $f(x, y) = x$ and let $A = \{(1, 0), (1, 1), (1, 2)\} \subset \mathbf{R}^2$, $B = \{1\} \subset \mathbf{R}$. Then $f(A)$ is linearly independent, but A is not; also, B is linearly independent, but $f^{-1}(B)$ is not.

16. Let V be a vector space over F, f a nonzero linear form on V (2.1.11), N the kernel of f, and y a vector such that $f(y) \neq 0$. Prove:

(1) $V = N \oplus Fy$ (in the sense of §1.6, Exercise 12). {Hint: §2.2, Exercise 2.}

(2) A linear form g on V is proportional to f if and only if $g(x) = 0$ for all $x \in N$. {Hint: Assuming $g = 0$ on N, the problem is to find a scalar c such that $g = cf$. Try $c = g(y)/f(y)$.}

17. Let f and g be the linear forms on $\mathbf{R}^3$ defined by

$$f(x_1, x_2, x_3) = x_1 + x_2 + x_3, \qquad g(x_1, x_2, x_3) = x_1 + x_2 - x_3.$$

True or false (explain): The elements f, g of the dual space $(\mathbf{R}^3)'$ (§2.3, Exercise 5) are linearly independent.

18. Let V and W be vector spaces over F and let M, N be linear subspaces of V such that $V = M \oplus N$ (§1.6, Exercise 12). Prove:

(i) If $R{:}M \to W$ and $S{:}N \to W$ are linear mappings, there exists a unique linear mapping $T{:}V \to W$ such that T agrees with R on M and with S on N.

(ii) Every linear mapping $M \to W$ may be extended to a linear mapping $V \to W$ (if we like, with the same range).

19. Let $M_1, \ldots, M_n$ be linear subspaces of a vector space V and let $M_1 \times \ldots \times M_n$ be the product vector space (1.3.11). The mapping $T{:}M_1 \times \ldots \times M_n \to V$ defined by $T(x_1, \ldots, x_n) = x_1 + \ldots + x_n$ is linear. If T is injective, the linear subspaces $M_1, \ldots, M_n$ are said to be (linearly) *independent*.

(i) M_1, M_2 are independent if and only if $M_1 \cap M_2 = \{\theta\}$.

(ii) M_1, M_2, M_3 are independent if and only if $M_1 \cap (M_2 + M_3) = M_2 \cap (M_1 + M_3) = M_3 \cap (M_1 + M_2) = \{\theta\}$.

(iii) Generalize the statement (ii).

(iv) If $x_1, \dots, x_n$ are nonzero vectors and M_i is the set of all scalar multiples of x_i, then the linear subspaces $M_1, \dots, M_n$ are independent if and only if the vectors $x_1, \dots, x_n$ are linearly independent.

20. With notations as in the preceding exercise, if $M_1, \dots, M_n$ are independent and $M_1 + \dots + M_n = V$, then V is said to be the *direct sum* (or 'internal direct sum') of the linear subspaces $M_1, \dots, M_n$, written $V = M_1 \oplus \dots \oplus M_n$. (The case $n = 2$ is covered by §1.6, Exercise 12.) This is equivalent to saying that T is bijective, and it implies that $M_1 \times \dots \times M_n \cong M_1 \oplus \dots \oplus M_n = V$.

21. Let $V_1, \dots, V_n$ be vector spaces over the same field of scalars and let $V = V_1 \times \dots \times V_n$ be the product vector space (Definition 1.3.11). For each index j, let M_j be the linear subspace of V consisting of all vectors $(x_1, \dots, x_n) \in V$ such that $x_i = \theta$ for all $i \neq j$. {For example, M_2 is the set of all vectors $(\theta, x_2, \theta, \dots, \theta)$ with $x_2 \in V_2$.} Prove:

(1) $M_j \cong V_j$ for $j = 1, \dots, n$.

(2) $V = M_1 \oplus \dots \oplus M_n$ in the sense of Exercise 20. {For this reason, $V_1 \times \dots \times V_n$ is also called the 'direct sum' of $V_1, \dots, V_n$ (or 'external direct sum', to distinguish it from the situation in Exercise 20).}

22. Suppose $V = M_1 \oplus \dots \oplus M_n$ as in Exercise 20. Prove:

(i) $V = (M_1 \oplus \dots \oplus M_{n-1}) \oplus M_n$.

(ii) $V/M_n \cong M_1 \oplus \dots \oplus M_{n-1}$.

23. Let $M_1, \dots, M_n$ be linear subspaces of V. Prove that the following conditions are equivalent:

(a) $M_1, \dots, M_n$ are independent (in the sense of Exercise 19);

(b) $(M_1 + \dots + M_i) \cap M_{i+1} = \{\theta\}$ for $i = 1, \dots, n - 1$.

24. Let V be a vector space over F and let $x_1, \dots, x_n$ be vectors in V. For $k = 1, \dots, n$ let M_k be the linear subspace of V generated by $\{x_1, \dots, x_k\}$ (3.1.4):

$$\begin{aligned} M_1 &= [\{x_1\}] = Fx_1 \\ M_2 &= [\{x_1, x_2\}] \\ &\dots \\ M_k &= [\{x_1, \dots, x_k\}] \\ &\dots \\ M_n &= [\{x_1, \dots, x_n\}]. \end{aligned}$$

Define $M_0 = \{\theta\}$. Thus

$$\{\theta\} = M_0 \subset M_1 \subset M_2 \dots \subset M_n.$$

Prove that the following conditions are equivalent:

(a) $x_1, \dots, x_n$ are linearly independent;

(b) $M_{k-1} \subset M_k$ properly, for $k = 1, \dots, n$.

3.4 Finitely generated vector spaces

3.4.1 Definition Let V be a vector space. A finite list of vectors $x_1, \dots, x_n$ in V is said to be **generating** (for V) if every vector in V is a linear combination of the vectors x_i.

This means that the set $\{x_1, \dots, x_n\}$ is generating in the sense of 3.1.4, that is, $[\{x_1, \dots, x_n\}] = V$ (3.1.5). The concept is also expressed by saying that 'the vectors $x_1, \dots, x_n$ *are* generating for V', or that '$x_1, \dots, x_n$ generate V'.

Although it's not obvious at the outset, the concepts 'generating' and 'independent' are natural companions. For example, the following theorem reminds us of 3.3.2:

3.4.2 Theorem *Let* V *be a vector space over the field* F, *let* $x_1, \dots, x_n$ *be a finite list of vectors in* V, *and let* $T:F^n \to V$ *be the linear mapping defined by* $T(a_1, \dots, a_n) = a_1x_1 + \dots + a_nx_n$ (2.1.3). *The following conditions are equivalent*:

(a) $x_1, \dots, x_n$ *are generating*;

(b) T *is surjective*.

Proof. This is immediate from the definitions. ■

3.4.3 Example In F^n, the list $e_1, \dots, e_n$ is generating (1.5.3).

3.4.4 Example In the vector space $\mathcal{P}$ of real polynomial functions (Example 1.3.5), no finite list $p_1, \dots, p_n$ can be generating. {Reason: If m is an integer larger than the degree of every p_i, then the monomial function $t \mapsto t^m$ is not a linear combination of $p_1, \dots, p_n$.

3.4.5 Example If $x_1, \dots, x_n$ are generating for V and x_{n+1} is any vector in V, then the list $x_1, \dots, x_n, x_{n+1}$ is dependent. {Reason: x_{n+1} is a linear combination of $x_1, \dots, x_n$; see 3.2.5.}

3.4.6 Example If $x_1, \dots, x_n$ are independent, then $x_1, \dots, x_{n-1}$ cannot be generating. {This is 3.4.5 in contrapositive form[1].}

The next theorem reminds us of 3.3.7:

3.4.7 Theorem *If* $T:V \to W$ *is a* **surjective** *linear mapping and if* $x_1, \dots, x_n$ *generate* V, *then* $Tx_1, \dots, Tx_n$ *generate* W.

[1] Appendix A.1.8.

Proof. By assumption, $W = T(V)$ and $V = [\{x_1, \ldots, x_n\}]$, therefore

$$W = T([\{x_1, \ldots, x_n\}]) = [T(\{x_1, \ldots, x_n\})]$$

by 3.1.6; but $T(\{x_1, \ldots, x_n\}) = \{Tx_1, \ldots, Tx_n\}$, so $W = [\{Tx_1, \ldots, Tx_n\}]$, in other words $Tx_1, \ldots, Tx_n$ generate W. ■

The analogue of Theorem 3.3.8:

3.4.8 Theorem *Let* V *be a vector space and* $x_1, \ldots, x_n$ *a finite list of vectors in* V. *Let* $1 \le r < n$, *let* $M = [\{x_{r+1}, \ldots, x_n\}]$ *be the linear subspace generated by* $\{x_{r+1}, \ldots, x_n\}$ (3.1.4), *and let* $Q: V \to V/M$ *be the quotient mapping* (2.6.1). *The following conditions are equivalent*:

(a) $x_1, \ldots, x_n$ *generate* V;

(b) $Qx_1, \ldots, Qx_r$ *generate* V/M.

Proof.

(a) ⇒ (b): Let $u \in V/M$; the problem is to express u as a linear combination of $Qx_1, \ldots, Qx_r$. Choose a vector $x \in V$ with $u = Qx$. By assumption, x is a linear combination of $x_1, \ldots, x_n$, say

$$x = c_1x_1 + \ldots + c_rx_r + c_{r+1}x_{r+1} + \ldots + c_nx_n;$$

applying the linear mapping Q and noting that $Qx_{r+1} = \ldots = Qx_n = Q\theta$ (because the kernel of Q is M), we see that $u = Qx = c_1Qx_1 + \ldots + c_rQx_r$.

(b) ⇒ (a): Let $x \in V$; we have to show that x is a linear combination of $x_1, \ldots, x_n$. By assumption, $Qx = c_1(Qx_1) + \ldots + c_r(Qx_r)$ for suitable scalars $c_1, \ldots, c_r$. Then $x - (c_1x_1 + \ldots + c_rx_r) \in \operatorname{Ker} Q = M$, say

$$x - (c_1x_1 + \ldots + c_rx_r) = c_{r+1}x_{r+1} + \ldots + c_nx_n,$$

so $x = c_1x_1 + \ldots + c_nx_n$. ■

3.4.9 Definition A vector space V over the field F is said to be **finitely generated** (over F) if there exists a finite list of vectors that is generating for V (in the sense of Definition 3.4.1).

3.4.10 Example The zero space $\{\theta\}$ is finitely generated (the list $x_1 = \theta$ is generating), as are the spaces F^n (3.4.3); we will see in the next section that *every* finitely generated vector space is isomorphic to one of these examples (3.5.7 and 3.5.6).

3.4.11 Theorem *A vector space* V *over* F *is finitely generated if and only if there exist a positive integer* n *and a surjective linear mapping* $T: F^n \to V$.

Proof. If $x_1, \ldots, x_n$ is a finite list of vectors that generates V, then there exists a linear surjection $F^n \to V$ by Theorem 3.4.2.

Conversely, suppose that for some positive integer n, there exists a linear surjection $T: F^n \to V$. Since F^n is finitely generated, so is V (3.4.7). ■

3.4.12 Remarks If T:V → W is a linear surjection and V is finitely generated, then W is also finitely generated (3.4.7).

In particular, if V is a finitely generated vector space and M is any linear subspace of V, then the quotient vector space V/M is finitely generated (let the quotient mapping V → V/M play the role of T). It is also true that M is finitely generated, but this is harder to see; the proof is given in the next section (3.5.14).

▶ ***Exercises***

1. True or false (explain): The vectors $x_1 = (1, 2, -3)$, $x_2 = (2, -1, 2)$, $x_3 = (5, 10, -15)$, $x_4 = (6, -3, 6)$ generate $\mathbf{R}^3$.
2. Prove that if the vectors x_1, x_2, x_3 generate a vector space V then the vectors x_1, $x_1 + x_2$, $x_1 + x_2 + x_3$ also generate V. Then state and prove a generalization.
3. Let V and W be vector spaces over F, V × W the product vector space (1.3.11). Prove: V × W is finitely generated if and only if both V and W are finitely generated. {Hint: Remark 3.4.12 and §3.1, Exercise 6.}
4. Let V be the vector space of all functions $\mathbf{R} \to \mathbf{R}$ having derivatives of all orders (see Example 2.3.8) and let
$$\mathrm{M} = \{f \in \mathrm{V}\colon f'' = f\},$$
where f'' is the second derivative of f. Prove that M is the linear subspace of V generated by the two functions $t \mapsto e^t$ and $t \mapsto e^{-t}$. {Hint: if $f'' = f$, look at $f = \frac{1}{2}(f + f') + \frac{1}{2}(f - f')$.}
5. With notations as in Exercise 4, M is also generated by the functions sinh, cosh.
6. With V as in Exercise 4, let
$$\mathrm{N} = \{f \in \mathrm{V}\colon f'' = -f\}.$$
Prove that N is generated by the functions sin, cos. {Hint: If $f'' = -f$, calculate the derivative of $f\sin + f'\cos$ and of $f\cos - f'\sin$.}
7. If V is a vector space and M is linear subspace such that both M and V/M are finitely generated, then V is finitely generated. More precisely, if $x_1, \ldots, x_m$ generate M and if $Qy_1, \ldots, Qy_r$ generate V/M (where Q:V → V/M is the quotient mapping), then V is generated by $x_1, \ldots, x_m, y_1, \ldots, y_r$.

3.5 Basis, dimension

In looking at properties of independent lists and generating lists, we have seen contrasts (3.4.6) and analogies (3.3.8 and 3.4.8); the decisive step is to merge the two concepts:

3.5.1 Definition Let V be a vector space. A finite list of vectors $x_1, \ldots, x_n$ in V is said to be a **basis** of V it if is both independent and generating.

This means that every vector $x \in \mathrm{V}$ is a linear combination $x = a_1x_1 + \ldots + a_nx_n$ (3.4.1) and that the coefficients $a_1, \ldots, a_n$ are uniquely determined by x (3.3.2 (b)).

3.5.2 Example In F^n, the list $e_1, \ldots, e_n$ is a basis (3.3.3 and 3.4.3); it is called the *canonical basis* (or 'natural basis') of F^n.

3.5.3 Example If $x_1, \ldots, x_n$ are independent in V and if $\mathrm{M} = [\{x_1, \ldots, x_n\}]$ is the linear subspace of V generated by the x_i, then $x_1, \ldots, x_n$ is a basis of M.

3.5.4 Example Suppose $x_1, \ldots, x_n$ is a basis of V and x_{n+1} is any vector in V. The augmented list $x_1, \ldots, x_n, x_{n+1}$ is not independent (3.4.5) and the diminished list $x_1, \ldots, x_{n-1}$ is not generating (3.4.6); so to speak, a basis is 'maximal-independent' and 'minimal-generating'.

3.5.5 Theorem *Let* V *be a vector space over the field* F, *let* $x_1, \ldots, x_n$ *be a finite list of vectors in* V, *and let* $T{:}\mathrm{F}^n \to \mathrm{V}$ *be the linear mapping defined by* $T(a_1, \ldots, a_n) = a_1x_1 + \ldots + a_nx_n$ (2.1.3). *The following conditions are equivalent*:

(a) $x_1, \ldots, x_n$ *is a basis of* V;

(b) T *is bijective*.

Proof. This is immediate from Theorems 3.3.2, 3.4.2, and the definitions. In slow motion:

(a) ⇒ (b): If $x_1, \ldots, x_n$ is a basis, then T is injective (3.3.2) and surjective (3.4.2), therefore bijective.

(b) ⇒ (a): If T is bijective then it is both injective and surjective, thus the list $x_1, \ldots, x_n$ is both independent (3.3.2) and generating (3.4.2), therefore is a basis. ■

3.5.6 Corollary *A vector space* V *over* F *has a basis* (*in the sense of* 3.5.1) *if and only if there exists a positive integer* n *such that* $\mathrm{V} \cong \mathrm{F}^n$.

Proof. Suppose $\mathrm{V} \cong \mathrm{F}^n$ and let $T{:}\mathrm{F}^n \to \mathrm{V}$ be a linear bijection (Theorem 2.4.6). Consider the basis $e_1, \ldots, e_n$ of F^n (3.5.2) and let $x_i = Te_i$ for $i = 1, \ldots, n$. The list $x_1, \ldots, x_n$ is independent (3.3.7) and generating (3.4.7), thus is a basis of V.

Conversely, if V has a basis $x_1, \ldots, x_n$ then $\mathrm{V} \cong \mathrm{F}^n$ by Theorem 3.5.5. ■

We will see in 3.5.10 that the integer n of the above corollary is *unique*.

It is obvious that if V has a basis in the sense of Definition 3.5.1, then it is finitely generated. Conversely:

3.5.7 Theorem *Every finitely generated vector space* $V \neq \{\theta\}$ *has a basis.*

Proof. By assumption, there exists a finite list of vectors that generates V. Among all such lists, choose one of minimal length. More formally, let

$$S = \{n \in \mathbf{P}:\ \exists \text{ list } x_1, \ldots, x_n \text{ generating } V\};$$

by assumption, the set S of positive integers is nonempty, therefore it has a smallest element, say m. Choose a list $x_1, \ldots, x_m$ of length m that generates V; it will suffice to prove that $x_1, \ldots, x_m$ are independent.

If $m = 1$ then $x_1 \neq \theta$ (because $V \neq \{\theta\}$) and we are done (3.3.1). Assuming $m > 1$, let us verify the criterion of 3.3.4: if j is any index, we must show that x_j is not a linear combination of the x_i with $i \neq j$. Consider the list obtained by omitting x_j:

$$(*) \qquad x_1, \ldots, x_{j-1}, x_{j+1}, \ldots, x_m$$

(if $j = 1$ we mean $x_2, \ldots, x_m$, and if $j = m$ we mean $x_1, \ldots, x_{m-1}$). The list $(*)$ has length $m - 1$, so it cannot generate V (m is the minimal length for a generating list); let W be the linear subspace of V generated by the vectors $(*)$. We know that $W \neq V$, therefore W cannot contain all of $x_1, \ldots, x_n$ (if it did, it would contain their linear span, hence would equal V); but W does contain the x_i for $i \neq j$, so it must fail to contain x_j. In other words, x_j is not a linear combination of the x_i with $i \neq j$. ■

We can be more specific: if $x_1, \ldots, x_n$ is any generating list for V, then some 'sublist' is a basis. The key to the proof: if some x_j is a linear combination of the remaining x_i, then x_j can be discarded from the list without changing the linear span, that is, the x_i for $i \neq j$ also generate V. Keep on discarding such 'redundant' vectors until none of the vectors that remain is a linear combination of its fellow survivors; this 'reduced' list is then both independent (3.3.4) and generating, thus is a basis.

The preceding argument shows that there can be many paths to a basis (start with any finite generating list, vary the order of suppressing redundant vectors, etc.). What is constant is the number of vectors left at the end of the process, that is, any two bases have the same number of elements; this is an easy consequence of the following theorem:

3.5.8 Theorem *Let* n *be a positive integer. If* V *is a vector space containing two lists of vectors* $x_1, \ldots, x_n$ *and* $y_1, \ldots, y_n$ *of the same length* n*, such that*

(1) $x_1, \ldots, x_n$ *generate* V, *and*

(2) $y_1, \ldots, y_n$ *are independent*,

then both lists are bases of V.

Proof. We have to show that list (1) is independent and list (2) is generating. The proof is by induction[1] on n. Let F be the field of scalars.

Suppose $n = 1$. By assumption, $V = Fx_1$ and $y_1 \neq \theta$. Say $y_1 = cx_1$. Since y_1 is nonzero, both c and x_1 are nonzero. From $x_1 \neq \theta$ we see that the list

[1] For the style of proof, see Appendix A.1.13.

x_1 is independent. Since every vector is a multiple of $x_1 = c^{-1}y_1$, and therefore of y_1, the list y_1 is generating. All's well for $n = 1$.

Let $n \geq 2$ and assume that the statement in the theorem is true for lists of length $n - 1$.

We assert first that $x_1, \ldots, x_n$ are independent. Assume to the contrary[1] that this list is linearly dependent, in other words, that one of the x_i is a linear combination of the others (Theorem 3.2.4). Suppose, for illustration, that x_n is a linear combination of $x_1, \ldots, x_{n-1}$. It then follows that the list $x_1, \ldots, x_{n-1}$ generates V; for, its linear span includes x_n as well as $x_1, \ldots, x_{n-1}$, so it must be all of V by (1). Since the list $y_1, \ldots, y_{n-1}$ is independent by (2), it follows from the induction hypothesis that $y_1, \ldots, y_{n-1}$ generate V. In particular, y_n is a linear combination of $y_1, \ldots, y_{n-1}$, which contradicts (2). The contradiction shows that $x_1, \ldots, x_n$ must be independent, as asserted.

The proof that $y_1, \ldots, y_n$ are generating will be accomplished by invoking the induction hypothesis in a suitable quotient space V/M; the first step is to construct an appropriate linear subspace M. Express y_n as a linear combination of $x_1, \ldots, x_n$ (possible by (1)), say

$$y_n = c_1x_1 + \ldots + c_nx_n.$$

Since $y_n \neq \theta$ (by (2)), one of the coefficients c_i must be nonzero; rearranging the x_i (which does not alter the fact that they generate V), we can suppose that $c_n \neq 0$. It follows that

$$x_n = (-c_1/c_n)x_1 + \ldots + (-c_{n-1}/c_n)x_{n-1} + (1/c_n)y_n,$$

thus the linear span of the list $x_1, \ldots, x_{n-1}, y_n$ includes all of the vectors $x_1, \ldots, x_n$; in view of (1), we conclude that

(∗) $\qquad x_1, \ldots, x_{n-1}, y_n$ generate V.

Let $M = Fy_n$ and let Q:V → V/M be the quotient mapping. Then

(3) $\qquad Qx_1, \ldots, Qx_{n-1}$ generate V/M

(by (∗) and 3.4.8) and

(4) $\qquad Qy_1, \ldots, Qy_{n-1}$ are independent

(by (2) and 3.3.8), so by the induction hypothesis, both of the lists (3) and (4) are bases of V/M.

In particular, the list (4) is generating for V/M; since M is generated by y_n, it follows from 3.4.8 that $y_1, \ldots, y_n$ generate V, as we wished to show. ■

No independent list can be longer than a generating list:

3.5.9 Corollary *If $x_1, \ldots, x_n$ are generating and $y_1, \ldots, y_m$ are independent, then $m \leq n$.*

Proof. Assume to the contrary that $n < m$, that is, $n + 1 \leq m$. Then $x_1, \ldots, x_n$ are generating and $y_1, \ldots, y_n$ are independent, so by the theorem $y_1, \ldots, y_n$ is a basis; in particular, y_{n+1} is a linear combination of $y_1, \ldots, y_n$, a contradiction (3.3.5). ■

The number of elements in a basis is unique:

3.5.10 Corollary *If* V *has a basis with* n *elements, then* ***every*** *basis of* V *has* n *elements.*

Proof. Suppose $x_1, \ldots, x_n$ and $y_1, \ldots, y_m$ are bases. Since the first list is generating and the second is independent, $m \le n$ by the preceding corollary; similarly $n \le m$. ■

3.5.11 Definition Let V be a finitely generated vector space. If $\mathrm{V} \neq \{\theta\}$ then V has a basis (3.5.7) and the number of basis vectors is unique (3.5.10); this number is called the **dimension** of V, written $\dim \mathrm{V}$. If $\mathrm{V} = \{\theta\}$, its dimension is defined to be 0.

Finitely generated vector spaces are also called **finite-dimensional**. More precisely, if $\dim \mathrm{V} = n$ we say that V is ***n*-dimensional**; if it is necessary to emphasize the scalar field F we say that V is n-dimensional **over** F and we write $\dim_{\mathrm{F}} \mathrm{V} = n$.

If a vector space is not finite-dimensional (that is, not finitely generated), it is said to be **infinite-dimensional**. Our main business is with finite-dimensional spaces.

The above definition assigns, to each finitely generated vector space V, a nonnegative integer $\dim \mathrm{V}$; up to isomorphism, this integer tells the story of V:

3.5.12 Theorem *Let* V *and* W *be vector spaces.*

(1) *If* V *is finite-dimensional and* $\mathrm{V} \cong \mathrm{W}$, *then* W *is also finite-dimensional and* $\dim \mathrm{V} = \dim \mathrm{W}$.

(2) *Conversely, if* V *and* W *are finite-dimensional and* $\dim \mathrm{V} = \dim \mathrm{W}$, *then* $\mathrm{V} \cong \mathrm{W}$.

(3) *Thus, if* V *and* W *are finite-dimensional, then*

$$\mathrm{V} \cong \mathrm{W} \Leftrightarrow \dim \mathrm{V} = \dim \mathrm{W}.$$

Proof. If $\mathrm{V} = \{\theta\}$ then $\dim \mathrm{V} = 0$ and all three assertions are obvious. Assume $\mathrm{V} \neq \{\theta\}$.

(1) Suppose $n = \dim \mathrm{V}$ and let $x_1, \ldots, x_n$ be a basis of V. By assumption, there exists a linear bijection $T\colon \mathrm{V} \to \mathrm{W}$. The list $Tx_1, \ldots, Tx_n$ generates W (3.4.7) and is independent (3.3.7), therefore is a basis of W; thus $\dim \mathrm{W} = n = \dim \mathrm{V}$.

(2) Let $n = \dim \mathrm{V} = \dim \mathrm{W}$ and choose a basis (necessarily of length n) in each of the spaces V and W. By Theorem 3.5.5, there exist isomorphisms $\mathrm{F}^n \cong \mathrm{V}$ and $\mathrm{F}^n \cong \mathrm{W}$, therefore $\mathrm{V} \cong \mathrm{W}$ (2.4.7).

(3) This is immediate from (1) and (2). ■

The next theorem restates earlier results in a convenient form for reference:

3.5.13 Theorem *In an n-dimensional vector space,*

(1) $y_1, \ldots, y_m$ *independent* $\Rightarrow m \le n$;

(2) $y_1, \ldots, y_m$ *generating* $\Rightarrow m \geq n$;

(3) $y_1, \ldots, y_n$ *independent* $\Rightarrow y_1, \ldots, y_n$ *is a basis*;

(4) $y_1, \ldots, y_n$ *generating* $\Rightarrow y_1, \ldots, y_n$ *is a basis*.

Proof. By assumption, there exists a list $x_1, \ldots, x_n$ that is both generating and independent; thus (1) and (2) are immediate from 3.5.9, while (3) and (4) are immediate from 3.5.8. ■

In words: in a space of dimension n, an independent list has length at most n, a generating list has length at least n, and an independent or generating list of the 'right length' is a basis.

3.5.14 Corollary *If* V *is a finite-dimensional vector space and* M *is a linear subspace of* V, *then*:

(i) M *is finite-dimensional and* $\dim M \leq \dim V$;

(ii) $M = V \Leftrightarrow \dim M = \dim V$.

Proof. Let $n = \dim V$. If $M = \{\theta\}$ then $\dim M = 0$ and the assertions are obvious. Assume henceforth that $M \neq \{\theta\}$ (therefore $V \neq \{\theta\}$ and $n \geq 1$).

(i) There exist independent lists in M (any nonzero vector in M defines an independent list of length 1), and every such list has length $\leq n$ (3.5.13). Let

$$x_1, \ldots, x_m$$

be an independent list in M whose length m is as large as possible. If x is any vector in M, then the list

$$x_1, \ldots, x_m, x$$

cannot be independent (too long!), therefore x is a linear combination of $x_1, \ldots, x_m$ (3.3.5). This proves that the list $x_1, \ldots, x_m$ generates M, hence is a basis of M (3.5.3), and $\dim M = m \leq n = \dim V$.

(ii) If $\dim M = \dim V = n$ then M has a basis $y_1, \ldots, y_n$ of length n; but $y_1, \ldots, y_n$ is also a basis of V (3.5.13), therefore

$$M = [\{y_1, \ldots, y_n\}] = V.$$

Thus, $\dim M = \dim V \Rightarrow M = V$. The converse is trivial. ■

In a finite-dimensional vector space, every independent list is part of a basis:

3.5.15 Theorem *If* V *is an n-dimensional vector space and* $x_1, \ldots, x_m$ *is a list of vectors in* V *that is independent but not generating, then* V *has a basis* $x_1, \ldots, x_m, x_{m+1}, \ldots, x_n$.

Proof. If $M = [\{x_1, \ldots, x_m\}]$ is the linear span of $x_1, \ldots, x_m$ then $\dim M = m$ (3.5.3); by assumption $M \neq V$, so $m < n$ (3.5.14).

The quotient vector space V/M is finitely generated (3.4.12) and nonzero (if $x \in V$ is a vector not in M, then $x + M$ is a nonzero element of V/M), so it has a basis (3.5.7), say $u_1, \ldots, u_r$. Let $Q: V \to V/M$ be the quotient mapping and choose vectors $y_1, \ldots, y_r$ in V such that $Qy_j = u_j$ for $j = 1, \ldots, r$; it will suffice to prove that the list

$$(*) \qquad x_1, \ldots, x_m, y_1, \ldots, y_r$$

is a basis of V. Indeed, since $x_1, \ldots, x_m$ is both independent and generating for M, the list $(*)$ is independent by 3.3.8 and it generates V by 3.4.8. ■

Implicit in the preceding proof:

3.5.16 Corollary *If* V *is a finite-dimensional vector space and* M *is a linear subspace of* V *with* $M \neq \{\theta\}$ *and* $M \neq V$, *then every basis of* M *can be expanded to basis of* V.

Proof. By 3.5.14, (i), M is finite-dimensional; apply 3.5.15 to a basis $x_1, \ldots, x_m$ of M. ■

▶ *Exercises*

1. Prove that $(1, 2, 3)$, $(2, 1, 0)$, $(-3, 2, 0)$ is a basis of $\mathbf{R}^3$. Then find a shorter proof.

2. Let F be any field. Explain why $\dim_F F^n = n$.

3. Prove that $F^m \cong F^n$ if and only if $m = n$.

4. Let V be a vector space, $V \neq \{\theta\}$. Prove that V is finitely generated if and only if $V \cong F^n$ for some positive integer n; if $V \cong F^n$ then $n = \dim_F V$ (in particular, n is unique).

5. Let α be a real number. Prove that the vectors

$$u = (\cos \alpha, \sin \alpha), \qquad v = (-\sin \alpha, \cos \alpha)$$

are a basis of $\mathbf{R}^2$.

6. True or false (explain): If x_1, x_2, x_3 is a basis of V, then so is x_1, $x_1 + x_2, x_1 + x_2 + x_3$.

7. Let $x_1, \ldots, x_n$ be a basis of V. For each $x \in V$, there exist *unique* scalars $a_1, \ldots, a_n$ such that $x = a_1x_1 + \ldots + a_nx_n$ (Theorem 3.5.5); one calls $a_1, \ldots, a_n$ the *coordinates* of x with respect to the basis $x_1, \ldots, x_n$.

(i) In $\mathbf{R}^2$, find the coordinates of the vector $(2, 3)$ with respect to the basis $(\frac{1}{2}\sqrt{3}, \frac{1}{2})$, $(-\frac{1}{2}, \frac{1}{2}\sqrt{3})$.

(ii) In $\mathbf{R}^n$, find the coordinates of the vector $(a_1, \ldots, a_n)$ with respect to the canonical basis $e_1, \ldots, e_n$ (3.5.2).

8. If $x_1 = (1, 2, 0)$, $x_2 = (2, 1, 0)$ and $x_3 = (a, b, 1)$, where a and b are any real numbers, prove that x_1, x_2, x_3 is a basis of $\mathbf{R}^3$.

9. Let V be a finite-dimensional vector space and let M be a linear subspace of V. Prove:

(i) V/M is finite-dimensional, and $\dim(V/M) \le \dim V$;
(ii) $\dim(V/M) = \dim V \Leftrightarrow M = \{\theta\}$;
(iii) $\dim(V/M) = 0 \Leftrightarrow M = V$.

10. Let $V = \mathcal{F}(\mathbf{R})$ be the real vector space of all functions $\mathbf{R} \to \mathbf{R}$ (1.3.4), let a, b, c be real numbers, and let x, y, z be the functions

$$x(t) = \sin(t + a), \qquad y = \sin(t + b), \qquad z = \sin(t + c).$$

Prove that x, y, z are linearly dependent. {Hint: Consider the linear subspace $M = [\{\sin, \cos\}]$ of V and look at Corollary 3.5.9.}

11. Let $x_1, \ldots, x_n$ be a basis of V and let $1 \le m < n$. Let M be the linear subspace of V generated by $x_1, \ldots, x_m$, and N the linear subspace generated by $x_{m+1}, \ldots, x_n$. Prove that $V = M \oplus N$ (in the sense of §1.6, Exercise 12).

12. Let V be a vector space over F, let $x_1, \ldots, x_n$ be nonzero vectors in V, and let $M_i = Fx_i$ for $i = 1, \ldots, n$. Prove that the following two conditions are equivalent:

(a) $x_1, \ldots, x_n$ is a basis of V;
(b) $V = M_1 \oplus \ldots \oplus M_n$ (in the sense of §3.3, Exercise 20).

13. Prove that if $x_1, \ldots, x_n$ is a basis of V and if $a_1, \ldots, a_n$ are nonzero scalars, then $a_1x_1, \ldots, a_nx_n$ is also a basis of V.

14. Prove that if x_1, x_2, x_3 is a basis of V and a_1, a_2, a_3 are nonzero scalars, then the list

$$a_1x_1, \qquad a_1x_1 + a_2x_2, \qquad a_1x_1 + a_2x_2 + a_3x_3$$

is also a basis of V. {Hint: Combine Exercises 13 and 6.}

15. Let $x_1, \ldots, x_n$ be a basis of V. For $k = 1, \ldots, n$, let M_k be the linear subspace of V generated by $x_1, \ldots, x_k$, and let $M_0 = \{\theta\}$ (cf. §3.3, Exercise 24). For $k = 1, \ldots, n$, choose a vector y_k such that $y_k \in M_k$ but $y_k \notin M_{k-1}$.

Prove that $y_1, \ldots, y_n$ is a basis of V. {Hint: For each $k = 1, \ldots, n$ write $y_k = a_kx_k + z_k$ with $z_k \in M_{k-1}$. Argue first that the a_k are nonzero, then show that the y_k are linearly independent.}

16. Let V be any vector space (not necessarily finitely generated). A subset B of V is said to be a **basis** of V if it has the following two properties:

(i) B is generating, that is $[B] = V$ (3.1.4);
(ii) B is linearly independent, that is, for every vector $x \in B$, $x \notin [B - \{x\}]$ (§3.3, Exercise 15).

It is shown in advanced courses (using some form of 'Zorn's Lemma') that (1) V has a basis, and (2) if B and C are any two bases of V, then there exists a bijection $B \to C$ (in this sense, any two bases of V 'have the same number of elements').

The exercise: Can you find this theorem in an advanced algebra book?

17. Prove that if B is a basis of V (in the sense of Exercise 16), then B is 'minimal-generating' and 'maximal-independent' in the following sense:

(1) if A is a proper subset of B (that is, $A \subset B$ and $A \neq B$) then A is not generating;

(2) if $B \subset C \subset V$ and B is a proper subset of C, then C is not linearly independent.

18. Let $\mathcal{P}$ be the vector space of real polynomial functions (Example 1.3.5). For $n = 0, 1, 2, \ldots$ let p_n be the monomial function $p_n(t) = t^n$, and let $B = \{p_0, p_1, p_2, \ldots\}$. Prove that B is a basis of $\mathcal{P}$ in the sense of Exercise 16.

19. Let V be the vector space of finitely nonzero sequences of real numbers (1.3.9). Find a basis B of V (in the sense of Exercise 16).

20. Let V be a finite-dimensional complex vector space. Regard V also as a real vector space by restriction of scalars (§1.3, Exercise 3).

Prove that $\dim_{\mathbf{R}} V = 2 \cdot \dim_{\mathbf{C}} V$. {Hint: $V_{\mathbf{C}} \cong \mathbf{C}^n$ for some n.}

21. If V is a finite-dimensional real vector space and W is its complexification (§1.3, Exercise 4), prove that $\dim_{\mathbf{C}} W = \dim_{\mathbf{R}} V$.

22. Let $x_1, \ldots, x_n$ be vectors in $\mathbf{R}^n$. Prove that $x_1, \ldots, x_n$ is a basis of the real vector space $\mathbf{R}^n$ if and only if it is a basis of the complex vector space $\mathbf{C}^n$.

23. Let V and W be finite-dimensional vector spaces, $V \times W$ the product vector space (1.3.11). Prove:

$$\dim(V \times W) = \dim V + \dim W.$$

{Hint: If $x_1, \ldots, x_n$ is a basis of V and $y_1, \ldots, y_m$ is a basis of W, show that the vectors

$$(x_1, \theta), \ldots, (x_n, \theta), (\theta, y_1), \ldots, (\theta, y_m)$$

are a basis of $V \times W$; cf. §3.1, Exercise 6.}

24. If V is a finite-dimensional vector space and M is a linear subspace of V, then there exists a linear subspace N of V such that $V = M \oplus N$ in the sense of §1.6, Exercise 12. {Hint: Corollary 3.5.16.}

25. As in Theorem 3.4.8, let V be a vector space, $x_1, \ldots, x_r, x_{r+1}, \ldots, x_n \in V$, $M = [\{x_{r+1}, \ldots, x_n\}]$ and $Q: V \to V/M$ the quotient mapping. The following conditions are equivalent:

(a) $x_1, \ldots, x_n$ is a basis of V;

(b) $Qx_1, \ldots, Qx_r$ is a basis of V/M and $x_{r+1}, \ldots, x_n$ is a basis of M.

26. Let V be a finite-dimensional vector space, M and N linear subspace of V such that $\dim M + \dim N = \dim V$. Prove: $M + N = V \Leftrightarrow M \cap N = \{\theta\}$.

{Hint: Corollary 3.5.14, (ii) and Theorem 3.5.13, (2).}

27. Let V be a finite-dimensional vector space over a field F, $V \times V$ the product vector space (1.3.11). Prove: If $\dim(V \times V) = (\dim V)^2$, then either $V = \{\theta\}$ or $V \cong F^2$.

3.6 Rank + nullity = dimension

Quotient vector spaces played a useful role in arriving at the concept of dimension (3.5.8); in turn, the concept of dimension applied to quotient spaces yields a powerful formula:

3.6.1 Theorem *Let* V *be a vector space,* M *a linear subspace of* V. *The following conditions are equivalent:*

(a) V *is finite-dimensional;*

(b) M *and* V/M *are finite-dimensional.*

When the conditions are verified,

$$\dim(V/M) = \dim V - \dim M.$$

Proof.

(a) ⇒ (b): The finite-dimensionality of M was proved in the preceding section (3.5.14), and that of V/M in the remarks at the end of §3.4.

(b) ⇒ (a): If $M = V$ the implication is trivial; moreover, $V/M = V/V$ consists of the single coset $\theta + V = V$, thus V/V is the zero vector space and the asserted equation reduces to $0 = \dim V - \dim V$.

If $M = \{\theta\}$ then the quotient mapping $V \to V/M$ is a linear surjection with kernel $M = \{\theta\}$, therefore is a linear bijection (2.2.9), so the finite-dimensionality of V follows from that of $V/\{\theta\}$ (Theorem 3.5.12) and the asserted equation reduces to $\dim(V/\{\theta\}) = \dim V = \dim V - 0$.

Finally, suppose $M \neq \{\theta\}$, $M \neq V$. Let $x_1, \ldots, x_m$ be a basis of M and $u_1, \ldots, u_r$ a basis of V/M. Say $u_k = y_k + M$ $(k = 1, \ldots, r)$. Then the list

$$(*) \qquad x_1, \ldots, x_m, y_1, \ldots, y_r$$

generates V (3.4.8), so V is finite-dimensional (3.5.11). Moreover, the list $(*)$ is independent (3.3.8), hence is a basis of V; in particular, $m + r = \dim V$, in other words $\dim M + \dim(V/M) = \dim V$. ■

The preceding theorem says that if V is finite-dimensional and $Q: V \to V/M$ is the quotient mapping, then the dimension of the domain of Q is equal to the sum of the dimensions of its range and kernel:

$$\dim V = \dim Q(V) + \dim(\operatorname{Ker} Q);$$

thanks to the First Isomorphism Theorem, the formula extends to arbitrary linear mappings with finite-dimensional domain:

3.6.2 Theorem *If* V *is a finite-dimensional vector space and* $T:\mathrm{V}\to\mathrm{W}$ *is a linear mapping, then*

$$\dim T(\mathrm{V}) + \dim(\operatorname{Ker} T) = \dim \mathrm{V}.$$

Proof. Since V is finite-dimensional, so are $T(\mathrm{V})$ (3.4.12) and $\operatorname{Ker} T$ (3.5.14). By the First Isomorphism Theorem (2.7.2), $\mathrm{V}/\operatorname{Ker} T \cong T(\mathrm{V})$, therefore (3.5.12)

$$\dim(\mathrm{V}/\operatorname{Ker} T) = \dim T(\mathrm{V});$$

in other words (3.6.1), $\dim \mathrm{V} - \dim(\operatorname{Ker} T) = \dim T(\mathrm{V})$. ■

3.6.3 Definition With notations as in the preceding theorem, the dimension of the range of T is called the **rank** of T, denoted $\rho(T)$, and the dimension of the kernel of T is called the **nullity** of T, denoted $\nu(T)$; thus,

$$\rho(T) = \dim T(\mathrm{V}),$$
$$\nu(T) = \dim(\operatorname{Ker} T).$$

The message of Theorem 3.6.2: $\rho(T) + \nu(T) = \dim \mathrm{V}$, more memorably

$$\text{rank} + \text{nullity} = \text{dimension},$$

ultimately compressed as

$$\mathrm{R} + \mathrm{N} = \mathrm{D}.$$

The applications of this powerful formula merit a section of their own (§3.7).
Another important application of Theorem 3.6.1:

3.6.4 Theorem *If* V *and* W *are finite-dimensional vector spaces over the same field and if* $\mathrm{V}\times\mathrm{W}$ *is the product vector space* (1.3.11), *then* $\mathrm{V}\times\mathrm{W}$ *is also finite-dimensional and* $\dim(\mathrm{V}\times\mathrm{W}) = \dim\mathrm{V} + \dim\mathrm{W}$.

Proof. Let $T:\mathrm{V}\times\mathrm{W}\to\mathrm{W}$ be the linear mapping defined by $T(x, y) = y$; clearly T is surjective and

$$\operatorname{Ker} T = \{(x, y)\colon y = \theta\} = \{(x, \theta)\colon x \in \mathrm{V}\} = \mathrm{V}\times\{\theta\},$$

therefore $(\mathrm{V}\times\mathrm{W})/(\mathrm{V}\times\{\theta\}) \cong \mathrm{W}$ by the first isomorphism theorem (2.7.2). Since V is finite-dimensional and $\mathrm{V} \cong \mathrm{V}\times\{\theta\}$ (via the mapping $x \mapsto (x, \theta)$), it follows from Theorem 3.5.12 that $\mathrm{V}\times\{\theta\}$ is finite-dimensional and $\dim(\mathrm{V}\times\{\theta\}) = \dim\mathrm{V}$. By the same reasoning, the quotient space $(\mathrm{V}\times\mathrm{W})/(\mathrm{V}\times\{\theta\})$, being isomorphic to W, is finite-dimensional with dimension $\dim\mathrm{W}$. It then follows from Theorem 3.6.1 that $\mathrm{V}\times\mathrm{W}$ is finite-dimensional and

$$\dim(\mathrm{V}\times\mathrm{W}) - \dim(\mathrm{V}\times\{\theta\}) = \dim[(\mathrm{V}\times\mathrm{W})/(\mathrm{V}\times\{\theta\})],$$

in other words, $\dim(\mathrm{V}\times\mathrm{W}) - \dim\mathrm{V} = \dim\mathrm{W}$. ■

3.6.5 Corollary *If* $\mathrm{V} = \mathrm{V}_1 \times \ldots \times \mathrm{V}_n$ *is a product of finite-dimensional vector spaces* $\mathrm{V}_1, \ldots, \mathrm{V}_n$ (1.3.11), *then* V *is also finite-dimensional and*

$$\dim V = \dim V_1 + \dots + \dim V_n.$$

Proof. There is a natural isomorphism

$$V_1 \times \dots \times V_n \cong (V_1 \times \dots \times V_{n-1}) \times V_n,$$

effected by the linear bijection $(x_1, \dots, x_n) \mapsto ((x_1, \dots, x_{n-1}), x_n)$, so the corollary follows by an easy induction (Theorem 3.6.4 is the case $n = 2$). ■

3.6.6 Corollary *If* W *is a finite-dimensional vector space and* $U = W \times \dots \times W$ (n factors) *is the product vector space, then* U *is also finite-dimensional and* $\dim U = n \dim W$.

Proof. Immediate from 3.6.5. ■

► ***Exercises***

1. Does there exist a linear mapping $T:\mathbf{R}^7 \to \mathbf{R}^3$ whose kernel is 3-dimensional?

2. Let V be a finite-dimensional vector space, M and N linear subspaces of V. Prove that

$$\dim M + \dim N = \dim (M + N) + \dim (M \cap N).$$

{Hint: §2.7, Exercise 1.}

3. Let M and N be linear subspaces of $\mathbf{R}^{10}$ such that $\dim M = 4$ and $\dim N = 7$. True or false (explain): M and N have a nonzero vector in common.

4. Let V be a finite-dimensional vector space, M and N linear subspaces of V such that the dimensions of

$$(M + N)/N, \qquad (M + N)/M, \qquad M \cap N$$

are 2, 3, 4, respectively. Find the dimensions of M, N and M + N.

5. Let V be a finite-dimensional vector space, M and N linear subspaces of V. Prove:

$$\dim (M + N) = \dim M + \dim N \Leftrightarrow M \cap N = \{\theta\}.$$

6. The converse of Theorem 3.6.4 is true: If a product space $V \times W$ is finite-dimensional, then so are the factor spaces V and W. {Hint: Remark 3.4.12.}

7. Let V be an n-dimensional vector space and let f be a nonzero linear form on V, that is, a nonzero element of the dual space V′ of V (§2.3, Exercise 5). Prove that the kernel of f is $(n - 1)$-dimensional. {Hint: §2.7, Exercise 3.} Conversely, every $(n - 1)$-dimensional linear subspace of V is the kernel of a linear form.

8. If V is an n-dimensional vector space and if f, g are linear forms on V that are not proportional (neither is a multiple of the other), prove that $(\operatorname{Ker} f) \cap (\operatorname{Ker} g)$ is $(n - 2)$-dimensional. {Hint: §3.3, Exercise 16.}

9. Let V be a vector space and suppose $f_1, \dots, f_n$ are linear forms on V such that

$$(\operatorname{Ker} f_1) \cap \ldots \cap (\operatorname{Ker} f_n) = \{\theta\}.$$

Prove: V is finite-dimensional and $\dim V \le n$. {Hint: If F is the field of scalars, discuss a suitable mapping $V \to F^n$.}

10. Let M be the set of all vectors $(x, y, z) \in \mathbf{R}^3$ such that $2x - 3y + z = 0$. Prove that M is a linear subspace of $\mathbf{R}^3$ and find its dimension. {Hint: Exercise 7.}

11. Let M and N be the linear subspaces of $\mathbf{R}^4$ defined by

$$M = \{(x_1, x_2, x_3, x_4):\ x_1 + x_2 + x_3 = 0\},$$
$$N = \{(x_1, x_2, x_3, x_4):\ x_1 + x_2 - x_4 = 0\}.$$

Find bases for M, N, M + N, M ∩ N. {Hint: First determine the dimensions of these subspaces (Exercises 8, 7 and 2).}

12. If M is the linear subspace of $\mathbf{R}^4$ defined by

$$M = \{(x_1, x_2, x_3, x_4): x_1 - x_4 = x_2 + x_3 = 0\},$$

find a basis of $\mathbf{R}^4/M$. {Hint: Exercise 8.}

13. With notations as in §2.1, Exercise 5, every relation among x_1, x_2, x_3 is a multiple of $(-2, 3, -1)$. Why?

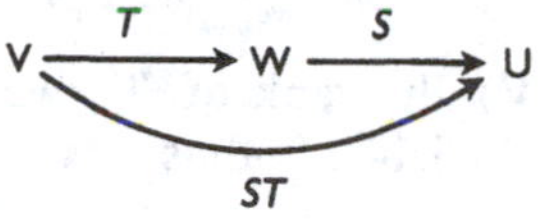

Fig. 13

14. Let T:V → W and S:W → U be linear mappings, with V finite-dimensional (Fig. 13).

(i) If S is injective, then $\operatorname{Ker} ST = \operatorname{Ker} T$ and $\rho(ST) = \rho(T)$.

(ii) If T is surjective, then $\operatorname{Im} ST = \operatorname{Im} S$ and $\nu(ST) - \nu(S) = \dim V - \dim W$.

15. Let V be a finite-dimensional vector space, $T \in \mathcal{L}(V)$. The followig conditions are equivalent:

(a) $V = \operatorname{Ker} T + \operatorname{Im} T$;

(b) $\operatorname{Ker} T \cap \operatorname{Im} T = \{\theta\}$;

(c) $V = \operatorname{Ker} T \oplus \operatorname{Im} T$.

{Hint: §3.5, Exercise 26.}

3.7 Applications of R + N = D

The full version of the formula in question (3.6.2): If V and W are finite-dimensional vector spaces and T:V → W is a linear mapping, then

$$\dim T(V) + \dim(\operatorname{Ker} T) = \dim V;$$

in words, the *rank* of a linear mapping plus its *nullity* equals the dimension of its domain.

For the first application, consider a system of m homogeneous linear equations in n unknowns $x_1, \ldots, x_n$:

$$(*)\begin{cases} a_{11}x_1 + a_{12}x_2 + \ldots + a_{1n}x_n = 0 \\ a_{21}x_1 + a_{22}x_2 + \ldots + a_{2n}x_n = 0 \\ \ldots \\ a_{m1}x_1 + a_{m2}x_2 + \ldots + a_{mn}x_n = 0 \end{cases}$$

where the coefficients a_{ij} are given elements of a field F. As in Example 2.1.9, we pose the question of the existence of *solutions* of the system, a solution being an n-ple $x = (x_1, \ldots, x_n) \in \mathrm{F}^n$ whose coordinates satisfy the equations $(*)$. There is always the *trivial solution* $x = \theta$ (that is, $x_1 = \ldots = x_n = 0$), and there may be no others (consider the 'system' $1x_1 = 0$, where $m = n = 1$ and $a_{11} = 1$). A solution x different from θ is called a *nontrivial solution*; when there are fewer equations than variables, nontrivial solutions are guaranteed to exist:

3.7.1 Theorem *If $m < n$ then the system $(*)$ has a nontrivial solution.*

Proof. As in 2.1.9, let $T{:}\mathrm{F}^n \to \mathrm{F}^m$ be the linear mapping defined by

$$Tx = (a_{11}x_1 + \ldots + a_{1n}x_n, \ldots, a_{m1}x_1 + \ldots + a_{mn}x_n).$$

The set of all solution-vectors $x = (x_1, \ldots, x_n)$ of $(*)$ is precisely the kernel of T. Assuming $m < n$, our problem is to show that $\mathrm{Ker}\, T \neq \{\theta\}$, in other words that the nullity of T is ≥ 1.

Let $\nu = \dim(\mathrm{Ker}\, T)$ be the nullity and $\rho = \dim T(\mathrm{V})$ the rank of T; we know that $\rho + \nu = n$ (3.6.2) and $\rho \leq m$ (3.5.14), thus $\nu = n - \rho \geq n - m > 0$. ■

3.7.2 Remarks (Optional) More precisely, assuming $m < n$, there exist $\nu = n - \rho$ linearly independent solution-vectors of the system $(*)$; to calculate ν, we need only calculate the rank ρ of T. This is a computational problem that the above 'existence proof' does not address; at any rate, here is a start on the problem. If $e_1, \ldots, e_n$ is the canonical basis of F^n (3.5.2), the range of T is the linear span of the vectors $Te_1, \ldots, Te_n$ in F^m (3.1.6 or 3.4.7). Explicitly,

$$Te_1 = (a_{11}, a_{21}, \ldots, a_{m1}),$$
$$Te_2 = (a_{12}, a_{22}, \ldots, a_{m2}),$$
$$\ldots$$
$$Te_n = (a_{1n}, a_{2n}, \ldots, a_{mn}),$$

which can be thought of as the 'column-vectors' in the $m \times n$ array (m rows, n columns) of coefficients of the system $(*)$:

$$(a_{ij}) = \begin{pmatrix} a_{11} & a_{12} & \ldots & a_{1n} \\ a_{21} & a_{22} & \ldots & a_{2n} \\ \ldots & & & \\ a_{m1} & a_{m2} & \ldots & a_{mn} \end{pmatrix}.$$

Writing $w_j = Te_j$ $(j = 1, \ldots, n)$, the range $T(\mathrm{F}^n)$ of T is the linear span of the vectors $w_1, \ldots, w_n$, and the rank $\rho = \dim T(\mathrm{V})$ is the largest number of linearly independent w's (remarks following Theorem 3.5.7). In principle, one can look at w_1, decide whether or not it is a linear combination of $w_2, \ldots, w_n$ (if it is, discard it; if not, keep it), then advance to w_2, etc. (At the last step, if $w_n = \theta$, think of it as a linear combination of the empty list and discard it.) In practice, there are efficient methods for calculating the rank; one of these methods is explained in the next chapter (§4.7), another in Chapter 10 (§10.4, Exercise 6).

The next applications (of $\mathrm{R} + \mathrm{N} = \mathrm{D}$) have to do with the interaction between the properties of a linear mapping $T:\mathrm{V} \to \mathrm{W}$ and the dimensions of V and W. They can be obtained more directly (as indicated in the exercises), but the proofs based on the formula $\mathrm{R} + \mathrm{N} = \mathrm{D}$ are elegant and easy to rediscover.

3.7.3 Theorem *Let* V *and* W *be finite-dimensional vector spaces,* $T:\mathrm{V} \to \mathrm{W}$ *a linear mapping.*

(1) *If* T *is injective, then* $\dim \mathrm{V} \le \dim \mathrm{W}$.

(2) *If* T *is surjective, then* $\dim \mathrm{V} \ge \dim \mathrm{W}$.

(3) *If* T *is bijective, then* $\dim \mathrm{V} = \dim \mathrm{W}$.

Proof. Let ρ and ν be the rank and nullity of T; thus, $\rho + \nu = \dim \mathrm{V}$ (3.6.2) and $\rho \le \dim \mathrm{W}$ (3.5.14).

(1) If T is injective, then $\operatorname{Ker} T = \{\theta\}$ (2.2.9), therefore $\nu = 0$ and $\dim \mathrm{V} = \rho + \nu = \rho \le \dim \mathrm{W}$.

(2) If T is surjective, then $\rho = \dim T(\mathrm{V}) = \dim \mathrm{W}$, thus $\dim \mathrm{V} = \dim \mathrm{W} + \nu \ge \dim \mathrm{W}$.

(3) This is immediate from (1) and (2). ■

For a linear mapping between spaces of the same dimension, the three 'jectivity' conditions coincide:

3.7.4 Theorem *Let* V *and* W *be finite-dimensional vector spaces* ***of the same dimension*** *and let* $T:\mathrm{V} \to \mathrm{W}$ *be a linear mapping. The following conditions are equivalent:*

(a) T *is injective;*

(b) T *is surjective;*

(c) T *is bijective.*

Proof. Since (c) ⇔ (a) & (b), that is, (c) is equivalent to the simultaneous assertion of (a) and (b), we need only show that (a) and (b) are equivalent.

(a) ⇔ (b): Let $n = \dim \mathrm{V} = \dim \mathrm{W}$, and let ρ and ν be the rank and nullity of T.

Injectivity means $\nu = 0$ (2.2.9) and surjectivity means $\rho = n$ (3.5.14); they mean the same thing (because $\nu = n - \rho$). ■

3.7.5 Corollary *Let* V *and* W *be finite-dimensional vector spaces* ***of the same dimension*** *and let* $T:\mathrm{V} \to \mathrm{W}$, $S:\mathrm{W} \to \mathrm{V}$ *be linear mappings. The following conditions are equivalent:*

(a) $ST = I_V$;
(b) $TS = I_W$;
(c) *T is bijective and $T^{-1} = S$.*

Proof. By I_V and I_W we mean the identity mappings on V and W (2.3.3). The implications (c) ⇒ (a) and (c) ⇒ (b) are trivial.

(a) ⇒ (c): From $ST = I_V$ it follows that T is injective; for, if $Tx = \theta$ then $S(Tx) = S\theta = \theta$, thus $\theta = (ST)x = I_V x = x$. By Theorem 3.7.4, T is bijective; right-multiplying $ST = I_V$ by T^{-1}, we see that $S = T^{-1}$.

(b) ⇒ (c): Assuming $TS = I_W$, the preceding paragraph shows that S is bijective and $S^{-1} = T$, therefore T is bijective and $T^{-1} = (S^{-1})^{-1} = S$. ■

The preceding corollary is of particular interest when V = W: if V is a finite-dimensional vector space and $T \in \mathcal{L}(V)$, the existence of an $S \in \mathcal{L}(V)$ with $ST = I$ is equivalent to the existence of an $S \in \mathcal{L}(V)$ with $TS = I$; this happens precisely when T is bijective, and $S = T^{-1}$ is the only candidate. *These assertions are false when* V *is infinite-dimensional* (see Exercise 5).

▶ ***Exercises***

1. Suppose that the homogeneous system

$$(*)\begin{cases} a_{11}x_1 + a_{12}x_2 + \dots + a_{1n}x_n = 0 \\ a_{21}x_1 + a_{22}x_2 + \dots + a_{2n}x_n = 0 \\ \dots \\ a_{n1}x_1 + a_{n2}x_2 + \dots + a_{nn}x_n = 0 \end{cases}$$

(n equations in n unknowns) has only the trivial solution. Prove that for any given scalars $c_1, \dots, c_n$, the system

$$(**)\begin{cases} a_{11}x_1 + a_{12}x_2 + \dots + a_{1n}x_n = c_1 \\ a_{21}x_1 + a_{22}x_2 + \dots + a_{2n}x_n = c_2 \\ \dots \\ a_{n1}x_1 + a_{n2}x_2 + \dots + a_{nn}x_n = c_n \end{cases}$$

has a unique solution. {Hint: Theorem 3.7.4.}

2. Prove that for any scalars a, b, c, the system

$$\begin{aligned} x_1 + 2x_2 - 3x_3 &= a \\ 2x_1 + x_2 + 2x_3 &= b \\ 3x_1 \qquad\qquad &= c \end{aligned}$$

has a unique solution. {Hint: §3.5, Exercise 1.}

3. Let V be a finite-dimensional vector space and let $S, T \in \mathcal{L}(V)$. Prove:

(i) $\rho(S + T) \le \rho(S) + \rho(T)$. {Hint: §3.6, Exercise 2.}
(ii) $\rho(ST) \le \rho(S)$ and $\rho(ST) \le \rho(T)$.
(iii) $\nu(S + T) \ge \nu(S) + \nu(T) - \dim V$.

(iv) $\nu(ST) \geq \nu(T)$ and $\nu(ST) \geq \nu(S)$.

4. Let V be a finite-dimensional vector space, $T:V \to V$ a linear mapping.

(i) Prove that if T and T^2 have the same rank, then $V = (\text{Ker } T) \oplus T(V)$ in the sense of §1.6, Exercise 12. {Hint: T and T^2 also have the same nullity; view $\text{Ker } T \subset \text{Ker } T^2$ in the light of Corollary 3.5.14.}

(ii) Prove that if k is a nonnegative integer such that T^k and T^{k+1} have the same rank, then $V = (\text{Ker } T^k) \oplus T^k(V)$.

(iii) Review §2.3, Exercise 13.

5. Let V be the vector space of infinite sequences of real numbers (Example 1.3.7). Define linear mappings $T:V \to V$ and $S:V \to V$ as follows: if $x = (a_1, a_2, a_3, \ldots) \in V$, then $Tx = (0, a_1, a_2, a_3, \ldots)$ and $Sx = (a_2, a_3, a_4, \ldots)$. Show that $ST = I$ but $TS \neq I$. (T is called the *shift operator* on V.)

6. Let $V = \mathcal{P}$ be the vector space of all real polynomial functions (1.3.5). As in Examples 2.1.4 and 2.1.5, let $S, T \in \mathcal{L}(V)$ be the linear mappings such that $Tp = p'$ and Sp is the prederivative of p with constant term 0. Then $TS = I$. Is T bijective?

7. Let V be any vector space (not necessarily finite-dimensional) and suppose $R, S, T \in \mathcal{L}(V)$ are such that $ST = I$ and $TR = I$. Prove that T is bijective and $R = S = T^{-1}$. {Hint: Look at $(ST)R$.}

8. With V as in Exercise 5, define $M \in \mathcal{L}(V)$ by $M(a_1, a_2, a_3, \ldots) = (a_1, 2a_2, 3a_3, \ldots)$. Show that M is bijective and find the formula for M^{-1}.

9. Let V be a finite-dimensional vector space and let $S, T \in \mathcal{L}(V)$. Prove:

(i) If there exists a nonzero vector x such that $STx = \theta$, then there exists a nonzero vector y such that $TSy = \theta$.

(ii) ST is bijective $\Leftrightarrow$ S and T are bijective $\Leftrightarrow$ TS is bijective.

{Hint: Exercise 3.}

3.8 Dimension of $\mathcal{L}(V, W)$

We will see in the next chapter that bases give us a powerful tool for *describing* linear mappings (via 'matrices'); the following theorem shows that they are also the ultimate tool for *constructing* linear mappings:

3.8.1. Theorem *Let* V *and* W *be vector spaces, with* V *finite-dimensional, say* $\dim V = n$.

If $x_1, \ldots, x_n$ *is a basis of* V *and if* $w_1, \ldots, w_n$ *are vectors in* W, *then there exists a unique linear mapping* $T:V \to W$ *such that* $Tx_i = w_i$ *for* $i = 1, \ldots, n$.

Proof. Note that the only restriction on the list $w_1, \ldots, w_n$ is its length; the vectors themselves can be arbitrary elements of W. Let F be the field of scalars.

Existence: Let $R:F^n \to V$ and $S:F^n \to W$ be the linear mappings defined by

$$R(a_1, \ldots, a_n) = a_1x_1 + \ldots + a_nx_n$$
$$S(a_1, \ldots, a_n) = a_1w_1 + \ldots + a_nw_n$$

(Theorem 2.1.3). In particular, if $e_1, \ldots, e_n$ is the canonical basis of F^n (3.5.2), then $Re_i = x_i$ and $Se_i = w_i$ for $i = 1, \ldots, n$. Since $x_1, \ldots, x_n$ is a basis of V, R is bijective (3.5.5); thus $e_i = R^{-1}x_i$ and $w_i = Se_i = SR^{-1}x_i$ for all i. Since R^{-1} is linear (2.4.6), the mapping $T = SR^{-1}$ meets the requirements (Fig. 14).

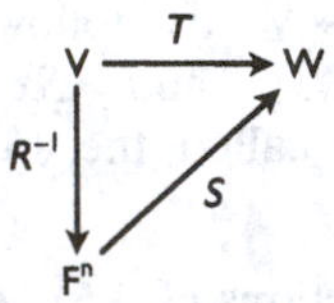

Fig. 14

Uniqueness: Suppose $T_1:V \to W$ and $T_2:V \to W$ are linear mappings such that $T_1x_i = w_i = T_2x_i$ for all i. Then $(T_1 - T_2)x_i = \theta$, thus the kernel of $T_1 - T_2$ contains every x_i; since the x_i generate V, it follows that $\text{Ker}(T_1 - T_2) = V$, in other words $T_1 - T_2 = 0$, thus $T_1 = T_2$. ■

The message of the theorem is that linear mappings can be constructed so as to have specified values at the vectors of a basis. The corollaries exploit this valuable idea.

3.8.2 Corollary *If* V *and* W *are finite-dimensional vector spaces, then* $\mathcal{L}(V, W)$ *is also finite-dimensional and*

$$\dim \mathcal{L}(V, W) = (\dim V)(\dim W).$$

Proof. Let $n = \dim V$, $m = \dim W$. Write $W^n = W \times \ldots \times W$ for the product vector space of n copies of W; by 3.6.6, W^n is nm-dimensional. Choose a basis $x_1, \ldots, x_n$ of V and let $\Phi:\mathcal{L}(V, W) \to W^n$ be the mapping defined by $\Phi(T) = (Tx_1, \ldots, Tx_n)$; Φ is linear by the definition of the linear operations in $\mathcal{L}(V, W)$, surjective by the 'existence' part of 3.8.1, and injective by the 'uniqueness' part. Thus Φ is a vector space isomorphism, therefore (by 3.5.12) $\mathcal{L}(V, W)$ is finite-dimensional and $\dim \mathcal{L}(V, W) = \dim W^n = nm = (\dim V)(\dim W)$. ■

3.8.3 Corollary *Let* V *be a finite-dimensional vector space,* $S \in \mathcal{L}(V)$ *a linear mapping such that* $ST = TS$ *for all* $T \in \mathcal{L}(V)$. *Then* S *is a scalar multiple of the identity.*

Proof. If $V = \{\theta\}$ then $I = 0$ is the only available linear mapping. Let us assume that $V \neq \{\theta\}$. The crux of the proof is to show that for every $x \in V$, Sx is proportional to x; the fact that the scalar factors can all be taken to be the same is an easy afterthought.

Let $x \in V$, $x \neq \theta$. We assert that there exists a scalar c such that $Sx = cx$. Assume to the contrary that no such scalar exists. The vectors x and Sx are then independent (3.3.5). Let $n = \dim V$ and construct a basis $x_1, \ldots, x_n$ of V for which $x_1 = x$ and $x_2 = Sx$ (3.5.15). By the theorem, there exists a linear mapping $T:V \to V$ such that $Tx_1 = x$ and $Tx_2 = x$ (and, if we like, $Tx_i = \theta$ for all $i > 2$). Thus, $Tx = x$ and $T(Sx) = x$. By assumption $TS = ST$, therefore

$$x = (TS)x = (ST)x = S(Tx) = Sx,$$

which contradicts the assumed independence of x and Sx. The contradiction establishes that there does indeed exist a scalar c with $Sx = cx$; since $x \neq \theta$, the scalar c is unique (1.4.7).

For every nonzero vector x, write $c(x)$ for the unique scalar such that $Sx = c(x)x$. We assert that $c(x) = c(y)$ for all nonzero vectors x and y. If x and y are dependent, say $y = ax$, then

$$c(y)y = Sy = S(ax) = a(Sx) = a \cdot c(x)x = c(x)ax = c(x)y,$$

therefore $c(y) = c(x)$. On the other hand, if x and y are independent, then $x + y \neq \theta$ and

$$c(x + y) \cdot (x + y) = S(x + y) = Sx + Sy = c(x)x + c(y)y;$$

comparing coefficients of the independent vectors x and y, we see that

$$c(x + y) = c(x) \quad \text{and} \quad c(x + y) = c(y),$$

so again $c(x) = c(y)$.

Write c for the common value of the $c(x)$ for $x \in \mathrm{V} - \{\theta\}$; then $Sx = cx$ for all $x \in \mathrm{V}$ (including $x = \theta$). ■

For use in the next section:

3.8.4 Corollary *Let* V *and* W *be vector spaces, with* V *finite-dimensional, and let* M *be a linear subspace of* V. *Every linear mapping* T_0:M → W *may be extended to a linear mapping* T:V → W.

Proof. Given $T_0 \in \mathcal{L}(\mathrm{M}, \mathrm{W})$, the assertion is that there exists a linear mapping $T \in \mathcal{L}(\mathrm{V}, \mathrm{W})$ whose restriction to M is T_0, that is, $T|\mathrm{M} = T_0$. If $\mathrm{M} = \{\theta\}$ or M = V, the assertion is trivial.

Suppose $\mathrm{M} \neq \{\theta\}$ and M ≠ V. Choose a basis $x_1, \dots, x_m$ of M, expand it to a basis $x_1, \dots, x_m, x_{m+1}, \dots, x_n$ of V (3.5.16), and let $T \in \mathcal{L}(\mathrm{V}, \mathrm{W})$ be the linear mapping such that $Tx_i = T_0 x_i$ for $i = 1, \dots, m$ and (for example) $Tx_i = 0$ for $i > m$ (such a T exists by Theorem 3.8.1); since $T|\mathrm{M}$ and T_0 agree on the basis vectors $x_1, \dots, x_m$ of M, they are identical on M. ■

▶ *Exercises*

1. Hiding in the proof of Corollary 3.8.3 is the proof that if V is 1-dimensional then every $S \in \mathcal{L}(\mathrm{V})$ is a scalar multiple of the identity. Can you find it?

2. True or false (explain): $\mathcal{L}(\mathbf{R}^2, \mathbf{R}^3) \cong \mathbf{R}^6$.

3. If V and W are finite-dimensional vector spaces, prove that $\mathcal{L}(\mathrm{V}, \mathrm{W}) \cong \mathcal{L}(\mathrm{W}, \mathrm{V})$. {Hint: *Don't* look for a mapping $\mathcal{L}(\mathrm{V}, \mathrm{W}) \to \mathcal{L}(\mathrm{W}, \mathrm{V})$.}

4. Let V be a finite-dimensional vector space, V′ its dual space (§2.3, Exercise 5). Prove: V′ is also finite-dimensional and dim V′ = dim V. {Hint: Corollary 3.8.2.}

5. Here is an alternate approach to Exercise 4: If F is the field of scalars and $n = \dim V$, prove that $V' \cong F^n$. {Hint: If $x_1, \ldots, x_n$ is a basis of V, then the mapping $V' \to F^n$ defined by $f \mapsto (f(x_1), \ldots, f(x_n))$ is surjective by Theorem 3.8.1.}

6. Here is a sketch of an alternate proof of Corollary 3.8.2, based on constructing an explicit basis for $\mathcal{L}(V, W)$. Let $x_1, \ldots, x_n$ be a basis of V and $y_1, \ldots, y_m$ a basis of W. For each $i \in \{1, \ldots, m\}$ and each $j \in \{1, \ldots, n\}$, there exists (by Theorem 3.8.1) a linear mapping $E_{ij}: V \to W$ such that

$$E_{ij}x_j = y_i \quad \text{and} \quad E_{ij}x_k = \theta \quad \text{for} \quad k \neq j.$$

Show that the mn linear mappings constructed in this way are a basis of $\mathcal{L}(V, W)$. {Verify that the E_{ij} are linearly independent and that every $T \in \mathcal{L}(V, W)$ is a linear combination of them.}

7. Prove: If V is a finite-dimensional vector space and $T \in \mathcal{L}(V)$, then there exists a finite list of scalars $a_0, a_1, \ldots, a_r$, not all 0, such that

$$a_0x + a_1Tx + a_2T^2x \ldots + a_rT^rx = \theta$$

for all $x \in V$. {Hint: The powers of T are defined following 2.3.13. Consider the sequence $I, T, T^2, T^3, \ldots$ in the finite-dimensional vector space $\mathcal{L}(V)$.}

8. Let $x_1, \ldots, x_n$ be a basis of V and, for each pair of indices $i, j \in \{1, \ldots, n\}$, let $E_{ij} \in \mathcal{L}(V)$ be the linear mapping such that $E_{ij}x_j = x_i$ and $E_{ij}x_k = \theta$ for $k \neq j$ (cf. Exercise 6). Show that (1) $E_{ij}E_{jk} = E_{ik}$, (2) $E_{ij}E_{hk} = 0$ when $j \neq h$, and (3) $E_{11} + E_{22} + \ldots + E_{nn} = I$.

9. Let $T: V \to W$ be a linear mapping. Suppose that W is finite-dimensional, say $\dim W = m$, and let $y_1, \ldots, y_m$ be a basis of W. For each $x \in V$, Tx is uniquely a linear combination of $y_1, \ldots, y_m$; denote the coefficient of y_i by $f_i(x)$, thus

$$Tx = \sum_{i=1}^{m} f_i(x)y_i \quad (x \in V).$$

Prove: The f_i are linear forms on V and, in the notation of Example 2.3.4, $T = \sum_{i=1}^{m} f_i \otimes y_i$.

10. If V and W are finite-dimensional vector spaces such that $\dim \mathcal{L}(V, W) = 7$, then $\mathcal{L}(V, W)$ is isomorphic to either V or W.

11. If U, V, W are finite-dimensional vector spaces over the same field, then there exists a natural isomorphism

$$\mathcal{L}(U, W) \times \mathcal{L}(V, W) \cong \mathcal{L}(U \times V, W).$$

{Hint: If $R: U \to W$ and $S: V \to W$ are linear mappings, consider the mapping $(x, y) \mapsto Rx + Sy$.}

12. If V is a finite-dimensional vector space and x is a nonzero vector in V, then there exists a linear form f on V such that $f(x) \neq 0$. {Hint:

Expand x to a basis of V (3.5.15) and apply Theorem 3.8.1 with the field of scalars F in the role of W. Alternate solution: If $\dim V = n$, compose an isomorphism $T: V \to F^n$ with an appropriate coordinate function $F^n \to F$.}

13. Let V be a finite-dimensional vector space, M a linear subspace of V, and x a vector in V such that $x \notin M$. Prove that there exists a linear form f on V such that $f(x) \neq 0$ and $f(y) = 0$ for all $y \in M$. {Hint: Adjoin x to a basis of M, expand to a basis of V (3.5.15), then apply Theorem 3.8.1. Alternate solution: apply Exercise 12 in the quotient space V/M and compose the linear form so obtained with the quotient mapping $V \to V/M$.}

3.9 Duality in vector spaces

One way to study a vector space V is to study linear mappings of V into other vector spaces W. Two particularly useful choices for W lie ready at hand: V itself and, a subtler choice, the field F of scalars. The first choice has to do with the space $\mathcal{L}(V)$ of linear mappings of V into itself (2.3.1), the second, with the space $\mathcal{L}(V, F)$ of all linear forms on V; the latter space is the subject of this section and we begin by giving it a name:

3.9.1 Definition If V is a vector space over the field F, the vector space $\mathcal{L}(V, F)$ of all linear forms on V (2.3.7) is called the **dual space** of V and is denoted V'. {There is no consensus on this notation: one also sees V^*, $\bar{V}$, V^d, etc.; for every notation, there is a context in which it is ill at ease.}

Finite-dimensionality is preserved under passage to the dual space, and so is the dimension:

3.9.2 Theorem *If V is a finite-dimensional vector space, then its dual space V' is also finite-dimensional and* $\dim V' = \dim V$.

Proof. The field F of scalars, regarded as a vector space over itself (cf. 2.1.10), is 1-dimensional (any nonzero scalar serves as a basis for F); by 3.8.2, $V' = \mathcal{L}(V, F)$ is finite-dimensional and $\dim V' = (\dim V)(\dim F) = \dim V$. ■

3.9.3 Corollary *Let V be a finite-dimensional vector space, $n = \dim V$. If $x_1, \ldots, x_n$ is a basis of V then there exists a basis $f_1, \ldots, f_n$ of the dual space V' such that $f_i(x_i) = 1$ for all i and $f_i(x_j) = 0$ when $i \neq j$.*

Proof. Define $\Phi: V' \to F^n$ by $\Phi(f) = (f(x_1), \ldots, f(x_n))$; from the proof of Corollary 3.8.2 (for the special case that $W = F$), we know that Φ is a linear bijection. Let $e_1, \ldots, e_n$ be the canonical basis of F^n and, for each i, let f_i be the element of V' such that $\Phi(f_i) = e_i$. The equation $\Phi(f_i) = e_i$ means that for $j = 1, \ldots, n$, the j'th coordinate of $\Phi(f_i)$—namely $f_i(x_j)$—is equal to 1 when $j = i$ and to 0 when $j \neq i$. Finally, the vectors $f_i = \Phi^{-1}(e_i)$ are a basis of V' by Theorems 3.3.7 and 3.4.7. ■

3.9.4 Definition With notations as in 3.9.3, one calls $f_1, \ldots, f_n$ the basis of V′ **dual** to the basis $x_1, \ldots, x_n$ of V.

The relation between the two bases can be expressed by the formula $f_i(x_j) = \delta_{ij}$, where $\delta_{ii} = 1$ and $\delta_{ij} = 0$ for $i \neq j$ (this interpretation of the symbol δ_{ij} is called the *Kronecker delta*).

3.9.5 Example If $e_1, \ldots, e_n$ is the canonical basis of F^n, then the dual basis of $(F^n)'$ is $f_1, \ldots, f_n$, where $f_i(a_1, \ldots, a_n) = a_i$; thus, f_i is the i'th *coordinate function* $F^n \to F$ that assigns to an n-ple its i'th coordinate.

Every linear mapping $V \to W$ induces a linear mapping $W' \to V'$ in a natural way:

3.9.6 Theorem *If* $T{:}V \to W$ *is a linear mapping then, for every linear form* g *on* W, *the composite mapping* gT *is a linear form on* V (Fig. 15); *the mapping* $T'{:}W' \to V'$ *defined by* $T'g = gT$ *is linear.*

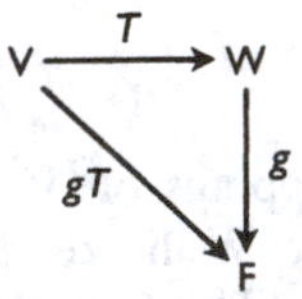

Fig. 15

Proof. The first statement is obvious (2.3.13). To see that T' is linear, let g and h be linear forms on W and let c be a scalar. For every vector $x \in V$,

$$\begin{aligned}(T'(g + h))(x) &= ((g + h)T)(x) = (g + h)(Tx)\\ &= g(Tx) + h(Tx) = (gT)(x) + (hT)(x)\\ &= (T'g)(x) + (T'h)(x) = (T'g + T'h)(x)\end{aligned}$$

(the justification for each equality is a definition: for the first equality, the definition of T'; what definitions are involved in the other equalities?), therefore $T'(g + h) = T'g + T'h$; similarly $T'(cg) = c(T'g)$ (details?). ■

3.9.7 Definition With notations as in 3.9.6, T' is called the **transpose** of T. {The reason for the name will become clear in the next chapter (§4.6).}

The next target is to prove that when V and W are finite-dimensional, T and T' have the same rank (3.9.12); but let's pause to look at an example.

3.9.8 Example Let $T{:}F^n \to V$ be a linear mapping and let $x_i = Te_i$, where $e_1, \ldots, e_n$ is the canonical basis of F^n. For every linear form f on V, $(T'f)(e_i) = f(Te_i) = f(x_i)$, therefore $(T'f)(a_1, \ldots, a_n) = a_1 f(x_1) + \ldots + a_n f(x_n)$; thus $T'f = f(x_1)f_1 + \ldots + f(x_n)f_n$, where $f_1, \ldots, f_n$ are the coordinate linear forms on F^n (3.9.5).

3.9.9 Theorem *If* $T{:}V \to W$ *is a linear mapping, then* $\operatorname{Ker} T' = \{g \in W' {:}\ g = 0$ *on* $T(V)\}$.

Proof. For $g \in W'$, $g = 0$ on $T(V) \Leftrightarrow g(T(V)) = 0 \Leftrightarrow (gT)(V) = 0 \Leftrightarrow (T'g)(V) = 0 \Leftrightarrow T'g = 0 \Leftrightarrow g \in \operatorname{Ker} T'$. ■

This prompts a useful definition.

3.9.10 Definition If M is linear subspace of the vector space V, the **annihilator** of M in V′, denoted M°, is the set of all linear forms f on V such that $f(x) = 0$ for all $x \in M$; thus

$$M^\circ = \{f \in V' : f = 0 \text{ on } M\}.$$

Note that M° is a linear subspace of V'. The message of Theorem 3.9.9: $\operatorname{Ker} T' = (T(V))^\circ$ for every linear mapping $T: V \to W$.

3.9.11 Theorem *If* V *is a finite-dimensional vector space,* M *is a linear subspace of* V, *and* M° *is the annihilator of* M *in* V' (3.9.10), *then* $\dim M^\circ = \dim V - \dim M$.

Proof. If f is a linear form on V, then the restriction $f|M$ of f to M (Appendix A.3.8) is a linear form on M. Define $R: V' \to M'$ by $Rf = f|M$. It is clear that R is linear and has kernel M°, therefore $V'/M^\circ \cong \operatorname{Im} R$ by the first isomorphism theorem. By Corollary 3.8.4, every $g \in M'$ is the restriction to M of some $f \in V'$, so that $g = f|M = Rf$; thus R is surjective and $V'/M^\circ \cong M'$. Then

$$\begin{aligned}\dim M &= \dim M' = \dim (V'/M^\circ)\\ &= \dim V' - \dim M^\circ = \dim V - \dim M^\circ\end{aligned}$$

(by 3.9.2, 3.5.12 and 3.6.1), thus $\dim M^\circ = \dim V - \dim M$. ■

The following theorem substantially simplifies the computation of rank (in the next chapter):

3.9.12 Theorem *If* V *and* W *are finite-dimensional vector spaces and* $T: V \to W$ *is any linear mapping, then* T *and* T' *have the same rank.*

Proof. By Theorem 3.9.9, $\operatorname{Ker} T' = (T(V))^\circ$, therefore

$$\begin{aligned}\nu(T') &= \dim (\operatorname{Ker} T') = \dim (T(V))^\circ\\ &= \dim W - \dim T(V) = \dim W - \rho(T),\end{aligned}$$

by the formula in 3.9.11; thus,

$$\rho(T) = \dim W - \nu(T') = \dim W' - \nu(T') = \rho(T')$$

by 3.9.2 and 3.6.2. {Caution: It does not follow that T and T' have the same nullity (Exercise 3).} ■

3.9.13 Corollary *If* $T: V \to W$ *is a linear mapping, where* V *and* W *are finite-dimensional vector spaces, then:*

(i) *T is surjective $\Leftrightarrow$ T' is injective;*

(ii) *T is injective $\Leftrightarrow$ T' is surjective.*

Proof.

(i) From the first set of equations in the proof of Theorem 3.9.12, we have

$$\nu(T') = 0 \Leftrightarrow \rho(T) = \dim W \Leftrightarrow T(V) = W$$

(the latter equivalence by (ii) of 3.5.14), thus $\operatorname{Ker} T' = \{\theta\} \Leftrightarrow T$ is surjective.

(ii) To say that T is injective means that $\nu(T) = 0$, that is, $\rho(T) = \dim V$, in other words $\rho(T') = \dim V'$, which is equivalent to $T'(W') = V'$. ■

At the beginning of the section we mentioned the possibility of studying V via either $\mathcal{L}(V)$ or V'. What in fact happens is a little of both: the study of a linear mapping $T \in \mathcal{L}(V)$ is enhanced by the study of its transpose $T' \in \mathcal{L}(V')$ (cf. Corollary 3.9.13 and Section 6 of the next chapter).

The rest of the section is for application in Chapter 13; it (and the exercises referring to it) can be omitted until that time. The basic reason for the symmetry in the statements (i), (ii) of 3.9.13 is that T can be regarded as the transpose of T' (well, almost; see Theorem 3.9.18 below). The key idea is to take the dual of the dual:

3.9.14 Definition If V is a vector space, the dual of its dual is called the **bidual** of V, denoted V''; thus $V'' = (V')'$. If $T:V \to W$ is a linear mapping, then $T':W' \to V'$, therefore $(T')':V'' \to W''$; one writes $T'' = (T')'$ (the 'bitranspose' of T).

Let V be a vector space over F and let $x \in V$. The mapping $V' \to F$ defined by $f \mapsto f(x)$ is a linear form on V', therefore an element of V''.

3.9.15 Definition If V is a vector space then, for each $x \in V$, we write x'' for the linear form on V' defined by $x''(f) = f(x)$. The mapping $S:V \to V''$ defined by $Sx = x''$ is called the **canonical mapping** of V into its bidual; more precisely, $S = S_V$ indicates the dependence of S on V.

3.9.16 Lemma *For every vector space* V, *the canonical mapping* $S:V \to V''$ *is linear.*

Proof. Given $x, y \in V$ and $c \in F$, the problem is to show that $(x+y)'' = x'' + y''$ and $(cx)'' = cx''$. For all $f \in V'$,

$$(x+y)''(f) = f(x+y) = f(x) + f(y)$$
$$= x''(f) + y''(f) = (x'' + y'')(f),$$
$$(cx)''(f) = f(cx) = cf(x) = cx''(f) = (cx'')(f). \blacksquare$$

3.9.17 Theorem (Principle of Duality) *If* V *is a finite-dimensional vector space, then the canonical mapping* $S:V \to V''$ *is a vector space isomorphism.*

Proof. If $V = \{\theta\}$ the assertion is trivial. Assuming $V \neq \{\theta\}$, let $n = \dim V$. Since V'' is also finite-dimensional and $\dim V'' = \dim V' = \dim V$, we need only show that S is injective. Assuming $x \in V$, $x \neq \theta$, we seek a linear form f on V such that $x''(f) \neq 0$, that is, $f(x) \neq 0$; the existence of such a form follows, for example, from composing an isomorphism $V \cong F^n$ with one of the coordinate linear forms on F^n (Example 3.9.5). {Alternatively, expand x to a basis of V, then cite Theorem 3.8.1}. ■

3.9.18 Theorem *If* V *and* W *are finite-dimensional vector spaces and* $T:V \to W$ *is a linear mapping, then* $(Tx)'' = T''x''$ *for all* $x \in V$.

Proof. For all $g \in W'$, $(Tx)''(g) = g(Tx)$, whereas $(T''x'')(g) = (x''T')(g) = x''(T'g) = (T'g)(x) = (gT)(x) = g(Tx)$. ■

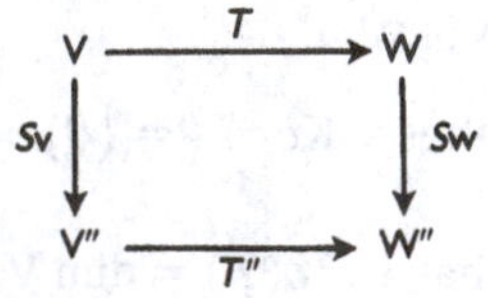

Fig. 16

Writing $S_V:V \to V''$ and $S_W:W \to W''$ for the canonical isomorphisms, the theorem says that $S_W(Tx) = T''(S_V x)$ for all $x \in V$, in other words $S_W T = T'' S_V$; this means that in Fig. 16 the two ways of getting from V to W'' agree.

▶ Exercises

1. Simplify the proof of Theorem 3.9.6 by citing §2.3, Exercise 2.

2. If V is a finite-dimensional vector space, M is a linear subspace of V, and $Q{:}\mathrm{V} \to \mathrm{V/M}$ is the quotient mapping, then $Q'{:}(\mathrm{V/M})' \to \mathrm{V}'$ is an injective linear mapping with range M°. In particular, $(\mathrm{V/M})' \cong \mathrm{M}^\circ$.

{Hint: By Corollary 3.9.13, Q' is injective. Show that $\operatorname{Im} Q'$ and M° have the same dimension (cf. Theorems 3.9.12, 3.9.11) and that $\operatorname{Im} Q' \subset \mathrm{M}^\circ$, therefore $\operatorname{Im} Q' = \mathrm{M}^\circ$.}

3. If V and W are finite-dimensional vector spaces and $T{:}\mathrm{V} \to \mathrm{W}$ is a linear mapping, prove that T and T' have the same nullity if and only if $\dim \mathrm{V} = \dim \mathrm{W}$.

4. If V and W are finite-dimensional vector spaces, then $(\mathrm{V} \times \mathrm{W})' \cong \mathrm{V}' \times \mathrm{W}'$ on grounds of dimensionalilty (Theorems 3.9.2 and 3.6.4). There is also a 'natural' isomorphism $T{:}\mathrm{V}' \times \mathrm{W}' \to (\mathrm{V} \times \mathrm{W})'$, defined as follows: if $f \in \mathrm{V}'$ and $g \in \mathrm{W}'$, let $T(f, g)$ be the linear form $(x, y) \mapsto f(x) + g(y)$ on $\mathrm{V} \times \mathrm{W}$. {Shortcut: §3.8, Exercise 11.}

5. Let $\mathrm{V} = \mathcal{P}$ be the vector space of real polynomial functions (1.3.5). Every nonnegative integer k defines a linear form φ_k on V by the formula $\varphi_k(p) = p^{(k)}(0)$, where $p^{(0)} = p$, $p^{(1)} = p'$, $p^{(2)} = p''$, etc. If $D{:}\mathrm{V} \to \mathrm{V}$ is the differentiation mapping $Dp = p'$ (2.1.4), show that $D'\varphi_k = \varphi_{k+1}$. {Hint: $\varphi_k(p) = (D^k p)(0)$.}

6. Let V be a vector space of dimension n and let $f_1, \ldots, f_n$ be a basis of V' (Theorem 3.9.2). Prove that the mapping $x \mapsto (f_1(x), \ldots, f_n(x))$ is a vector space isomorphism $\mathrm{V} \to \mathrm{F}^n$. {Hint: §3.8, Exercise 12.}

7. If V and W are finite-dimensional vector spaces, $T{:}\mathrm{V} \to \mathrm{W}$ is a linear mapping, and $T'{:}\mathrm{W}' \to \mathrm{V}'$ is the transpose of T, then T is bijective if and only if T' is bijective. {Hint: Corollary 3.9.13.}

8. It is an exercise in concentration to calculate the transpose of a linear form $f{:}\mathrm{V} \to \mathrm{F}$.

(i) For each $a \in \mathrm{F}$, the formula $a'(c) = ac$ defines a linear form a' on F, and $a \mapsto a'$ is an isomorphism $\mathrm{F} \to \mathrm{F}'$.
(ii) $a'f$ (the composite of a' and f) is equal to af (the scalar multiple of f by the scalar a).
(iii) $f'a' = af$ for all $a \in \mathrm{F}$.

9. Let $x_1, \ldots, x_n$ be a basis of V and let $f_1, \ldots, f_n$ be the basis of V' dual to $x_1, \ldots, x_n$. Show that $x_1'', \ldots, x_n''$ is the basis of V'' dual to $f_1, \ldots, f_n$. {See Definition 3.9.15 for the notations.}

10. Let V and W be finite-dimensional vector spaces. Show that the mapping $\mathcal{L}(\mathrm{V}, \mathrm{W}) \to \mathcal{L}(\mathrm{W}', \mathrm{V}')$ defined by $T \mapsto T'$ is linear and bijective.

11. If V and W are finite-dimensional, then the mapping $\mathcal{L}(\mathrm{V}, \mathrm{W}) \to \mathcal{L}(\mathrm{V}'', \mathrm{W}'')$ defined by $T \mapsto T''$ is linear and bijective.

12. Let V be a finite-dimensional vector space and let $f_1, \ldots, f_n$ be a basis of V′ (3.9.2). Prove that there exists a basis $x_1, \ldots, x_n$ of V such that $f_1, \ldots, f_n$ is the basis of V′ dual to $x_1, \ldots, x_n$. {Hint: Apply 3.9.3 with V and V′ replaced by V′ and V″, then cite 3.9.17. Alternate proof: cf. Exercise 6 and the proof of 3.9.3.}

13. Let V be a finite-dimensional vector space, M a linear subspace of V, M° the annihilator of M in V′ (3.9.10). Prove:

(i) $M^\circ = \{0\} \Leftrightarrow M = V$.

(ii) $M^\circ = V' \Leftrightarrow M = \{\theta\}$.

14. Let V be a finite-dimensional vector space, V′ its dual, N a linear subspace of V′. Define

$$^\circ N = \{x \in V: f(x) = 0 \quad \text{for all} \quad f \in N\},$$

that is, °N is the intersection of the kernels of all the $f \in N$. {This is a sort of 'backward annihilator', analogous to the 'forward annihilator' M° of Definition 3.9.10; the effect on subspaces is that $M \mapsto M^\circ$ moves us forward in the list V, V′, V″, V‴, . . . and $N \mapsto {}^\circ N$ moves us backward.}

Prove that $^\circ(M^\circ) = M$ for every linear subspace M of V.

{Hint: The inclusion $M \subset {}^\circ(M^\circ)$ is trivial; for the reverse inclusion, see §3.8, Exercise 13. Alternate proof: With $S: V \to V''$ as in 3.9.15, let $(M^\circ)^\circ$ be the annihilator of M° in V″; show that $S(M) \subset (M^\circ)^\circ$ then argue that $S(M) = (M^\circ)^\circ$ on the grounds of dimensionality (3.9.11), and finally observe that $^\circ(M^\circ) = S^{-1}((M^\circ)^\circ)$.}

15. Let V be a finite-dimensional vector space, M and N linear subspaces of V, and M°, N° their annihilators in V′ (3.9.10). Prove:

(i) If $M^\circ = N^\circ$ then $M = N$. {Hint: $S(M) = S(N)$ by the hint for Exercise 14.}

(ii) $(M + N)^\circ = M^\circ \cap N^\circ$.

(iii) $(M \cap N)^\circ = M^\circ + N^\circ$. {Hint: $S(M \cap N) = S(M) \cap S(N)$, that is, $((M \cap N)^\circ)^\circ = (M^\circ)^\circ \cap (N^\circ)^\circ$; cite (ii) to conclude that $((M \cap N)^\circ)^\circ = (M^\circ + N^\circ)^\circ$, then cite (i).}

16. Let V be a finite-dimensional vector space, N a linear subspace of V′. The 'backward annihilator' °N of N is a linear subspace of V (Exercise 14); the 'forward annihilator' N° of N is a linear subspace of V″ (3.9.10). Prove:

(i) $S(^\circ N) = N^\circ$.

(ii) $(^\circ N)^\circ = N$. {Hint: If $M = {}^\circ N$ then $N^\circ = S(M) = (M^\circ)^\circ$; cite (i) of Exercise 15.}

17. Let V be a finite-dimensional vector space and let f and $f_1, \ldots, f_n$ be linear forms on V. The following conditions are equivalent:

(a) f is a linear combination of $f_1, \ldots, f_n$;

(b) f vanishes on the intersection of the null spaces of the f_i, that is, if

x is a vector in V such that $f_i(x) = 0$ for $i = 1, \ldots, n$ then $f(x) = 0$.

{Hint: Let N be the linear span of $f_1, \ldots, f_n$ and look at Exercise 16.}

18. Let $T:\mathrm{V} \to \mathrm{W}$ be a linear mapping, M a linear subspace of V, N a linear subspace of W. Prove:

(i) If $T(\mathrm{M}) \subset \mathrm{N}$ then $T'(\mathrm{N}^\circ) \subset \mathrm{M}^\circ$.

(ii) $T'(\mathrm{W}') \subset (\mathrm{Ker}\, T)^\circ$. {Hint: In (i), consider $\mathrm{M} = \mathrm{Ker}\, T$, $\mathrm{N} = \{\theta\}$.}

(iii) If V and W are finite-dimensional, then $T'(\mathrm{W}') = (\mathrm{Ker}\, T)^\circ$. {Hint: In (ii), calculate dimensions.}

19. Let V be a finite-dimensional vector space, M a linear subspace of V, and $T:\mathrm{M} \to \mathrm{V}$ the insertion mapping $Tx = x$ $(x \in \mathrm{M})$. Show that $T':\mathrm{V}' \to M'$ is the mapping R in the proof of Theorem 3.9.11.

20. If V is a vector space and $f_1, \ldots, f_n$ are linearly independent linear forms on V such that

$$(\mathrm{Ker}\, f_1) \cap \ldots \cap (\mathrm{Ker}\, f_n) = \{\theta\}$$

then V is n-dimensional. {Hint: Cite §3.6, Exercise 9, then argue that $\dim \mathrm{V} \geq n$. Shortcut: Exercise 17.}

21. Let V and W be finite-dimensional vector spaces, $T:\mathrm{V} \to \mathrm{W}$ a linear mapping. If W is 12-dimensional and the kernel of T' is 4-dimensional, find the rank of T.

4

Matrices

4.1 Matrices

If V is an n-dimensional vector space over the field F, choosing a basis for V sets up a vector space isomorphism $F^n \to V$ (3.5.5), whereby each vector $x \in V$ acquires a numerical description ('coordinates'), namely, the n-ple of scalars $(a_1, \ldots, a_n) \in F^n$ that corresponds to x under the isomorphism.

Linear mappings may be described numerically in a similar way; let's begin with an example.

4.1.1 Example Let $T:\mathbf{R}^3 \to \mathbf{R}^2$ be the linear mapping defined by

$$T(x, y, z) = (2x - 3y + 4z, 5x - 2z).$$

If e_1, e_2, e_3 is the canonical basis of $\mathbf{R}^3$ and f_1, f_2 is the canonical basis of $\mathbf{R}^2$, then

$$\begin{aligned} Te_1 &= T(1, 0, 0) = (2, 5) = 2f_1 + 5f_2 \\ Te_2 &= T(0, 1, 0) = (-3, 0) = -3f_1 + 0f_2 \\ Te_3 &= T(0, 0, 1) = (4, -2) = 4f_1 - 2f_2. \end{aligned}$$

If $u = (x, y, z)$ is an arbitrary vector in $\mathbf{R}^3$, then $u = xe_1 + ye_2 + ze_3$ and

$$Tu = x(Te_1) + y(Te_2) + z(Te_3),$$

thus T is completely determined by its values on the basis vectors e_1, e_2, e_3; in turn, the vectors Te_1, Te_2, Te_3 are completely determined by the 6 coefficients needed to express them as linear combinations of the basis vectors f_1, f_2. These 6 coefficients are in a sense the 'coordinates' of T with respect to the canonical bases of $\mathbf{R}^3$ and $\mathbf{R}^2$; it proves to be notationally convenient to organize the coefficients into the rectangular array

$$\begin{pmatrix} 2 & -3 & 4 \\ 5 & 0 & -2 \end{pmatrix}.$$

The array is said to be 2×3 (read '2 by 3'), with 2 rows and 3 columns; Te_1 contributes the first column, Te_2 the second, Te_3 the third, whereas the first row consists of the coefficients of f_1 in the expressions for Te_1, Te_2, Te_3 and the second row consists of the coefficients of f_2.

Now let's generalize the example.

4.1.2 Example Let V and W be finite-dimensional vector spaces over F, $T: \mathrm{V} \to \mathrm{W}$ a linear mapping. Let $n = \dim \mathrm{V}$, $m = \dim \mathrm{W}$ and choose bases $x_1, \ldots, x_n$ and $y_1, \ldots, y_m$ of V and W, respectively. For each $j \in \{1, \ldots, n\}$ express Tx_j as a linear combination of $y_1, \ldots, y_m$:

$$(*) \qquad Tx_j = a_{1j}y_1 + a_{2j}y_2 + \ldots + a_{mj}y_m = \sum_{i=1}^{m} a_{ij}y_i.$$

It is worth writing this out in slow motion the first time:

$$\begin{aligned} Tx_1 &= a_{11}y_1 + a_{21}y_2 + \ldots + a_{m1}y_m \\ Tx_2 &= a_{12}y_1 + a_{22}y_2 + \ldots + a_{m2}y_m \\ &\ldots \\ Tx_n &= a_{1n}y_1 + a_{2n}y_2 + \ldots + a_{mn}y_m. \end{aligned}$$

The data of $(*)$ is a collection of mn scalars $a_{ij} \in \mathrm{F}$, where $i \in \{1, \ldots, m\}$ and $j \in \{1, \ldots, n\}$, mostly conveniently organized into a rectangular array with m rows and n columns,

$$\begin{pmatrix} a_{11} & a_{12} & \ldots & a_{1j} & \ldots & a_{1n} \\ a_{21} & a_{22} & \ldots & a_{2j} & \ldots & a_{2n} \\ \ldots & & & & & \\ a_{i1} & a_{i2} & \ldots & a_{ij} & \ldots & a_{in} \\ \ldots & & & & & \\ a_{m1} & a_{m2} & \ldots & a_{mj} & \ldots & a_{mn} \end{pmatrix},$$

with a_{ij} appearing in row i and column j. The coefficients for Tx_j in $(*)$ contribute the j'th column, whereas the i'th row consists of the coefficients of y_i for the vectors $Tx_1, Tx_2, \ldots, Tx_n$.

What is really going on here? For each ordered pair (i, j) with $i \in \{1,\ldots,m\}$ and $j \in \{1,\ldots,n\}$ we are given a scalar a_{ij}. Writing $\mathrm{I} = \{1,\ldots,m\}$, $\mathrm{J} = \{1,\ldots,n\}$, for each $(i, j) \in \mathrm{I} \times \mathrm{J}$ we have an element a_{ij} of F; this is nothing more nor less than a function $\mathrm{I} \times \mathrm{J} \to \mathrm{F}$.

There is some economy in reversing the order of appearance of $m \times n$ rectangular arrays and functions on $\mathrm{I} \times \mathrm{J}$:

4.1.3 Definition Let m and n be positive integers and let $\mathrm{I} = \{1,\ldots,m\}$, $\mathrm{J} = \{1,\ldots,n\}$. An $m \times n$ **matrix** over F is a function $\mathrm{I} \times \mathrm{J} \to \mathrm{F}$.

4.1.4 Remarks With notations as in 4.1.3, let $A:\mathrm{I} \times \mathrm{J} \to \mathrm{F}$ and let $a_{ij} = A(i, j)$ be the value of A at the point $(i, j) \in \mathrm{I} \times \mathrm{J}$. The values of A can be organized into an $m \times n$ rectangular array

$$\begin{pmatrix} a_{11} & a_{12} & \cdots & a_{1n} \\ a_{21} & a_{22} & \cdots & a_{2n} \\ \cdots & & & \\ a_{m1} & a_{m2} & \cdots & a_{mn} \end{pmatrix};$$

we regard this array as just another notation for the function A (a notation that explicitly shows the effect of A on the points of its domain), that is, we regard the array as being *equal* to A and we write

$$A = \begin{pmatrix} a_{11} & a_{12} & \cdots & a_{1n} \\ a_{21} & a_{22} & \cdots & a_{2n} \\ \cdots & & & \\ a_{m1} & a_{m2} & \cdots & a_{mn} \end{pmatrix}$$

or, more concisely,

$$A = (a_{ij})_{1\le i\le m, 1\le j\le n}$$

and even more concisely

$$A = (a_{ij}).$$

The scalars a_{ij} are also called the *entries* of A.

The set of all such functions A is precisely the set $\mathcal{F}(\mathrm{I} \times \mathrm{J}, \mathrm{F})$ (1.3.4). If $B = (b_{ij})$ is another $m \times n$ matrix over F, that is, if $B \in \mathcal{F}(\mathrm{I} \times \mathrm{J}, \mathrm{F})$ and $b_{ij} = B(i, j)$, then the equality $A = B$ (as functions) means that

$$A(i, j) = B(i, j) \quad \text{for all} \quad (i, j) \in \mathrm{I} \times \mathrm{J};$$

in other words,

$$(a_{ij}) = (b_{ij}) \quad \Leftrightarrow \quad a_{ij} = b_{ij} \text{ for all } i \text{ and } j.$$

It is time to replace the ponderous notation $\mathcal{F}(\mathrm{I} \times \mathrm{J}, \mathrm{F})$ by something more compact:

4.1.5 Definition The set of all $m \times n$ matrices over F will be denoted $\mathrm{M}_{m,n}(\mathrm{F})$; when $m = n$ we abbreviate $\mathrm{M}_{n,n}(\mathrm{F})$ to $\mathrm{M}_n(\mathrm{F})$ and call its elements **square** matrices of **order** n. {The notations $\mathrm{F}_{m,n}$ and F_n are also used.}

For example, $\mathrm{M}_{2,3}(\mathbf{R})$ is the set of all 2×3 matrices over the real field **R**, and $\mathrm{M}_5(\mathbf{C})$ is the set of all 5×5 matrices (i.e., square matrices of order 5) over the complex field **C**.

When $\mathrm{F} = \mathbf{R}$ we speak of **real** matrices, and when $\mathrm{F} = \mathbf{C}$, of **complex** matrices.

When $n = 1$ we speak of **column** matrices

$$\begin{pmatrix} a_{11} \\ a_{21} \\ \vdots \\ a_{m1} \end{pmatrix}$$

and when $m = 1$ we speak of **row** matrices

$$(a_{11} \quad a_{12} \quad \ldots \quad a_{1n}).$$

The set $\mathrm{M}_{1,n}(\mathrm{F})$ of all row matrices with n entries is (give or take some commas) indistinguishable from F^n; the set $\mathrm{M}_{m,1}(\mathrm{F})$ of all column matrices with m entries can similarly be identified with F^m.

In particular, the rows and columns of a matrix can be thought of as vectors: if $A = (a_{ij})$ is an $m \times n$ matrix we call

$$(a_{i1} \quad a_{i2} \quad \ldots \quad a_{in})$$

the i'th **row vector** of A (thought of as an element of F^n), and we call

$$\begin{pmatrix} a_{1j} \\ a_{2j} \\ \vdots \\ a_{mj} \end{pmatrix}$$

the j'th **column vector** of A (thought of as an element of F^m).

4.1.6 Theorem *The set* $\mathrm{M}_{m,n}(\mathrm{F})$ *of* $m \times n$ *matrices over* F *is a vector space for the operations*

$$(a_{ij}) + (b_{ij}) = (a_{ij} + b_{ij}), \qquad c(a_{ij}) = (ca_{ij}).$$

Proof. There is essentially nothing to prove, since $\mathrm{M}_{m,n}(\mathrm{F})$ is just another notation for the vector space of functions $\mathcal{F}(\mathrm{I} \times \mathrm{J}, \mathrm{F})$ with the pointwise linear operations, where $\mathrm{I} = \{1, \ldots, m\}$, $\mathrm{J} = \{1, \ldots, n\}$ (4.1.5, 4.1.3). It is nevertheless reassuring to write $A = (a_{ij})$, $B = (b_{ij})$ and compute:

$$(A + B)(i, j) = A(i, j) + B(i, j) = a_{ij} + b_{ij},$$
$$(cA)(i, j) = cA(i, j) = ca_{ij}$$

for all i and j, therefore $A + B = (a_{ij} + b_{ij})$ and $cA = (ca_{ij})$; thus the operations indicated in the statement of the theorem are just the linear operations in the vector space $\mathcal{F}(\mathrm{I} \times \mathrm{J}, \mathrm{F})$. ■

In particular, $-A = (-a_{ij})$ (called the *negative* of the matrix A), and the zero element $0 \in M_{m,n}(F)$ is represented by the $m \times n$ array with the scalar 0 in every position (called the *$m \times n$ zero matrix*).

To summarize, the description of linear mappings in terms of bases leads to the idea of a matrix (a function defined on a finite cartesian product $I \times J$, with a dual personality as a rectangular array). So far, matrices are just a notation; in the next section we look at the interplay between linear mappings and matrices (the dominant theme of the rest of the book).

▶ Exercises

1. If

$$A = \begin{pmatrix} 3 & 2 & 4 \\ -1 & 0 & -2 \end{pmatrix}, \quad B = \begin{pmatrix} 5 & 3 & -1 \\ 1 & 2 & 0 \end{pmatrix}$$

calculate $A + B$, $-A$, $2A + 3B$.

2. Solve for w, x, y, z:

$$\begin{pmatrix} 4w & 3x+2 \\ y+1 & 2-z \end{pmatrix} = \begin{pmatrix} 1-w & x+1 \\ 4z & 3y \end{pmatrix}.$$

3. If the column vectors of the matrix

$$A = \begin{pmatrix} a & c \\ b & d \end{pmatrix}$$

are linearly dependent, prove that $ad - bc = 0$. Then prove the converse. What about rows?

4.2 Matrices of linear mappings

4.2.1 Definition

Let $T: V \rightarrow W$ be a linear mapping, where V and W are finite-dimensional vector spaces over the field F. Say $\dim V = n$, $\dim W = m$. Choose bases $x_1, \ldots, x_n$ of V and $y_1, \ldots, y_m$ of W and, for each $j \in \{1, \ldots, n\}$, express Tx_j as a linear combination of $y_1, \ldots, y_m$:

$$Tx_j = \sum_{i=1}^{m} a_{ij} y_i \qquad (j = 1, \ldots, n).$$

The $m \times n$ matrix (a_{ij}) is called the *matrix of T relative to the bases $x_1, \ldots, x_n$ and $y_1, \ldots, y_m$*.

It is important to note that a linear mapping doesn't have a unique matrix; there's one for each choice of bases. (An exception is the zero linear mapping, all of whose matrices are the zero matrix.)

When $V = W$ we have the option (not always exercised) of choosing the same basis for V as initial space and as final space:

4.2.2 Definition

If $T: V \rightarrow V$ is a linear mapping and $x_1, \ldots, x_n$ is a basis of V, and if

$$Tx_j = \sum_{i=1}^{n} a_{ij} x_i$$

for all j, then the square matrix (a_{ij}) of order n is called the *matrix of* T *relative to the basis* $x_1, \ldots, x_n$.

4.2.3 Example If $T:\mathbf{R}^3 \to \mathbf{R}^2$ is the linear mapping defined by

$$T(x, y, z) = (2x - 3y + 4z, 5x - 2z),$$

then, by the discussion in Example 4.1.1, the matrix of T relative to the canonical bases of $\mathbf{R}^3$ and $\mathbf{R}^2$ is

$$\begin{pmatrix} 2 & -3 & 4 \\ 5 & 0 & -2 \end{pmatrix}.$$

4.2.4 Example Let V be a vector space of dimension n, $I:\mathrm{V} \to \mathrm{V}$ the identity linear mapping, and $x_1, \ldots, x_n$ any basis of V. Then

$$Ix_j = x_j = \sum_{i=1}^{n} \delta_{ij} x_i,$$

where δ_{ij} is the Kronecker delta (3.9.4), thus the matrix of I relative to $x_1, \ldots, x_n$ is the $n \times n$ matrix

$$(\delta_{ij}) = \begin{pmatrix} 1 & 0 & \ldots & 0 \\ 0 & 1 & 0 \ \ldots & 0 \\ \vdots & \vdots & \ddots & \vdots \\ 0 & 0 & \ldots & 1 \end{pmatrix}.$$

This matrix is called the $n \times n$ *identity matrix* and is also denoted I. For example, the 3×3 identity matrix is

$$\begin{pmatrix} 1 & 0 & 0 \\ 0 & 1 & 0 \\ 0 & 0 & 1 \end{pmatrix}.$$

4.2.5 Example Let $I:\mathbf{R}^2 \to \mathbf{R}^2$ be the identity mapping, $e_1 = (1, 0)$, $e_2 = (0, 1)$. The matrix of I relative to the pair of bases e_1, e_2 and e_2, e_1 is

$$\begin{pmatrix} 0 & 1 \\ 1 & 0 \end{pmatrix},$$

because

$$Ie_1 = e_1 = 0e_2 + 1e_1,$$
$$Ie_2 = e_2 = 1e_2 + 0e_1.$$

This reminds us that a basis is a finite *list* (not just a finite set); more precisely, a basis of an n-dimensional vector space is an *ordered n-ple* of vectors.

Definition 4.2.1 is a two-way street:

4.2.6 Theorem *If $A = (a_{ij})$ is an $m \times n$ matrix over* F *and if* V *and* W *are vector spaces over* F *with* $\dim V = n$, $\dim W = m$, *then for each choice of bases of* V *and* W, *there exists a unique linear mapping* $T: V \to W$ *whose matrix relative to the chosen bases is* A.

Proof. Let $x_1, \ldots, x_n$ be a basis of V and $y_1, \ldots, y_m$ a basis of W. We seek a linear mapping $T: V \to W$ such that

$$Tx_j = \sum_{i=1}^{m} a_{ij} y_i$$

for $j = 1, \ldots, n$. Let

$$w_j = \sum_{i=1}^{m} a_{ij} y_i \qquad (j = 1, \ldots, n);$$

the existence and uniqueness of a linear mapping $T: V \to W$ such that $Tx_j = w_j$ for all j are assured by Theorem 3.8.1. ■

It follows that matrices reflect faithfully the vector space structure of $\mathcal{L}(V, W)$:

4.2.7 Corollary *Let* V *and* W *be finite-dimensional vector spaces over* F, *choose bases of* V *and* W, *and, for each linear mapping* $T: V \to W$, *let* M_T *be the matrix of* T *relative to the chosen bases* (4.2.1). *Then the mapping* $T \mapsto M_T$ *is a linear bijection* $\mathcal{L}(V, W) \mapsto M_{m,n}(F)$, *where* $m = \dim W$ *and* $n = \dim V$.

Proof. The mapping is surjective by the 'existence' part of 4.2.6, injective by the 'uniqueness' part. It remains to prove that if $S, T \in \mathcal{L}(V, W)$ and $c \in F$, then

$$M_{S+T} = M_S + M_T \quad \text{and} \quad M_{cS} = cM_S.$$

Let $x_1, \ldots, x_n$ and $y_1, \ldots, y_m$ be the chosen bases of V and W. Thus, if $M_S = (a_{ij})$ and $M_T = (b_{ij})$, then

$$Sx_j = \sum_{i=1}^{m} a_{ij} y_i, \qquad Tx_j = \sum_{i=1}^{m} b_{ij} y_i$$

for all j. It follows from the definitions of the linear operations in $\mathcal{L}(V, W)$ (2.3.9) that

$$(S + T)x_j = Sx_j + Tx_j = \sum_{i=1}^{m} (a_{ij} + b_{ij}) y_i,$$

$$(cS)x_j = c(Sx_j) = \sum_{i=1}^{m} (ca_{ij}) y_i,$$

therefore

$$M_{S+T} = (a_{ij} + b_{ij}) = M_S + M_T,$$
$$M_{cS} = (ca_{ij}) = cM_S,$$

by the definition of the linear operations in $M_{m,n}(F)$ (4.1.6) ■

Matrices can be enlisted in the computation of the rank of a linear mapping. This will be treated in detail later on (§4.7), but it's not too soon to see the relevance of matrices to the task:

4.2.8 Lemma *Let* V *be an n-dimensional vector space over* F, $T{:}\mathrm{V} \to \mathrm{F}^{\mathrm{m}}$ *a linear mapping. Let* $x_1, \ldots, x_n$ *be a basis of* V, *let* $f_1, \ldots, f_m$ *be the canonical basis of* F^{m}, *and let* A *be the matrix of* T *relative to these bases. Then* $T(\mathrm{V})$ *is the linear span of the column vectors of* A.

Proof. Say $A = (a_{ij})$. The j'th column vector of A is

$$\begin{pmatrix} a_{1j} \\ a_{2j} \\ \vdots \\ a_{mj} \end{pmatrix}$$

(remarks following Definition 4.1.5); written horizontally, with commas, it is the vector

$$(a_{1j}, a_{2j}, \ldots, a_{mj}) = \sum_{i=1}^{m} a_{ij} f_i = Tx_j$$

of F^{m}. Since V is the linear span of the x_j, $T(\mathrm{V})$ is the linear span of the Tx_j (3.1.6); in other words, the range of T is the linear span of the column vectors of A. ■

This prompts a burst of terminology:

4.2.9 Definition Let A be an $m \times n$ matrix over the field F. The **column space** of A is the linear subspace of F^{m} generated by the column vectors of A; its dimension is called the **column rank** of A. Similarly, the **row space** of A is the linear subspace of F^{n} generated by the row vectors of A, and its dimension is called the **row rank** of A.

Although the row space and column space are subspaces of different vector spaces (F^{n} and F^{m}, respectively), it turns out that they have the same dimension, in other words, *the row rank and column rank are equal*; this miracle is explained in §4.6.

It follows from 4.2.8 that the rank of the linear mapping T is equal to the column rank of one of its matricial representations. In fact, the rank of a linear mapping between any two finite-dimensional vector spaces is equal to the column rank of its matrix relative to any pair of bases:

4.2.10 Theorem *If* $T{:}\mathrm{V} \to \mathrm{W}$ *is a linear mapping and* A *is the matrix of* T *relative to any pair of bases of* V *and* W, *then the rank of* T *is equal to the column rank of* A.

Proof. We can suppose that the notations of Definition 4.2.1 are in force; thus $A = (a_{ij})$ and

$$Tx_j = \sum_{i=1}^{m} a_{ij} y_i \qquad (j = 1, \ldots, n),$$

where $x_1, \ldots, x_n$ and $y_1, \ldots, y_m$ are the bases of V and W in question.

Let $f_1, \ldots, f_m$ be the canonical basis of F^m and let $S{:}W \to F^m$ be the vector space isomorphism such that $Sy_i = f_i$ for all i (Theorem 3.5.5). Then

$$(ST)x_j = S(Tx_j) = \sum_{i=1}^{m} a_{ij}(Sy_i) = \sum_{i=1}^{m} a_{ij}f_i,$$

thus A is also the matrix of ST relative to the bases $x_1, \ldots, x_n$ and $f_1, \ldots, f_m$. By Lemma 4.2.8, the range of ST is the column space of A, consequently the rank of ST is equal to the column rank of A.

Since S is linear and injective, it defines a vector space isomorphism of $T(V)$ onto $S(T(V)) = (ST)(V)$, therefore

$$\dim T(V) = \dim (ST)(V)$$

by Theorem 3.5.12; thus T and ST have the same rank, in other words, the rank of T is equal to the column rank of A. ■

We conclude with more examples of matrices of linear mappings and some useful definitions they suggest.

4.2.11 Example Let V be a vector space of dimension n, let $c_1, \ldots, c_n$ be scalars and let $x_1, \ldots, x_n$ be a basis of V. If $T{:}V \to V$ is the linear mapping such that $Tx_j = c_j x_j$ for all j (3.8.1), then the matrix of T for the basis $x_1, \ldots, x_n$ is

$$\begin{pmatrix} c_1 & & & \text{\Large 0} \\ & c_2 & & \\ & & \ddots & \\ \text{\Large 0} & & & c_n \end{pmatrix}$$

where the oversized 'zeros' indicate that every position not filled by the c_j is occupied by a 0. Such a matrix is called a **diagonal matrix** and is also denoted

$$\operatorname{diag}(c_1, c_2, \ldots, c_n).$$

4.2.12 Definition If $A = (a_{ij})$ is a square matrix of order n, then $a_{11}, a_{22}, \ldots, a_{nn}$ are called the elements of the **principal diagonal** (or 'main diagonal') of A.

4.2.13 Example Let V be a finite-dimensional vector space, c a fixed scalar, and $T{:}V \to V$ the scalar linear mapping defined by c (2.3.5), that is, $Tx = cx$ for all vectors x. The matrix of T relative to every basis of V is the $n \times n$ diagonal matrix

$$\operatorname{diag}(c, c, \ldots, c) = \begin{pmatrix} c & & & \text{\Large 0} \\ & c & & \\ & & \ddots & \\ \text{\Large 0} & & & c \end{pmatrix}.$$

{Reason: $T = cI_V$, therefore the matrix of T is cI by 4.2.7 and 4.2.4.} Such a matrix is called a **scalar matrix**. There are no other linear mappings for which the matrix is independent of basis (Exercise 6).

▶ Exercises

1. Let V be a 3-dimensional vector space, let x_1,x_2,x_3 be a basis of V and let $T \in \mathcal{L}(V)$ be the linear mapping such that

$$Tx_1 = x_1, \quad Tx_2 = x_1 + x_2, \quad Tx_3 = x_1 + x_2 + x_3.$$

Find the matrix of T and of T^{-1} relative to the basis x_1,x_2,x_3.

2. Let E_{ij} be the linear mappings of §3.8, Exercise 6, that is, $E_{ij}x_k = \delta_{jk}y_i$ for all $i \in \{1,\ldots,m\}$ and all $j, k \in \{1,\ldots,n\}$. Describe the matrix of E_{ij} relative to the bases $x_1,\ldots,x_n$ and $y_1,\ldots,y_m$.

3. Let $T:\mathbf{R}^2 \to \mathbf{R}^3$ be the linear mapping defined by

$$T(x, y) = (2x - y, x + 4y, 3x - 2y).$$

Find the matrix of T relative to the canonical bases of $\mathbf{R}^2$ and $\mathbf{R}^3$.

4. Let $V = \mathcal{P}_3$ be the vector space of real polynomials of degree ≤ 3 (1.3.6) and let $D:V \to V$ be the linear mapping $Dp = p'$ (the derivative of p). Find the matrix of D relative to the basis $1, t, t^2, t^3$ of V.

5. Fix $y \in \mathbf{R}^3$ and let $T:\mathbf{R}^3 \to \mathbf{R}^3$ be the linear mapping $Tx = x \times y$ (§2.1, Exercise 1). Find the matrix of T relative to the canonical basis of $\mathbf{R}^3$.

6. Let V be a finite-dimensional vector space, $S \in \mathcal{L}(V)$ a linear mapping whose matrix is the same for every basis of V. Prove that $S = cI$ for some scalar c.

{Hint: Let A be the matrix of S (the same for all bases). As in the proof of Corollary 3.8.3, it suffices to show that if x is any nonzero vector, then Sx is proportional to x. The alternative is that x, Sx are independent, therefore V has a basis $x, Sx, x_3,\ldots,x_n$ (Theorem 3.5.15) and $Sx = 0x + 1(Sx) + 0x_3 + \ldots + 0x_n$ shows that the first column of A is

$$\begin{pmatrix} 0 \\ 1 \\ 0 \\ \vdots \\ 0 \end{pmatrix}.$$

Infer that if y, z are any two independent vectors then $Sy = z$; then apply this to the independent vectors $y, y + z$ to arrive at the absurdity $z = Sy = y + z$.}

7. If $A = (a_{ij})$ is an $n \times n$ matrix, the *trace* of A, denoted $\operatorname{tr} A$, is defined to be the sum of the (principal) diagonal elements of A:

$$\operatorname{tr} A = \sum_{i=1}^{n} a_{ii}.$$

Prove that tr: $M_n(F) \to F$ is a linear form on the vector space $M_n(F)$ of all $n \times n$ matrices over the field F.

8. Let V be an n-dimensional vector space over F and let $f: V \to F$ be a linear form on V. Describe the matrix of f relative to a basis $x_1, \ldots, x_n$ of V and the basis 1 of F.

9. If $A = (a_{ij})$ is an $m \times n$ matrix over F, the *transpose* of A, denoted A', is in the $n \times m$ matrix (b_{ji}) for which $b_{ji} = a_{ij}$. (Thus, the columns of A' are the rows of A.) Prove that $A \mapsto A'$ is a vector space isomorphism $M_{m,n}(F) \to M_{n,m}(F)$.

10. Find the matrix of the identity linear mapping $I: \mathbf{R}^2 \to \mathbf{R}^2$ relative to the bases $x_1 = (2, 3)$, $x_2 = (4, -5)$ and $y_1 = (1, 1)$, $y_2 = (-3, 4)$.

11. Find the column rank of each of the matrices in Exercises 3–5.

12. (i) Let V be a finite-dimensional vector space, $T \in \mathscr{L}(V)$. Prove that T is bijective if and only if there exists a pair of bases of V relative to which the matrix of T is the identity matrix I. {Hint: If $x_1, \ldots, x_n$ and $Tx_1, \ldots, Tx_n$ are both bases of V, what is the matrix of T relative to these bases?}

(ii) Generalize (i) to the case that $T \in \mathscr{L}(V, W)$, where V and W have the same dimension.

4.3 Matrix multiplication

It is convenient to explain *how* matrices are multiplied before explaining *why* they are multiplied that way.

You may have met 'dot products' in physics or in multidimensional calculus: if $x = (x_1, x_2, x_3)$ and $y = (y_1, y_2, y_3)$ are vectors in $\mathbf{R}^3$, their *dot product* is defined to be the real number

$$x \cdot y = x_1 y_1 + x_2 y_2 + x_3 y_3.$$

It's useful to carry over this idea to vectors x, y in F^n (F any field): if $x = (x_1, \ldots, x_n)$ and $y = (y_1, \ldots, y_n)$, the dot product of the vectors x and y is the scalar

$$x \cdot y = x_1 y_1 + x_2 y_2 + \ldots + x_n y_n.$$

With this concept in hand, the definition of matrix products is straightforward: the entry in row i and column j of a matrix product AB is the dot product of the i'th row of A with the j'th column of B.

4.3.1 Example If

$$A = \begin{pmatrix} 3 & -1 \\ 2 & 0 \\ 4 & -2 \end{pmatrix}, \quad B = \begin{pmatrix} 3 & 2 & 0 & 1 \\ 5 & -3 & 1 & 2 \end{pmatrix}$$

then

$$AB = \begin{pmatrix} 4 & 9 & -1 & 1 \\ 6 & 4 & 0 & 2 \\ 2 & 14 & -2 & 0 \end{pmatrix}.$$

In particular, the (3, 2)-entry of AB is

$$(4 \quad -2)\cdot\begin{pmatrix} 2 \\ -3 \end{pmatrix} = (4)(2) + (-2)(-3) = 8 + 6 = 14.$$

Not all matrices A and B can be multiplied: for the required dot products to make sense, a row of A must have the same length as a column of B; in other words, the number of columns of A must equal the number of rows of B. Thus if A is an $m \times n$ matrix then B must be $n \times p$ for some p; there are then m choices for the row of A and p choices for the column of B, so the product matrix AB will be an $m \times p$ matrix. The formal definition is as follows:

4.3.2 Definition If $A = (a_{ij})$ is an $m \times n$ matrix and $B = (b_{jk})$ an $n \times p$ matrix over the field F, then the **product** of A and B, denoted AB, is the $m \times p$ matrix (c_{ik}), where

$$c_{ik} = (a_{i1} \; a_{i2} \ldots a_{in}) \cdot \begin{pmatrix} b_{1k} \\ b_{2k} \\ \vdots \\ b_{nk} \end{pmatrix}$$

$$= a_{i1}b_{1k} + a_{i2}b_{2k} + \ldots + a_{in}b_{nk},$$

concisely,

$$c_{ik} = \sum_{j=1}^{n} a_{ij}b_{jk}$$

for $i = 1, \ldots, m$ and $k = 1, \ldots, p$.

We can keep track of the number of rows and columns symbolically by writing

$$(m \times n)(n \times p) = m \times p;$$

note that the 'inner dimensions' are equal (both are n). In the example given above, $(3 \times 2)(2 \times 4) = 3 \times 4$. The fact that the product of an $m \times n$ matrix A with an $n \times p$ matrix B is an $m \times p$ matrix AB is also expressed by $\mathrm{M}_{m,n}(\mathrm{F}) \cdot \mathrm{M}_{n,p}(\mathrm{F}) \subset \mathrm{M}_{m,p}(\mathrm{F})$.

That's *how* matrices are multiplied; here's *why*:

4.3.3 Theorem *Let* U, V, W *be finite-dimensional vector spaces over the field* F *and let* T:U → V, S:V → W *be linear mappings,* ST:U → W *the composite mapping* (Fig. 17).

Choose a basis in each of the spaces U, V, W *and let* A *be the matrix*

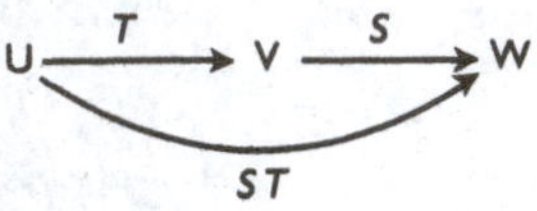

Fig. 17

of S, B the matrix of T, relative to these bases. Then the matrix of ST is AB relative to the chosen bases of U *and* W. *Suggestively,*

$$\operatorname{mat}(ST) = (\operatorname{mat} S)(\operatorname{mat} T)$$

relative to the chosen bases of U, V, W.

Proof. Let $p = \dim \mathrm{U}$, $n = \dim \mathrm{V}$, $m = \dim \mathrm{W}$. Choose bases

$$\begin{aligned} &z_1, \ldots, z_p \quad \text{of } \mathrm{U}, \\ &x_1, \ldots, x_n \quad \text{of } \mathrm{V}, \\ &y_1, \ldots, y_m \quad \text{of } \mathrm{W}, \end{aligned}$$

and suppose

$$\begin{aligned} Sx_j &= \sum_{i=1}^{m} a_{ij} y_i \qquad && (j = 1, \ldots, n), \\ Tz_k &= \sum_{j=1}^{n} b_{jk} x_j && (k = 1, \ldots, p), \\ (ST)z_k &= \sum_{i=1}^{m} c_{ik} y_i && (k = 1, \ldots, p); \end{aligned}$$

the matrices of S, T, and ST relative to those bases are

$$A = (a_{ij}), \quad B = (b_{jk}), \quad C = (c_{ik}),$$

respectively, and the assertion of the theorem is that $C = AB$. Note that A is $m \times n$ and B is $n \times p$, so the product AB is defined and is $m \times p$; C is also $m \times p$, so it is eligible to be compared with AB. For every k $(k = 1, \ldots, p)$,

$$\begin{aligned} \sum_{i=1}^{m} c_{ik} y_i &= (ST)z_k = S(Tz_k) \\ &= S\left(\sum_{j=1}^{n} b_{jk} x_j\right) = \sum_{j=1}^{n} b_{jk}(Sx_j) \\ &= \sum_{j=1}^{n} b_{jk}\left(\sum_{i=1}^{m} a_{ij} y_i\right) \\ &= \sum_{j=1}^{n} \sum_{i=1}^{m} b_{jk} a_{ij} y_i \\ &= \sum_{i=1}^{m} \sum_{j=1}^{n} a_{ij} b_{jk} y_i \\ &= \sum_{i=1}^{m} \left(\sum_{j=1}^{n} a_{ij} b_{jk}\right) y_i; \end{aligned}$$

comparing coefficients of the independent vectors $y_1, \ldots, y_m$, we see that

$$c_{ik} = \sum_{j=1}^{n} a_{ij} b_{jk}$$

for all i (and for each k), therefore $C = AB$. ■

4.3.4 Remark Let U, V, W and their bases be as in the proof of 4.3.3. By 4.2.7, there are linear bijections $\mathscr{L}(V, W) \to M_{m,n}(F)$, $\mathscr{L}(U, V) \to M_{n,p}(F)$, $\mathscr{L}(U, W) \to M_{m,p}(F)$ defined, respectively, by

$$S \mapsto M_S, \quad T \mapsto M_T, \quad R \mapsto M_R,$$

where M_S, M_T, M_R are the matrices of S, T, R relative to the chosen bases. In the context of these notations, the message of 4.3.3 is that

$$M_{ST} = M_S M_T$$

for all $S \in \mathscr{L}(V, W)$ and $T \in \mathscr{L}(U, V)$.

► ***Exercises***

1. Prove that the inclusion $M_{2,1}(F) \cdot M_{1,2}(F) \subset M_2(F)$ is proper. {Hint: Show that the 2×2 identity matrix is not a product AB of the indicated sort.}

2. (i) Check that

$$\begin{pmatrix} 1 & 0 & 0 \\ 0 & 1 & 0 \end{pmatrix} \begin{pmatrix} 1 & 0 \\ 0 & 1 \\ 0 & 0 \end{pmatrix} = \begin{pmatrix} 1 & 0 \\ 0 & 1 \end{pmatrix}$$

but the product in the reverse order is *not* the 3×3 identity matrix.

(ii) Let A be an $m \times n$ matrix, B an $n \times p$ matrix, and suppose that $AB = I$ with I an identity matrix. Prove that $m = p \le n$. {Hint: Let U, V, W and their bases be as in the proof of Theorem 4.3.3 and let S:V → W, T:U → V be the linear mappings with matrices A and B, respectively, relative to these bases. From $m = p$ we can suppose that U = W with the same basis. Infer that $ST = I$.}

(iii) If, as in (ii), $AB = I$, and if $m = n$, conclude that $BA = I$. {Hint: With the preceding notations, cite Corollary 3.7.5.}

3. If A is an $m \times n$ matrix and I is the $n \times n$ identity matrix, show that $AI = A$. {Hint: Refer the matter back to linear mappings, as in the hints for Exercise 2.} Show similarly that $IA = A$, where this time I stands for the $m \times m$ identity matrix. (So to speak, the symbol I is 'neutral' for matrix multiplication.)

4. Let $A \in M_{m,n}(F)$ and $B \in M_{n,m}(F)$ (thus both AB and BA are defined). Say $AB = (c_{hi})$, $BA = (d_{jk})$. Prove that

$$\sum_{i=1}^{m} c_{ii} = \sum_{j=1}^{n} d_{jj},$$

that is, $\operatorname{tr}(AB) = \operatorname{tr}(BA)$ (for the notation, see §4.2, Exercise 7).

4.4 Algebra of matrices

If A and B are $m \times n$ matrices over a field F and if $c \in F$, then $c(A + B) = cA + cB$; this is immediate from Theorem 4.1.6, as are the other identities that express the fact that $M_{m,n}(F)$ is a vector space over F.

The identities for multiplication take a little more work. It is possible to prove them by direct computation but we learn more by translating them into questions about linear mappings, using the 'dictionary' established in the preceding two sections; it is precisely this technique that makes matrices and linear mappings so useful to each other.

4.4.1 Theorem *Let A, B, C be matrices over* F. *Consider the equations*

(1) $$(AB)C = A(BC),$$

(2) $$A(B + C) = AB + AC,$$

(3) $$(B + C)A = BA + CA.$$

If one side of an equation is defined, then so is the other side and the two sides are equal.

Proof. (2) Suppose $A(B + C)$ is defined. Say A is an $m \times n$ matrix. Then $B + C$ must be $n \times p$ for some p (4.3.2), therefore B and C must both be $n \times p$ (4.1.6). It follows that AB and AC are both defined (and are $m \times p$ matrices) and so is $AB + AC$. Thus, if the left side of (2) is defined then so is the right side, and the converse of this assertion is proved similarly.

Assuming now that both sides are defined, let us show that they are equal. With m, n, p as in the preceding paragraph, let U, V, W be vector spaces over F of dimensions p, n, m, respectively, and choose a basis for each of these spaces. (For example, let U, V, W be F^p, F^n, F^m and choose the canonical bases.) Let $S, T \in \mathscr{L}(\mathrm{U}, \mathrm{V})$ be the linear mappings whose matrices (relative to the chosen bases) are B, C, respectively, and let $R \in \mathscr{L}(\mathrm{V}, \mathrm{W})$ be the linear mapping with matrix A:

$$\mathrm{U} \xrightarrow{S,T} \mathrm{V} \xrightarrow{R} \mathrm{W}.$$

With notations as in 4.3.4, we have

$$A(B + C) = \mathrm{M}_R(\mathrm{M}_S + \mathrm{M}_T) = \mathrm{M}_R\mathrm{M}_{S+T} = \mathrm{M}_{R(S+T)},$$
$$AB + AC = \mathrm{M}_R\mathrm{M}_S + \mathrm{M}_R\mathrm{M}_T = \mathrm{M}_{RS} + \mathrm{M}_{RT} = \mathrm{M}_{RS+RT},$$

so we are reduced to showing that $R(S + T) = RS + RT$.

That's easy: for all $x \in \mathrm{U}$,

$$\begin{aligned}(R(S + T))x &= R((S + T)x) = R(Sx + Tx)\\ &= R(Sx) + R(Tx) = (RS)x + (RT)x\\ &= (RS + RT)x.\end{aligned}$$

(3) The proof is similar to that of (2).

(1) If, for example, $A(BC)$ is defined, then so is $A(BC)$ by an argument similar to that for (2). As in the proof of (2), set up linear mappings

$$\mathrm{U} \xrightarrow{T} \mathrm{V} \xrightarrow{S} \mathrm{W} \xrightarrow{R} \mathrm{X}$$

so that R, S, T have matrices A, B, C, respectively, relative to chosen bases of U, V, W, X. Then (cf. 4.3.4)

$$(AB)C = (\mathrm{M}_R\mathrm{M}_S)\mathrm{M}_T = \mathrm{M}_{RS}\mathrm{M}_T = \mathrm{M}_{(RS)T}$$

and similarly $A(BC) = \mathrm{M}_{R(ST)}$, thus $(AB)C = A(BC)$ is a consequence of the associative law for the composition of functions: $(RS)T = R(ST)$. {Either side, applied to $x \in \mathrm{U}$, leads to $R(S(Tx))$.} ■

In a similar vein,

4.4.2 Theorem *If A and B are matrices over* F *such that AB is defined, and if $c \in \mathrm{F}$, then*

$$(cA)B = c(AB) = A(cB).$$

Proof. With notations as in the proof of (2) of Theorem 4.4.1, it all boils down to the equalities

$$(cR)S = c(RS) = R(cS),$$

which are easily verified. {Each of the three members, applied to $x \in \mathrm{U}$, leads to $c(R(Sx))$. ■

Finally (tedious but necessary):

4.4.3 Theorem *If A is any matrix over* F, *then*

$$AI = A, \quad IA = A, \quad A0 = 0, \quad 0A = 0,$$

where I and 0 denote the identity matrices and zero matrices of the appropriate sizes.

Proof. Say A is $m \times n$. In the equality $A0 = 0$, the zero on the left side is $n \times p$ for some p and the zero on the right side is then $m \times p$; the equality is immediate by direct computation, as is the equality $0A = 0$ (with analogous remarks about sizes).

In the expression AI, I stands for the $n \times n$ identity matrix; in IA, it is the $m \times m$ identity matrix. As in the proof of Theorem 4.4.1, let $R\!:\!\mathrm{V} \to \mathrm{W}$ be a linear mapping whose matrix is A relative to chosen bases of V and W, and look at the diagram

$$\mathrm{V} \xrightarrow{I_{\mathrm{V}}} \mathrm{V} \xrightarrow{R} \mathrm{W} \xrightarrow{I_{\mathrm{W}}} \mathrm{W};$$

the equalities $AI = A = IA$ then follow from the obvious equalities

$$RI_{\mathrm{V}} = R = I_{\mathrm{W}}R$$

(cf. 4.2.4 and 4.3.4). ■

4.4.4 Definition If A is a square matrix over F, the **powers** of A are defined recursively by the formulas

$$A^0 = I, \quad A^{i+1} = A^iA.$$

{In particular, $A^1 = A$, $A^2 = A^1A = AA$, $A^3 = A^2A = (AA)A = A(AA) = AA^2$, etc.} Additional important information on powers is given in the exercises.

▶ Exercises

1. If $T \in \mathcal{L}(\mathrm{V})$ and A is the matrix of T relative to a basis of V then, for $i = 0, 1, 2, \ldots, A^i$ is the matrix of T^i relative to the same basis.

2. If A is a square matrix, prove that there exists a finite list of scalars c_0, $c_1, c_2, \ldots, c_r$, not all 0, such that $c_0I + c_1A + c_2A^2 + \ldots + c_rA^r = 0$. {Hint: §3.8, Exercise 7.}

3. Check the correctness of (1) of Theorem 4.4.1 by trying it out on the matrices

$$A = \begin{pmatrix} 2 & -1 & 3 \\ 1 & 4 & 2 \end{pmatrix}, \qquad B = \begin{pmatrix} 4 & 3 \\ 1 & 0 \\ 3 & -2 \end{pmatrix}, \qquad C = \begin{pmatrix} 5 & 0 & 2 & 3 \\ 4 & 2 & 1 & 0 \end{pmatrix}.$$

4. Prove Theorems 4.4.1–4.4.3 by direct computation with the matrix elements (that is, without the intervention of linear mappings).

5. Let $J \in M_2(\mathbf{R})$ be the matrix

$$J = \begin{pmatrix} 0 & -1 \\ 1 & 0 \end{pmatrix}$$

and let $\mathcal{C}$ be the set of all matrices of the form

$$aI + bJ = \begin{pmatrix} a & 0 \\ 0 & a \end{pmatrix} + \begin{pmatrix} 0 & -b \\ b & 0 \end{pmatrix} = \begin{pmatrix} a & -b \\ b & a \end{pmatrix},$$

where $a, b \in \mathbf{R}$; thus $\mathcal{C}$ is the (2-dimensional) linear span of the set $\{I, J\}$ in the 4-dimensional vector space $M_2(\mathbf{R})$.

Show that $J^2 = -I$ and that the product of two elements of $\mathcal{C}$ belongs to $\mathcal{C}$; persuade yourself that sums and products of such matrices $aI + bJ$ mimic the familiar sum and product for complex numbers $a + bi$, so that the correspondence $aI + bJ \mapsto a + bi$ can be used to 'identify' $\mathcal{C}$ with the field $\mathbf{C}$ of complex numbers.

*4.5 A model for linear mappings

Let V and W be finite-dimensional vector spaces over a field F, $T: \mathrm{V} \to \mathrm{W}$ a linear mapping. Say $\dim \mathrm{V} = n$, $\dim \mathrm{W} = m$. Let's show that T can be replaced by a linear mapping $\mathrm{F}^n \to \mathrm{F}^m$ that 'preserves' a given matricial representation of T.

More precisely, let $x_1, \ldots, x_n$ be a basis of V and $y_1, \ldots, y_m$ a basis of W, and let $A = (a_{ij})$ be the matrix of T relative to these bases (4.2.1):

$$Tx_j = \sum_{i=1}^{m} a_{ij} y_i \qquad (j = 1, \ldots, n).$$

Let $e_1, \ldots, e_n$ be the canonical basis of F^n, $R: \mathrm{F}^n \to \mathrm{V}$ the isomorphism such that $Re_j = x_j$ for all j (3.5.5); similarly, let $f_1, \ldots, f_m$ be the canonical basis of F^m and $S: \mathrm{F}^m \to \mathrm{W}$ the isomorphism such that $Sf_i = y_i$ for all i. Consider the diagram:

* Optional reading; instructive, but omissible.

$$\begin{array}{ccc} & T & \\ V & \rightarrow & W \\ R \uparrow & & \uparrow S \\ F^n & \cdots & F^m \end{array}$$

We propose to fill in the dots with a linear mapping $T^{\#}:F^n \to F^m$ whose properties faithfully reflect the properties of T. The diagram itself suggests the definition:

$$T^{\#} = S^{-1}TR.$$

{Caution: $T^{\#}$ depends not only on T but also on the specific isomorphisms R, S that arise from a choice of bases.}

4.5.1 Theorem *With the preceding notations, the matrix of $T^{\#}$ relative to the canonical bases is also A.*

Proof. From the definition of $T^{\#}$, we have

$$ST^{\#} = TR,$$

therefore

$$(\text{mat}\, S)(\text{mat}\, T^{\#}) = (\text{mat}\, T)(\text{mat}\, R)$$

relative to the chosen bases (4.3.3). By assumption, $\text{mat}\, T = A$; from $Re_j = x_j$ we see that $\text{mat}\, R = I$ (the $n \times n$ identity matrix) and similarly $\text{mat}\, S = I$ (the $m \times m$ matrix), so the above matrix equation reduces to

$$I(\text{mat}\, T^{\#}) = AI,$$

in other words, $\text{mat}\, T^{\#} = A$ (4.4.3). ■

We can study T or we can study $T^{\#}$; the formulas $T^{\#} = S^{-1}TR$ and $T = ST^{\#}R^{-1}$ show that each mirrors faithfully the properties of the other. But $T^{\#}$ has some computational advantages:

4.5.2 Corollary *With the preceding notations, if $x = (c_1, \ldots, c_n) \in F^n$ then $T^{\#}x$ is the vector in F^m obtained by regarding x as a column vector (or $n \times 1$ matrix) and computing the matrix product*

$$Ax = \begin{pmatrix} a_{11} & a_{12} & \cdots & a_{1n} \\ a_{21} & a_{22} & \cdots & a_{2n} \\ \cdots & & & \\ a_{m1} & a_{m2} & \cdots & a_{mn} \end{pmatrix} \begin{pmatrix} c_1 \\ c_2 \\ \vdots \\ c_n \end{pmatrix}.$$

Proof. From the theorem, we know that

$$T^{\#}e_j = \sum_{i=1}^{m} a_{ij}f_i = (a_{1j}, a_{2j}, \ldots, a_{mj})$$

for $j = 1, \ldots, n$; thus, if

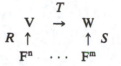

$$x = (c_1, \ldots, c_n) = \sum_{j=1}^{n} c_j e_j,$$

then

$$\begin{aligned} T^{\#}x &= \sum_{j=1}^{n} c_j(T^{\#}e_j) \\ &= \sum_{j=1}^{n} c_j\left(\sum_{i=1}^{m} a_{ij} f_i\right) \\ &= \sum_{i=1}^{m}\left(\sum_{j=1}^{n} a_{ij} c_j\right) f_i \\ &= \left(\sum_{j=1}^{n} a_{1j}c_j, \sum_{j=1}^{n} a_{2j}c_j, \ldots, \sum_{j=1}^{n} a_{mj}c_j\right). \end{aligned}$$

Thus, the i'th component of $T^{\#}x$ is the dot product (§4.3)

$$\sum_{j=1}^{n} a_{ij}c_j$$

of the i'th row of A with the vector x, which is just what the corollary asserts. ■

So we have a choice: study linear mappings $T{:}\mathrm{V} \to \mathrm{W}$ with $\dim \mathrm{V} = n$, $\dim \mathrm{W} = m$; or study $m \times n$ matrices $A = (a_{ij})$ acting on the elements of F^n (regarded as column vectors of length n), via left multiplication, to produce elements of F^m (regarded as column vectors of length m).

Does this mean that it was a waste of time to study general vector spaces? Not at all: if we are interested in the rank of a linear mapping $T{:}\mathrm{F}^n \to \mathrm{F}^m$ (and we are), we have to be prepared to talk about the dimension of its range $T(\mathrm{F}^n)$, which may be a proper subspace of F^m (hence not of the tidy form F^k for some k).

► ***Exercises***

1. Let x_1, x_2, x_3 be a basis of a 3-dimensional vector space V, $T \in \mathcal{L}(\mathrm{V})$ the linear mapping such that

$$Tx_1 = x_1, \qquad Tx_2 = x_1 + x_2, \qquad Tx_3 = x_1 + x_2 + x_3,$$

and A the matrix of T relative to the basis x_1, x_2, x_3. Find the formula for the linear mapping $T^{\#}$ of Theorem 4.5.1 (constructed for this choice of basis for V).

2. Suppose $m = n$ in Corollary 4.5.2. Prove that T is bijective if and only if the column vectors of A are linearly independent.

4.6 Transpose of a matrix

As in the case of matrix products, it is convenient to explain the 'how' before the 'why':

4.6.1 Definition If $A = (a_{ij})_{1\le i\le m, 1\le j\le n}$ is an $m \times n$ matrix, the **transpose** of A, denoted A' (also ${}^tA$), is the $n \times m$ matrix $(b_{ji})_{1\le j\le n, 1\le i\le m}$ for which $b_{ji} = a_{ij}$. Thus, the element in row j and column i of A' is the element in row i and column j of A. Viewing A and A' as functions (4.1.3), $A'(j, i) = A(i, j)$.

4.6.2 Example If

$$A = \begin{pmatrix} 2 & 1 & -3 \\ 4 & 0 & 5 \end{pmatrix},$$

then

$$A' = \begin{pmatrix} 2 & 4 \\ 1 & 0 \\ -3 & 5 \end{pmatrix}.$$

Thus, the rows of A are the columns of A', and the columns of A are the rows of A'. This observation merits a little fanfare:

4.6.3 Remark If A is an $m \times n$ matrix over a field F then, in the terminology of Definition 4.2.9,

(i) the row space of A is equal to the column space of A' (a linear subspace of F^n),

(ii) the column space of A is equal to the row space of A' (a linear subspace of F^m), consequently

(iii) the row rank of A equals the column rank of A', and the column rank of A equals the row rank of A' (we see in 4.6.7 that all of these numbers are equal to each other).

4.6.4 Example Consider the square matrix

$$A = \begin{pmatrix} 2 & 1 & -3 \\ 4 & 0 & 5 \\ 1 & -7 & 6 \end{pmatrix};$$

then

$$A' = \begin{pmatrix} 2 & 4 & 1 \\ 1 & 0 & -7 \\ -3 & 5 & 6 \end{pmatrix}.$$

4.6.5 Remarks Note, for example, that the (2, 1) *position* of a square matrix (irrespective of the element in that position) is the mirror image of the (1, 2) position, with the principal diagonal (Definition 4.2.12) serving as mirror. We obtain A' by reflecting the elements of A across the principal diagonal (in the process, the elements of the principal diagonal itself remain fixed). {Exercise: Cover up the matrix A' of Example 4.6.4, then reconstruct it from A by reflecting each nondiagonal element across the principal diagonal.} A similar remark applies to *matrices that are not necessarily square*; again, the elements a_{11}, $a_{22}, \ldots$ serve as the 'principal diagonal' (as long as they hold out; the principal diagonal of the matrices A and A' in Example 4.6.2 consists of the elements 2, 0).

That's what the transpose *is*; here's where it comes from:

4.6.6 Theorem *Let* V *and* W *be finite-dimensional vector spaces,* $T:\mathrm{V} \to \mathrm{W}$ *a linear mapping,* $T':\mathrm{W}' \to \mathrm{V}'$ *the transpose of* T (3.9.7). *Choose bases of* V *and* W, *and construct the bases of* V′ *and* W′ *dual to them* (3.9.4).

Relative to these bases, the matrix of T' *is the transpose of the matrix of* T.

Proof. A suggestive way of expressing the conclusion is that

$$\mathrm{mat}\,(T') = (\mathrm{mat}\,T)'$$

relative to the bases in question.

Let $x_1, \ldots, x_n$ be the chosen basis of V and $y_1, \ldots, y_m$ the chosen basis of W. Let $f_1, \ldots, f_n$ be the basis of V′ dual to $x_1, \ldots, x_n$, and let $g_1, \ldots, g_m$ be the basis of W′ dual to $y_1, \ldots, y_m$; thus

$$f_j(x_k) = \delta_{jk} \quad \text{for} \quad j, k = 1, \ldots, n,$$
$$g_i(y_h) = \delta_{ih} \quad \text{for} \quad i, h = 1, \ldots, m.$$

(The δ notation is explained in 3.9.4.) Suppose

(1) $$Tx_j = \sum_{i=1}^{m} a_{ij} y_i \qquad (j = 1, \ldots, n),$$

(2) $$T'g_i = \sum_{j=1}^{n} b_{ji} f_j \qquad (i = 1, \ldots, m);$$

the problem is to show that $b_{ji} = a_{ij}$ for all i and j.

For any $k = 1, \ldots, n$, evaluate the linear form of (2) at the vector x_k and compute:

$$(T'g_i)(x_k) = \sum_{j=1}^{n} b_{ji} f_j(x_k),$$

$$g_i(Tx_k) = \sum_{j=1}^{n} b_{ji} \delta_{jk},$$

$$g_i(Tx_k) = b_{ki}.$$

We can also calculate $g_i(Tx_k)$ using (1):

$$g_i(Tx_k) = g_i\left(\sum_{h=1}^{m} a_{hk} y_h\right) = \sum_{h=1}^{m} a_{hk} g_i(y_h)$$
$$= \sum_{h=1}^{m} a_{hk} \delta_{ih} = a_{ik}.$$

Thus, $b_{ki} = g_i(Tx_k) = a_{ik}$ for all $i = 1, \ldots, m$ and all $k = 1, \ldots, n$. In other words, if $A = (a_{ij})$ is the matrix of T relative to the chosen bases of V and W, then A' is the matrix of T' relative to the dual bases of W′ and V′. ■

4.6.7 Corollary *The row rank of a matrix is equal to its column rank.*

Proof. Suppose $A = (a_{ij})$ is an $m \times n$ matrix over the field F. With notations as in 4.2.6, let $T: V \to W$ be a linear mapping whose matrix relative to chosen bases is A; then A' is the matrix of T' relative to suitable bases of W′ and V′ (4.6.6). Citing 4.2.10, we have

$$\text{rank } T = \text{column rank of } A,$$
$$\text{rank } T' = \text{column rank of } A';$$

since $\text{rank } T = \text{rank } T'$ (Theorem 3.9.12), we conclude that

$$\text{column rank of } A = \text{column rank of } A',$$

in other words (4.6.3) the column rank of A is equal to the row rank of A. ■

We can therefore simplify the terminology:

4.6.8 Definition The number described in 4.6.7 is called the **rank** of the matrix.

The closeness of the interplay between matrices and linear mappings can be appreciated by reflecting on the following question: Did we use linear mappings to calculate the rank of a matrix (4.6.7), or matrices to calculate the rank of a linear mapping (4.2.10)?

▶ ***Exercises***

1. (i) With notations as in Definition 4.1.5, prove that the mapping $M_{m,n}(F) \to M_{n,m}(F)$ defined by $A \mapsto A'$ is linear and bijective, and that $(A')' = A$ for all $A \in M_{m,n}(F)$.

(ii) Prove that if A and B are matrices over F such that the product AB is defined, then so is $B'A'$, and $(AB)' = B'A'$. {Hint: First show that $(ST)' = T'S'$ for linear mappings S, T, then exploit Theorems 4.6.6 and 4.3.3.}

(iii) For every matrix A, the products $A'A$ and AA' are defined and are square matrices.

2. Let $A \in M_{2,3}(F)$, say

$$A = \begin{pmatrix} a & b & c \\ d & e & f \end{pmatrix}.$$

(i) Calculate $A'A$, then calculate its trace (§4.3, Exercise 7).

(ii) Prove that if $F = \mathbf{R}$ and $A'A = 0$ (the zero matrix) then $A = 0$. What if $AA' = 0$?

(iii) If $F = \mathbf{C}$ show that $A'A = 0$ is possible even if $A \neq 0$.

(iv) Generalize (ii) to real matrices of any size.

3. If $A = (a_{ij}) \in M_{m,n}(\mathbf{C})$ one defines

$$A^* = (b_{ji}) \in M_{n,m}(\mathbf{C}),$$

where $b_{ji} = \bar{a}_{ij}$ is the complex conjugate of a_{ij}. {Review: If $a = 4 + 5i$ then $\bar{a} = 4 - 5i$ and $\bar{a}a = 4^2 + 5^2 = |a|^2$.} The matrix A^* is called the *conjugate–transpose* of A.

(i) Prove that the mapping $M_{m,n}(C) \to M_{n,m}(C)$ defined by $A \mapsto A^*$ has the properties

$$(A + B)^* = A^* + B^*, \qquad (cA)^* = \bar{c}A^*, \qquad (A^*)^* = A.$$

(ii) If A and B are matrices over C such that the product AB is defined, then so is B^*A^* and $(AB)^* = B^*A^*$.

(iii) For every $A \in M_{m,n}(C)$, the products A^*A and AA^* are defined.

4. (i) Suppose $A \in M_{2,3}(C)$, say

$$A = \begin{pmatrix} a & b & c \\ d & e & f \end{pmatrix}.$$

Calculate A^*A, then calculate its trace.

(ii) Prove that if $A^*A = 0$ then $A = 0$.

(iii) If $A \in M_{m,n}(C)$, find a formula for $\operatorname{tr}(A^*A)$.

(iv) Generalize (ii) to complex matrices of any size.

4.7 Calculating the rank

The rank of a linear mapping between finite-dimensional vector spaces is equal to the rank of any of its matrix representations (4.2.10, 4.6.8). In this section we see how to calculate the rank of a matrix. The idea is to keep simplifying the matrix, taking care to leave the rank unchanged, until we arrive at a matrix so simple that its rank is *obvious*. As usual, F is the field of scalars, from which the entries of the matrices are drawn.

First, a matrix whose rank is obvious:

4.7.1 Lemma *If I is the $r \times r$ identity matrix, then the rank of I is equal to r.*

Proof. The column vectors of I are the canonical basis $e_1, \ldots, e_r$ of the r-dimensional coordinate space F^r (3.5.2). ■

Another:

4.7.2 Lemma *If I is the $r \times r$ identity matrix, then a matrix,*

$$C = \left(\begin{array}{c|c} I & O \\ \hline O & O \end{array}\right),$$

with I appearing in the 'northwest corner' and zeros elsewhere, has rank r.

Proof. Say C is an $m \times n$ matrix. The linear span of the column vectors of C is the same as the linear span of the first r columns, which are the first r vectors of the canonical basis of F^m. ■

For example, the 5×7 matrix

$$\begin{pmatrix} 1 & 0 & 0 & 0 & 0 & 0 & 0 \\ 0 & 1 & 0 & 0 & 0 & 0 & 0 \\ 0 & 0 & 1 & 0 & 0 & 0 & 0 \\ 0 & 0 & 0 & 0 & 0 & 0 & 0 \\ 0 & 0 & 0 & 0 & 0 & 0 & 0 \end{pmatrix} = \left(\begin{array}{ccc|c} 1 & & 0 & \\ & 1 & & 0 \\ 0 & & 1 & \\ \hline & 0 & & 0 \end{array}\right) = \left(\begin{array}{c|c} I & 0 \\ \hline 0 & 0 \end{array}\right)$$

has rank 3.

The strategy now is as follows: Given any $m \times n$ matrix A other than the zero matrix, we show that a finite number of rank-preserving operations on A will bring it into the form of the matrix in 4.7.2 (or a simplification thereof, in which rows of zeros or columns of zeros may be absent). These operations are of three basic types.

4.7.3 Lemma (Operations of type I) *If the matrix B is obtained from A by permuting the columns (or the rows) of A, then* $\operatorname{rank} B = \operatorname{rank} A$.

Proof. A permutation of the column vectors (for instance) does not change their linear span in F^m. {Thought: This might change the linear span of the rows—but not its dimension (4.6.7).} ■

For example, if

$$A = \begin{pmatrix} a & b & c \\ d & e & f \end{pmatrix}$$

then the matrix

$$B = \begin{pmatrix} b & c & a \\ e & f & d \end{pmatrix}$$

has the same rank as A.

4.7.4 Lemma (Operations of type II) *If B is obtained from A by adding to a column of A a scalar multiple of one of the other columns of A, then* $\operatorname{rank} B = \operatorname{rank} A$ (*and similarly for rows*).

For example, if

$$A = \begin{pmatrix} a & b & c \\ d & e & f \end{pmatrix}$$

and k is any scalar, then the matrix

$$B = \begin{pmatrix} a + kc & b & c \\ d + kf & e & f \end{pmatrix}$$

has the same rank as A (B is the result of adding the k'th multiple of the third column of A to the first column).

Proof of 4.7.4. The crux of the matter is that the linear span of vectors

$$x_1, x_2, x_3, \ldots, x_n$$

is the same as the linear span of the vectors

$$x_1 + kx_2, x_2, x_3, \ldots, x_n.$$

For, every vector in the second list is obviously a linear combination of vectors in the first. Conversely, since

$$x_1 = (x_1 + kx_2) - kx_2 = 1(x_1 + kx_2) + (-k)x_2,$$

every vector in the first list is a linear combination of vectors in the second. Thus, the two lists have the same linear span. ■

On performing a finite number of operations of type II, we see that if B is obtained from A by adding to a column of A a linear combination of other columns of A, then $\operatorname{rank} B = \operatorname{rank} A$ (and similarly for rows).

4.7.5 Lemma (Operations of type III) *If B is obtained from A by multiplying all of the elements of a column (or of a row) of A by the same nonzero scalar, then* $\operatorname{rank} B = \operatorname{rank} A$.

Proof. The linear span of the column vectors (for example) is unchanged if one of the vectors is replaced by a nonzero scalar multiple of itself. ■

For example, if k is a nonzero scalar and

$$A = \begin{pmatrix} a & b & c \\ d & e & f \end{pmatrix},$$

then the matrix

$$B = \begin{pmatrix} ka & kb & kc \\ d & e & f \end{pmatrix}$$

has the same rank as A.

4.7.6 Example Consider the real matrix

$$A = \begin{pmatrix} 2 & 0 & -5 \\ 1 & 3 & 4 \end{pmatrix}.$$

Interchange the rows, obtaining

$$\begin{pmatrix} 1 & 3 & 4 \\ 2 & 0 & -5 \end{pmatrix};$$

To the second row, add -2 times the first row, obtaining

$$\begin{pmatrix} 1 & 3 & 4 \\ 0 & -6 & -13 \end{pmatrix};$$

from the second column, subtract 3 times the first column, obtaining

$$\begin{pmatrix} 1 & 0 & 4 \\ 0 & -6 & -13 \end{pmatrix};$$

from the third column, subtract 4 times the first column, obtaining

$$\begin{pmatrix} 1 & 0 & 0 \\ 0 & -6 & -13 \end{pmatrix};$$

multiply the second column by $-1/6$, obtaining

$$\begin{pmatrix} 1 & 0 & 0 \\ 0 & 1 & -13 \end{pmatrix};$$

to the third column, add 13 times the second column, obtaining

$$\begin{pmatrix} 1 & 0 & 0 \\ 0 & 1 & 0 \end{pmatrix}.$$

Conclusion: the rank of A is 2. This was obvious from the outset, since the rows of A are not proportional; the merit of the mechanical procedure illustrated is that it works for *all* matrices:

4.7.7 Theorem *If A is not the zero matrix then, by a finite number of operations of types* I, II *and* III, *A can be brought into the form*

$$\left(\begin{array}{c|c} I & 0 \\ \hline 0 & 0 \end{array}\right)$$

where I is the $r \times r$ identity matrix for some r, and 0 abbreviates the zero matrices of the appropriate sizes (with the zero columns or the zero rows—or both—possibly absent).

The rank of A is r.

Proof. If A is an $m \times n$ matrix of rank r, the assertion of the theorem is that A can be brought (by operations of types I, II, III) to one of the forms

$$\left(\begin{array}{c|c} I & O \\ \hline O & O \end{array}\right), \quad \begin{pmatrix} I \\ O \end{pmatrix}, \quad (I \; O)$$

where I is the $r \times r$ identity matrix. The first case will occur when $r < m$ and $r < n$, the second when $r = n < m$, and the third when $r = m < n$.

To illustrate the general method, let's assume that $m \geq 2$ and $n \geq 2$ (the argument for the case that $m = 1$ or $n = 1$ is an easy simplification of what follows).

By assumption, A has a nonzero element. Permuting the rows, we can suppose there is a nonzero element in the first row; a permutation of the columns then brings this element to the $(1, 1)$ position. Dividing the first row (or the first column) by this element, we can suppose that the $(1, 1)$ entry is 1. We now have a matrix

$$\left(\begin{array}{c|ccc} 1 & b_{12} & \dots & b_{1n} \\ \hline b_{21} & & & \\ \vdots & & B_1 & \\ b_{m1} & & & \end{array}\right)$$

where B_1 abbreviates the rest of the matrix (it is a matrix with $m - 1$ rows and $n - 1$ columns).

Next, we 'sweep out' the rest of the first row: from the second column, subtract b_{12} times the first column, producing a 0 in the $(1, 2)$ position; from the third column, subtract b_{13} times the first column, producing a 0 in the $(1, 3)$ position, and so on. In this way we arrive at a matrix

$$\left(\begin{array}{c|ccc} 1 & 0 & \dots & 0 \\ \hline b_{21} & & & \\ \vdots & & B_2 & \\ b_{m1} & & & \end{array}\right)$$

where B_2 is an $(m-1) \times (n-1)$ matrix.

Similarly 'sweep out' the first column (using row operations), arriving at the matrix

$$\left(\begin{array}{c|ccc} 1 & 0 & \dots & 0 \\ \hline 0 & & & \\ \vdots & & B_2 & \\ 0 & & & \end{array}\right).$$

(Yes, B_2 again; the subtraction of various multiples of the first row leaves the entries of B_2 unchanged.)

If $B_2 = 0$ we are through (and the rank of A is 1); otherwise, we apply the same procedure to B_2, and so on.

The 'and so on' is easily replaced by a formal induction, say on m:

If $m = 1$ then, after a finite number of obvious operations, we reach a row matrix

$$(1\ 0\ \dots\ 0);$$

this is of the form $(I\ 0)$, where I is the 1×1 identity matrix and 0 stands for the $1 \times (n-1)$ zero matrix. All's well with $m = 1$. The case that $n = 1$ is treated similarly, so we can suppose that $n \geq 2$.

Assume inductively that $m \geq 2$ and that all is well with $m - 1$. By the argument at the beginning of the proof, a finite number of the allowable operations produces a matrix

$$B = \left(\begin{array}{c|c} 1 & 0 \\ \hline 0 & B_2 \end{array}\right)$$

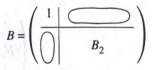

where B_2 is an $(m-1) \times (n-1)$ matrix. The rank of B is r (the rank of A) and its first column is not in the linear span of the other columns; it follows that B_2 has rank $r - 1$. If $B_2 = 0$ we are done. Otherwise, by the inductive hypothesis, a finite number of the allowable operations brings B_2 to one of the three desired forms

$$\left(\begin{array}{c|c} J & O \\ \hline O & O \end{array}\right), \quad \left(\begin{array}{c} J \\ \hline O \end{array}\right), \quad (J \quad O),$$

where J is the identity matrix of order $r-1$; if the row and column operations that bring this about are performed on the corresponding rows and columns of B, then B is transformed into one of the matrices

$$\left(\begin{array}{c|cc} 1 & O & \\ \hline O & J & O \\ & O & O \end{array}\right), \quad \left(\begin{array}{c|c} 1 & O \\ \hline O & J \\ & O \end{array}\right), \quad \left(\begin{array}{c|cc} 1 & O & \\ \hline O & J & O \end{array}\right)$$

and the proof is concluded by the observation that the $r \times r$ northwest corner is the identity matrix I of order r. ■

All of this is easier done than said (try the exercises!).

▶ Exercises

1. Find the rank of the matrix

$$\begin{pmatrix} 2 & 1 & -5 & -3 \\ 1 & 2 & 4 & 8 \\ 3 & 3 & -1 & 5 \end{pmatrix}.$$

2. Find the rank of the matrix

$$\begin{pmatrix} 2 & 4 & -3 \\ 0 & 5 & 6 \\ 0 & 0 & 4 \end{pmatrix}.$$

3. Prove that if a, b, c are distinct scalars, then the matrix

$$\begin{pmatrix} 1 & a & a^2 & d \\ 1 & b & b^2 & e \\ 1 & c & c^2 & f \end{pmatrix}$$

has rank 3.

4. Why does the word 'other' appear in the statement of Lemma 4.7.4? {Hint: There are fields in which $2 = 0$ (cf. Appendix A.4.2). Imagine the effect of adding a column to itself.}

5. Call *operation of type* II′ the following operation on a matrix: add one column (or row) to another. An operation of type II involving a nonzero scalar k can be obtained by performing three successive operations, of types III, II′ and III, where the scalars in the first and third operations are k and k^{-1}, respectively.

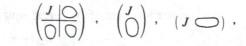

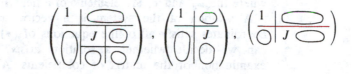

4.8 When is a linear system solvable?

Consider a system of m linear equations in n unknowns $x_1, \ldots, x_n$,

$$(*) \qquad \begin{aligned} a_{11}x_1 + a_{12}x_2 + \ldots + a_{1n}x_n &= c_1 \\ a_{21}x_1 + a_{22}x_2 + \ldots + a_{2n}x_n &= c_2 \\ &\ldots \\ a_{m1}x_1 + a_{m2}x_2 + \ldots + a_{mn}x_n &= c_m \end{aligned}$$

where the a_{ij} and c_i are elements of a field F.

A **solution** of the system is a vector $x = (x_1, \ldots, x_n)$ in F^n whose coordinates satisfy all of the equations of $(*)$ (Examples 2.1.9, 2.2.4). How can we decide whether a solution *exists*? The answer depends on an examination of the matrix of coefficients $A = (a_{ij})$ and of the **augmented matrix**

$$B = \begin{pmatrix} a_{11} & a_{12} & \ldots & a_{1n} & c_1 \\ a_{21} & a_{22} & \ldots & a_{2n} & c_2 \\ \ldots & & & & \\ a_{m1} & a_{m2} & \ldots & a_{mn} & c_m \end{pmatrix}$$

obtained by adjoining to A the 'column of constants' (the column of right members of $(*)$). Writing $A_1, \ldots, A_n$ for the column vectors of A, and C for the column of constants, the system $(*)$ can be written in the form

$$x_1A_1 + \ldots + x_nA_n = C;$$

thus, the problem is to decide whether C can be expressed as a linear combination of the columns of A. There is an easy special case that puts us on the right track:

4.8.1 Theorem *If the $m \times n$ matrix of coefficients A has rank m, then the system $(*)$ has at least one solution.*

Proof. With the preceding notations, the assumption is that the linear span of the column vectors $A_1, \ldots, A_n$ is m-dimensional, so it must be all of F^m (3.5.14); in particular, C is a linear combination of $A_1, \ldots, A_n$. ■

Wasn't that easy? The general case, equally easy, depends on the following lemma:

4.8.2 Lemma *Let $u_1, \ldots, u_n, v$ be elements of a vector space, let*

$$M = [\{u_1, \ldots, u_n\}]$$

be the linear span of $u_1, \ldots, u_n$ and let

$$N = [\{u_1, \ldots, u_n, v\}]$$

be the linear span of $u_1, \ldots, u_n, v$. The following conditions are equivalent:

(a) *v is a linear combination of $u_1, \ldots, u_n$;*

(b) $\dim M = \dim N$.

Proof. In any case $M \subset N$, so (a) is equivalent to the condition $M = N$, which is in turn equivalent to (b) by 3.5.14. ■

4.8.3 Theorem *The system (*) is solvable if and only if the rank of the coefficient matrix is equal to the rank of the augmented matrix.*

Proof. In the notations introduced earlier, the claim is that (*) has a solution if and only if the linear span of the columns of the coefficient matrix A has the same dimension as the linear span of the columns of the augmented matrix B. The latter condition means, by the lemma, that C is a linear combination of $A_1, \ldots, A_n$, in other words that (*) has a solution. ■

There remains the problem of computing the ranks of A and B, but this is straightforward (§4.7). There's a shortcut: in effect, it's enough to compute the rank of B 'the right way'. The idea is as follows.

Let r be the rank of A. This means (4.7.7) that if certain row and column operations are performed on A, one reaches the $m \times n$ matrix

$$D = \left(\begin{array}{c|c} I & O \\ \hline O & O \end{array}\right)$$

where I is the $r \times r$ identity matrix (and the zero rows, or zero columns, or both, may be absent). If the *same* operations are performed on the augmented matrix B, one obtains a matrix

$$E = \left(\begin{array}{c|c|c} I & O & \begin{matrix} d_1 \\ \vdots \\ d_r \end{matrix} \\ \hline O & O & \begin{matrix} d_{r+1} \\ \vdots \\ d_m \end{matrix} \end{array}\right) = \left(\begin{array}{c|c} D & \begin{matrix} d_1 \\ \vdots \\ d_r \\ d_{r+1} \\ \vdots \\ d_m \end{matrix} \end{array}\right)$$

consisting of D augmented by the column that results from subjecting the original column of constants to the *row* operations used in arriving at D. The obvious column operations on E yield the $m \times (n+1)$ matrix

$$G = \left(\begin{array}{c|c|c} I & O & \begin{matrix} 0 \\ \vdots \\ 0 \end{matrix} \\ \hline O & O & \begin{matrix} d_{r+1} \\ \vdots \\ d_m \end{matrix} \end{array}\right)$$

whose rank is either r (if $d_{r+1} = \ldots = d_m = 0$) or $r + 1$ (if one of these d_i is nonzero).

Thus, the rank of B (i.e. the rank of G) is equal to the rank of A (namely r) if and only if $d_{r+1} = \ldots = d_m = 0$. This says, in view of 4.8.3, that *the system* $(*)$ *is solvable if and only if* $d_{r+1} = \ldots = d_m = 0$ (a condition that is vacuously satisfied when $r = m$, as in 4.8.1). To sum up:

4.8.4 Scholium Transform the augmented matrix B into the form E by performing the row and column operations needed to put A into the form D, and observe what has happened to the last column: if its last $m - r$ entries are 0 then the system $(*)$ is solvable, otherwise not. {Caution: In the course of transforming B into E, the elements of the last column will be affected by the row operations performed, but the last column is not to be used in column operations.}

There remains the task of actually computing the solutions when they exist. The crux of the matter is that one should stick to row operations and let the chips fall where they may (column operations would amount to unauthorized tampering with the variables, that is, with the solutions). Again, this is easy to do, tedious to explain (unbearable to read an explanation of); we compromise by working out an example in the next section.

▶ Exercises

1. If an $m \times n$ matrix A has rank m, necessarily $m \leq n$. Why? What if A has rank n?

2. Suppose, as in Theorem 4.8.1, that the $m \times n$ matrix A has rank m (hence $m \leq n$ and the system $(*)$ is solvable).

   Prove: The system $(*)$ has a unique solution if and only if $m = n$. {Hint: §3.3, Exercise 3; see also Theorem 3.7.1.}

3. If the coefficient matrix A of the system $(*)$ has rank n, does it follow that there exists a solution?

4.9 An example

For illustration, let's apply the method of the preceding section to the system of equations

$$\begin{aligned} 3x_1 + 6x_2 + 13x_3 + 25x_4 &= 15 \\ 2x_1 + 4x_2 + 9x_3 + 15x_4 &= 10 \\ x_1 + 2x_2 + 3x_3 + 13x_4 &= 9 \\ 4x_1 + 8x_2 + 17x_3 + 35x_4 &= 20. \end{aligned}$$

The augmented matrix is

$$\begin{pmatrix} 3 & 6 & 13 & 25 & 15 \\ 2 & 4 & 9 & 15 & 10 \\ 1 & 2 & 3 & 13 & 9 \\ 4 & 8 & 17 & 35 & 20 \end{pmatrix}.$$

Moving the third row to the top places a 1 in the (1, 1) position:

$$\begin{pmatrix} 1 & 2 & 3 & 13 & 9 \\ 3 & 6 & 13 & 25 & 15 \\ 2 & 4 & 9 & 15 & 10 \\ 4 & 8 & 17 & 35 & 20 \end{pmatrix}.$$

Now use this 1 to sweep out the rest of the first column (for example, subtract 3 times the first row from the second row):

$$\begin{pmatrix} 1 & 2 & 3 & 13 & 9 \\ 0 & 0 & 4 & -14 & -12 \\ 0 & 0 & 3 & -11 & -8 \\ 0 & 0 & 5 & -17 & -16 \end{pmatrix}.$$

The first two columns are in final form; nothing more can, or should, be done to them by means of row operations.

The next target is the 4 in the (2, 3) position; let's make it a 1. This could be done by dividing the second row by 4 (thereby introducing fractions); it is simpler to subtract the third row from the second:

$$\begin{pmatrix} 1 & 2 & 3 & 13 & 9 \\ 0 & 0 & 1 & -3 & -4 \\ 0 & 0 & 3 & -11 & -8 \\ 0 & 0 & 5 & -17 & -16 \end{pmatrix}.$$

Now use this 1 to sweep out the rest of the third column (for example, subtract 3 times the second row from the first):

$$\begin{pmatrix} 1 & 2 & 0 & 22 & 21 \\ 0 & 0 & 1 & -3 & -4 \\ 0 & 0 & 0 & -2 & 4 \\ 0 & 0 & 0 & -2 & 4 \end{pmatrix}.$$

The first three columns are now in final form. The last two rows happen to be equal (the luck of the draw); subtract the third row from the fourth:

$$\begin{pmatrix} 1 & 2 & 0 & 22 & 21 \\ 0 & 0 & 1 & -3 & -4 \\ 0 & 0 & 0 & -2 & 4 \\ 0 & 0 & 0 & 0 & 0 \end{pmatrix}.$$

The next target is the -2 in the (3, 4) position; let's make it a 1 (by dividing through the third row by -2):

$$\begin{pmatrix} 1 & 2 & 0 & 22 & 21 \\ 0 & 0 & 1 & -3 & -4 \\ 0 & 0 & 0 & 1 & -2 \\ 0 & 0 & 0 & 0 & 0 \end{pmatrix}.$$

Now use this 1 to sweep out the rest of the fourth column (for example, subtract 22 times the third row from the first):

$$\begin{pmatrix} 1 & 2 & 0 & 0 & 65 \\ 0 & 0 & 1 & 0 & -10 \\ 0 & 0 & 0 & 1 & -2 \\ 0 & 0 & 0 & 0 & 0 \end{pmatrix}.$$

This is the end of the line; it is called the 'row-echelon form' of the augmented matrix. The good news is that the element in the $(4, 5)$ position is 0, which assures the rank of the augmented matrix is equal to the rank of the coefficient matrix (namely 3); thus, *a solution exists*.

If these (row) operations had been performed on the linear equations of the system instead of on the augmented matrix, we would have arrived at the 'equivalent' (same solutions) system

$$\begin{aligned} x_1 + 2x_2 \qquad\quad &= 65 \\ x_3 \quad &= -10 \\ x_4 &= -2 \end{aligned}$$

We can choose x_2 to be anything we like; then $x_1 = 65 - 2x_2$ and we have a solution

$$x = (65 - 2x_2, x_2, -10, -2).$$

For example, setting $x_2 = 1$ we have a solution $(63, 1, -10, -2)$ of the original system. (See if it checks!)

That's how the method works. It is not easy to state *formally* what the method does, let alone prove that it does it.[1]

► ***Exercises***

1. Find the solutions (if any) of the system

$$\begin{aligned} 5x_1 + 2x_2 + x_3 + x_4 &= 14 \\ x_1 + 2x_2 + x_3 \qquad &= 3 \\ 3x_1 + 6x_2 + 3x_3 + x_4 &= 12 \\ 4x_1 + 2x_2 + x_3 + x_4 &= 12. \end{aligned}$$

2. Find the solutions (if any) of the system

$$\begin{aligned} 2x_1 + 4x_2 &= 9 \\ x_1 + x_2 &= 3 \\ 3x_1 + 4x_2 &= 11. \end{aligned}$$

[1] For a lucid description of what row-echelon form is, see D. Zelinsky, *A first course in linear algebra*, p. 107, (2nd edn), Academic Press, New York, 1973. A formal proof that row-echelon form can always be attained is given in G. D. Mostow and J. H. Sampson, *Linear algebra*, p. 128, McGraw-Hill, New York, 1969; it is not pretty.

3. Let

$$A = \begin{pmatrix} 3 & 6 & 13 & 25 \\ 2 & 4 & 9 & 15 \\ 1 & 2 & 3 & 13 \\ 4 & 8 & 17 & 35 \end{pmatrix}$$

be the matrix of coefficients of the linear system solved in this section and let $T:\mathbf{R}^4 \to \mathbf{R}^4$ be the linear mapping whose matrix relative to the canonical basis of $\mathbf{R}^4$ is A.

(i) The kernel of T is the set of solutions of the homogeneous system

$$\begin{aligned} x_1 + 2x_2 \qquad\qquad &= 0 \\ x_3 \qquad &= 0 \\ x_4 &= 0; \end{aligned}$$

this is the set of all scalar multiples of the vector $(2, -1, 0, 0)$. In particular, T has nullity 1 and rank 3.

(ii) Using only column operations, the matrix A can be brought to the form

$$C = \begin{pmatrix} 1 & 0 & 0 & 0 \\ 0 & 1 & 0 & 0 \\ 0 & 0 & 1 & 0 \\ 2 & -1 & 0 & 0 \end{pmatrix}.$$

Writing C_1, C_2, C_3 for the first three columns of C, we see that the column space of A is the linear span of C_1, C_2, C_3 (this is also the range of T by Lemma 4.2.8). For example,

$$15C_1 + 10C_2 + 9C_3 = \begin{pmatrix} 15 \\ 10 \\ 9 \\ 20 \end{pmatrix}.$$

4.10 Change of basis, similar matrices

Let V be a finite-dimensional vector space. A linear mapping $T:\mathrm{V} \to \mathrm{V}$ has a matrix relative to each basis of V (4.2.2). If we change the basis, how does the matrix of T change? The answer involves matrices that have a 'reciprocal' (the official word is 'inverse') in the sense suggested by the following theorem:

4.10.1 Theorem *Let $x_1, \ldots, x_n$ and $y_1, \ldots, y_n$ be two bases of an n-dimensional vector space. Express the vectors of each basis in terms of the other basis, say*

$$y_j = \sum_{i=1}^{n} a_{ij}x_i, \quad x_j = \sum_{i=1}^{n} b_{ij}y_i \quad (j = 1, \ldots, n).$$

Let $A = (a_{ij})$ *and* $B = (b_{ij})$ *be the matrices of coefficients. Then* $AB = BA = I$ *(the* $n \times n$ *identity matrix).*

Proof. If V is the vector space in question, write $J = I_V$ for the identity mapping of V (to distinguish it from the identity matrix I). The equations

$$Jy_j = y_j = \sum_{i=1}^{n} a_{ij}x_i \quad (j = 1, \ldots, n)$$

show that A is the matrix of J relative to the pair of bases $y_1, \ldots, y_n$ and $x_1, \ldots, x_n$ (4.2.1). Similarly, B is the matrix of J relative to the bases $x_1, \ldots, x_n$ and $y_1, \ldots, y_n$. All of this is compactly expressed by the diagram

$$\underset{(x_i)}{V} \xrightarrow[B]{J} \underset{(y_i)}{V} \xrightarrow[A]{J} \underset{(x_i)}{V}$$

where, below an arrow, we write the matrix, relative to the indicated bases, of the linear mapping above the arrow. By Theorem 4.3.3, AB is the matrix of $JJ = J$ relative to the basis $x_1, \ldots, x_n$, thus $AB = I$ (4.2.4). Similarly, $BA = I$. ■

Think of $x_1, \ldots, x_n$ as the 'old' basis and $y_1, \ldots, y_n$ as the 'new' basis; A is called the *change-of-basis matrix* that expresses the new basis in terms of the old.

4.10.2 Definition An $n \times n$ matrix A over a field F is said to be **invertible** if there exists an $n \times n$ matrix B over F such that $AB = BA = I$.

If such a matrix B exists then it is unique. {If also $AC = I$ then $C = IC = (BA)C = B(AC) = BI = B$.} One calls B the **inverse** of A, written $B = A^{-1}$; thus

$$AA^{-1} = A^{-1}A = I$$

for an invertible matrix A.

The relation $BA = I$ in 4.10.1 is in fact a *consequence* of $AB = I$:

4.10.3 Theorem *Let* A *be an* $n \times n$ *matrix over a field* F, *that is,* $A \in M_n(F)$ (4.1.5). *The following conditions are equivalent:*

(a) A *is invertible;*

(b) $AB = I$ *for some* $B \in M_n(F)$;

(c) $CA = I$ *for some* $C \in M_n(F)$.

If these conditions are satisfied, necessarily $B = C = A^{-1}$.

Proof.

(a) ⇒ (b): This is obvious from the definition of invertibility.

(b) ⇒ (a): Assuming $AB = I$, let's show that $BA = I$. Let V be an n-dimensional vector space over F, choose a basis of V, and let

$S, T \in \mathcal{L}(V)$ be the linear mappings such that

$$A = \text{mat}\, S, \quad B = \text{mat}\, T$$

relative to the chosen basis (4.2.6). By 4.3.3,

$$\text{mat}\,(ST) = (\text{mat}\, S)(\text{mat}\, T) = AB = I = \text{mat}\, I_V,$$

therefore $ST = I_V$. It follows from the finite-dimensionality of V that $TS = I_V$ (3.7.5), therefore $BA = I$ (4.3.3), thus A is invertible and $B = A^{-1}$.

This proves that (a) $\Leftrightarrow$ (b); the proof that (a) $\Leftrightarrow$ (c) is similar. Finally, if $AB = I = CA$, the foregoing proof shows that A is invertible and that $B = A^{-1} = C$. {Alternatively, $B = IB = (CA)B = C(AB) = CI = C$.} ■

Invertibility is to matrices what bijectivity is to linear mappings:

4.10.4 Theorem *If A is the matrix of a linear mapping T:V $\to$ V relative to a basis of* V, *then A is invertible if and only if T is bijective.*

Proof. 'If': Suppose T is bijective and $S = T^{-1}$. From $ST = TS = I_V$ we infer (4.3.3) that, relative to the given basis,

$$(\text{mat}\, S)(\text{mat}\, T) = (\text{mat}\, T)(\text{mat}\, S) = \text{mat}\, I_V,$$

that is,

$$(\text{mat}\, S)A = A(\text{mat}\, S) = I;$$

thus A is invertible, with $A^{-1} = \text{mat}\, S$.

'Only if': Suppose A is invertible and $B = A^{-1}$. Let $S \in \mathcal{L}(V)$ be the linear mapping such that $\text{mat}\, S = B$ relative to the given basis of V. The matrix relations $AB = BA = I$ translate (4.3.4) into the relations $TS = ST = I_V$ in $\mathcal{L}(V)$, consequently T is bijective. ■

We now answer the question posed at the beginning of the section:

4.10.5 Theorem *Let* V *be an n-dimensional vector space, T:V $\to$ V a linear mapping.*

Let $x_1, \ldots, x_n$ and $y_1, \ldots, y_n$ be two bases of V; *let A be the matrix of T relative to the basis $x_1, \ldots, x_n$ and let B be the matrix of T relative to the basis $y_1, \ldots, y_n$.*

Finally, let C be the 'change-of-basis matrix' that expresses the y's in terms of the x's (4.10.1).

Then $B = C^{-1}AC$.

Proof. If $C = (c_{ij})$, the assumption is that

$$y_j = \sum_{i=1}^{n} c_{ij} x_i \qquad (j = 1, \ldots, n).$$

As noted in the proof of 4.10.1, C is the matrix of the identity mapping $J = I_V$ relative to the pair of bases $y_1, \ldots, y_n$ and $x_1, \ldots, x_n$. Consider the diagram

$$\underset{(y_i)}{V} \xrightarrow[C]{J} \underset{(x_i)}{V} \xrightarrow[A]{T} \underset{(x_i)}{V},$$

where C and A are the matrices of J and T relative to the indicated bases. By 4.3.3, AC is the matrix of $TJ = T$ relative to the bases $y_1, \ldots, y_n$ and $x_1, \ldots, x_n$.

On the other hand, the diagram

$$\underset{(y_i)}{V} \xrightarrow[B]{T} \underset{(y_i)}{V} \xrightarrow[C]{J} \underset{(x_i)}{V}$$

shows that CB is the matrix of $JT = T$ relative to the bases $y_1, \ldots, y_n$ and $x_1, \ldots, x_n$.

Conclusion: $CB = AC$. Since C is invertible (4.10.1, 4.10.2), left-multiplication by C^{-1} yields $B = C^{-1}AC$. ■

The formula in 4.10.5 prompts a definition:

4.10.6 Definition Matrices A and B are said to be **similar** if there exists an invertible matrix C such that $B = C^{-1}AC$. (It is tacit from Definition 4.10.2 that we are talking about square matrices of the same order with entries in a field.)

In applications, Theorem 4.10.5 is often used in the following way (expressed informally): If A is the matrix of a linear mapping $T \in \mathcal{L}(V)$ relative to some basis of V, the problem of finding a basis of V relative to which the matrix of T is 'nicer' than A amounts to finding an invertible matrix C such that $C^{-1}AC$ is 'nicer' than A. A prerequisite for success is that T itself be 'nice' in an appropriate sense; the work of the next two chapters culminates in the analysis of just such a class of linear mappings, with a pretty application to analytic geometry (and deeper results to come in Chapters 9 and 12).

► ***Exercises***

1. Let A be an $n \times n$ matrix over a field F. Prove that A is invertible if and only if its column vectors are linearly independent. {Hint: Theorems 4.2.10, 4.10.4.} What about rows?

2. Let A be an $n \times n$ matrix over a field F. Prove that A is invertible if and only if A is the matrix of the identity mapping on F^n relative to a suitable pair of bases. {Hint: Theorem 4.10.1.}

3. In the set $M_n(F)$ of $n \times n$ matrices over F, write $A \sim B$ if A is similar to B. Prove that $\sim$ is an equivalence relation in $M_n(F)$ (2.5.3).

4. If $A \in M_n(F)$ is such that $C^{-1}AC = A$ for every invertible matrix C, then A is a scalar matrix (4.2.13). {Hint: §4.2, Exercise 6.}

5. If A and B are similar matrices, then $\operatorname{tr} A = \operatorname{tr} B$. {Hint: §4.3, Exercise 4.}

6. Prove that similar matrices have the same rank.

7. Let A and B be $n \times n$ matrices over F, V an n-dimensional vector space over F. The following conditions are equivalent:

   (a) A and B are similar;

   (b) there exist a linear mapping $T \in \mathcal{L}(V)$ and bases $x_1, \ldots, x_n$ and

$y_1, \ldots, y_n$ of V, such that A is the matrix of T relative to $x_1, \ldots, x_n$ and B is the matrix of T relative to $y_1, \ldots, y_n$.

{Hint: Theorem 4.10.5 is half the battle; a rearrangement of its proof yields the other half.}

8. If $T{:}\mathrm{V} \to \mathrm{V}$ is a bijective linear mapping and A is the matrix of T relative to a pair of bases $x_1, \ldots, x_n$ and $y_1, \ldots, y_n$ of V, prove that A^{-1} is the matrix of T^{-1} relative to the bases $y_1, \ldots, y_n$ and $x_1, \ldots, x_n$. Concisely,

$$\operatorname{mat}(T^{-1}) = (\operatorname{mat} T)^{-1}$$

relative to the indicated pairs of bases. State and prove a similar result for a bijective linear mapping $T{:}\mathrm{V} \to \mathrm{W}$ between vector spaces of the same dimension.

9. Let $A \in \mathrm{M}_n(\mathrm{F})$ be invertible and suppose B is a matrix obtained from A by a permutation of its rows. True or false (explain): There exists a matrix C such that $BC = I$.

10. If $A, B \in \mathrm{M}_n(\mathrm{F})$ are invertible, prove that AB is invertible and find a formula for $(AB)^{-1}$.

11. If $A \in \mathrm{M}_n(\mathrm{F})$ is invertible, prove that the transpose A' of A is invertible and find a formula for $(A')^{-1}$.

12. Suppose $T{:}\mathrm{V} \to \mathrm{W}$, where $\dim \mathrm{V} = \dim \mathrm{W} = n$. Let $x_1, \ldots, x_n$ be a basis of V, $y_1, \ldots, y_n$ a basis of W, and $A \in \mathrm{M}_n(\mathrm{F})$ the matrix of T relative to the bases $x_1, \ldots, x_n$ and $y_1, \ldots, y_n$. Prove: T is bijective if and only if A is invertible.

{Hint: The case that $\mathrm{V} = \mathrm{W}$ and $y_j = x_j$ for all j is covered by Theorem 4.10.4. In the general case, let $S{:}\mathrm{W} \to \mathrm{V}$ be the linear mapping such that $Sy_j = x_j$ for all j (what is its matrix?) and apply the preceding case to the linear mapping $ST{:}\mathrm{V} \to \mathrm{V}$.}

13. If $A, B \in \mathrm{M}_n(\mathrm{F})$, the following conditions are equivalent: (a) AB is invertible; (b) A and B are invertible; (b) BA is invertible.

{Hint: §3.7, Exercise 9.}

5

Inner product spaces

In this and the next chapter, we consider only real vector spaces in the text proper. (The modifications needed to treat the complex case are sketched in the exercises.)

Our discussion of vector spaces began with the intuitive example of vectors in 3-space (§1.1). The example has two aspects: *geometric* (arrows, length, parallelograms) and *algebraic* (numerical components, coordinatewise linear operations). The first four chapters developed the algebraic aspect; what mainly happened is that (1) the basic vectorial ideas were extended to spaces of arbitrary finite dimension, and (2) we made friends with linear mappings.

We now carry over the geometric ideas in the 3-dimensional example to the algebraic terrain prepared in the first four chapters. The present chapter treats geometric matters in dimension n, and the next chapter gets down to brass tacks in dimensions 2 and 3. (The themes of these two chapters are pursued in greater depth in Chapters 9 and 12.)

5.1 Inner product spaces, Euclidean spaces

Inner product spaces are vector spaces with extra structure of a geometric nature (enabling us to define length, perpendicularity, etc.). Duality (linear forms) plays a key role in the introduction of this extra structure:

5.1.1 Definition

A (real) **inner product space** is a real vector space E such that, for every pair of vectors x, y in E, there is determined a real number $(x|y)$, called the **inner product** of x and y, subject to the following axioms:

(1) $(x|x) > 0$ for all $x \neq \theta$;

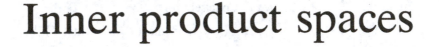

(2) $(x|y) = (y|x)$ for all vectors x, y;
(3) $(x + y|z) = (x|z) + (y|z)$ for all vectors x, y, z;
(4) $(cx|y) = c(x|y)$ for all vectors x, y and all real numbers c.

A finite-dimensional inner product space is called a **Euclidean space**.

Convention: $\{\theta\}$ is admitted as a 0-dimensional Euclidean space, with $(\theta|\theta) = 0$. {The implication «$x \neq \theta \Rightarrow (x|x) > 0$» is vacuously true.}

Property (1) is expressed by saying that the inner product is *positive definite*; property (2) is called *symmetry*. Properties (3) and (4) say that for each fixed vector y, the mapping $x \mapsto (x|y)$ is a linear form on E (2.1.11); it is in this sense that 'duality' is built into the definition of an inner product space.

What we have here is a function $\mathrm{E} \times \mathrm{E} \to \mathbf{R}$, given by $(x, y) \mapsto (x|y)$; this function is also called the 'inner product' (on the space E). If M is a linear subspace of E, then the restriction of the inner product of E to $\mathrm{M} \times \mathrm{M}$ is an inner product on M; thus, every linear subspace of an inner product space (Euclidean space) is an inner product space (Euclidean space).

5.1.2 Example An inner product on $\mathbf{R}^n$ (called the **canonical inner product**) is defined by the formula

$$(x|y) = \sum_{i=1}^{n} a_i b_i$$

for all $x = (a_1, \ldots, a_n)$ and $y = (b_1, \ldots, b_n)$. {Properties (1)–(4) are easily checked. {For example, if $x \neq \theta$ then $(x|x) = \sum_{k=1}^{n}(a_i)^2 > 0$ because not every a_i is 0.} Thus $\mathbf{R}^n$ is a Euclidean space for this inner product. We also write E^n for $\mathbf{R}^n$ equipped with the canonical inner product.

5.1.3 Example Let $[a, b]$ be a closed interval in $\mathbf{R}$, with $a < b$, and let E be the vector space of all continuous functions $x:[a, b] \to \mathbf{R}$, with the pointwise linear operations (cf. 1.6.5) and with inner products defined by the formula

$$(x|y) = \int_a^b x(t)y(t)dt.$$

Property (2) of 5.1.1 is obvious. Properties (3) and (4) follow from the linearity of integration (familiar, but not obvious). Since

$$(x|x) = \int_a^b (x(t))^2 dt,$$

property (1) essentially restates the fact that a nonnegative continuous function with zero integral must be identically zero.[1] In contrast with E^n (5.1.2), this space is not finite-dimensional (cf. 3.4.4), hence is not a Euclidean space.

[1] Even if properties (3) and (4) are trivialized by taking E to be the vector space of polynomial functions on $[a, b]$ (cf. Example 1.6.7) and defining integration in terms of prederivatives, property (1) remains a serious piece of busines.

Repeating a pattern familiar from §1.4, let's explore the consequences of the axioms (1)–(4).

5.1.4 Theorem *In an inner product space,*

(5) $(x|y + z) = (x|y) + (x|z)$,

(6) $(x|cy) = c(x|y)$,

(7) $(x|\theta) = (\theta|y) = 0$,

for all vectors x, y, z and all real numbers c.

Proof. Property (5) follows from (2) and (3):

$$(x|y + z) = (y + z|x) = (y|x) + (z|x) = (x|y) + (x|z).$$

Similarly, (6) follows from (2) and (4), and (7) follows from the fact that every linear form vanishes at the zero vector (cf. 2.1.2). ■

Enter geometry:

5.1.5 Definition If E is an inner product space and $x \in \mathrm{E}$, the **norm** (or 'length') of x, written $\|x\|$, is defined by the formula

$$\|x\| = (x|x)^{\frac{1}{2}}.$$

A vector of length 1 is called a **unit vector**. The **distance** between vectors x, y is defined to be $\|x - y\|$.

The function $\mathrm{E} \to \mathbf{R}$ given by $x \mapsto \|x\|$ is also called the 'norm' (on the space E).

5.1.6 Examples In the Euclidean space E^{n} (Example 5.1.2),

$$\|x\| = (a_1^2 + \ldots + a_n^2)^{\frac{1}{2}}$$

and

$$\|x - y\| = [(a_1 - b_1)^2 + \ldots + (a_n - b_n)^2]^{\frac{1}{2}};$$

for $n = 1, 2, 3$, these are the usual Euclidean length and distance. In the inner product space of Example 5.1.3,

$$\|x\| = \left(\int_a^b x(t)^2 dt\right)^{\frac{1}{2}}.$$

Some general properties of the norm:

5.1.7 Theorem *In an inner product space,*

(8) $\|\theta\| = 0$ *and* $\|x\| > 0$ *when* $x \neq \theta$;

(9) $\|cx\| = |c| \cdot \|x\|$ *for all vectors x and real numbers c*;

(10) $(x|y) = \frac{1}{4}\{\|x + y\|^2 - \|x - y\|^2\}$ *for all vectors x and y* (Polarization identity);

(11) $\|x + y\|^2 + \|x - y\|^2 = 2\|x\|^2 + 2\|y\|^2$ *for all vectors x and y* (Parallelogram law).

Proof.
(8) This is obvious from the definitions.

(9) $\|cx\|^2 = (cx|cx) = c^2(x|x) = |c|^2\|x\|^2$.

(10), (11) From earlier properties,

$$\begin{aligned}\|x + y\|^2 &= (x + y|x + y) = (x|x + y) + (y|x + y)\\ &= (x|x) + (x|y) + (y|x) + (y|y)\\ &= \|x\|^2 + \|y\|^2 + 2(x|y);\end{aligned}$$

replacing y by $-y = (-1)y$,

$$\|x - y\|^2 = \|x\|^2 + \|y\|^2 - 2(x|y).$$

Subtraction of these expressions yields the polarization identity

$$\|x + y\|^2 - \|x - y\|^2 = 4(x|y),$$

whereas addition yields the parallelogram law. ■

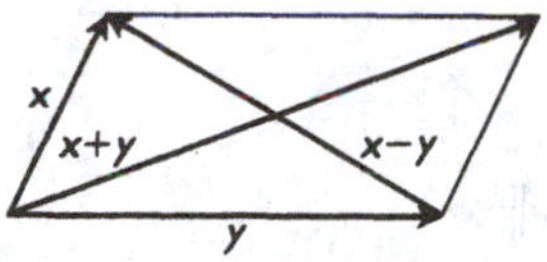

Fig. 18

A fast perspective on (10): norm is *defined* in terms of a special inner product, while the polarization identity expresses general inner products in terms of the norm.[2] Formula (11) comes to life in the Euclidean plane E^2: the sum of the squares of the two diagonals of a parallelogram is equal to the sum of the squares of the four sides (Fig. 18).

The polarization identity and the parallelogram law acting in concert yield an interesting and useful inequality:

5.1.8 Theorem. (Cauchy–Schwarz inequality) *In an inner product space,*

(12) $$|(x|y)| \le \|x\|\,\|y\|$$

for all vectors x and y; the equality $|(x|y)| = \|x\|\,\|y\|$ holds if and only if x and y are linearly dependent.

Proof. If $x = \theta$ or $y = \theta$, the inequality (12) holds trivially by (7). Assuming x and y nonzero, let

$$u = \|x\|^{-1}x, \qquad v = \|y\|^{-1}y.$$

By (9), u and v are unit vectors (obtained by 'normalizing' x and y), and

$$(x|y) = (\|x\|u|\|y\|v) = \|x\|\,\|y\|(u|v),$$

so it will suffice to show that $|(u|v)| \le 1$; indeed, by (10) and (11) we have

$$\begin{aligned}|(u|v)| &\le \tfrac{1}{4}\{\|u + v\|^2 + \|u - v\|^2\}\\ &= \tfrac{1}{4}\{2\|u\|^2 + 2\|v\|^2\} = 1.\end{aligned}$$

Suppose x and y are linearly dependent. If, for example, $y = cx$, then $|(x|y)| = |c(x|x)| = |c|\,\|x\|^2 = \|x\|\,\|y\|$.

Conversely, suppose $|(x|y)| = \|x\|\,\|y\|$. If $x = \theta$ or $y = \theta$, then x, y are trivially dependent. Assuming x and y nonzero, let $u = \|x\|^{-1}x$, $v = \|y\|^{-1}y$;

[2] This is interesting, but we shouldn't be overwhelmed: $ab = \frac{1}{4}\{(a + b)^2 - (a - b)^2\}$ for all real numbers a, b.

u and v are unit vectors, and by assumption $|(u|v)| = 1$, that is, $(u|v) = \pm 1$. If $(u|v) = 1$ then

$$\|u - v\|^2 = \|u\|^2 + \|v\|^2 - 2(u|v) = 1 + 1 - 2 = 0,$$

so $u - v = \theta$, that is, $\|x\|^{-1}x - \|y\|^{-1}y = \theta$. Similarly, $(u|v) = -1$ implies $u + v = \theta$. Either way, x and y are linearly dependent. ■

5.1.9 Corollary. (Triangle inequality) *In an inner product space,*

$$(13) \qquad \|x + y\| \le \|x\| + \|y\|$$

for all vectors x and y.

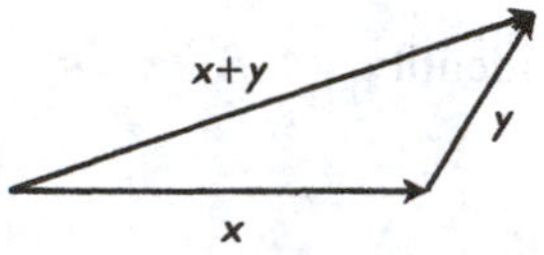

Fig. 19

Proof. Citing the Cauchy–Schwarz inequality at the appropriate step,

$$\|x + y\|^2 = \|x\|^2 + \|y\|^2 + 2(x|y) \le$$
$$\|x\|^2 + \|y\|^2 + 2|(x|y)| \le \|x\|^2 + \|y\|^2 + 2\|x\|\,\|y\| = (\|x\| + \|y\|)^2,$$

whence (13). {The reason for the name is suggested by Fig. 19.} ■

5.1.10 Theorem *For vectors x, y in an inner product space,*

$$(x|y) = 0 \Leftrightarrow \|x + y\|^2 = \|x\|^2 + \|y\|^2.$$

Proof. This is immediate from the identity

$$\|x + y\|^2 = \|x\|^2 + \|y\|^2 + 2(x|y). \quad ■$$

The preceding theorem reminds us of the theorem of Pythagoras and suggests a plausible definition of orthogonality in inner product spaces:

5.1.11 Definition Vectors x, y of an inner product space are said to be **orthogonal** (or 'perpendicular'), written $x \perp y$, if $(x|y) = 0$.

Note that $x \perp y \Leftrightarrow y \perp x$; $x \perp x \Leftrightarrow x = \theta$; and if $x \perp y_i$ for $i = 1, \ldots, n$ then $x \perp y$ for every linear combination $y = c_1 y_1 + \ldots + c_n y_n$.

Geometry and algebra co-operate admirably in the following theorem:

5.1.12 Theorem *If $x_1, \ldots, x_n$ are pairwise orthogonal nonzero vectors in an inner product space, then they are linearly independent.*

Proof. The hypothesis is that $(x_i|x_j) = 0$ when $i \ne j$ and that $\|x_i\| > 0$ for all i. Assuming $c_1 x_1 + \ldots + c_n x_n = \theta$, we have to show that every c_j is 0; this follows from the computation

$$0 = (\theta|x_j) = \left(\sum_{i=1}^{n} c_i x_i \Big| x_j\right) = \sum_{i=1}^{n} c_i (x_i|x_j) = c_j \|x_j\|^2. \quad ■$$

The concept of orthogonality can be extended from individuals to crowds:

5.1.13 Definition Subsets A and B of an inner product space E are said to be **orthogonal**, written $A \perp B$, if $x \perp y$ for all $x \in A$ and $y \in B$. If $\{x\} \perp B$ we also write $x \perp B$; thus $x \perp B$ means that x is orthogonal to every vector in B.

The set of all vectors $x \in E$ such that $x \perp A$ is denoted $A^\perp$ (verbalized

'A perp') and is called the (orthogonal) **annihilator** of A in E; the annihilator of $A^{\perp}$ is denoted $A^{\perp\perp}$, thus $A^{\perp\perp} = (A^{\perp})^{\perp}$.

Convention: $\varnothing^{\perp} = E$. {The implication «$x \in \varnothing$, $y \in E \Rightarrow x \perp y$» is vacuously true.}

In essence, we have a correspondence $A \mapsto A^{\perp}$ that assigns to each subset A of E its orthogonal annihilator $A^{\perp}$; the formal properties of this correspondence, which will be used frequently (beginning with the next section), are as follows:

5.1.14 Theorem *For subsets* A *and* B *of an inner product space* E,

(i) $A^{\perp}$ *is a linear subspace of* E,

(ii) $A \subset A^{\perp\perp}$,

(iii) $A \subset B \Rightarrow B^{\perp} \subset A^{\perp}$,

(iv) $(A^{\perp\perp})^{\perp} = (A^{\perp})^{\perp\perp} = A^{\perp}$.

(v) $A \cap A^{\perp} \subset \{\theta\}$.

Proof.

(i) If x and y are orthogonal to every vector in A, then the same is true of every linear combination of x and y (remarks following 5.1.11).

(ii) If $x \in A$ then, for every $y \in A^{\perp}$, $x \perp y$ by the definition of $A^{\perp}$, thus $x \in (A^{\perp})^{\perp}$.

(iii) If $A \subset B$ and $x \perp B$ then also $x \perp A$.

(iv) By (ii), $A \subset A^{\perp\perp}$, therefore $(A^{\perp\perp})^{\perp} \subset A^{\perp}$ by (iii). The reverse inclusion follows from (ii): $A^{\perp} \subset (A^{\perp})^{\perp\perp} = ((A^{\perp})^{\perp})^{\perp} = (A^{\perp\perp})^{\perp}$.

(v) The inclusion is trivial if $A = \varnothing$. If $x \in A$ and $x \in A^{\perp}$ then in particular $x \perp x$, therefore $x = \theta$. ∎

To say that x and y are orthogonal means that x is in the kernel of the linear form defined by y (remarks following 5.1.1); it is this circumstance that makes the theory of duality (§3.9) so useful in inner product spaces, as we see in the next section.

▶ *Exercises*

1. Show that if x and y are vectors in an inner product space such that $\|x\|^2 = \|y\|^2 = (x|y)$ then $x = y$. What does this mean for the inner product space of Example 5.1.3?

2. (*Bessel's inequality*) In an inner product space, if $x_1, \ldots, x_n$ are pairwise orthogonal unit vectors ($\|x_i\| = 1$ for all i, and $x_i \perp x_j$ when $i \neq j$), then

$$\sum_{i=1}^{n} |(x|x_i)|^2 \leq \|x\|^2.$$

(For $n = 1$ this is essentially the Cauchy–Schwarz inequality.)
{Hint: Define $c_i = (x|x_i)$, $y = c_1x_1 + \ldots + c_nx_n$ and $z = x - y$. Show that $y \perp z$ and apply the Pythagorean formula to $x = y + z$ to conclude that $\|x\|^2 \geq \|y\|^2$.}

3. For nonzero vectors x, y in an inner product space, $\|x+y\| = \|x\| + \|y\|$ if and only if $y = cx$ with $c > 0$. {Hint: Inspect the proof of Corollary 5.1.9, then apply the second assertion of Theorem 5.1.8.}

4. Vectors x, y in a (real) inner product space are orthogonal if and only if $\|x+y\| = \|x-y\|$.

5. In an inner product space, let $x_1, \dots, x_n$ be vectors that are 'pairwise orthogonal', that is, $(x_i|x_j) = 0$ whenever $i \neq j$.

(i) Prove (by induction) that $\|\sum_{i=1}^{n} x_i\|^2 = \sum_{i=1}^{n} \|x_i\|^2$.

(ii) Deduce an alternate proof of Theorem 5.1.12.

6. If $r_1, \dots, r_n$ are real numbers >0, then the formula

$$[x|y] = \sum_{i=1}^{n} r_i a_i b_i,$$

for $x = (a_1, \dots, a_n)$, $y = (b_1, \dots, b_n)$, defines an inner product on $\mathbf{R}^n$ (different from the canonical inner product, unless $r_1 = \dots = r_n = 1$).

7. In the inner product space of Example 5.1.3, if x and y are continuously differentiable (that is, the derivative functions x' and y' exist and are continuous) then

$$(x|y') + (x'|y) = x(b)y(b) - x(a)y(a).$$

{Hint: 'Integration by parts'.}

8. Prove that, in a (real) inner product space, $\|x\| = \|y\|$ if and only if $x + y$ and $x - y$ are orthogonal. Geometric interpretation? {Hint: Calculate $(x+y|x-y)$. Fancy solution: In the polarization identity, replace x, y by $x+y$, $x-y$.}

9. If T is a linear mapping in a (real) inner product space, then

$$(Tx|Ty) = \tfrac{1}{4}\{\|T(x+y)\|^2 - \|T(x-y)\|^2\}$$

for all vectors x and y. {Hint: Theorem 5.1.7.}

10. For subsets A, B of an inner product space, (i) $\mathrm{A} \perp \mathrm{B} \Leftrightarrow \mathrm{A} \subset \mathrm{B}^\perp$, and (ii) $(\mathrm{A}+\mathrm{B})^\perp = (\mathrm{A} \cup \mathrm{B})^\perp = \mathrm{A}^\perp \cap \mathrm{B}^\perp$. (iii) In general, $(\mathrm{A} \cap \mathrm{B})^\perp \neq \mathrm{A}^\perp + \mathrm{B}^\perp$. {Hint: In the Euclidean space E^2 (5.1.2), consider $\mathrm{A} = \{e_1, e_2\}$, $\mathrm{B} = \{e_1, e_1 + e_2\}$.}

11. For vectors of $\mathbf{C}^n$, inner products are defined by the formula

$$(x|y) = \sum_{k=1}^{n} a_k \bar{b}_k,$$

where $\bar{b}_k$ is the complex conjugate of b_k. The result is called a *complex Euclidean space* (or 'unitary space'). The exercise: explore its properties.

5.2 Duality in inner product spaces

Linear forms are 'present at the creation' of an inner product space (5.1.1), so it's not surprising that duality looms large in such spaces. The two key results in this section, stated informally:

(i) a finite-dimensional inner product space is essentially equal to its dual (5.2.5);

(ii) an n-dimensional inner product space is essentially equal to the Euclidean space E^n (5.2.17).

5.2.1 Definition If E is an inner product space and $y \in E$, we write y' for the linear form on E defined by

$$y'(x) = (x|y) \quad \text{for all} \quad x \in E.$$

Thus $y': E \to \mathbf{R}$ is linear, that is, $y' \in E' = \mathcal{L}(E, \mathbf{R})$.

5.2.2 Theorem *If* E *is an inner product space, then the mapping* $E \to E'$ *defined by* $y \mapsto y'$ *is linear and injective.*

Proof. The formula $(x|y+z) = (x|y) + (x|z)$ shows that $(y+z)' = y' + z'$, and $(x|cy) = c(x|y)$ shows that $(cy)' = cy'$. If $y' = 0$ then $0 = y'(y) = (y|y)$, therefore $y = \theta$; thus the linear mapping $y \mapsto y'$ has kernel $\{\theta\}$, hence is injective. ■

5.2.3 Corollary *If* T *is a nonempty set,* E *is an inner product space, and* $f: T \to E$, $g: T \to E$ *are functions such that* $(f(t)|x) = (g(t)|x)$ *for all* $x \in E$ *and* $t \in T$, *then* $f = g$.

Proof. The validity of $(x|f(t)) = (x|g(t))$ for all x and t says, in the notation of 5.2.1, that $(f(t))' = (g(t))'$ for all t, therefore $f(t) = g(t)$ for all t by 5.2.2. ■

5.2.4 Remarks If E and F are inner product spaces and $S, T: E \to F$ are mappings such that $(Sx|y) = (Tx|y)$ for all $x \in E$, $y \in F$, then $S = T$ by 5.2.3. However, there exist *linear* mappings $T: \mathbf{R}^2 \to \mathbf{R}^2$ such that $(Tx|x) = 0$ for all vectors x but $T \neq 0$ (consider rotation by 90°).

A more dramatic consequence of Theorem 5.2.2:

5.2.5 Theorem. (Self-duality of Euclidean spaces) *If* E *is a Euclidean space* (5.1.1), *then the mapping* $y \mapsto y'$ *of* Theorem 5.2.2 *is a vector space isomorphism* $E \to E'$.

Proof. Since $\dim E' = \dim E$ (3.9.2) the linear mapping $y \mapsto y'$, which is injective by Theorem 5.2.2, is necessarily bijective (3.7.4). ■

5.2.6 Definition With notations as in 5.2.5, the linear bijection $J: E \to E'$ defined by $Jy = y'$ is called the **canonical isomorphism** of the Euclidean space E onto its dual space. We write $J = J_E$ when it is necessary to indicate the dependence of J on E (for example, in contexts where several Euclidean spaces appear).

There is a harvest of consequences of the self-duality of Euclidean spaces:

5.2.7 Theorem *If* M *is a finite-dimensional linear subspace of an inner product space* E, *then* $E = M + M^\perp$ *and* $M \cap M^\perp = \{\theta\}$.

Proof. The zero vector belongs to both M and $M^{\perp}$ (they are linear subspaces), thus $\{\theta\} \subset M \cap M^{\perp}$; the reverse inclusion is (v) of Theorem 5.1.14.

To prove that $E = M + M^{\perp}$, given any $x \in E$ we must show that $x = y + z$ for suitable $y \in M$ and $z \in M^{\perp}$. Let x' be the linear form on E defined by x as in 5.2.1 and let $f = x'|M$ be the restriction of f to M; then f is a linear form on the Euclidean space M, so by Theorem 5.2.5 there is a vector $y \in M$ such that $f(w) = (w|y)$ for all $w \in M$, that is, $(w|x) = (w|y)$ for all $w \in M$. Then $(w|x - y) = 0$ for all $w \in M$, so the vector $z = x - y$ belongs to $M^{\perp}$. {In the notation of §1.6, Exercise 12, $E = M \oplus M^{\perp}$.} ■

5.2.8 Corollary *If* E *is a Euclidean space and* M *is a linear subspace of* E, *then* $\dim M^{\perp} = \dim E - \dim M$.

Proof. If $y_1, \dots, y_m$ is a basis of M and $z_1, \dots, z_r$ is a basis of $M^{\perp}$, then the list $y_1, \dots, y_m, z_1, \dots, z_r$ generates E (because $M + M^{\perp} = E$) and is linearly independent (because $M \cap M^{\perp} = \{\theta\}$), hence is a basis of E; thus $\dim E = m + r$, that is, $\dim M^{\perp} = r = \dim E - m = \dim E - \dim M$.

Alternate proof: Let $J : E \to E'$ be the canonical isomorphism (5.2.6). Writing $M^{\circ} = \{f \in E' : f = 0 \text{ on } M\}$ as in 3.9.10, we have

$$x \in M^{\perp} \Leftrightarrow x' = 0 \text{ on } M \Leftrightarrow Jx \in M^{\circ},$$

thus $M^{\perp} = J^{-1}(M^{\circ})$, in other words $J(M^{\perp}) = M^{\circ}$; since J is injective, $M^{\perp} \cong M^{\circ}$ and the dimension formula follows from Theorems 3.9.11 and 3.5.12. ■

5.2.9 Definition With notations as in 5.2.8, $M^{\perp}$ is called the **orthogonal complement** of M in E.

5.2.10 Examples If M is the set of multiples of a vector y in a Euclidean space, then the orthogonal complement of M is the kernel of the linear form y'. In the Euclidean space E^3 (5.1.2), the orthogonal complement of a line through the origin is a plane through the origin.

5.2.11 Corollary *If* M *is a linear subspace of a Euclidean space, then* $M = M^{\perp\perp}$.

Proof. Applying Corollary 5.2.8 twice, we see that M and $M^{\perp\perp} = (M^{\perp})^{\perp}$ have the same dimension; since $M \subset M^{\perp\perp}$ (5.1.14), it follows from 3.5.14 that $M = M^{\perp\perp}$. ■

5.2.12 Corollary *If* A *is a subset of a Euclidean space, then the linear span of* A *is* $A^{\perp\perp}$, *that is,* $[A] = A^{\perp\perp}$.

Proof. By Theorem 5.1.14, $A^{\perp\perp}$ is a linear subspace containing A, so $A^{\perp\perp} \supset [A]$. On the other hand, $A \subset [A]$, therefore $A^{\perp} \supset [A]^{\perp}$, $A^{\perp\perp} \subset [A]^{\perp\perp} = [A]$ by 5.2.11. ■

We will now deduce from Corollary 5.2.8 that there is essentially only one

Euclidean space of dimension n (5.2.17 below); this requires some preparation:

5.2.13 Definition Vectors $x_1, \ldots, x_n$ in an inner product space are said to be **orthonormal** if they are pairwise orthogonal unit vectors, that is, if $\|x_i\| = 1$ for all i and $(x_i|x_j) = 0$ when $i \neq j$; in other words, $(x_i|x_j) = \delta_{ij}$ for all i and j (3.9.4). {It follows that the x_i are linearly independent (5.1.12).}

Convention: If x_1 is a unit vector, the list x_1 of length 1 is declared to be orthonormal (any two vectors in the list are orthogonal 'by default').

5.2.14 Example In the Euclidean space E^n (5.1.2), the canonical basis vectors $e_1, \ldots, e_n$ are orthonormal.

5.2.15 Example In the Euclidean plane E^2, the vectors $(\cos\alpha, \sin\alpha)$, $(-\sin\alpha, \cos\alpha)$ form an orthonormal basis, for each real number α.

5.2.16 Theorem *Every Euclidean space $\neq \{\theta\}$ has an orthonormal basis.*

Proof. Assuming E is an n-dimensional inner product space, we seek a basis $x_1, \ldots, x_n$ of E consisting of pairwise orthogonal unit vectors.

If $n = 1$, any unit vector will do (Definition 5.2.13). We proceed by induction: let $n \geq 2$ and assume the theorem true for inner product spaces of dimension $n - 1$. Let x_n be any unit vector in E (choose a nonzero vector $x \in E$ and let $x_n = \|x\|^{-1}x$) and let $M = \mathbf{R}x_n$ be the 1-dimensional linear subspace spanned by x_n. By Corollary 5.2.8, $M^\perp$ is $(n-1)$-dimensional, so by the induction hypothesis it has an orthonormal basis $x_1, \ldots, x_{n-1}$. The vectors $x_1, \ldots, x_{n-1}, x_n$ are orthonormal and their linear span is $M^\perp + M = E$ (5.2.7), so they must be a basis of E (3.5.13). {An alternate path to linear independence is Theorem 5.1.12.} ■

5.2.17 Corollary *If E is an n-dimensional Euclidean space and E^n is the n-dimensional Euclidean space of 5.1.2, then there exists a bijective linear mapping $T:E \rightarrow E^n$ such that*

$$(Tx|Ty) = (x|y)$$

for all vectors x, y in E.

Proof. Let $x_1, \ldots, x_n$ be an orthonormal basis of E (5.2.16), let $e_1, \ldots, e_n$ be the canonical orthonomal basis of E^n (5.2.14) and let $S:E^n \rightarrow E$ be the mapping

$$S(a_1, \ldots, a_n) = a_1x_1 + \ldots + a_nx_n;$$

then S is a linear bijection with $Se_i = x_i$ for all i (3.5.5), therefore $T = S^{-1}$ is a linear bijection $T:E \rightarrow E^n$ with $Tx_i = e_i$ for all i.

Let $x, y \in E$, say

$$x = \sum_{i=1}^{n} a_i x_i, \qquad y = \sum_{i=1}^{n} b_i x_i;$$

then

$$Tx = \sum_{i=1}^{n} a_i Tx_i = \sum_{i=1}^{n} a_i e_i = (a_1, \ldots, a_n),$$

and similarly $Ty = (b_1, \ldots, b_n)$, so

$$(Tx|Ty) = \sum_{i=1}^{n} a_i b_i.$$

On the other hand,

$$\begin{aligned}(x|y) &= (\sum_{i=1}^{n} a_i x_i | \sum_{i=1}^{n} b_i x_i) \\ &= \sum_{i,j} a_i b_j (x_i|x_j) \\ &= \sum_{i,j} a_i b_j \delta_{ij} = \sum_{i=1}^{n} a_i b_i,\end{aligned}$$

thus $(Tx|Ty) = (x|y)$. ■

5.2.18 Definition Two inner product spaces[1] E, F are said to be **isomorphic** (as inner product spaces) if there exists a bijective linear mapping $T:\mathrm{E} \to \mathrm{F}$ such that $(Tx|Ty) = (x|y)$ for all $x, y \in \mathrm{E}$ (so to speak, T preserves inner products).

The message of Corollary 5.2.17: up to isomorphism (in the sense of 5.2.18) there is only one n-dimensional Euclidean space.

► **Exercises**

1. Let E be the inner product space of 5.1.3 (the space of continuous real-valued functions on a closed interval $[a, b]$, with inner products defined by means of the integral).

(i) Fix a point $c \in [a, b]$ and let f be the linear form on E defined by $f(x) = x(c)$ for all $x \in \mathrm{E}$ (f is 'pointwise evaluation at c'). Prove that there does *not* exist a function $y \in \mathrm{E}$ such that $f = y'$ in the sense of Definition 5.2.1. {Hint: Note that $f \neq 0$. Show that if $y \in \mathrm{E}$, $y \neq 0$, then there exists $x \in \mathrm{E}$ such that $x(c) = 0$ but $\int_a^b x(t)y(t)dt > 0$.}

(ii) Let f be the linear form on E defined by

$$f(x) = \int_a^b x(t)dt \quad \text{for} \quad x \in \mathrm{E}.$$

True or false (explain): There exists $y \in \mathrm{E}$ such that $f = y'$.

(iii) Fix a point c, $a < c < b$, and let f be the linear form on E defined by

$$f(x) = \int_a^c x(t)dt \quad \text{for} \quad x \in \mathrm{E}.$$

Same question as in (ii). (Different answer!)

[1] Since we are now sticking to real vector spaces, the letter F, previously reserved for the field of scalars, is again liberated for general duty.

2. Let V be the vector space of all functions $x:\mathbf{P}\to\mathbf{R}$ with the pointwise linear operations (cf. 1.3.7) and let E be the set of all functions $x \in \mathrm{V}$ whose 'support'

$$\{n \in \mathbf{P}: x(n) \neq 0\}$$

is a finite subset of **P**. Prove:

(i) E is a linear subspace of V;

(ii) the formula

$$(x|y) = \sum_n x(n)y(n)$$

defines an inner product on E (n runs over **P**, but the sum has at most finitely many nonzero terms).

(iii) The formula

$$f(x) = \sum_n x(n)$$

defines a linear form on E.

(iv) There does not exist a vector $y \in \mathrm{E}$ such that $f = y'$.

3. Let E be a Euclidean space, M and N linear subspaces of E such that $\mathrm{M} \subset \mathrm{N}$. Prove:

(i) $\mathrm{N} = \mathrm{M} + (\mathrm{N} \cap \mathrm{M}^{\perp})$. {Hint: Theorem 5.2.7 and the 'modular law' (§1.6, Exercise 5).}

(ii) If $\mathrm{M} \subset \mathrm{N}$ properly, then $\mathrm{N} \cap \mathrm{M}^{\perp} \neq \{\theta\}$.

4. If E is any inner product space (not necessarily finite-dimensional), then $\mathrm{M} = \mathrm{M}^{\perp\perp}$ for every finite-dimensional linear subspace M. {Hint: The proof of Corollary 5.2.11 is not applicable here. The problem is to show that $\mathrm{M}^{\perp\perp} \subset \mathrm{M}$. If $x \in \mathrm{M}^{\perp\perp}$ and $x = y + z$ as in the proof of Theorem 5.2.7, then $z = x - y \in \mathrm{M}^{\perp\perp}$, so $z \perp z$.}

5. It can be shown that in the inner product space of continuous functions on $[a, b]$ (Example 5.1.3), the polynomial functions form a linear subspace M such that $\mathrm{M}^{\perp} = \{0\}$ (the proof is not easy!), therefore $\mathrm{M} \neq \mathrm{M}^{\perp\perp}$.

6. If M and N are linear subspaces of a Euclidean space, then $(\mathrm{M} + \mathrm{N})^{\perp} = \mathrm{M}^{\perp} \cap \mathrm{N}^{\perp}$ and $(\mathrm{M} \cap \mathrm{N})^{\perp} = \mathrm{M}^{\perp} + \mathrm{N}^{\perp}$. {Hint: Corollary 5.2.11.}

7. Let e_1, e_2, e_3 be the canonical orthonormal basis of the Euclidean space $\mathbf{R}^3$, let $x_1 = e_1 + e_2 + e_3$, $x_2 = e_1 - e_2 + e_3$, and let $\mathrm{M} = [\{x_1, x_2\}]$ be the linear span of x_1 and x_2. Find $\mathrm{M}^{\perp}$.

8. (i) Let E be an inner product space and let $x_1, \ldots, x_n, x$ be vectors such that x is not a linear combination of $x_1, \ldots, x_n$. Prove that there exists a vector y such that

$$[\{x_1, \ldots, x_n, x\}] = [\{x_1, \ldots, x_n, y\}]$$

and such that y is orthogonal to every x_i. {Hint: Exercise 3, part (ii).}

(ii) If $x_1, \ldots, x_n$ are linearly independent vectors in E, use the idea of (i) to construct orthonormal vectors $y_1, \ldots, y_n$ such that $[\{x_1, \ldots, x_k\}] = [\{y_1, \ldots, y_k\}]$ for $k = 1, \ldots, n$. (See Exercise 12 for a more explicit construction.)

9. Given a system of homogeneous linear equations

$$a_{11}x_1 + a_{12}x_2 + \ldots + a_{1n}x_n = 0$$

$$a_{21}x_1 + a_{22}x_2 + \ldots + a_{2n}x_n = 0$$

$$\ldots$$

$$a_{m1}x_1 + a_{m2}x_2 + \ldots + a_{mn}x_n = 0$$

with real coefficients a_{ij}. If $m < n$, we know from Theorem 3.7.1 that there exist nonzero solution-vectors $x = (x_1, \ldots, x_n)$; deduce another proof of this from Corollary 5.2.8. {Hint: Let $A = (a_{ij})$ be the $m \times n$ matrix of coefficients and let M be the row space of A (4.2.9).}

10. A *unitary space* is a finite-dimensional complex vector space E with an 'inner product' $(x, y) \mapsto (x|y)$ taking complex values, satisfying (1) and (3) of Definition 5.1.1, (4) for complex scalars, and, instead of (2), the identity $\overline{(x|y)} = (y|x)$, where the overbar denotes complex-conjugate (cf. §5.1, Exercise 11).

(i) Show that $(x|cy) = \bar{c}(x|y)$ for all vectors x, y and all complex numbers c.

(ii) If $y \in$ E define $y'(x) = (x|y)$ for all $x \in$ E. Show that $y' \in \mathrm{E}' = \mathscr{L}(\mathrm{E}, \mathrm{C})$ and that $(y + z)' = y' + z'$, $(cy)' = \bar{c}y'$.

(iii) Prove the unitary space analogue of Theorem 5.2.5, with 'isomorphism' replaced by 'conjugate-isomorphism' in the sense suggested by (ii).

(iv) Prove the unitary space analogues of the remaining results in this section.

(v) Verify the 'Polarization identity'

$$(x|y) = \tfrac{1}{4}\{\|x + y\|^2 - \|x - y\|^2 + i\|x + iy\|^2 - i\|x - iy\|^2\}.$$

11. If E is a unitary space (Exercise 10) and T:E$\rightarrow$E is a linear mapping such that $(Tx|x) = 0$ for all vectors x, then (in contrast with the example in 5.2.4) $T = 0$. {Hint: Adapt the argument for (v) of Exercise 10 to express $(Tx|y)$ as a linear combination of numbers $(Tz|z)$.}

12. Let E be a Euclidean space.

(i) If u, x are vectors in E such that $\|u\| = 1$ and x is not a multiple of u, find a nonzero vector z such that $u \perp z$ and $x = au + z$ for a suitable scalar a.

(ii) If x_1, x_2 are linearly independent vectors, find orthonormal vectors u_1, u_2 such that u_1 is a multiple of x_1, and u_2 is a linear combination of x_1, x_2.

(iii) If $x_1, \ldots, x_n$ are linearly independent, then there exist orthonormal vectors $u_1, \ldots, u_n$ such that $[\{x_1, \ldots, x_k\}] = [\{u_1, \ldots, u_k\}]$ for $k = 1, \ldots, n$. (*Gram–Schmidt orthogonalization process*)

(iv) Infer an alternate proof of 5.2.16: Every Euclidean space has an orthonormal basis.

{Hints:

(i) If such a, z exist, then $(x|u) = a$ and $z = x - au$.

(ii) Let $u_1 = \|x_1\|^{-1}x_1$, apply (i) with $u = u_1$, $x = x_2$ and normalize the resulting vector z.

(iii) Induction.}

13. If $u_1, \ldots, u_k$ are orthonormal vectors in a Euclidean space E of dimension $n > k$, then E has an orthonormal basis $u_1, \ldots, u_k$, $u_{k+1}, \ldots, u_n$. {Hint: Consider $\mathrm{M}^\perp$ for the appropriate linear subspace M.}

14. If M is a linear subspace of a Euclidean space E then $\dim \mathrm{M} + \dim \mathrm{M}^\perp = \dim \mathrm{E}$ by the 'alternate proof' of Corollary 5.2.8. Infer from $\mathrm{M} \cap \mathrm{M}^\perp = \{\theta\}$ that $\mathrm{E} = \mathrm{M} + \mathrm{M}^\perp$.
{Hint: §3.5, Exercise 26 or §3.6, Exercise 5.}

5.3 The adjoint of a linear mapping

If T is a linear mapping in vector space, its transpose T' acts in the dual space. A Euclidean space is 'self-dual' (5.2.5), so we expect something *like* the transpose to act in the space itself (instead of the dual space); that 'something' is the subject of this section.

More generally, consider a pair of Euclidean spaces E, F and a linear mapping T:E$\to$F. We are looking for a mapping F$\to$E 'like' the transpose (but acting in the given spaces rather than their duals). As in Definition 5.2.6, we write J_{E}:E$\to$E$'$ for the canonical isomorphism of E onto its dual ($J_{\mathrm{E}}x = x'$ for all $x \in \mathrm{E}$) and similarly for J_{F}:F$\to$F$'$. The following diagram offers a roundabout way of getting from F to E and suggests the definition of the desired mapping:

$$\begin{array}{ccc} \mathrm{E} & \xrightarrow{\;T\;} & \mathrm{F} \\ J_{\mathrm{E}} \downarrow & & \downarrow J_{\mathrm{F}} \\ \mathrm{E}' & \xleftarrow[\;T'\;]{} & \mathrm{F}' \end{array}$$

5.3.1 Definition With the preceding notations, the **adjoint** of T:E$\to$F is the mapping T^*:F$\to$E defined by the formula $T^* = J_{\mathrm{E}}^{-1}T'J_{\mathrm{F}}$.

Since T^* is the composite of linear mappings, it is also linear. The diagram

$$\begin{array}{ccc} & T^* & \\ \mathrm{E} & \longleftarrow & \mathrm{F} \\ J_{\mathrm{E}} \downarrow & & \downarrow J_{\mathrm{F}} \\ \mathrm{E}' & \longleftarrow & \mathrm{F}' \\ & T' & \end{array}$$

is 'commutative' in the sense that the two ways of getting from F to E′ are equal: $J_{\mathrm{E}}T^* = T'J_{\mathrm{F}}$.

5.3.2 Theorem *Let* E *and* F *be Euclidean spaces,* $T:\mathrm{E}\rightarrow\mathrm{F}$ *a linear mapping,* $T^*:\mathrm{F}\rightarrow\mathrm{E}$ *the adjoint of* T (5.3.1). *Then:*

(1) T^* *is a linear mapping.*

(2) $(Tx|y) = (x|T^*y)$ *for all* $x \in \mathrm{E}$, $y \in \mathrm{F}$.

(3) $(T^*y|x) = (y|Tx)$ *for all* $y \in \mathrm{F}$, $x \in \mathrm{E}$.

(4) *If* $S:\mathrm{F}\rightarrow\mathrm{E}$ *is a mapping such that* $(Tx|y) = (x|Sy)$ *for all* $x \in \mathrm{E}$, $y \in \mathrm{F}$, *then necessarily* $S = T^*$.

Proof.

(1) Remarked above.

(2) For all $x \in \mathrm{E}$ and $y \in \mathrm{F}$,

$$\begin{aligned}(Tx|y) &= y'(Tx) = (T'y')(x) = (T'J_{\mathrm{F}}y)(x)\\ &= (J_{\mathrm{E}}T^*y)(x) = (T^*y)'(x) = (x|T^*y).\end{aligned}$$

(3) In view of the symmetry of the inner product, this is a restatement of (2).

(4) By hypothesis, $(Sy|x) = (y|Tx) = (T^*y|x)$ for all x and y, therefore $S = T^*$ by 5.2.3. ■

Here are some useful algebraic properties of adjuction:

5.3.3 Theorem *Let* E, F, G *be Euclidean spaces.*

(5) $(T^*)^* = T$ *for all* $T \in \mathscr{L}(\mathrm{E}, \mathrm{F})$.

(6) $(S + T)^* = S^* + T^*$ *for all* $S, T \in \mathscr{L}(\mathrm{E}, \mathrm{F})$.

(7) $(cT)^* = cT^*$ *for all* $c \in \mathbf{R}$ *and* $T \in \mathscr{L}(\mathrm{E}, \mathrm{F})$.

(8) $(ST)^* = T^*S^*$ *for all* $T \in \mathscr{L}(\mathrm{E}, \mathrm{F})$ *and* $S \in \mathscr{L}(\mathrm{F}, \mathrm{G})$.

(9) $I^* = I$, *where* $I \in \mathscr{L}(\mathrm{E})$ *is the identity mapping.*

(10) $0^* = 0$, *where the* 0*'s denote the zero mappings of* E *into* F *and of* F *into* E.

(11) *If* $T \in \mathscr{L}(\mathrm{E}, \mathrm{F})$ *and* $T^*T = 0$, *then* $T = 0$; *similarly,* $TT^* = 0$ *implies* $T = 0$.

Proof.

(5) For all $x \in \mathrm{E}$ and $y \in \mathrm{F}$, $((T^*)^*x|y) = (x|T^*y) = (Tx|y)$, therefore $(T^*)^* = T$ by 5.2.3.

(6) For all $y \in \mathrm{F}$ and $x \in \mathrm{E}$,

$$\begin{aligned}((S + T)^*y|x) &= (y|(S + T)x) = (y|Sx + Tx)\\ &= (y|Sx) + (y|Tx) = (S^*y|x) + (T^*y|x)\\ &= (S^*y + T^*y|x) = ((S^* + T^*)y|x),\end{aligned}$$

therefore $(S + T)^* = S^* + T^*$ by 5.2.3.

(7) For all x and y, $((cT)^*y|x) = (y|(cT)x) = (y|c(Tx)) = c(y|Tx) = c(T^*y|x) = (c(T^*y)|x) = ((cT^*)y|x)$.

(8) The picture is as follows:

$$\begin{array}{c} \mathrm{E} \xrightarrow{T} \mathrm{F} \xrightarrow{S} \mathrm{G} \\ \mathrm{E} \xleftarrow[T^*]{} \mathrm{F} \xleftarrow[S^*]{} \mathrm{G} \end{array}$$

For all $x \in \mathrm{E}$ and $z \in \mathrm{G}$, $((ST)^*z|x) = (z|(ST)x) = (z|S(Tx)) = (S^*z|Tx) = (T^*(S^*z)|x) = ((T^*S^*)z|x)$.

(9), (10) The proofs are straightforward.

(11) For all $x \in \mathrm{E}$,

$$\|Tx\|^2 = (Tx|Tx) = (T^*(Tx)|x) = ((T^*T)x|x),$$

so $T^*T = 0$ implies that $Tx = \theta$ for all vectors x, that is, $T = 0$. The second assertion of (11) follows from applying the first to T^*. ■

5.3.4 Corollary *If* E, F *are Euclidean spaces and* $T{:}\mathrm{E} \to \mathrm{F}$ *is a linear bijection, then* $T^*{:}\mathrm{F} \to \mathrm{E}$ *is also bijective and* $(T^*)^{-1} = (T^{-1})^*$.

Proof. Taking adjoints in the equation $T^{-1}T = I$, we have $T^*(T^{-1})^* = I$. Similarly $(T^{-1})^*T^* = I$ (the identity mapping of F), thus T^* is bijective, with inverse $(T^{-1})^*$. ■

Now let's tie in adjoints with matrix theory:

5.3.5 Theorem *Let* E *and* F *be Euclidean spaces,* $T{:}\mathrm{E} \to \mathrm{F}$ *a linear mapping. Choose* ***orthonormal*** *bases of* E *and* F, *and let A be the matrix of T relative to these bases. Then the matrix of* $T^*{:}\mathrm{F} \to \mathrm{E}$ *relative to these bases is the transpose A' of A.*

Proof. Let $x_1, \ldots, x_n$ and $y_1, \ldots, y_m$ be the given orthonormal bases of E and F, respectively. Assuming A is the matrix of T with respect to $x_1, \ldots, x_n$ and $y_1, \ldots, y_m$, we have to show that the matrix of T^* with respect to $y_1, \ldots, y_m$ and $x_1, \ldots, x_n$ is A'. {Caution: Even if $\mathrm{E} = \mathrm{F}$, the order of appearance of the two bases has to be reversed in calculating the matrix of T^*.} A straightforward proof of the theorem based on the relation (3) of Theorem 5.3.2 makes a nice exercise, but the following 'conceptual' proof teaches us more.

The strategy is to look at the diagram of Definition 5.3.1 through the appropriate lens:

$$\begin{array}{ccc} & T^* & \\ \mathrm{E}' & \longleftarrow & \mathrm{F} \\ J_{\mathrm{E}} \downarrow & & \downarrow J_{\mathrm{F}} \\ \mathrm{E}' & \longleftarrow & \mathrm{F}' \\ & T' & \end{array}$$

Regard E and F as equipped with the bases $x_1, \ldots, x_n$ and $y_1, \ldots, y_m$. Equip E′ and F′ with the bases

$$J_E x_1, \ldots, J_E x_n \quad \text{and} \quad J_F y_1, \ldots, J_F y_m,$$

respectively (3.3.7, 3.4.7). The matrix of J_E relative to the bases $x_1, \ldots, x_n$ and $J_E x_1, \ldots, J_E x_n$ is the identity matrix, and similarly for J_F. By the definition of T^*, $J_E T^* = T' J_F$. Relative to the chosen bases, this equation translates (by 4.3.3) into the matrix equation

$$(\text{mat } J_E)(\text{mat } T^*) = (\text{mat } T')(\text{mat } J_F);$$

since $\text{mat } J_E = I$ and $\text{mat } J_F = I$, this simplifies to

$$\text{mat } T^* = \text{mat } T'.$$

We are thus reduced to calculating the matrix of $T':F' \to E'$ relative to the bases in question. Now comes the key observation: $J_E x_1, \ldots, J_E x_n$ is the basis of E′ *dual* to the basis $x_1, \ldots, x_n$ (3.9.4):

$$(x_j)'(x_k) = (x_k | x_j) = \delta_{jk}.$$

Similarly, $J_F y_1, \ldots, J_F y_m$ is the basis of F′ dual to $y_1, \ldots, y_m$, consequently $\text{mat } T' = A'$ by Theorem 4.6.6. ■

Finally, a useful theorem about the effect of adjoints on linear subspaces:

5.3.6 Theorem *Let* E *and* F *be Euclidean spaces,* $T:E \to F$ *a linear mapping,* M *a linear subspace of* E, *and* N *a linear subspace of* F.

If $T(M) \subset N$ *then* $T^*(N^\perp) \subset M^\perp$.

Proof. Assuming $T(M) \subset N$ and $y \in N^\perp$, we have to show that T^*y is orthogonal to every vector in M. For all $x \in M$,

$$(T^*y | x) = (y | Tx) = 0$$

because $Tx \in N$ and $y \in N^\perp$. ■

▶ Exercises

1. Let $T:\mathbf{R}^3 \to \mathbf{R}^2$ be the linear mapping defined by

$$T(a, b, c) = (2a - c, 3b + 4c).$$

(i) Viewing $\mathbf{R}^3$ and $\mathbf{R}^2$ as Euclidean spaces (Example 5.1.2), find the matrix of T^* relative to the canonical orthonormal bases, then deduce a formula for T^*.

(ii) Find the matrix of T^* relative to the basis $\frac{1}{2}(1, \sqrt{3})$, $\frac{1}{2}(-\sqrt{3}, 1)$ of $\mathbf{R}^2$ and the canonical basis e_1, e_2, e_3 of $\mathbf{R}^3$.

2. If E and F are Euclidean spaces, then the mapping $T \mapsto T^*$ is a linear bijection $\mathcal{L}(E, F) \to \mathcal{L}(F, E)$. What is the inverse mapping?

3. If E and F are inner product spaces and if $T:E \to F$, $S:F \to E$ are mappings such that $(Tx|y) = (x|Sy)$ for all $x \in E$ and $y \in F$, then S and T are linear (and $S = T^*$ when E and F are Euclidean spaces).

4. For A, B in the vector space $M_n(\mathbf{R})$ of $n \times n$ real matrices (4.1.5), define

$$(A|B) = \text{tr}(AB')$$

(cf. §4.2, Exercise 7). Prove that $M_n(\mathbf{R})$ is a Euclidean space (of dimension n^2) and find the formula for $\|A\|$ if $A = (a_{ij})$.

5. The converse of Theorem 5.3.6 is also true, thus $T(\mathrm{M}) \subset \mathrm{N}$ if and only if $T^*(\mathrm{N}^\perp) \subset \mathrm{M}^\perp$.

6. Theorem 5.3.6 looks suspiciously like §3.9, Exercise 18 in disguise. Can you unmask it?

7. Let E and F be Euclidean spaces, $T:\mathrm{E} \to \mathrm{F}$ a linear mapping, A a subset of E, and B a subset of F. Prove:

(i) $(T(\mathrm{A}))^\perp = (T^*)^{-1}(\mathrm{A}^\perp)$;
(i′) $(T^*(\mathrm{B}))^\perp = T^{-1}(\mathrm{B}^\perp)$;
(ii) $(T(\mathrm{E}))^\perp = \operatorname{Ker} T^*$, therefore $\mathrm{F} = T(\mathrm{E}) \oplus \operatorname{Ker} T^*$;
(ii′) $(T^*(\mathrm{F}))^\perp = \operatorname{Ker} T$, therefore $\mathrm{E} = T^*(\mathrm{F}) \oplus \operatorname{Ker} T$;
(iii) $T(\mathrm{A}) \subset \mathrm{B} \Rightarrow T^*(\mathrm{B}^\perp) \subset \mathrm{A}^\perp$.

8. Extend the results of this section to unitary spaces (§5.2, Exercise 10). For example, in the definition analogous to 5.3.1, J_{E} and J_{F} are both 'conjugate-linear' (see (ii) of the cited exercise), so T^* ends up being linear. In (7) of 5.3.3, expect $(cT)^* = \bar{c}T^*$. In 5.3.5, expect the conjugate-transpose of A (§4.6, Exercise 3).

5.4 Orthogonal mappings and matrices

An *isomorphism* of inner product spaces is a bijective linear mapping between them that preserves the inner product (5.2.18); these were the mappings used to identify finite-dimensional inner product spaces with the Euclidean spaces E^n (5.2.17). There is a special name for such mappings when the two spaces are equal:

5.4.1 Definition Let E be a Euclidean space. A linear mapping $T \in \mathcal{L}(\mathrm{E})$ is said to be **orthogonal** if $(Tx|Ty) = (x|y)$ for all x, y in E.

Initially defined in terms of inner products, orthogonality also has useful characterizations in terms of norm and in terms of adjunction:

5.4.2 Theorem *If* E *is a Euclidean space and* $T \in \mathcal{L}(\mathrm{E})$, *the following conditions are equivalent:*

(a) T *is orthogonal* (5.4.1);
(b) $\|Tx\| = \|x\|$ *for all vectors* $x \in \mathrm{E}$;
(c) $T^*T = I$;
(d) $TT^* = I$.

Proof.
(a) ⇒ (b): Put $y = x$ in 5.4.1: $\|Tx\|^2 = (Tx|Tx) = (x|x) = \|x\|^2$ for all vectors x.

(b) ⇒ (c): By the polarization identity (5.1.7) and the condition (b),

$$\begin{aligned}4(Tx|Ty) &= \|Tx + Ty\|^2 - \|Tx - Ty\|^2\\ &= \|T(x + y)\|^2 - \|T(x - y)\|^2\\ &= \|x + y\|^2 - \|x - y\|^2\\ &= 4(x|y),\end{aligned}$$

thus $(T^*Tx|y) = (Tx|Ty) = (x|y) = (Ix|y)$ for all x and y, so $T^*T = I$ by 5.2.3.

(c) ⇔ (d): This is a special case of 3.7.5.

(c) ⇒ (a): For all vectors x and y,

$$(Tx|Ty) = (T^*Tx|y) = (Ix|y) = (x|y),$$

thus T is orthogonal (5.4.1). ■

The reason for the term 'orthogonal' becomes clearer when we say it with matrices:

5.4.3 Theorem *Let* E *be a Euclidean space, let* $T \in \mathcal{L}(\mathrm{E})$ *and let* A *be the matrix of* T *relative to an orthonormal basis of* E. *The following conditions are equivalent:*

(a) *T is orthogonal*;

(b) $A'A = I$.

Proof. Condition (b) says that, relative to the given orthonormal basis of E, $(\text{mat } T^*)(\text{mat } T) = I$ (5.3.5), in other words $\text{mat}(T^*T) = \text{mat } I$ (4.3.3), which is equivalent to $T^*T = I$ (4.2.7), in other words (by 5.4.2), to (a). ■

5.4.4 Definition An $n \times n$ real matrix A is said to be **orthogonal** if $A'A = I$.

5.4.5 Remarks If, relative to *some* orthonormal basis, the matrix of $T \in \mathcal{L}(\mathrm{E})$ is orthogonal, then T is orthogonal ('(b) ⇒ (a)' of 5.4.3), therefore the matrix of T relative to *every* orthonormal basis is orthogonal ('(a) ⇒ (b)' of 5.4.3).

If $A = (a_{ij})$ is an $n \times n$ real matrix, the condition $A'A = I$ says that

$$\sum_{k=1}^{n} a_{ki}a_{kj} = \delta_{ij}$$

for all i and j; this means that the column vectors of A are orthonormal in the Euclidean space E^n, hence form an orthonormal basis of E^n (5.1.12). It follows, free of charge, that the *rows* of A are also orthonormal:

5.4.6 Theorem *The following conditions on an $n \times n$ real matrix A are equivalent:*

(a) *A is orthogonal;*

(b) *A is invertible and $A^{-1} = A'$;*

(c) *A' is orthogonal.*

Proof. Condition (a) says that $A'A = I$, condition (c) says that $AA' = (A')'A' = I$, and the three conditions are equivalent by Theorem 4.10.3. ■

The orthogonal matrices are precisely the 'change-of-basis' matrices between pairs of orthonormal bases (cf. 4.10.1):

5.4.7 Theorem *Let* E *be a Euclidean space,* $x_1, \ldots, x_n$ *an orthonormal basis of* E, *and* $A = (a_{ij})$ *an* $n \times n$ *real matrix. Let*

$$y_j = \sum_{i=1}^{n} a_{ij} x_i \quad \text{for} \quad j = 1, \ldots, n.$$

The following conditions are equivalent:

(a) *A is orthogonal;*

(b) *$y_1, \ldots, y_n$ is an orthonormal basis of* E.

Proof. Let $T \in \mathcal{L}(\mathrm{E})$ be the linear mapping such that $Tx_i = y_i$ for all i (3.8.1); A is the matrix of T relative to the basis $x_1, \ldots, x_n$. In view of Theorem 5.4.3, the problem is to show that T is orthogonal if and only if $y_1, \ldots, y_n$ is an orthonormal basis.

If T is orthogonal, then

$$(y_i|y_j) = (Tx_i|Tx_j) = (x_i|x_j) = \delta_{ij},$$

thus $y_1, \ldots, y_n$ are orthonormal, hence are a basis (5.1.12, 3.5.13).

Conversely, if the vectors $y_1, \ldots, y_n$ are orthonormal, then

$$(Tx_i|Tx_j) = (y_i|y_j) = \delta_{ij} = (x_i|x_j)$$

for all i and j. Since the x_i generate E, it follows by linearity that

$$(Tx|Tx_j) = (x|x_j)$$

for every vector x and for all j, and a repetition of this argument shows that $(Tx|Ty) = (x|y)$ for all vectors x and y; in other words, T is orthogonal. ■

5.4.8 Corollary *Let* E *be an n-dimensional Euclidean space,* $T \in \mathcal{L}(\mathrm{E})$, *$A$ the matrix of T relative to an orthonormal basis of* E. *Let B be another $n \times n$ real matrix. The following conditions are equivalent:*

(a) *B is the matrix of T relative to some orthonormal basis of* E;

(b) *there exists an orthogonal matrix C such that $B = C'AC$.*

Proof. Say A is the matrix of T relative to the orthonormal basis $x_1, \ldots, x_n$.

(a) ⇒ (b): Suppose B is the matrix of T relative to an orthonormal basis $y_1, \ldots, y_n$. Let C be the change-of-basis matrix that expresses the y's in terms of the x's (cf. 4.10.1). By 4.10.5, $B = C^{-1}AC$; but C is orthogonal (5.4.7), so $C^{-1} = C'$ and $B = C'AC$.

(b) ⇒ (a): Let $C = (c_{ij})$ be an orthogonal matrix such that $B = C'AC$. Defining

$$y_j = \sum_{i=1}^{n} c_{ij} x_i \quad \text{for} \quad j = 1, \ldots, n,$$

we know from 5.4.7 that $y_1, \ldots, y_n$ is an orthonormal basis of E, and by 4.10.5 the matrix of T relative to the basis $y_1, \ldots, y_n$ is $C^{-1}AC = C'AC = B$. ■

5.4.9 Definition With notations as in (b) of 5.4.8, A and B are said to be **orthogonally similar** (cf. 4.10.6).

5.4.10 Example Let's determine all of the 2×2 orthogonal matrices. Suppose

$$A = \begin{pmatrix} a & c \\ b & d \end{pmatrix}$$

is orthogonal. From $A'A = I$ we have

(1) $a^2 + b^2 = c^2 + d^2 = 1$,

(2) $ac + bd = 0$,

and from $AA' = I$ we have

(3) $a^2 + c^2 = b^2 + d^2 = 1$,

(4) $ab + cd = 0$.

From (1) and (3),

$$d^2 = 1 - c^2 = a^2,$$

therefore $d = \pm a$. Similarly, $c^2 = 1 - d^2 = b^2$, so $c = \pm b$.

First let's dispose of a special case. If $a = 0$ then also $d = 0$ and we have four possiblities for A:

(5) $$\begin{pmatrix} 0 & 1 \\ 1 & 0 \end{pmatrix}, \begin{pmatrix} 0 & -1 \\ -1 & 0 \end{pmatrix}, \begin{pmatrix} 0 & -1 \\ 1 & 0 \end{pmatrix}, \begin{pmatrix} 0 & 1 \\ -1 & 0 \end{pmatrix}.$$

Now assume $a \neq 0$ (hence $d \neq 0$). If $d = a$ then (2) yields $a(c + b) = 0$, so $c = -b$ and A is the matrix

(6) $$\begin{pmatrix} a & -b \\ b & a \end{pmatrix}.$$

On the other hand, if $d = -a$ then (2) yields $a(c - b) = 0$, so $c = b$ and A is the matrix

(7) $$\begin{pmatrix} a & b \\ b & -a \end{pmatrix}.$$

In any case, $a^2 + b^2 = 1$ shows that (a, b) is a point on the unit circle of the Euclidean plane. If α is an angle in 'standard position' whose terminal ray passes through (a, b) (as in Fig. 20), then $(a, b) = (\cos\alpha, \sin\alpha)$ and we see that the matrices

$$(8) \qquad \begin{pmatrix} \cos\alpha & -\sin\alpha \\ \sin\alpha & \cos\alpha \end{pmatrix}, \quad \begin{pmatrix} \cos\alpha & \sin\alpha \\ \sin\alpha & -\cos\alpha \end{pmatrix}$$

cover all possiblities for A. In particular, the matrices in (5) are obtained by choosing $\alpha = \pi/2$ or $\alpha = 3\pi/2$ in (8).

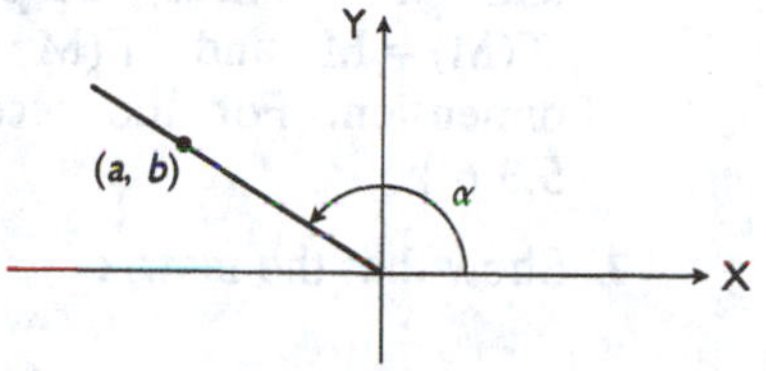

Fig. 20

► Exercises

1. Let E be a Euclidean space, $T \in \mathcal{L}(\mathrm{E})$. The following conditions are equivalent:

(a) T is orthogonal (that is, T preserves inner products);

(b) $\|Tx - Ty\| = \|x - y\|$ for all vectors x, y (that is, T preserves distances).
{Hint: §5.1, Exercise 9.}

2. Let $x_1, \ldots, x_n$ be an orthonormal basis of a Euclidean space E, let $r_1, \ldots, r_n$ be real numbers whose squares are equal (for example, let $r_i = \pm 2$ for all i), and let $T \in \mathcal{L}(\mathrm{E})$ be the linear mapping such that $Tx_i = r_i x_i$ for all i. Prove that T preserves orthogonality, in the sense that $(x|y) = 0 \Rightarrow (Tx|Ty) = 0$, and that T is a scalar multiple of an orthogonal mapping, but T is not orthogonal unless $r_i = \pm 1$ for all i.

3 If E is the Euclidean space and $T \in \mathcal{L}(\mathrm{E})$ is a linear mapping that preserves orthogonality (in the sense of Exercise 2), then T is a scalar multiple of an orthogonal mapping. Here's a sketch of the proof (not difficult):

(i) If $\|x\| = \|y\|$ then $\|Tx\| = \|Ty\|$. {Hint: §5.1, Exercise 8.}

(ii) Let $x_1, \ldots, x_n$ be an orthonormal basis of E and let r be the common value of the $\|Tx_i\|$ (part (i)). Show that $T^*T = r^2 I$. {Hint: $(T^*Tx_i|x_j) = (r^2 x_i|x_j)$ for all i and j; cf. the proof of Theorem 5.4.7.}

(iii) If $r \neq 0$ then $r^{-1}T$ is orthogonal.

4. Let A and B be $n \times n$ real matrices, I the $n \times n$ identity matrix. Prove:

(i) I is an orthogonal matrix;

(ii) if A and B are orthogonal, then so is AB;

(iii) if A is orthogonal then it is invertible and A^{-1} is orthogonal.

5. State and prove the analogue of Exercise 4 for linear mappings.

6. If $T \in \mathcal{L}(\mathbf{R}^2)$ is orthogonal, then T may be interpreted as either a rotation or a reflection in the Euclidean plane. {Hint: Example 5.4.10.}

7. Show by elementary computations that equations (1),(2) of Example 5.4.10 imply the equations (3), (4).

8. Let E be a Euclidean space, $T \in \mathcal{L}(\mathrm{E})$ an orthogonal linear mapping, and M a linear subspace of E such that $T(\mathrm{M}) \subset \mathrm{M}$. Prove that $T(\mathrm{M}) = \mathrm{M}$ and $T(\mathrm{M}^{\perp}) = \mathrm{M}^{\perp}$. {Hint: For the first equality, think dimension. For the second, recall that $T^{-1} = T^*$ and cite Theorem 5.3.6.}

9. Show that the matrix

$$\frac{1}{3}\begin{pmatrix} 1 & -2 & 2 \\ 2 & 2 & 1 \\ 2 & -1 & -2 \end{pmatrix}$$

is orthogonal.

10. Show that the matrix

$$\frac{1}{5}\begin{pmatrix} 3 & -4 & 0 \\ 4 & 3 & 0 \\ 0 & 0 & 5 \end{pmatrix}$$

is orthogonal, then invent another one like it. {Hint: $5^2 + 12^2 = 13^2$.}

11. In unitary spaces (§5.2, Exercise 10), linear mappings satisfying the condition of Definition 5.4.1 are called *unitary*. A square complex matrix A is said to be *unitary* if $A^*A = I$, where A^* is the conjugate-transpose of A. Extend the results of this section to unitary spaces.

12. Let e_1, e_2 be the canonical basis of E^2, $x_1 = e_1 + e_2$, $x_2 = e_1 - e_2$, and let $\mathrm{A} = \{x_1, x_2, -x_1, -x_2\}$; interpret A as the set of vertices of a square, centered at the origin, with edge length 2. Let $O = \{T \in \mathcal{L}(\mathrm{E}^2)\colon T(\mathrm{A}) = \mathrm{A}\}$, the set of all linear mappings of Euclidean 2-space that preserve the vertices of the square. Prove:

(i) Every $T \in O$ is bijective.

(ii) O contains 8 elements.

(iii) $I \in O$; if $S, T \in O$ then $ST \in O$ and $T^{-1} \in O$.

(iv) Every $T \in O$ is orthogonal.

(v) Find the matrices of the $T \in O$ relative to the canonical basis; relative to the basis x_1, x_2.

(vi) Interpret the elements of O as $I, R, R^2, R^3, H, V, D, E$, where R is counter-clockwise rotation of 90°, H is reflection in the x-axis, V is reflection in the y-axis, D is reflection in the diagonal containing $\pm x_1$, and E is reflection in the other diagonal. (O is called the *group of symmetries of the square,* or the *octic group.*)

{Hint: (i) A is generating, so T is surjective. (ii) There are four choices for Tx_1, each of which determines $T(-x_1)$; once Tx_1 is chosen, there are two possibilities ($\pm$) for Tx_2. (iv) Note that $x_1 \perp x_2$, $\|x_1\| = \|x_2\|$, $Tx_1 \perp Tx_2$ and $\|Tx_i\| = \|x_i\|$, then calculate $\|T(ax_1 + bx_2)\|^2$.}

6

Determinants (2×2 and 3×3)

This chapter is intended to be accessible to a reader with no prior background in the theory of determinants.[1] To this end, we make three concessions:

(1) The discussion is limited to determinants of 'small' matrices (2×2 and 3×3).

(2) We restrict attention to real scalars. This allows us to use inner product spaces to simplify the discussion of 3×3 determinants.

(3) We refrain from proving everything that can be proved about small matrices. (In particular, we make do without the 'product rule' for determinants, thereby avoiding a messy computation in the 3×3 case.)

The guidelines for the chapter, in brief: small matrices, real scalars, and let's not get carried away.

The reader who already feels comfortable with small-order determinants can go directly to §6.4; the reader who is not interested in the geometric applications (§6.7) can omit the present chapter altogether and advance either to determinants of order n (Ch. 7) or to the determinant-free treatment of diagonalization in dimension n (Ch. 9).

[1] Determinants are not to be conquered in a day. It is easy to learn the *rules* for calculating determinants of any size, but the theory underlying the rules—the key to understanding and applying them—is intricate and challenging (cf. Ch. 7).

6.1 Determinant of a 2×2 matrix

A simple test for linear independence in $\mathbf{R}^2$ leads us straight to determinants:

6.1.1 Theorem *The vectors $x = (a, b)$ and $y = (c, d)$ in $\mathbf{R}^2$ are linearly independent if and only if $ad - bc \neq 0$.*

Proof. Arguing contrapositively, it is the same to prove that x and y are linearly dependent if and only if $ad - bc = 0$.

If x and y are dependent, then one of them is a scalar multiple of the other (3.2.3). If, for example, $y = kx$, then $(c, d) = (ka, kb)$ and

$$ad - bc = a(kb) - b(ka) = k(ab - ba) = 0.$$

Conversely, assuming $ad - bc = 0$, we are to show that one of x, y is a multiple of the other. If $x = \theta$ this is obvious ($x = 0y$); suppose $x \neq \theta$, that is, either $a \neq 0$ or $b \neq 0$. If $a \neq 0$ then $d = (c/a)b$ and

$$y = (c, d) = (c, (c/a)b) = (c/a)(a, b) = kx$$

with $k = c/a$; similarly, if $b \neq 0$ then $y = kx$ with $k = d/b$. ■

6.1.2 Corollary *A 2×2 real matrix*

$$A = \begin{pmatrix} a & c \\ b & d \end{pmatrix}$$

is invertible if and only if $ad - bc \neq 0$.

Proof. Recall that A is said to be invertible if there exists a matrix $B \in \mathrm{M}_2(\mathbf{R})$ such that $AB = BA = I$ (4.10.2). Let $T \in \mathcal{L}(\mathbf{R}^2)$ be the linear mapping whose matrix relative to the canonical basis of $\mathbf{R}^2$ is A (4.2.6). The proof is most conveniently organized as a chain of equivalences:

A invertible $\Leftrightarrow$ T bijective	(4.10.4)
$\Leftrightarrow$ T surjective	(3.7.4)
$\Leftrightarrow$ the columns of A span $\mathbf{R}^2$	(4.2.8)
$\Leftrightarrow$ the columns of A are independent	(3.5.13)
$\Leftrightarrow$ $ad - bc \neq 0$	(6.1.1) ■

6.1.3 Definition With notations as in Corollary 6.1.2, the scalar $ad - bc$ is called the **determinant** of the matrix A and is denoted $|A|$ or $\det A$. When the first of these notations is used, it is customary to omit the parentheses that 'contain' the elements of A and to write

$$\begin{vmatrix} a & c \\ b & d \end{vmatrix} = ad - bc.$$

The determinant of a 2×2 matrix (before it has been calculated) is called a

'2 × 2 determinant' (or a 'determinant of order 2'); nevertheless, it is just a scalar.

6.1.4 Example Determinants come up in connection with the solution of systems of linear equations

$$ax + cy = r,$$
$$bx + dy = s.$$

When $ad - bc \neq 0$, a solution-pair (x, y) exists and is given by the formulas

$$x = \frac{\begin{vmatrix} r & c \\ s & d \end{vmatrix}}{\begin{vmatrix} a & c \\ b & d \end{vmatrix}}, \quad y = \frac{\begin{vmatrix} a & r \\ b & s \end{vmatrix}}{\begin{vmatrix} a & c \\ b & d \end{vmatrix}}.$$

For expressing the *properties* of determinants, it is more convenient to think of a determinant as a function of two vectors (rather than of one matrix):

6.1.5 Definition For vectors $x = (a, b)$ and $y = (c, d)$ in $\mathbf{R}^2$, we define

$$\det(x, y) = ad - bc = \begin{vmatrix} a & c \\ b & d \end{vmatrix}.$$

From this perspective, the determinant of a 2×2 matrix

$$A = \begin{pmatrix} a & c \\ b & d \end{pmatrix}$$

is a scalar-valued function of its two column vectors. Note that the transpose of A has the same determinant as A,

$$\begin{vmatrix} a & b \\ c & d \end{vmatrix} = \begin{vmatrix} a & c \\ b & d \end{vmatrix},$$

thus we can regard the determinant of A equally well as a function of its row vectors (a, c), (b, d).

The function $\det : \mathbf{R}^2 \times \mathbf{R}^2 \to \mathbf{R}$ defined in this way has the following properties:

6.1.6 Theorem *If $x, y, z \in \mathbf{R}^2$ and $c \in \mathbf{R}$, then*

(1) $\det(x, x) = 0$,
(2) $\det(y, x) = -\det(x, y)$,
(3) $\det(x + y, z) = \det(x, z) + \det(y, z)$,
(4) $\det(x, y + z) = \det(x, y) + \det(x, z)$,
(5) $\det(cx, y) = c \det(x, y) = \det(x, cy)$,
(6) $\det(e_1, e_2) = 1$,

where e_1, e_2 is the canonical basis of $\mathbf{R}^2$ (3.5.2).

Proof. (1) Obvious (a 2×2 determinant with equal rows is zero).

(3) If $x = (x_1, x_2)$, $y = (y_1, y_2)$, $z = (z_1, z_2)$, then

$$\begin{aligned}\det(x + y, z) &= \begin{vmatrix} x_1 + y_1 & x_2 + y_2 \\ z_1 & z_2 \end{vmatrix} \\ &= (x_1 + y_1)z_2 - z_1(x_2 + y_2) \\ &= (x_1 z_2 - z_1 x_2) + (y_1 z_2 - z_1 y_2) \\ &= \begin{vmatrix} x_1 & x_2 \\ z_1 & z_2 \end{vmatrix} + \begin{vmatrix} y_1 & y_2 \\ z_1 & z_2 \end{vmatrix} \\ &= \det(x, z) + \det(y, z).\end{aligned}$$

(4) Similar to (3).

(2) By (1), (3) and (4), we have

$$\begin{aligned}0 = \det(x + y, x + y) &= \det(x, x + y) + \det(y, x + y) \\ &= \det(x, x) + \det(x, y) + \det(y, x) + \det(y, y) \\ &= 0 + \det(x, y) + \det(y, x) + 0,\end{aligned}$$

whence (2). {There is a shorter proof!}

(5) For example,

$$\begin{aligned}\det(cx, y) = \begin{vmatrix} cx_1 & cx_2 \\ y_1 & y_2 \end{vmatrix} &= (cx_1)y_2 - y_1(cx_2) \\ &= c(x_1 y_2 - y_1 x_2) = c\begin{vmatrix} x_1 & x_2 \\ y_1 & y_2 \end{vmatrix} \\ &= c\det(x, y).\end{aligned}$$

(6) $$\det(e_1, e_2) = \begin{vmatrix} 1 & 0 \\ 0 & 1 \end{vmatrix} = 1 \cdot 1 - 0 \cdot 0 = 1. \blacksquare$$

Properties (3)–(5) can be expressed by saying that the determinant of a 2×2 matrix is a linear function of each row (or of each column). Property (6) says that $|I| = 1$, where I is the 2×2 identity matrix.

▶ Exercises

1. If $A = \begin{pmatrix} 2 & 4 \\ 3 & 5 \end{pmatrix}$ find A^{-1}.

2. Give a direct elementary proof of 6.1.2 by solving the equation $AB = I$, regarding the four entries of B as unknowns.

3. If

$$A = \begin{pmatrix} a & c \\ b & d \end{pmatrix}, \quad \text{define } B = \begin{pmatrix} d & -c \\ -b & a \end{pmatrix}.$$

Show that $AB = |A| \cdot I$. If A is invertible, deduce a formula for A^{-1}.

4. If A is a 2×2 matrix and k is a scalar, prove that $|kA| = k^2|A|$.

5. Let A and B be 2×2 real matrices. Prove:

(i) $|AB| = |A| \cdot |B|$ (the 'product rule').
(ii) If A and B are similar (4.10.6), then $|A| = |B|$.

Show by means of an example that the converse of the implication in (ii) is false.

6.2 Cross product of vectors in $\mathbf{R}^3$

The determinant test for proportionality (6.1.1) extends to a pair of vectors in $\mathbf{R}^3$, at the cost of considering three determinants instead of one:

6.2.1 Theorem *The vectors $x = (x_1, x_2, x_3)$ and $y = (y_1, y_2, y_3)$ in $\mathbf{R}^3$ are linearly independent if and only if at least one of the three determinants*

$$(*) \qquad \begin{vmatrix} x_2 & x_3 \\ y_2 & y_3 \end{vmatrix}, \quad \begin{vmatrix} x_1 & x_3 \\ y_1 & y_3 \end{vmatrix}, \quad \begin{vmatrix} x_1 & x_2 \\ y_1 & y_2 \end{vmatrix}$$

is nonzero.

Proof. These determinants are formed by considering the matrix

$$A = \begin{pmatrix} x_1 & x_2 & x_3 \\ y_1 & y_2 & y_3 \end{pmatrix},$$

omitting a column and taking the determinant of what's left; in view of 6.1.2, the assertion of the theorem is that A has rank 2 if and only if one of its 2×2 'submatrices' is invertible.

Stated contrapositively, our problem is to show that the vectors x, y are dependent if and only if all three of the determinants $(*)$ are zero.

Suppose x, y are dependent. If, for example, $y = kx$, then in particular $(y_2, y_3) = (kx_2, kx_3)$, so the first of the determinants $(*)$ is zero; similarly for the other two.

Conversely, suppose that all three of the determinants $(*)$ are zero. If $x = \theta$ then $x = 0y$ and we're done; let's assume $x \neq \theta$. Not all of x_1, x_2, x_3 are zero, so there are two possibilities: (1) exactly one of them is nonzero, or (2) at least two of them are nonzero.

case 1: exactly one of x_1, x_2, x_3 is nonzero.

If, for example, $x_3 \neq 0$, then $x_1 = x_2 = 0$ by assumption. Since the determinants $(*)$ are zero, in particular $x_1y_3 - y_1x_3 = 0$, we have

$$y_1x_3 = x_1y_3 = 0y_3 = 0,$$

therefore $y_1 = 0$; similarly, from $x_2y_3 - y_2x_3 = 0$ we see that $y_2 = 0$. Thus,

$$y = (0, 0, y_3) = (y_3/x_3)(0, 0, x_3) = kx$$

with $k = y_3/x_3$.

case 2: at least two of x_1, x_2, x_3 are nonzero.

Then $(x_2, x_3) \neq (0, 0)$, $(x_1, x_3) \neq (0, 0)$ and $(x_1, x_2) \neq (0, 0)$. Since

$(x_2, x_3) \neq (0, 0)$ and the first determinant of $(*)$ is zero, it follows from Theorem 6.1.1 that (y_2, y_3) is proportional to (x_2, x_3), say

$$(y_2, y_3) = r(x_2, x_3),$$

r a suitable scalar. Similarly, there are scalars s and t such that

$$(y_1, y_3) = s(x_1, x_3), \quad (y_1, y_2) = t(x_1, x_2);$$

if we show that $r = s = t$, then it will follow that $y = rx$.

By assumption, at least two of the x_i are nonzero; suppose, for example, that $x_1 \neq 0$ and $x_2 \neq 0$. From the proportionalities just derived, we see that

$$sx_1 = y_1 = tx_1 \quad \text{and} \quad rx_2 = y_2 = tx_2;$$

cancelling x_1 and x_2, we have $s = t$ and $r = t$. ■

If you've met 'cross products' in physics or multidimensional calculus, the determinants $(*)$ will ring a bell (with one sour note in the middle term):

6.2.2 Definition Let $x = (x_1, x_2, x_3)$ and $y = (y_1, y_2, y_3)$ be vectors in $\mathbf{R}^3$. The **cross product** of x and y (in that order) is the vector, denoted $x \times y$, whose components are (in order)

$$\begin{vmatrix} x_2 & x_3 \\ y_2 & y_3 \end{vmatrix}, \quad \begin{vmatrix} x_3 & x_1 \\ y_3 & y_1 \end{vmatrix}, \quad \begin{vmatrix} x_1 & x_2 \\ y_1 & y_2 \end{vmatrix};$$

in other words,

$$x \times y = (x_2y_3 - y_2x_3, x_3y_1 - y_3x_1, x_1y_2 - y_1x_2).$$

Notice that the second component of $x \times y$ is the *negative* of the second determinant in the list $(*)$ of 6.2.1; in forming the determinants that make up the components of $x \times y$, the columns of the matrix

$$\begin{pmatrix} x_1 & x_2 & x_3 \\ y_1 & y_2 & y_3 \end{pmatrix}$$

are to be taken in 'cyclic order': column 1 is followed by column 2, column 2 is followed by column 3, and *column* 3 *is followed by column* 1.

6.2.3 Corollary *Vectors x, y in $\mathbf{R}^3$ are linearly independent if and only if $x \times y \neq \theta$.*

Proof. Immediate from 6.2.1 and 6.2.2. ■

The basic formal properties of the cross product are as follows:

6.2.4 Theorem *If x, y, z are vectors in $\mathbf{R}^3$ and $c \in \mathbf{R}$, then*

(1) $x \times x = \theta$,

(2) $y \times x = -(x \times y)$,

(3) $(x + y) \times z = x \times z + y \times z$,

(4) $x \times (y + z) = x \times y + x \times z$,

(5) $(cx) \times y = c(x \times y) = x \times (cy)$.

Proof. Direct computation, citing the corresponding properties of 2×2 determinants (Theorem 6.1.6). ■

Properties (3)–(5) of the cross product remind us of some of the properties of inner products (Definition 5.1.1), but (1) and (2) are at opposite poles from the 'positive definiteness' and 'symmetry' of an inner product. Of course the most vivid difference is that the cross product of two vectors in $\mathbf{R}^3$ is a vector, whereas the inner product of two vectors is a scalar. While we're on the subject, the canonical inner product on $\mathbf{R}^3$ (5.1.2) is also called the **dot product** and is written $x \cdot y$ instead of $(x|y)$.

One property of the cross product is so striking that it merits a theorem of its own:

6.2.5 Theorem *If* $x, y \in \mathbf{R}^3$ *then* $x \times y$ *is orthogonal to both* x *and* y.

Proof. The orthogonality in question refers to the canonical inner product on $\mathbf{R}^3$ (5.1.2). Our problem is to show that $(x \times y|x) = 0$ and $(x \times y|y) = 0$; since $x \times y = -(y \times x)$, it is enough to show that $(x \times y|x) = 0$.

Suppose $y = (a, b, c)$. If e_1, e_2, e_3 is the canonical basis of $\mathbf{R}^3$ (3.5.2), then $y = ae_1 + be_2 + ce_3$, so

$$x \times y = a(x \times e_1) + b(x \times e_2) + c(x \times e_3)$$

by 6.2.4, therefore

$$(x \times y|x) = a(x \times e_1|x) + b(x \times e_2|x) + c(x \times e_3|x);$$

we are thus reduced to proving that $(x \times e_i|x) = 0$ for $i = 1, 2, 3$.

Say $x = (x_1, x_2, x_3)$. From the matrix

$$\begin{pmatrix} x_1 & x_2 & x_3 \\ 1 & 0 & 0 \end{pmatrix}$$

we see that the components of $x \times e_1$ are

$$\begin{vmatrix} x_2 & x_3 \\ 0 & 0 \end{vmatrix}, \quad \begin{vmatrix} x_3 & x_1 \\ 0 & 1 \end{vmatrix}, \quad \begin{vmatrix} x_1 & x_2 \\ 1 & 0 \end{vmatrix}$$

thus $x \times e_1 = (0, x_3, -x_2)$ and

$$(x \times e_1|x) = 0x_1 + x_3x_2 + (-x_2)x_3 = 0.$$

Similarly, $x \times e_2$ has components

$$\begin{vmatrix} x_2 & x_3 \\ 1 & 0 \end{vmatrix}, \quad \begin{vmatrix} x_3 & x_1 \\ 0 & 0 \end{vmatrix}, \quad \begin{vmatrix} x_1 & x_2 \\ 0 & 1 \end{vmatrix}$$

thus $x \times e_2 = (-x_3, 0, x_1)$ and

$$(x \times e_2|x) = (-x_3)x_1 + 0x_2 + x_1x_3 = 0.$$

Exercise: Check that $(x \times e_3|x) = 0$. ■

6.2.6 Corollary *If $x, y \in \mathbf{R}^3$ are linearly independent, then $x, y, x \times y$ is a basis of $\mathbf{R}^3$.*

Proof. Since $x \times y$ is orthogonal to each of x and y (6.2.5), it is orthogonal to every linear combination of x and y:

$$(x \times y|ax + by) = a(x \times y|x) + b(x \times y|y) = a0 + b0 = 0.$$

But $x \times y$ is nonzero (6.2.3), so it is not orthogonal to itself ((1) of 5.1.1). Conclusion: $x \times y$ is not a linear combination of x and y. It follows that x, y, $x \times y$ are independent (3.3.5), hence are a basis of $\mathbf{R}^3$ (3.5.13). ■

6.2.7 Remark With notations as in 6.2.6, let $\mathrm{M} = \mathbf{R}x + \mathbf{R}y$ be the linear span of x and y. Since M is 2-dimensional, its orthogonal complement $\mathrm{M}^\perp$ is 1-dimensional (5.2.8); but $x \times y$ is a nonzero vector (6.2.3) belonging to $\mathrm{M}^\perp$ (proof of 6.2.6), consequently $\mathrm{M}^\perp = \mathbf{R}(x \times y)$. Thus, in the notation of 3.1.1,

$$[\{x, y\}]^\perp = [\{x \times y\}]$$

(provided x and y are linearly independent).

The final theorem is for application in the next section (Theorem 6.3.3):

6.2.8 Theorem *Let $x \in \mathbf{R}^3$ and let $T:\mathbf{R}^3 \to \mathbf{R}^3$ be the mapping defined by $Ty = x \times y$. Then T is linear and its adjoint is given by the formula $T^*y = y \times x$.*

Proof. The mapping T is linear by 6.2.4. Let e_1, e_2, e_3 be the canonical basis of $\mathbf{R}^3$. By the proof of 6.2.5,

$$\begin{aligned} Te_1 &= x \times e_1 = (0, x_3, -x_2) = 0e_1 + x_3e_2 - x_2e_3, \\ Te_2 &= x \times e_2 = (-x_3, 0, x_1) = -x_3e_1 + 0e_2 + x_1e_3, \\ Te_3 &= x \times e_3 = (x_2, -x_1, 0) = x_2e_1 - x_1e_2 + 0e_3, \end{aligned}$$

thus the matrix of T relative to the canonical basis is

$$A = \begin{pmatrix} 0 & -x_3 & x_2 \\ x_3 & 0 & -x_1 \\ -x_2 & x_1 & 0 \end{pmatrix}.$$

Since the basis is orthonormal, the matrix of T^* is the transposed matrix A' (5.3.5); but

$$A' = \begin{pmatrix} 0 & x_3 & -x_2 \\ -x_3 & 0 & x_1 \\ x_2 & -x_1 & 0 \end{pmatrix} = -A,$$

therefore $T^* = -T$ (4.2.7), thus $T^*y = -Ty = -(x \times y) = y \times x$ for all $y \in \mathbf{R}^3$. ■

▶ Exercises

1. Show that $\|x \times y\|^2 = \|x\|^2\|y\|^2 - (x|y)^2$ for all vectors x, y in $\mathbf{R}^3$ (*Lagrange's identity*).

2. If u_1, u_2 are orthonormal vectors in $\mathbf{R}^3$ (5.2.13), prove that u_1, u_2, $u_1 \times u_2$ is an orthonormal basis of $\mathbf{R}^3$. {Hint: Theorem 6.2.5 and Exercise 1.}

3. Calculate the 9 vectors $e_i \times e_j$, where e_1, e_2, e_3 is the canonical basis of $\mathbf{R}^3$. {For example, $e_1 \times e_2 = e_3$.}

4. If $x, y \in \mathbf{R}^3$ are linearly dependent, show that $[\{x, y\}]^\perp \neq [\{x \times y\}]$ (cf. 6.2.7). Conclude that $[\{x, y\}]^\perp = [\{x \times y\}]$ if and only if x, y are independent.

5. Vectors x, y in an inner product space are linearly dependent if and only if $|(x|y)| = \|x\| \cdot \|y\|$ (Theorem 5.1.8). Give a lightning proof for vectors in $\mathbf{R}^3$. {Hint: Exercise 1.}

6.3 Determinant of a 3 × 3 matrix

As in §6.1, we begin with a criterion for linear independence:

6.3.1 Theorem *The vectors x, y, z in $\mathbf{R}^3$ are linearly independent if and only if* $(x \times y|z) \neq 0$.

Proof. Let $M = \mathbf{R}x + \mathbf{R}y$ be the linear span of x and y; note that if x and y are independent then $M^\perp = [\{x \times y\}]$ (6.2.7), therefore $M = [\{x \times y\}]^\perp$ (5.2.11).

If x, y, z are independent then $M = [\{x \times y\}]^\perp$ and $z \notin M$ (3.3.5), therefore z is not orthogonal to $x \times y$.

Conversely, if $(x \times y|z) \neq 0$ then $x \times y \neq \theta$ and $z \notin [\{x \times y\}]^\perp$, therefore x, y are independent (6.2.3) and $z \notin M$; consequently x, y, z are independent (3.3.5). ■

A less formal, more intuitive way of seeing the equivalence

$$x,\ y,\ z \text{ independent} \Leftrightarrow (x \times y|z) \neq 0$$

is as follows. The right side says that x, y are independent $(x \times y \neq \theta)$ and z is not perpendicular to $x \times y$; that is, x, y are independent and z is not in the plane spanned by x and y. That is what the left side says, too.

6.3.2 Definition If $x, y, z \in \mathbf{R}^3$ we define

$$\det(x, y, z) = (x \times y|z),$$

called the **determinant** of the ordered triple of vectors x, y, z.

The message of 6.3.1 is that vectors x, y, z in $\mathbf{R}^3$ are independent if and only if $\det(x, y, z) \neq 0$. Thus, as in 6.1.1, the issue of independence hangs on a single determinant. The three 2 × 2 determinants considered earlier (6.2.1) are of course alive and well:

$$\begin{aligned}\det(x, y, z) &= (x \times y|z)\\ &= z_1\begin{vmatrix} x_2 & x_3 \\ y_2 & y_3 \end{vmatrix} + z_2\begin{vmatrix} x_3 & x_1 \\ y_3 & y_1 \end{vmatrix} + z_3\begin{vmatrix} x_1 & x_2 \\ y_1 & y_2 \end{vmatrix}\\ &= z_1\begin{vmatrix} x_2 & x_3 \\ y_2 & y_3 \end{vmatrix} - z_2\begin{vmatrix} x_1 & x_3 \\ y_1 & y_3 \end{vmatrix} + z_3\begin{vmatrix} x_1 & x_2 \\ y_1 & y_2 \end{vmatrix}.\end{aligned}$$

The minus sign arises from interchanging two columns of a determinant (6.1.6); when this is done, the columns appear in their 'natural' order rather than their 'cyclic' order. One also writes

$$\begin{vmatrix} x_1 & x_2 & x_3 \\ y_1 & y_2 & y_3 \\ z_1 & z_2 & z_3 \end{vmatrix}$$

for $\det(x, y, z)$, and the preceding formula is called the 'expansion' of the determinant by its third row.

The basic properties of the mapping $\det:\mathbf{R}^3 \times \mathbf{R}^3 \times \mathbf{R}^3 \to \mathbf{R}$ are as follows:

6.3.3 Theorem *Let $x, y, z \in \mathbf{R}^3$ and $c \in \mathbf{R}$. Then:*

(1) *For fixed y and z, the mapping $x \mapsto \det(x, y, z)$ is a linear form on $\mathbf{R}^3$.*

(2) $\det(y, x, z) = -\det(x, y, z)$.

(3) $\det(x, z, y) = -\det(x, y, z)$.

(4) $\det(e_1, e_2, e_3) = 1$, *where e_1, e_2, e_3 is the canonical basis of $\mathbf{R}^3$.*

Proof.

(1) If any two of the vectors x, y, z are fixed, then $\det(x, y, z)$ is a linear function of the remaining vector; this is clear from the 'bilinearity' of cross products (6.2.4) and inner products (5.1.1, 5.1.4).

(4) $\det(e_1, e_2, e_3) = (e_1 \times e_2|e_3) = (e_3|e_3) = 1$.

(2) $\det(y, x, z) = (y \times x|z) = -(x \times y|z) = -\det(x, y, z)$.

(3) Regard $x \in \mathbf{R}^3$ as fixed and, as in 6.2.8, write $Ty = x \times y$ for $y \in \mathbf{R}^3$; then (by 6.2.8)

$$\begin{aligned}\det(x, y, z) &= (x \times y|z) = (Ty|z)\\ &= (y|T^*z) = (y|z \times x)\\ &= (z \times x|y) = -(x \times z|y)\\ &= -\det(x, z, y). \quad \blacksquare\end{aligned}$$

6.3.4 Remarks The meaning of (2) and (3) is that if adjacent rows of the determinant

$$\begin{vmatrix} x_1 & x_2 & x_3 \\ y_1 & y_2 & y_3 \\ z_1 & z_2 & z_3 \end{vmatrix}$$

are interchanged, then the determinant changes sign. The effect of *any* permutation of the rows is then easily computed: if the permutation is achieved by k successive interchanges of adjacent rows, the net effect is to multiply the original determinant by $(-1)^k$.

We see from (2) that $\det(x, x, z) = 0$ (also clear from the definition). More generally, if any two rows are proportional, then the determinant is 0; for example,

$$\det(x, y, cx) = c\det(x, y, x) = -c\det(x, x, y) = 0.$$

Other properties are noted in the exercises; in the text proper, we're going to make do with the bare minimum.

The determinant notation can be carried over to 3×3 matrices:

6.3.5 Definition Let $A = (a_{ij})$ be a 3×3 real matrix. Denote the row vectors of A by x, y, z:

$$x = (a_{11}, a_{12}, a_{13}),$$
$$y = (a_{21}, a_{22}, a_{23}),$$
$$z = (a_{31}, a_{32}, a_{33}).$$

The **determinant** of A, denoted $|A|$ (or $\det A$), is defined to be $\det(x, y, z)$; thus, in the notation following 6.3.2,

$$|A| = \begin{vmatrix} a_{11} & a_{12} & a_{13} \\ a_{21} & a_{22} & a_{23} \\ a_{31} & a_{32} & a_{33} \end{vmatrix}.$$

6.3.6 Theorem *A 3×3 real matrix A is invertible if and only if $|A| \neq 0$.*

Proof. With 2 replaced by 3 in the proof of 6.1.2, we see that A is invertible if and only if its column vectors are linearly independent, in other words, A has rank 3; this is equivalent to the independence of the row vectors of A (4.6.7), which is in turn equivalent to $|A| \neq 0$ (remark following 6.3.2). ■

6.3.7 Remarks (Optional) One of the deeper properties of determinants is the 'product rule': $|AB| = |A| \cdot |B|$ for all A, B in $M_3(\mathbf{R})$, that is, the determinant of a product is the product of the determinants of the factors.[1] We won't be proving this here, but, for a moment, let's take it for granted; it would follow that if $C \in M_3(\mathbf{R})$ is invertible, then

$$|C^{-1}AC| = |C^{-1}| \cdot |AC| = |AC| \cdot |C^{-1}| = |ACC^{-1}| = |AI| = |A|,$$

thus similar matrices (4.10.6) have the same determinant.

In particular, if A and B are similar, then $|A| \neq 0$ if and only if $|B| \neq 0$; we do need this fact, but we can prove it directly without resorting to the product rule:

6.3.8 Corollary *If A and B are similar real matrices, then $|A| \neq 0 \Leftrightarrow |B| \neq 0$.*

[1] The word 'deep' is certainly appropriate for $n \times n$ determinants (cf. Theorem 7.3.6); for $n = 3$, the proof of the product rule is just a messy computation, but the property is still deep (sure it's true, but *why* is it true?).

Proof. By assumption, there is an invertible matrix C such that $B = C^{-1}AC$. If $|A| \neq 0$ then A is invertible (6.3.6, therefore so is B (with inverse $C^{-1}A^{-1}C$), therefore $|B| \neq 0$ (6.3.6). Conversely, since $A = CBC^{-1}$, $|B| \neq 0$ implies $|A| \neq 0$. ■

6.3.9 Remark With notations as in Definition 6.3.5, $\det(-x, -y, -z) = (-1)^3 \det(x, y, z)$, in other words $|-A| = -|A|$ for a 3×3 matrix A.

▶ *Exercises*

1. Are the following vectors linearly independent?:

$$x = (2, -1, 3), \qquad y = (3, 2, 4), \qquad z = (5, 8, 6).$$

2. Calculate the determinant

$$\begin{vmatrix} 1 & -2 & 3 \\ 4 & 5 & 2 \\ 2 & -3 & 6 \end{vmatrix}.$$

3. Prove that $\det(x, y, z)$ is a linear function of y (for fixed x and z) and a linear function of z (for fixed x and y).

4. If A is a 3×3 real matrix, prove by direct computation that $|A'| = |A|$, where A' is the transpose of A.

5. If A and B are 3×3 real matrices, prove by direct computation (in two hours or less) that $|AB| = |A| \cdot |B|$. Deduce (in thirty seconds or less) that $|A| = |B|$ when A and B are similar.

6. If A is a 3×3 real matrix and c is a real number, prove that $|cA| = c^3|A|$. {Hint: Exercise 3.}

7. Calculate the determinant of the matrix

$$A = \begin{pmatrix} 2 & 3 & 7 \\ 0 & 4 & 5 \\ 0 & 0 & -1 \end{pmatrix}.$$

8. If x, y, z are vectors in $\mathbf{R}^3$, prove that $\det(z, y, x) = -\det(x, y, z)$.

9. Find a scheme for expanding a 3×3 determinant by any row or column. {Hint: In the formula given in the remarks following Definition 6.3.2, the coefficients z_1, z_2, z_3 are preceded by the signs $+ - +$; when using other rows or columns, you may need to use the pattern $- + -$.}

6.4 Characteristic polynomial of a matrix (2 × 2 or 3 × 3)

The determinant of a matrix gives numerical information about the matrix, but the information is in a sense inconclusive because very different matrices can have the same determinant; for example, the matrices

$$\begin{pmatrix} 2 & 0 \\ 0 & 2 \end{pmatrix}, \begin{pmatrix} 2 & 1 \\ 0 & 2 \end{pmatrix}, \begin{pmatrix} 5 & 1 \\ 0 & 4/5 \end{pmatrix}, \begin{pmatrix} 5 & 1 \\ 1 & 1 \end{pmatrix}$$

all have determinant 4, but no two of them are similar to each other in the sense of §4.10. {The first is similar only to itself; the other three have different traces, so are ineligible to be similar (§4.10, Exercise 5).}

How can we squeeze more information out of the determinant? By introducing variation into the problem[1]:

6.4.1 Definition Let

$$A = \begin{pmatrix} a & c \\ b & d \end{pmatrix}$$

be a 2×2 real matrix. For each real number r, the matrix

$$A - rI = \begin{pmatrix} a - r & c \\ b & d - r \end{pmatrix}$$

has determinant

$$\begin{aligned} |A - rI| &= (a - r)(d - r) - bc \\ &= r^2 - (a + d)r + (ad - bc), \end{aligned}$$

which is a quadratic polynomial function of r. The **characteristic polynomial** of A, denoted p_A, is the quadratic polynomial

$$p_A = t^2 - (a + d)t + (ad - bc)$$

(here t is an indeterminate[2]). Allowing the entries of a matrix to be polynomials, we have formally

$$p_A = |A - tI| = \begin{vmatrix} a - t & c \\ b & d - t \end{vmatrix};$$

since $|A - tI| = |tI - A|$ (because $|-B| = |B|$ for every 2×2 matrix B), the characteristic polynomial of A is also given by the formula

$$p_A = |tI - A|.$$

The roots[3] of the polynomial p_A are called the **characteristic roots** of the matrix A.

6.4.2 Examples The characteristic polynomials of the four matrices considered earlier are, respectively,

$$(t - 2)^2, \ (t - 2)^2, \ (t - 5)(t - 4/5), \ t^2 - 6t + 4.$$

For the first three, the characteristic roots are obvious; for the fourth, they are $3 \pm \sqrt{5}$ (quadratic formula!). For the matrix

$$\begin{pmatrix} 0 & 1 \\ -1 & 0 \end{pmatrix}$$

[1] The fundamental strategy of calculus!

[2] Think of an 'indeterminate' as an abstract place-holder for an honest number, that is, a symbol that is manipulated algebraically as if it were a number.

[3] A *root* (or 'zero') of a polynomial p is a number c such that $p(c) = 0$.

the characteristic polynomial is $t^2 + 1$ and the characteristic roots are $\pm i$.

In general, the characteristic roots are complex numbers, but there's an important case in which they are guaranteed to be real:

6.4.3 Theorem *If A is a 2×2 real matrix such that $A' = A$, then the characteristic roots of A are real.*

Proof. To say that A equals its transpose means that

$$A = \begin{pmatrix} a & b \\ b & d \end{pmatrix}$$

for suitable real numbers a, b, d. The characteristic polynomial of A is

$$p_A = t^2 - (a + d)t + (ad - b^2);$$

its discriminant

$$\Delta = (a + d)^2 - 4(ad - b^2) = (a - d)^2 + 4b^2$$

is ≥ 0, so the roots are real numbers (namely, $(a + d \pm \sqrt{\Delta})/2$). ■

6.4.4 Definition A square matrix A is said to be **symmetric** if $A' = A$.

For example, the fourth matrix given at the beginning of the section is symmetric; we now know why its characteristic roots turned out to be real numbers (Theorem 6.4.3).

The following theorem is the tool for studying symmetry by means of linear mappings:

6.4.5 Theorem *Let E be a Euclidean space, $T \in \mathcal{L}(\mathrm{E})$, and let A be the matrix of T relative to an orthonormal basis of E. Then $T^* = T$ if and only if A is symmetric.*

Proof. The matrix of T^* relative to the given orthonormal basis is A' (5.3.5), so $T^* = T$ if and only if $A' = A$ (4.2.7). ■

6.4.6 Definition A linear mapping T in a Euclidean space is said to be **self-adjoint** if $T^* = T$.

6.4.7 Remark If the matrix of $T \in \mathcal{L}(\mathrm{E})$ relative to *some* orthonormal basis is symmetric, then T is self-adjoint (the 'if' part of 6.4.5), therefore its matrix relative to *every* orthonormal basis is symmetric (the 'only if' part of 6.4.5).

Let's advance to 3×3 matrices:

6.4.8 Definition If $A = (a_{ij})$ is a 3×3 real matrix, the **characteristic polynomial** of A, denoted p_A, is defined by the formula $p_A = |tI - A|$. (As in Definition 6.4.1, we accept polynomials as matrix entries and compute as if they were numbers.[4])

Note that $p_A = |-(A - tI)| = -|A - tI|$ (6.3.9), thus

[4] This is not really so mysterious; the ratios p/q of real polynomials p, q ($q \neq 0$) form a field (Appendix A.4) and we are back in the framework of matrices with entries in a field.

$$-p_A = |A - tI| = \begin{vmatrix} a_{11} - t & a_{12} & a_{13} \\ a_{21} & a_{22} - t & a_{23} \\ a_{31} & a_{32} & a_{33} - t \end{vmatrix}.$$

By the formula of Definition 6.3.2,

$-p_A =$

$$a_{31}\begin{vmatrix} a_{12} & a_{13} \\ a_{22} - t & a_{23} \end{vmatrix} - a_{32}\begin{vmatrix} a_{11} - t & a_{13} \\ a_{21} & a_{23} \end{vmatrix} + (a_{33} - t)\begin{vmatrix} a_{11} - t & a_{12} \\ a_{21} & a_{22} - t \end{vmatrix},$$

thus p_A is a cubic polynomial with t^3 the term of highest degree. The roots of p_A are called the **characteristic roots** of the matrix A.

6.4.9 Remarks When A is a 2×2 matrix, $|A - tI| = |tI - A|$; when A is a 3×3 matrix, $|A - tI| = -|tI - A|$. In either case, the characteristic roots of A are also the roots of $|A - tI|$.

When the concept of characteristic polynomial is extended to square matrices of order n (as in Chapter 8), the definition $p_A = |tI - A|$ proves to be more convenient notationally than $|A - tI|$. One possible reason: $|tI - A|$ reliably leads off with t^n (that's not earthshaking, but it's nice), whereas the leading term of $|A - tI|$ oscillates between t^n and $-t^n$ according as n is even or odd (not the end of the world, but it introduces an unwelcome variable in the discussion of factorization). The advantage of $|A - tI|$ is that it is easier to calculate in actual examples: one simply subtracts t from the diagonal elements of A and takes determinant (whereas $tI - A$ generates a swarm of minus signs).

Our compromise: the official definition of the characteristic polynomial is $p_A = |tI - A|$, but we *calculate* $|A - tI|$ whenever it makes life easier. {In particular, when A is 3×3, it is easier to calculate p_A by calculating $|A - tI|$, then multiplying by -1.}

As in the case of 2×2 matrices, there is no assurance that the characteristic roots of a 3×3 matrix are real; but, unlike the 2×2 case, we can be sure that at least one of them is real:

6.4.10 Theorem *A 3×3 real matrix has at least one real characteristic root.*

Proof.[5] If A is a 3×3 real matrix, its characteristic polynomial $p = p_A$ has the form

$$p = t^3 + at^2 + bt + c$$

for suitable real numbers a, b, c. For nonzero $r \in \mathbf{R}$, we have a factorization

$$p(r) = r^3(1 + a/r + b/r^2 + c/r^3) = r^3 f(r),$$

where $f(r)$ can be made near 1 (hence > 0) by taking $|r|$ sufficiently large. Choose $r_1 < 0$ with $|r_1|$ large enough that $f(r_1) > 0$, and choose $r_2 > 0$

[5] If you believe that the graph of a real polynomial of odd degree must cross the x-axis, you can skip this proof.

large enough that $f(r_2) > 0$; then $p(r_1) < 0$ and $p(r_2) > 0$, so p must be 0 at some point of the interval (r_1, r_2) by the 'Intermediate value theorem' of Calculus.[6] ∎

The significance of characteristic roots for linear mappings is as follows:

6.4.11 Theorem *Let* V *be a real vector space of dimension* 2 *or* 3, *let* $T \in \mathcal{L}(\mathrm{V})$, *and let* A *be the matrix of* T *relative to some basis of* V. *Let* r *be a real number. The following conditions on* r *are equivalent*:

(a) *r is a characteristic root of A*;

(b) *$Tx = rx$ for some nonzero vector $x \in \mathrm{V}$.*

Proof. Condition (a) says that $|A - rI| = 0$ and condition (b) says that $\mathrm{Ker}(T - rI) \neq \{\theta\}$. Relative to the given basis, $B = A - rI$ is the matrix of the linear mapping $S = T - rI$, so our problem is to show that $|B| = 0$ if and only if $\mathrm{Ker}\, S \neq \{\theta\}$. Stated contrapositively, we have to show that $|B| \neq 0$ if and only if $\mathrm{Ker}\, S = \{\theta\}$. Consider the statements

(1) $|B| \neq 0$,

(2) B is invertible,

(3) S is bijective,

(4) S is injective,

(5) $\mathrm{Ker}\, S = \{\theta\}$.

We have (1) ⇔ (2) by 6.1.2 and 6.3.6; (2) ⇔ (3) by 4.10.4; (3) ⇔ (4) by 3.7.4; and (4) ⇔ (5) by 2.2.9. Thus (1) ⇔ (5) and we are done. ∎

Condition (b) of the theorem generates some terminology:

6.4.12 Definition Let V be a real vector space, $T \in \mathcal{L}(\mathrm{V})$, $r \in \mathbf{R}$, and x a *nonzero* vector in V. If $Tx = rx$, we say that x is an **eigenvector** of T and that r is an **eigenvalue**[7] of T. Thus,

(i) the real number r is an eigenvalue of T if and only if $\mathrm{Ker}(T - rI) \neq \{\theta\}$, and

(ii) the nonzero vector x is an eigenvector of T if and only if Tx is a scalar multiple of x.

6.4.13 Remark With notations as in Theorem 6.4.11, we see that the eigenvalues of T are the real characteristic roots of A.

6.4.14 Example Let $T \in \mathcal{L}(\mathbf{R}^2)$ be the linear mapping $T(a, b) = (b, -a)$. The matrix of T relative to the canonical basis of $\mathbf{R}^2$ is

[6] Plausible as it appears, the intermediate value theorem is not easy to prove; it involves fundamental assumptions about the field of real numbers and the concept of continuity for a function. (See, for example, K. A. Ross *Elementary analysis: The theory of calculus*, Springer, 1980, p. 96.)

[7] The meaning of this hybrid terminology can be appreciated by a short tour of dictionaries: *eigen* (Ger.), *propre* (Fr.), proper; in fact, the terms 'proper vector' and 'proper value' are also used (in German, 'Eigenvektor' and 'Eigenwerte').

$$A = \begin{pmatrix} 0 & 1 \\ -1 & 0 \end{pmatrix};$$

the characteristic polynomial of A is $t^2 + 1$, which has no real roots. Conclusion: T has no eigenvalues (or eigenvectors).

6.4.15 Example Every $T \in \mathcal{L}(\mathbf{R}^3)$ has at least one eigenvalue (6.4.10, 6.4.11).

6.4.16 Example If $T \in \mathcal{L}(\mathbf{R}^2)$ and if, relative to some basis of $\mathbf{R}^2$, the matrix of T is symmetric, then T has at least one eigenvalue (6.4.3, 6.4.11).

6.4.17 Theorem *Let* V *be a finite-dimensional real vector space,* $T \in \mathcal{L}(\mathrm{V})$, *and* A *the matrix of* T *relative to a basis* $x_1, \ldots, x_n$ *of* V. *For* A *to be a diagonal matrix, it is necessary and sufficient that the* x_i *be eigenvectors of* T.

Proof. Assuming $A = (a_{ij})$, we have

$$Tx_j = \sum_{i=1}^{n} a_{ij} x_i$$

for $j = 1, \ldots, n$. For Tx_j to be proportional to x_j, it is necessary and sufficient that $a_{ij} = 0$ for all $i \neq j$; this is true for every j precisely when $a_{ij} \neq 0$ for all $i \neq j$, in other words, when A is a diagonal matrix (4.2.11). ■

6.4.18 Theorem *If* B *is a* 2×2 *or* 3×3 *diagonal matrix over* $\mathbf{R}$, *then the diagonal elements of* B *are its characteristic roots.*

Proof. If, for example,

$$B = \begin{pmatrix} b_{11} & 0 & 0 \\ 0 & b_{22} & 0 \\ 0 & 0 & b_{33} \end{pmatrix},$$

the formula of 6.4.8 shows that $-p_B = |B - tI|$ is equal to

$$(b_{33} - t)\begin{vmatrix} b_{11} - t & 0 \\ 0 & b_{22} - t \end{vmatrix} = (b_{11} - t)(b_{22} - t)(b_{33} - t). \ ■$$

6.4.19 Theorem *Let* A *and* B *be similar matrices over* $\mathbf{R}$ (*both* 2×2 *or both* 3×3) *and let* $r \in \mathbf{R}$. *Then* r *is a characteristic root of* A *if and only if it is a characteristic root of* B.

Proof. Suppose, to be definite, that A and B are 3×3 matrices. Let V be a 3-dimensional real vector space, choose a basis of V and let $T \in \mathcal{L}(\mathrm{V})$ be the linear mapping whose matrix relative to the chosen basis is A (4.2.6). By hypothesis, there exists an invertible real matrix C such that $B = C^{-1}AC$. This means that B is the matrix of T relative to a suitable basis of V (proof of 4.10.5). Then r is a characteristic root of A if and only if it is an eigenvalue of T (6.4.13), if and only if it is a characteristic root of B (6.4.13 again). ■

6.4.20 Corollary *Let A be a 2×2 or 3×3 real matrix. If A is similar to a diagonal matrix B, then the diagonal elements of B are characteristic roots of A.*

Proof. The diagonal elements of B are characteristic roots of B (6.4.18), hence of A (6.4.19). {Alternate proof: Show that similar matrices have the same characteristic polynomial (Exercise 5), then cite 6.4.18.} ■

▶ *Exercises*

1. Find the characteristic polynomial and characteristic roots of each of the following matrices A:

$$\text{(i) } A = \begin{pmatrix} 2 & 6 \\ 6 & -3 \end{pmatrix}, \quad \text{(ii) } A = \begin{pmatrix} 3 & 0 \\ 0 & 5 \end{pmatrix},$$

$$\text{(iii) } A = \begin{pmatrix} 3 & 0 \\ 0 & 3 \end{pmatrix}, \quad \text{(iv) } A = \begin{pmatrix} 3 & 1 \\ 0 & 3 \end{pmatrix}.$$

2. Find the characteristic polynomial and characteristic roots of the matrix

$$A = \begin{pmatrix} 2 & 3 & 7 \\ 0 & 4 & 5 \\ 0 & 0 & -1 \end{pmatrix}.$$

3. Find the characteristic polynomial and characteristic roots of the matrix

$$A = \begin{pmatrix} 2 & -1 & 3 \\ 3 & 2 & 4 \\ 5 & 8 & 6 \end{pmatrix}.$$

4. Give an example of a 3×3 matrix over $\mathbf{R}$ that has one real and two non-real characteristic roots.

5. If A and B are similar matrices over $\mathbf{R}$ (both 2×2 or both 3×3), prove that A and B have the same characteristic polynomial. {Hint: §6.1, Exercise 5 and §6.3, Exercise 5.}

6. Show that the matrices

$$A = \begin{pmatrix} 1 & 0 \\ 0 & 1 \end{pmatrix}, \quad B = \begin{pmatrix} 1 & 1 \\ 0 & 1 \end{pmatrix}$$

have the same characteristic polynomial, but are not similar.

7. If A is a 3×3 real matrix and p is its characteristic polynomial, show that the constant term of p is $-|A|$ and the coefficient of the term of second degree is $-\operatorname{tr} A$ (§4.2, Exercise 7).

8. Let $T \in \mathcal{L}(\mathbf{R}^3)$ be an orthogonal linear mapping in Euclidean 3-space (5.4.1) and let A be the matrix of T relative to some orthonormal basis of $\mathbf{R}^3$, so that A is an orthogonal matrix (Theorem 5.4.3). Prove:

(i) There exists a nonzero vector u such that $Tu = \pm u$. {Hint: Example 6.4.15; an orthogonal linear mapping preserves norm.}

(ii) The matrix A has either $+1$ or -1 as a characteristic root.

(iii) If $|A| = 1$ then the orthogonal matrix A has $+1$ as a characteristic root. {Hint: Let r_1, r_2, r_3 be the characteristic roots (not necessarily real). If they are all real, they are eigenvalues of T hence must be ± 1; their product is 1 (Exercise 7), so they can't all be -1. If only one of them is real, say r_1, then the other two are complex conjugates of each other, thus $1 = r_1 r_2 r_3 = r_1 |r_2|^2$, therefore $r_1 = 1$.}

(iv) If $|A| = 1$ then T may be regarded as a rotation in $\mathbf{R}^3$. {Hint: By (iii) there is a unit vector u_1 such that $Tu_1 = u_1$. Let $\mathrm{M} = \mathbf{R}u_1$; the vectors of M are left fixed by T, so M will turn out to be the axis of the rotation (think of it as a line through the origin). Then $T(\mathrm{M}^\perp) = \mathrm{M}^\perp$ (§5.4, Exercise 8), so the restriction of T to $\mathrm{M}^\perp$ is an orthogonal linear mapping $S = T|\mathrm{M}^\perp$ in the 2-dimensional Euclidean space $\mathrm{M}^\perp$. Let u_2, u_3 be an orthonormal basis of $\mathrm{M}^\perp$ and let

$$C = \begin{pmatrix} a & c \\ b & d \end{pmatrix}$$

be the matrix of S relative to the basis u_2, u_3. Then the matrix of T relative to u_1, u_2, u_3 is

$$B = \begin{pmatrix} 1 & 0 & 0 \\ 0 & a & c \\ 0 & b & d \end{pmatrix}.$$

Since B is similar to A (4.10.5), $|A| = |B| = 1$ (§6.3, Exercise 5); clearly $|B| = |C|$, so $|C| = 1$ and S is a rotation in the plane $\mathrm{M}^\perp$ through the origin perpendicular to the line M (5.4.10).}

(v) $|A| = \pm 1$. {Hint: §6.3, Exercises 4, 5.}

(vi) If $|A| = -1$ then $-T$ is a rotation, so $T = (-1)(-T)$ is composed of a rotation followed by reflection in the origin. {Hint: 6.3.9.}

9. In the Euclidean plane $\mathbf{R}^2$, let $x_1 = (1, 0)$, $x_2 = (1, 1)$, and let $T \in \mathcal{L}(\mathbf{R}^2)$ be the linear mapping such that $Tx_1 = x_2$, $Tx_2 = x_1$. Prove that the matrix of T relative to the basis x_1, x_2 is symmetric, but T is not self-adjoint. Does this contradict Theorem 6.4.5? {Hint: Calculate the matrix of T relative to the canonical basis e_1, e_2.}

10. Prove that the matrix

$$A = \begin{pmatrix} 1 & 1 \\ 0 & 1 \end{pmatrix}$$

is not similar to a diagonal matrix. {Hint: Corollary 6.4.20 and Exercise 6.}

11. Let T be a self-adjoint linear mapping in a Euclidean space.

(i) Show that

$$(T(x+y)|x+y) - (T(x-y)|x-y) = 4(Tx|y)$$

for all vectors x and y.

(ii) If $(Tx|x) = 0$ for all vectors x, then $T = 0$ (cf. §5.2, Exercise 11).

6.5 Diagonalizing 2 × 2 symmetric real matrices

Given a symmetric 2×2 matrix A over $\mathbf{R}$, we seek a 2×2 orthogonal matrix C such that the matrix $B = C^{-1}AC = C'AC$ is diagonal, that is, of the form

$$B = \begin{pmatrix} r_1 & 0 \\ 0 & r_2 \end{pmatrix}.$$

Why does this result interest us? First, because there is a pretty application to analytic geometry (§6.7); without going into details, if we think of A as the matrix of a linear mapping in the Euclidean plane $\mathbf{R}^2$ (relative to the canonical orthonormal basis), the orthogonal matrix C accomplishes a rotation of axes (5.4.10) that dramatically simplifies the matrix of the linear mapping. Secondly, there's a generalization to $n \times n$ symmetric real matrices[1], a result with many important applications; the details for dimension n are substantially harder, but dimension 3 is within easy reach and is treated in the next section.

It's useful to have some concise terminology for what we're trying to do:

6.5.1 Definition Let A be a square matrix. If we can find an invertible matrix C such that $C^{-1}AC$ is diagonal[2], we say that we have **diagonalized** A under similarity; if we can accomplish this with an orthogonal matrix C, we say that we have diagonalized A under **orthogonal similarity** (5.4.9).

Stated in terms of linear mappings, the problem of diagonalizing matrices amounts to finding bases consisting of eigenvectors (6.4.17, 4.10.5); to ensure that the matrix C that transforms A to diagonal form is orthogonal, we have to stick to orthonormal bases (5.4.8). The crux of the matrix diagonalization problem posed at the beginning of the section is the following theorem about linear mappings:

6.5.2 Theorem *If* E *is a 2-dimensional Euclidean space and* $T \in \mathcal{L}(\mathrm{E})$ *is a self-adjoint linear mapping, then* E *has an orthonormal basis consisting of eigenvectors of* T.

[1] The generalization to dimension n is known as the 'spectral theorem' or 'principal axis theorem' (§9.3).

[2] Not always possible (§6.4, Exercise 10).

Proof. Let x_1, x_2 be an orthonormal basis of E (Theorem 5.2.16) and let A be the matrix of T relative to this basis. We are assuming that $T^* = T$, so $A' = A$ by Theorem 6.4.5. Our problem is to construct an orthonormal basis y_1, y_2 such that Ty_i is a multiple of y_i for $i = 1, 2$.

Since A is symmetric, its characteristic roots are real numbers (6.4.3), therefore they are eigenvalues of T (6.4.13). Let r be a characteristic root of A and choose a nonzero vector y such that $Ty = ry$; then $y_1 = \|y\|^{-1}y$ is a unit vector with $Ty_1 = ry_1$.

Let $\mathrm{M} = \mathbf{R}y_1$ be the 1-dimensional linear subspace spanned by y_1. From $Ty_1 = ry_1$ we see that $Tx = rx$ for all $x \in \mathrm{M}$; in particular, $T(\mathrm{M}) \subset \mathrm{M}$, therefore $T^*(\mathrm{M}^\perp) \subset \mathrm{M}^\perp$ (5.3.6), in other words $T(\mathrm{M}^\perp) \subset \mathrm{M}^\perp$. Since $\mathrm{M}^\perp$ is also 1-dimensional (5.2.8), $\mathrm{M}^\perp = \mathbf{R}z$ for a suitable nonzero vector z. From $Tz \in T(\mathrm{M}^\perp) \subset \mathrm{M}^\perp = \mathbf{R}z$, we see that Tz is a multiple of z; in other words, z is an eigenvector of T, therefore $y_2 = \|z\|^{-1}z$ is an eigenvector of norm 1.

Finally, from $y_1 \in \mathrm{M}$, $y_2 \in \mathrm{M}^\perp$ and $\|y_1\| = \|y_2\| = 1$, we see that y_1, y_2 is an orthonormal basis of E. ■

6.5.3 Corollary *If A is a symmetric 2×2 matrix over $\mathbf{R}$, then there exists a 2×2 orthogonal matrix C such that $C'AC$ is diagonal and such that $|C| = 1$.*

Proof. Let $T \in \mathscr{L}(\mathbf{R}^2)$ be the linear mapping, in the Euclidean plane, whose matrix relative to the canonical orthonormal basis e_1, e_2 is A. Since A is symmetric, T is self-adjoint (6.4.5); by the theorem, there is an orthonormal basis y_1, y_2 of $\mathbf{R}^2$ consisting of eigenvectors of T, say $Ty_1 = r_1y_1$, $Ty_2 = r_2y_2$. The matrix of T relative to y_1, y_2 is the diagonal matrix

$$B = \begin{pmatrix} r_1 & 0 \\ 0 & r_2 \end{pmatrix}.$$

If C is the change-of-basis matrix that expresses the y's in terms of the e's, then $B = C^{-1}AC$ (4.10.5); since the bases are orthonormal, C is orthogonal (5.4.7), that is, $C^{-1} = C'$.

From 5.4.10, we know that $|C| = \pm 1$. If $|C| = 1$, we're done. If $|C| = -1$ then replacing y_1 by $-y_1$ has the effect of changing the signs of the entries in the first column of C, which changes the determinant from -1 to $+1$. ■

6.5.4 Remarks With notations as in the proof of Corollary 6.5.3, we know that the characteristic polynomial of the diagonal matrix B is $p_B = (t - r_1)(t - r_2)$ (6.4.18). Let's show that this is also the characteristic polynomial of A (cf. §6.4, Exercise 5). At any rate, p_A and p_B are quadratic polynomials, with leading term t^2, such that $p_A(r_i) = p_B(r_i) = 0$ for $i = 1, 2$ (6.4.20). If $r_1 \neq r_2$, then $p_A = (t - r_1)(t - r_2)$ by the factor theorem, thus $p_A = p_B$; on the other hand, if $r_1 = r_2$ then $B = r_1 I$, therefore $A = CBC^{-1} = r_1 CIC^{-1} = r_1 I = B$ and there is nothing to prove.

The significance of $p_A = p_B$ is that, by merely calculating the roots of p_A, *we can predict what the diagonal matrix of* 6.5.3 *will be*.

6.5.5 Example The characteristic polynomial of the symmetric matrix

$$A = \begin{pmatrix} 2 & 6 \\ 6 & -3 \end{pmatrix}$$

is $t^2 + t - 42 = (t-6)(t+7)$, so the characteristic roots are 6 and -7. Conclusion: there exists an orthogonal matrix C such that

$$C'AC = \begin{pmatrix} 6 & 0 \\ 0 & -7 \end{pmatrix}.$$

Let's find such a matrix C.

Let $T \in \mathcal{L}(\mathbf{R}^2)$ be the linear mapping whose matrix is A relative to the canonical orthonormal basis e_1, e_2 of $\mathbf{R}^2$; thus

$$Te_1 = (2, 6), \quad Te_2 = (6, -3)$$

and, in general, $T(a, b) = (2a + 6b, 6a - 3b)$. For (a, b) to be an eigenvector corresponding to the eigenvalue 6, we require that $T(a, b) = 6(a, b)$, that is,

$$\begin{aligned} 2a + 6b &= 6a, \\ 6a - 3b &= 6b. \end{aligned}$$

Both equations reduce to $2a = 3b$, thus $a = 3$, $b = 2$ is a solution and we have an eigenvector $y = (3, 2)$. Let z be the vector obtained by interchanging the coordinates of y and changing the sign of the new first coordinate: $z = (-2, 3)$ is orthogonal to y, and the proof of 6.5.2 shows that z is also an eigenvector. We don't have to take 6.5.2's word for it (and anyway, we should check!):

$$\begin{aligned} T(-2, 3) &= (2(-2) + 6(3), 6(-2) - 3(3)) \\ &= (14, -21) = -7(-2, 3). \end{aligned}$$

Normalizing y and z, we have an orthonormal basis

$$\begin{aligned} y_1 &= (1/\sqrt{13})(3, 2) \\ y_2 &= (1/\sqrt{13})(-2, 3) \end{aligned}$$

consisting of eigenvectors of T. The change-of-basis matrix in question is

$$C = (1/\sqrt{13}) \begin{pmatrix} 3 & -2 \\ 2 & 3 \end{pmatrix},$$

clearly orthogonal and of determinant 1. The claim is that $C'AC$ is diagonal, more precisely

$$(1/13) \begin{pmatrix} 3 & 2 \\ -2 & 3 \end{pmatrix} \begin{pmatrix} 2 & 6 \\ 6 & -3 \end{pmatrix} \begin{pmatrix} 3 & -2 \\ 2 & 3 \end{pmatrix} = \begin{pmatrix} 6 & 0 \\ 0 & -7 \end{pmatrix},$$

a claim easily verified by computing the left side. As noted in Example 5.4.10,

$$C = \begin{pmatrix} \cos\alpha & -\sin\alpha \\ \sin\alpha & \cos\alpha \end{pmatrix}$$

for a suitable angle α (in the present example, $\alpha = \text{Arcsin}(2/\sqrt{13})$), thus the matrix of T has been diagonalized by a 'rotation of axes'.

Another way to organize the computation of eigenvectors is to use the 'model' idea of §4.5: the matrix equation

$$\begin{pmatrix} 2 & 6 \\ 6 & -3 \end{pmatrix}\begin{pmatrix} a \\ b \end{pmatrix} = 6\begin{pmatrix} a \\ b \end{pmatrix}$$

leads to the scalar equations

$$\begin{aligned} 2a + 6b &= 6a \\ 6a - 3b &= 6b \end{aligned}$$

as before.

6.5.6 Example Here is a useful machine for making up examples of symmetric matrices

$$A = \begin{pmatrix} a & b \\ b & d \end{pmatrix}$$

whose characteristic roots are *integers*: choose integers m, n, d and define

$$(*) \qquad a = d + (m^2 - n^2), \qquad b = mn.$$

{For example, putting $m = 3$, $n = 2$, $d = -3$, we get the example of 6.5.5.} Why does it work? The characteristic polynomial of such a matrix is

$$t^2 - (a + d)t + (ad - b^2),$$

and its discriminant is

$$\Delta = (a + d)^2 - 4(ad - b^2) = (a - d)^2 + (2b)^2.$$

The equations $(*)$ may be written

$$a - d = m^2 - n^2, \qquad 2b = 2mn,$$

therefore $\Delta = (m^2 - n^2)^2 + (2mn)^2 = (m^2 + n^2)^2$ is the square of an integer and the characteristic roots are

$$\begin{aligned} r &= [a + d \pm \sqrt{\Delta}]/2 \\ &= [2d + m^2 - n^2 \pm (m^2 + n^2)]/2 \\ &= d + m^2 \quad \text{or} \quad d - n^2 \end{aligned}$$

(also integers). For the example of 6.5.5, $r = -3 + 3^2 = 6$ or $r = -3 - 2^2 = -7$.

▶ Exercises

1. For each of the following symmetric matrices A, find an orthogonal matrix C such that $C'AC$ is diagonal:

$$\text{(i) } A = \begin{pmatrix} 2 & 1 \\ 1 & 2 \end{pmatrix}, \qquad \text{(ii) } A = \begin{pmatrix} 5 & 4 \\ 4 & -1 \end{pmatrix},$$

$$\text{(iii) } A = \begin{pmatrix} 1 & 2 \\ 2 & 4 \end{pmatrix}, \qquad \text{(iv) } A = \begin{pmatrix} 5 & 6 \\ 6 & 0 \end{pmatrix}.$$

2. If $T \in \mathcal{L}(\mathbf{R}^2)$ and if, relative to some basis of $\mathbf{R}^2$ (not necessarily orthonormal), the matrix of T is symmetric, prove that there exists a basis of $\mathbf{R}^2$ consisting of eigenvectors of T. {Hint: Corollary 6.5.3, Theorems 4.10.5, 6.4.17.}

6.6 Diagonalizing 3×3 symmetric real matrices

The agenda is similar to that of the preceding section: given a 3×3 symmetric real matrix A, we seek an orthogonal matrix C such that the matrix $B = C^{-1}AC = C'AC$ is diagonal, that is, of the form

$$B = \begin{pmatrix} r_1 & 0 & 0 \\ 0 & r_2 & 0 \\ 0 & 0 & r_3 \end{pmatrix}.$$

Again, the strategy is to reformulate the problem in terms of a self-adjoint linear mapping (cf. the proof of 6.5.3).

6.6.1 Theorem *If* E *is a 3-dimensional Euclidean space and* $T \in \mathcal{L}(\mathrm{E})$ *is a self-adjoint linear mapping, then* E *has an orthonormal basis consisting of eigenvectors of* T.

Proof. Let x_1, x_2, x_3 be an orthonormal basis of E (5.2.16) and let A be the matrix of T relative to this basis. Since $T^* = T$, the matrix A is symmetric (6.4.5).

By 6.4.10, A has a real characteristic root r, and r is an eigenvalue of T by 6.4.11; choose a unit vector y_1 such that $Ty_1 = ry_1$ and let $\mathrm{M} = \mathbf{R}y_1$ be the 1-dimensional linear subspace spanned by y_1. The orthogonal complement $\mathrm{M}^\perp$ of M is a 2-dimensional Euclidean space (5.2.8) and, as argued in the proof of 6.5.2, $T(\mathrm{M}^\perp) \subset \mathrm{M}^\perp$. It follows that the restriction $S = T|\mathrm{M}^\perp$ of T to $\mathrm{M}^\perp$ is a self-adjoint linear mapping in $\mathrm{M}^\perp$. {Proof: Since $T^* = T$, we have $(Tu|v) = (u|Tv)$ for all u, v in E, in particular for all u, v in $\mathrm{M}^\perp$. Thus $(Su|v) = (u|Sv)$ for all u, v in $\mathrm{M}^\perp$, consequently $S^* = S$ (5.3.2).}

Apply 6.5.2 to the 2-dimensional Euclidean space $\mathrm{M}^\perp$: there exists an orthonormal basis y_2, y_3 of $\mathrm{M}^\perp$ consisting of eigenvectors of S (hence of T). The orthonormal basis y_1, y_2, y_3 of E meets the requirements of the theorem. ■

6.6.2 Corollary *If* A *is a symmetric* 3×3 *matrix over* $\mathbf{R}$, *then there exists a* 3×3 *orthogonal matrix* C *(of determinant* 1, *if we like) such that* $C'AC$ *is diagonal.*

Proof. The proof follows the same format as 6.5.3, with the Euclidean plane $\mathbf{R}^2$ replaced by Euclidean 3-space $\mathbf{R}^3$. ■

6.6.3 Corollary *The characteristic roots of a* 3×3 *symmetric real matrix are all real.*

Proof. With notations as in 6.6.2, let $B = C'AC$; thus,

$$B = \begin{pmatrix} r_1 & 0 & 0 \\ 0 & r_2 & 0 \\ 0 & 0 & r_3 \end{pmatrix}$$

with r_1, r_2, r_3 real numbers. We know from 6.4.20 that the r_i are characteristic roots of A.

If $r_1 = r_2 = r_3$ then $B = r_1 I$, therefore $A = r_1 I$ (cf. 6.5.4) and r_1 is the unique characteristic root of A.

If the r_i are not all equal, then A has at least two real characteristic roots; since the characteristic polynomial is a real cubic, its remaining root has no choice but to be real. ■

6.6.4 Example The symmetric matrix

$$A = \begin{pmatrix} 1 & 0 & 1 \\ 0 & -1 & 0 \\ 1 & 0 & 2 \end{pmatrix}$$

has lots of zero entries, so its characteristic polynomial $p = p_A$ is easy to compute:

$$p = t^3 - 2t^2 - 2t + 1.$$

{To learn more from the example, let's pretend we can't see that $p(-1) = 0$.} The derivative polynomial is $p' = 3t^2 - 4t - 2$ and its roots are $r = (2 \pm \sqrt{10})/3$. Let's show that p' and p have no roots in common. If r is a root of p', then

$$0 = p'(r) = 3r^2 - 4r - 2,$$

thus

(i) $$3r^2 = 4r + 2,$$

(ii) $$3r^3 = 4r^2 + 2r.$$

These formulas are useful for calculating $p(r)$: citing (ii), we have

$$\begin{aligned} 3p(r) &= 3r^3 - 6r^2 - 6r + 3 \\ &= (4r^2 + 2r) - 6r^2 - 6r + 3 = -2r^2 - 4r + 3, \end{aligned}$$

therefore (citing (i))

$$\begin{aligned} 9p(r) &= -2(3r^2) - 12r + 9 \\ &= -2(4r + 2) - 12r + 9 = -20r + 5, \end{aligned}$$

thus $p(r) \neq 0$ (because $r \neq 1/4$). Since p and p' have no root in common, p can't have a repeated root. In other words, the roots of p (real numbers, by 6.6.3) must be distinct.

We know from Corollary 6.6.2 that A is orthogonally similar to a diagonal matrix

$$B = \begin{pmatrix} r_1 & 0 & 0 \\ 0 & r_2 & 0 \\ 0 & 0 & r_3 \end{pmatrix}.$$

The three distinct real roots of p are among the r_i (Theorem 6.4.19), therefore the r_i *are* the three roots of p and we have $p = (t - r_1)(t - r_2)(t - r_3)$ by the factor theorem.

The bottom line: We can predict the diagonal form by calculating the roots of p. In general, this is a messy computation (roots of cubics!); in the example at hand, $p = (t + 1)(t^2 - 3t + 1)$, so the roots are -1 and $(3 \pm \sqrt{5})/2$.

6.6.5 Example The symmetric matrix

$$A = \begin{pmatrix} 0 & 0 & 1 \\ 0 & -1 & 0 \\ 1 & 0 & 0 \end{pmatrix}$$

has characteristic polynomial $p = t^3 + t^2 - t - 1$. The derivative polynomial is

$$p' = 3t^2 + 2t - 1 = (t + 1)(3t - 1),$$

with roots -1 and $1/3$. Evaluate p at the roots of p': from $p(-1) = 0$ we know that -1 is a repeated root of p; in fact,

$$p = (t + 1)^2(t - 1).$$

By Corollary 6.6.2, A is orthogonally similar to a diagonal matrix B, and the only possible diagonal elements of B are -1 and $+1$ (6.4.20). Apart from permutations of the diagonal elements, there are just two candidates for B:

$$B_1 = \begin{pmatrix} 1 & 0 & 0 \\ 0 & 1 & 0 \\ 0 & 0 & -1 \end{pmatrix} \quad \text{or} \quad B_2 = \begin{pmatrix} 1 & 0 & 0 \\ 0 & -1 & 0 \\ 0 & 0 & -1 \end{pmatrix}.$$

{If we had verified that similar matrices have the same determinant (§6.3, Exercise 5), we could decide between B_1 and B_2 on the basis of $|A| = 1$.}

Notice that

$$B_1 - (-1)I = \begin{pmatrix} 2 & 0 & 0 \\ 0 & 2 & 0 \\ 0 & 0 & 0 \end{pmatrix}$$

has rank 2, whereas

$$B_2 - (-1)I = \begin{pmatrix} 2 & 0 & 0 \\ 0 & 0 & 0 \\ 0 & 0 & 0 \end{pmatrix}$$

has rank 1. Since similar matrices have the same rank (they are the matrices of the same linear mapping on $\mathbf{R}^3$ relative to two different bases; cf. the proof of Theorem 4.10.5), $B - (-1)I$ has the same rank as $A - (-1)I$; but

$$A - (-1)I = \begin{pmatrix} 1 & 0 & 1 \\ 0 & 0 & 0 \\ 1 & 0 & 1 \end{pmatrix}$$

has rank 1, so the diagonal form must be B_2.

6.6.6 Armed with what we've learned from the preceding two examples, we're ready to *predict the diagonal form of an arbitrary* 3×3 *symmetric real matrix* A.

From Corollary 6.6.2, we know that A is orthogonally similar to a diagonal matrix B; the problem is to predict (with a minimum of computation) what B will be.

If A is a scalar matrix $A = cI$, then A is already diagonal and there is nothing to predict; let's assume that A is not a scalar matrix.

Let $p = p_A$ be the characteristic polynomial of A, p' the derivative polynomial. Calculate the roots of p' (quadratic formula) and test them in p to decide whether p has a repeated root (cf. Exercise 3).

case 1: If neither root of p' is a root of p, then the analysis of Example 6.6.4 shows that p has three distinct real roots r_1, r_2, r_3 and A is orthogonally similar to the diagonal matrix

$$B = \begin{pmatrix} r_1 & 0 & 0 \\ 0 & r_2 & 0 \\ 0 & 0 & r_3 \end{pmatrix}.$$

There remains the problem (in general messy) of actually calculating the r_i, then finding an orthogonal matrix C such that $C'AC = B$.

case 2: If p vanishes at one of the roots of p', say $p(r_1) = p'(r_1) = 0$, then r_1 is a multiple root of p. The multiplicity of r_1 in p can't be 3, because $p = (t - r_1)^3$ would imply that $B = r_1 I$ (6.4.20), therefore $A = r_1 I$, contrary to our assumption that A is not scalar. Thus

$$p = (t - r_1)^2(t - r_2)$$

with r_1, r_2 distinct real numbers. By the analysis of Example 6.6.5, A is orthogonally similar to one of the matrices

$$B_1 = \begin{pmatrix} r_1 & 0 & 0 \\ 0 & r_1 & 0 \\ 0 & 0 & r_2 \end{pmatrix} \quad \text{or} \quad B_2 = \begin{pmatrix} r_1 & 0 & 0 \\ 0 & r_2 & 0 \\ 0 & 0 & r_2 \end{pmatrix},$$

and we can decide between them by calculating the rank of $A - r_1 I$.

The rank of $A - r_1 I$ is easily calculated by inspection: the rank is not 3 ($A - r_1 I$ has determinant 0, so it is not invertible), and it is not 0 (A is not a scalar matrix), so it must be either 1 or 2. If the rank is 1, we can *see* that all rows of $A - r_1 I$ are multiples of one of them; otherwise, the rank is 2. {In the present case, the rank will turn out to be 1, so the diagonal form is

B_1. No *theoretical* proof of this has been given, but none is needed; we calculate the rank and decide.}

6.6.7 Here is a way of constructing examples of 3×3 symmetric real matrices A whose diagonal form $B = C'AC$ has only integer entries. The idea is to start with such a matrix B and let $A = CBC'$; for this, we need a 'nice' orthogonal matrix C, for example the matrix

$$C = \frac{1}{3}\begin{pmatrix} 1 & 2 & -2 \\ 2 & 1 & 2 \\ -2 & 2 & 1 \end{pmatrix}.$$

If you want A to have diagonal form

$$B = \begin{pmatrix} 1 & 0 & 0 \\ 0 & -1 & 0 \\ 0 & 0 & 2 \end{pmatrix},$$

define it to be the matrix

$$A = CBC' = CBC = \frac{1}{9}\begin{pmatrix} 5 & -8 & -10 \\ -8 & 11 & -2 \\ -10 & -2 & 2 \end{pmatrix}.$$

{Exercise: Check the computation CBC and check that $|tI - A| = (t-1)(t+1)(t-2)$.}

▶ Exercises

1. Find the characteristic roots of the following matrices:

$$\text{(i)}\quad \frac{1}{9}\begin{pmatrix} 5 & 10 & -8 \\ 10 & 2 & 2 \\ -8 & 2 & 11 \end{pmatrix};\qquad \text{(ii)}\quad \frac{1}{9}\begin{pmatrix} 5 & 4 & 2 \\ 4 & 5 & -2 \\ 2 & -2 & 8 \end{pmatrix}.$$

2. If $T \in \mathscr{L}(\mathbf{R}^3)$ and if, relative to some basis of $\mathbf{R}^3$ (not necessarily orthonormal), the matrix of T is symmetric, prove that $\mathbf{R}^3$ has a basis consisting of eigenvectors of T. {Hint: Cf. §6.5, Exercise 2.}

3. If you're familiar with 'Euclid's algorithm', here's a shortcut for deciding in 6.6.6 whether p has a repeated root. Calculate the greatest common divisor d of p and p' (using Euclid's algorithm); p has a repeated root if and only if d is not a constant (Appendix B.5.4). The virtue of this method is that it avoids the quadratic formula and the radicals that usually come with it.

{Hint: The only possible degrees for d are 0, 1, 2. Degree 0 means d is a constant (p has no repeated roots); degree 2 means that p' is a divisor of p (possible only when p has a single root of multiplicity 3); degree 1 means that p has two distinct roots, of multiplicities 2 and 1, respectively, the root of multiplicity 2 being the one it shares with p'.}

6.7 A geometric application (conic sections)

6.7.1 Conic sections A remarkable theorem of plane analytic geometry asserts that the graph of every equation of the second degree

(1) $$ax^2 + bxy + cy^2 + dx + ey + f = 0$$

is a conic section, that is, an ellipse, hyperbola or parabola[1], or a degenerate case of one of these (including the possibility that the graph is empty[2]).

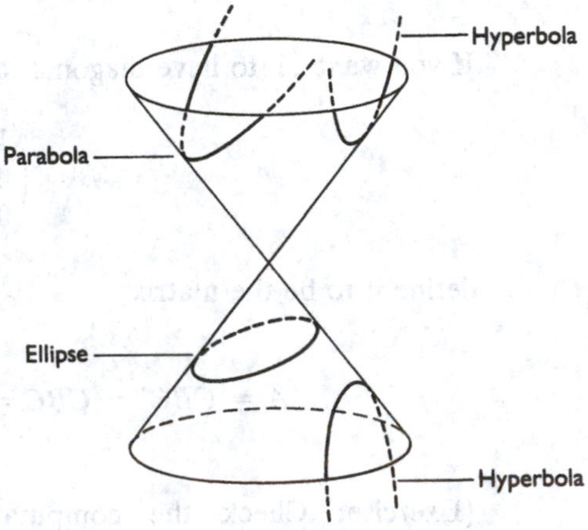

Fig. 21

This is proved by finding a new rectangular coordinate system, in which the coordinates x', y' of the points of the graph satisfy a simplified equation

(1′) $$a'(x')^2 + c'(y')^2 + d'x' + e'y' + f' = 0$$

(with $x'y'$ missing), the identification of the possible graphs of the simpler equation being relatively easy. The new coordinate axes can be obtained by rotating the old axes through a suitable angle α (Fig. 22), and the coordinates (x, y) and (x', y') of the same point P relative to the two coordinate systems are related by the equations

(2) $$\begin{aligned} x' &= x\cos\alpha + y\sin\alpha, \\ y' &= -x\sin\alpha + y\cos\alpha, \end{aligned}$$

[1] Why 'conic section'? Imagine (Fig. 21) taking a planar cross-section of a right circular cone (with two nappes). If the plane intersects only one nappe, the result is either an ellipse (a circle when the plane is perpendicular to the axis of the cone) or a parabola (if the plane is parallel to an element of the cone). If the plane intersects both nappes, the result is a hyperbola (two 'branches'), degenerating to a pair of intersecting lines when the plane passes through the vertex of the cone.

[2] For example, the equation $x^2 + y^2 + 1 = 0$.

in matrix form

$$(3)\qquad \begin{pmatrix} x' \\ y' \end{pmatrix} = \begin{pmatrix} \cos\alpha & \sin\alpha \\ -\sin\alpha & \cos\alpha \end{pmatrix} \begin{pmatrix} x \\ y \end{pmatrix}.$$

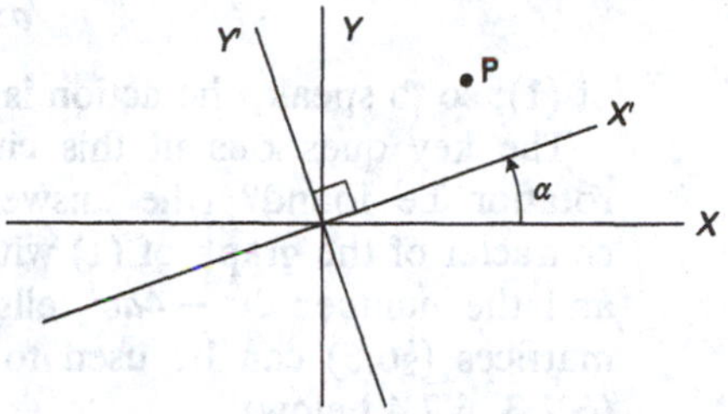

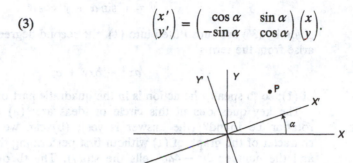

Fig. 22

The 2×2 matrix of coefficients in (3) is invertible (its determinant is 1), in fact orthogonal (5.4.10), so (3) can also be written

$$(3')\qquad \begin{pmatrix} x \\ y \end{pmatrix} = \begin{pmatrix} \cos\alpha & -\sin\alpha \\ \sin\alpha & \cos\alpha \end{pmatrix} \begin{pmatrix} x' \\ y' \end{pmatrix},$$

the matrix of coefficients in (3′) being the transpose of the matrix in (3). Thus, associated with the rotation of axes, we have an orthogonal matrix

$$(4)\qquad C = \begin{pmatrix} \cos\alpha & -\sin\alpha \\ \sin\alpha & \cos\alpha \end{pmatrix},$$

and the equations

$$\begin{pmatrix} x \\ y \end{pmatrix} = C \begin{pmatrix} x' \\ y' \end{pmatrix}, \qquad \begin{pmatrix} x' \\ y' \end{pmatrix} = C' \begin{pmatrix} x \\ y \end{pmatrix}$$

link the 'old coordinates' (x, y) of a point with its 'new coordinates' (x', y'). If S is the linear mapping in $\mathbf{R}^2$ whose matrix relative to the canonical basis e_1, e_2 is C, we see from (4) that

$$Se_1 = (\cos\alpha, \sin\alpha), \quad Se_2 = (-\sin\alpha, \cos\alpha),$$

thus the effect of S on e_1 and e_2 is to rotate them through the angle α; it follows that S rotates every vector $(x, y) = xe_1 + ye_2$ through the angle α (rotation of the sides xe_1, ye_2 of a rectangle implies a rotation of its diagonal).

The geometric character of the graph of (1′) is governed by the terms of second degree,

$$a'(x')^2 + c'(y')^2;$$

setting aside degenerate cases, the graph is elliptic when $a'c' > 0$ (circular if $a' = c'$), hyperbolic when $a'c' < 0$, parabolic when $a'c' = 0$ (that is, $a' = 0$ or $c' = 0$). If the formulas

$$x = x' \cos \alpha - y' \sin \alpha$$
$$y = x' \sin \alpha + y' \cos \alpha$$

given by (3′) are substituted into (1), the second-degree terms of (1′) can only arise from the terms

$$ax^2 + bxy + cy^2$$

of (1); so to speak, the action is in the quadratic part of (1).

The key questions in this circle of ideas are: (a) can such a simplifying rotation be found? (the answer is yes); (b) can we predict the geometric character of the graph of (1) without first performing the rotation? (yes again, and the number $b^2 - 4ac$ tells the story). The theory of 2×2 symmetric matrices (§6.5) can be used to answer both questions simply and elegantly (6.7.3, 6.7.4 below).

6.7.2 Quadratic forms associated with linear mappings

Every linear mapping in $\mathbf{R}^2$ leads to expressions of the form

$$ax^2 + bxy + cy^2$$

(the action end of a conic section) in the following way. Let $T \in \mathscr{L}(\mathbf{R}^2)$, let u_1, u_2 be an orthonormal basis of $\mathbf{R}^2$, and let

$$A = \begin{pmatrix} a & c \\ b & d \end{pmatrix}$$

be the matrix of T relative to this basis, in other words

$$Tu_1 = au_1 + bu_2,$$
$$Tu_2 = cu_1 + du_2.$$

For each vector $u \in \mathbf{R}^2$ there are unique scalars x, y such that

$$u = xu_1 + yu_2;$$

x and y are called the *coordinates* of u relative to the basis u_1, u_2. By linearity,

$$\begin{aligned} Tu &= x(Tu_1) + y(Tu_2) \\ &= x(au_1 + bu_2) + y(cu_1 + du_2) \\ &= (ax + cy)u_1 + (bx + dy)u_2, \end{aligned}$$

so the coordinates of Tu (relative to the basis u_1, u_2) can be obtained from the matrix computation

$$A \begin{pmatrix} x \\ y \end{pmatrix} = \begin{pmatrix} a & c \\ b & d \end{pmatrix} \begin{pmatrix} x \\ y \end{pmatrix} = \begin{pmatrix} ax + cy \\ bx + dy \end{pmatrix}.$$

From the orthonormality of u_1, u_2, we see that

$$\begin{aligned} (Tu|u) &= (ax + cy)x + (bx + dy)y \\ &= ax^2 + (b + c)xy + dy^2. \end{aligned}$$

The function $u \mapsto (Tu|u)$ is called the *quadratic form*[3] associated with T; its formula in terms of coordinates depends on the matrix of T, thus on the particular orthonormal basis chosen for $\mathbf{R}^2$.

6.7.3 The self-adjoint linear mapping associated with a quadratic form

Given a quadratic form

$$(*) \qquad ax^2 + bxy + cy^2$$

with real coefficients $a, b, c,$ there are many linear mappings $T \in \mathcal{L}(\mathbf{R}^2)$ such that

$$(Tu|u) = ax^2 + bxy + cy^2 \quad \text{for all} \quad u = (x, y);$$

for example, the linear mapping with matrix

$$\begin{pmatrix} a & 0 \\ b & c \end{pmatrix}$$

relative to the canonical orthonormal basis e_1, e_2 of $\mathbf{R}^2$, or the linear mapping with matrix

$$\begin{pmatrix} a & 1 \\ b-1 & c \end{pmatrix},$$

and so on. If we want the linear mapping to be self-adjoint (and we do), we'll have to take the matrix to be symmetric (6.4.5). For this, the only possible choice is the matrix

$$A = \begin{pmatrix} a & b/2 \\ b/2 & c \end{pmatrix};$$

it is called *the matrix of the quadratic form* $(*)$. The linear mapping $T \in \mathcal{L}(\mathbf{R}^2)$ whose matrix relative to the canonical orthonormal basis e_1, e_2 is A is called *the linear mapping associated with the quadratic form* $(*)$; since A is symmetric and the basis is orthonormal, T is *self-adjoint*.

For example, the matrix of the quadratic form

$$2x^2 + 12xy - 3y^2$$

is the (symmetric) matrix

$$\begin{pmatrix} 2 & 6 \\ 6 & -3 \end{pmatrix}$$

and the linear mapping associated with the form is the (self-adjoint) linear mapping $T \in \mathcal{L}(\mathbf{R}^2)$ such that

$$Te_1 = 2e_1 + 6e_2,$$
$$Te_2 = 6e_1 - 3e_2.$$

[3] The name 'quadratic form' refers to the fact that each nonzero term of the expression $ax^2 + (b + c)xy + dy^2$ has the same total degree (the sum of the degree in x and the degree in y) and that degree is 2. Similarly, $x^3 - 4x^2y + 5y^3$ is a 'cubic form'. A 'linear form' is an expression $ax + by$, which is consistent with the earlier use of the term since the linear forms on $\mathbf{R}^2$ in the sense of duality (Definition 2.1.11) are precisely the functions $(x, y) \mapsto ax + by$ (cf. Theorem 5.2.5).

Returning to the self-adjoint linear mapping T associated with the quadratic form (*), let's apply to it the diagonalization theory of §6.5: there exists an orthonormal basis u_1, u_2 of $\mathbf{R}^2$ consisting of eigenvectors of T, say

$$Tu_1 = r_1u_1,$$
$$Tu_2 = r_2u_2.$$

Every vector $u = (x, y) \in \mathbf{R}^2$ has a unique representation

$$u = x'u_1 + y'u_2,$$

and

$$\begin{aligned}(Tu|u) &= (x'r_1u_1 + y'r_2u_2|x'u_1 + y'u_2)\\ &= r_1(x')^2 + r_2(y')^2.\end{aligned}$$

We thus have two formulas for the quadratic form associated with T:

$$ax^2 + bxy + cy^2 = (Tu|u) = r_1(x')^2 + r_2(y')^2.$$

If $C = (c_{ij})$ is the change-of-basis matrix (4.10.1) that expresses u_1, u_2 in terms of e_1, e_2,

$$u_1 = c_{11}e_1 + c_{21}e_2,$$
$$u_2 = c_{12}e_1 + c_{22}e_2,$$

then C is orthogonal (5.4.7) and

$$C'AC = \begin{pmatrix} r_1 & 0 \\ 0 & r_2 \end{pmatrix}$$

by Theorem 4.10.5. We can suppose (as in 6.5.3) that $|C| = 1$, so that

$$C = \begin{pmatrix} \cos\alpha & -\sin\alpha \\ \sin\alpha & \cos\alpha \end{pmatrix}$$

for a suitable angle α (5.4.10), thus

$$u_1 = (\cos\alpha)e_1 + (\sin\alpha)e_2,$$
$$u_2 = (-\sin\alpha)e_1 + (\cos\alpha)e_2.$$

For every vector $u = (x, y) \in \mathbf{R}^2$, we have (for suitable x', y')

$$\begin{aligned}xe_1 + ye_2 = u &= x'u_1 + y'u_2\\ &= x'(c_{11}e_1 + c_{21}e_2) + y'(c_{12}e_1 + c_{22}e_2)\\ &= (c_{11}x' + c_{12}y')e_1 + (c_{21}x' + c_{22}y')e_2;\end{aligned}$$

thus, the coordinates of u (relative to the two bases) are related by

$$\begin{pmatrix} x \\ y \end{pmatrix} = \begin{pmatrix} c_{11} & c_{12} \\ c_{21} & c_{22} \end{pmatrix}\begin{pmatrix} x' \\ y' \end{pmatrix} = C\begin{pmatrix} x' \\ y' \end{pmatrix},$$

therefore

$$\begin{pmatrix} x' \\ y' \end{pmatrix} = C^{-1}\begin{pmatrix} x \\ y \end{pmatrix} = C'\begin{pmatrix} x \\ y \end{pmatrix} = \begin{pmatrix} \cos\alpha & \sin\alpha \\ -\sin\alpha & \cos\alpha \end{pmatrix}\begin{pmatrix} x \\ y \end{pmatrix}.$$

The resulting formulas

$$x' = x\cos\alpha + y\sin\alpha,$$
$$y' = -x\sin\alpha + y\cos\alpha,$$

show that the change of orthonormal basis amounts to a rotation of axes (6.7.1).

Conclusion: The quadratic form $(*)$ can be brought to the diagonal form

$$r_1(x')^2 + r_2(y')^2$$

by a suitable rotation of axes.

6.7.4 Predicting the geometric character of the graph Continuing the notations of 6.7.3, the characteristic polynomial $p = p_A$ of the symmetric matrix A is

$$p = |tI - A| = t^2 - (a + c)t + (ac - b^2/4);$$

from 6.5.4, we know that

$$p = (t - r_1)(t - r_2) = t^2 - (r_1 + r_2)t + r_1 r_2.$$

In particular, the determinant of A is

$$|A| = ac - b^2/4 = r_1 r_2,$$

thus $b^2 - 4ac = -4r_1 r_2$. The geometric character of the graph of the equation (1) can therefore be predicted from the coefficients a, b, c:

$$\begin{aligned} &\text{elliptic:} && b^2 - 4ac < 0,\\ &\text{hyperbolic:} && b^2 - 4ac > 0,\\ &\text{parabolic:} && b^2 - 4ac = 0, \end{aligned}$$

corresponding, respectively, to the conditions $r_1 r_2 > 0$, $r_1 r_2 < 0$, $r_1 r_2 = 0$.

6.7.5 An example Consider the equation

$$2x^2 + 12xy - 3y^2 - 1 = 0.$$

The matrix of the quadratic part (6.7.3) is

$$A = \begin{pmatrix} 2 & 6 \\ 6 & -3 \end{pmatrix}.$$

From $|A| = -42 < 0$ we know at once that the graph is hyperbolic. The characteristic polynomial of A is

$$p_A = t^2 + t - 42 = (t - 6)(t + 7),$$

so the characteristic roots are 6 and -7. A suitable rotation of axes will bring the original equation to the form

$$6(x')^2 - 7(y')^2 - 1 = 0;$$

a rotation that does it is calculated in Example 6.5.5.

6.7.6 Quadric surfaces

In dimension 3, conic sections graduate to quadric surfaces (ellipsoids, hyperboloids, paraboloids, cones and cylinders) and, for a quadratic form in three variables x, y, z,

$$ax^2 + by^2 + cz^2 + dxy + eyz + fxz,$$

the associated 3×3 symmetric matrix is

$$A = \begin{pmatrix} a & d/2 & f/2 \\ d/2 & b & e/2 \\ f/2 & e/2 & c \end{pmatrix}.$$

The geometry of quadric surfaces is much more complicated than that of the conic sections—and solving cubics much harder than solving quadratics—but it is clear that the methods of the preceding paragraphs are, in principle, applicable. A further complication is that the sign of $|A| = r_1 r_2 r_3$ can be positive (say) in more than one way (for example, $r_1 < 0, r_2 < 0, r_3 > 0$), so the art of predicting geometric character is more subtle. Finally, the connection between change of orthonormal bases and rotation of axes is not trivial (§6.4, Exercise 8).

For an extraordinarily thorough discussion of quadric surfaces (with all the pictures) see the classic book of Ali R. Amir-Moéz and A. L. Fass, *Elements of linear spaces*, Pergamon, New York, 1962.[4]

▶ Exercises

1. For each of the following equations, determine in advance the geometric character of the graph, then find a rotation of axes that eliminates the xy term:

   (i) $2x^2 + 2xy + 2y^2 - 1 = 0$;

   (ii) $5x^2 + 8xy - y^2 + 2x - 3y + 5 = 0$;

   (iii) $x^2 + 4xy + 4y^2 - 7x + 3y - 2 = 0$.

2.

   (i) For a real matrix

$$A = \begin{pmatrix} a & c \\ b & d \end{pmatrix}$$

   let

$$B = \begin{pmatrix} a & (b+c)/2 \\ (b+c)/2 & d \end{pmatrix},$$

$$C = \begin{pmatrix} 0 & (c-b)/2 \\ (b-c)/2 & 0 \end{pmatrix},$$

[4] See also H. Anton, *Elementary linear algebra*, 5th edn, Wiley, 1987; H. W. Brinkmann and E. A. Klotz, *Linear algebra and analytic geometry*, Addison-Wesley, 1971; R. C. McCann, *Introduction to linear algebra*, Harcourt Brace Jovanovich, 1984; S. Venit and W. Bishop, *Elementary linear algebra*, 2nd edn, Prindle, Weber, and Schmidt, 1985; and N. V. Yefimov, *Quadratic forms and matrices* (translated from the Russian by A. Shenitzer), Academic Press, 1964.

that is, $B = \frac{1}{2}(A + A')$ and $C = \frac{1}{2}(A - A')$. Then $A = B + C$, with $B' = B$ and $C' = -C$. (Such matrices C are called *skew-symmetric*.)

(ii) Show that the characteristic roots of C are of the form $\pm ri$ with r real (that is, they are 'pure imaginary').

(iii) If $AA' = A'A$, show that A is one of the two matrices

$$\begin{pmatrix} a & b \\ b & d \end{pmatrix}, \quad \begin{pmatrix} a & -b \\ b & a \end{pmatrix} = aI + \begin{pmatrix} 0 & -b \\ b & 0 \end{pmatrix},$$

thus either A is symmetric or $A - aI$ is skew-symmetric. A matrix A of the latter type is a scalar multiple of a rotation (5.4.10). (See also §4.4, Exercise 5.)

(iv) If either A is symmetric or $A - aI$ is skew-symmetric, show that $AA' = A'A$. (An $n \times n$ real matrix that commutes with its transpose is said to be *normal*.)

(v) Guess the definition of 'normal' for a linear mapping in Euclidean space; for a complex matrix; for a linear mapping in unitary space.

Part two

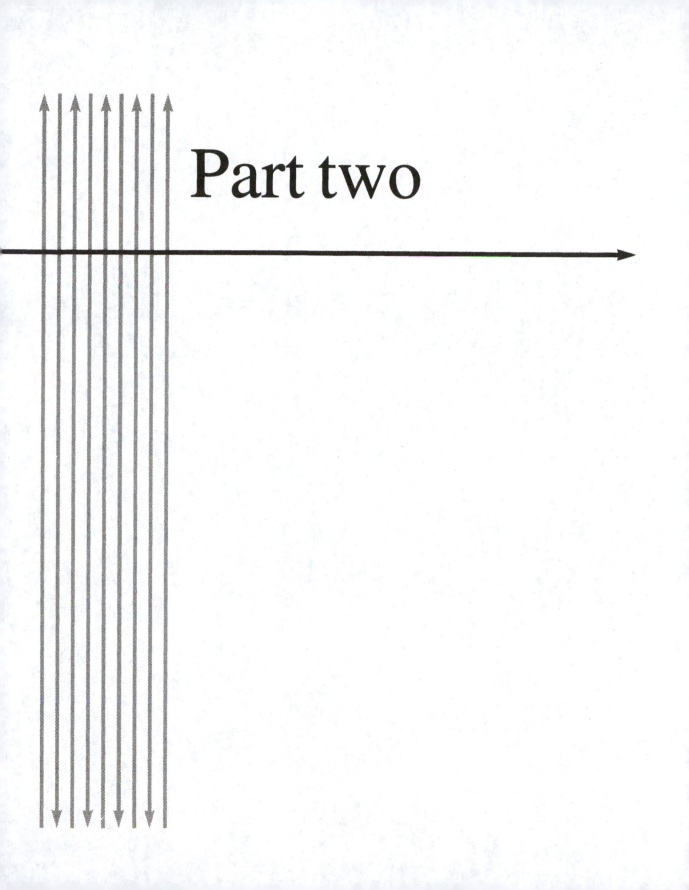

7

Determinants $(n \times n)$

Notations fixed for the chapter:

F a field

n an integer $\geqslant 2$

V an n-dimensional vector space over F

$\mathrm{V}^{\mathrm{n}} = \mathrm{V} \times \ldots \times \mathrm{V}$ (n factors) the product set

No vector space structure on V^{n} is assumed; it is just the **set** of all ordered n-ples $x = (x_1, \ldots, x_{\mathrm{n}})$ of vectors in V.

7.1 Alternate multilinear forms

7.1.1 Definition A function $f: \mathrm{V}^{\mathrm{n}} \to \mathrm{F}$ is called a **multilinear form** if it is a linear function of each coordinate of V^{n}; that is, for each fixed element $y = (y_1, \ldots, y_{\mathrm{n}}) \in \mathrm{V}^{\mathrm{n}}$ and for each index i, the function

$$x_{\mathrm{i}} \mapsto f(y_1, \ldots, y_{\mathrm{i}-1}, x_i, y_{\mathrm{i}+1}, \ldots, y_{\mathrm{n}}) \quad (x_{\mathrm{i}} \in \mathrm{V})$$

is a linear form on V. {When $i = 1$ the notation is to be interpreted as $x_1 \mapsto f(x_1, y_2, \ldots, y_{\mathrm{n}})$; when $i = n$, as $x_{\mathrm{n}} \mapsto f(y_1, \ldots, y_{\mathrm{n}-1}, x_{\mathrm{n}})$.} We write $\mathscr{M}(\mathrm{V})$ for the set of all multilinear forms on V. Note that the definition makes sense for V an arbitrary vector space (not necessarily n-dimensional). The term **bilinear** is used when $n = 2$.

7.1.2 Theorem *$\mathscr{M}(\mathrm{V})$ is a linear subspace of the vector space $\mathcal{F}(\mathrm{V}^{\mathrm{n}}, \mathrm{F})$, hence is a vector space for the pointwise linear operations.*

Proof. Of course the zero function is multilinear. Let $f, g \in \mathcal{M}(\mathrm{V})$, $c \in \mathrm{F}$; we have to show that $f+g$ and cf are multilinear. Fix an n-ple $(y_1, \ldots, y_n) \in \mathrm{V}^n$. For every $h \in \mathcal{M}(\mathrm{V})$ and each index i, define $h_i:\mathrm{V} \to \mathrm{F}$ by the formula

$$h_i(x_i) = h(y_1, \ldots, y_{i-1}, x_i, y_{i+1}, \ldots, y_n) \quad (x_i \in \mathrm{V});$$

since h is multilinear, $h_i, \ldots, h_n$ are linear forms on V. For each i,

$$(f+g)_i = f_i + g_i \quad \text{and} \quad (cf)_i = cf_i$$

by the definitions of the linear operations on $\mathcal{F}(\mathrm{V}^n, \mathrm{F})$; f_i and g_i are linear forms on V (because f and g are multilinear), therefore so are $f_i + g_i$ and cf_i, in other words, $(f+g)_i$ and $(cf)_i$; since this is true for every choice of $(y_1, \ldots, y_n)$ and for each index i, it follows that $f+g$ and cf are multilinear. ∎

7.1.3 Definition A multilinear form f on V is said to be **alternate** if $f(x_1, \ldots, x_n) = 0$ whenever $x_i = x_{i+1}$ for some index i (cf. Corollary 7.1.10). We write $\mathcal{A}(\mathrm{V})$ for the set of all alternate multilinear forms on V. {Again, the definition of alternate form makes sense for V an arbitrary vector space, but the notation $\mathcal{A}(\mathrm{V})$ is reserved for the case that V is n-dimensional.}

7.1.4 Theorem *$\mathcal{A}(\mathrm{V})$ is a linear subspace of $\mathcal{M}(\mathrm{V})$, hence is a vector space for the pointwise linear operations.*

Proof. Suppose $f, g \in \mathcal{A}(\mathrm{V})$ and $c \in \mathrm{F}$. If $(x_1, \ldots, x_n) \in \mathrm{V}^n$ and $x_i = x_{i+1}$ for some i, then both f and g vanish at $(x_1, \ldots, x_n)$, therefore so do $f+g$ and cf. ∎

The key to the chapter: $\mathcal{A}(\mathrm{V})$ *is* 1-*dimensional* (Theorem 7.1.14 below).

7.1.5 Definition We write $\mathbf{S}_n$ for the symmetric group of degree n, that is, the group of all permutations σ of the set $\{1, \ldots, n\}$. In other words, $\mathbf{S}_n$ is the group of all bijections $\sigma:\{1, \ldots, n\} \to \{1, \ldots, n\}$ with composition as the group law. The identity element of $\mathbf{S}_n$ is the identity permutation ι. For every function $f:\mathrm{V}^n \to \mathrm{F}$ and every permutation $\sigma \in \mathbf{S}_n$, we define $\sigma \cdot f:\mathrm{V}^n \to \mathrm{F}$ by the formula $(\sigma \cdot f)(x_1, \ldots, x_n) = f(x_{\sigma(1)}, \ldots, x_{\sigma(n)})$.

7.1.6 Theorem *With notations as in* 7.1.5, $\iota \cdot f = f$ *and* $(\sigma\tau)\cdot f = \sigma \cdot (\tau \cdot f)$ *for all permutations* σ *and* τ.

Proof. The equality $\iota \cdot f = f$ is obvious. Given $(x_1, \ldots, x_n) \in \mathrm{V}^n$, write $y_i = x_{\sigma(i)}$ for $i = 1, \ldots, n$. Then

$$\begin{aligned} [\sigma \cdot (\tau \cdot f)](x_1, \ldots, x_n) &= (\tau \cdot f)(x_{\sigma(1)}, \ldots, x_{\sigma(n)}) \\ &= (\tau \cdot f)(y_1, \ldots, y_n) \\ &= f(y_{\tau(1)}, \ldots, y_{\tau(n)}) \\ &= f(x_{\sigma(\tau(1))}, \ldots, x_{\sigma(\tau(n))}) \\ &= f(x_{(\sigma\tau)(1)}, \ldots, x_{(\sigma\tau)(n)}) \\ &= [(\sigma\tau)\cdot f](x_1, \ldots, x_n). \ ∎ \end{aligned}$$

7.1.7 Lemma *If $f \in \mathcal{A}(V)$ and $\tau = (i \;\; i+1)$ is a transposition that interchanges two consecutive indices, then $\tau \cdot f = -f$.*

Proof. The problem is to show that if $x = (x_1, \ldots, x_n)$ then $(\tau \cdot f)(x_1, \ldots, x_n) + f(x_1, \ldots, x_n) = 0$, that is,

$$(*) \qquad \begin{aligned} &f(x_1, \ldots, x_{i-1}, x_{i+1}, x_i, x_{i+2}, \ldots, x_n) \\ &\quad + f(x_1, \ldots, x_{i-1}, x_i, x_{i+1}, x_{i+2}, \ldots, x_n) = 0. \end{aligned}$$

The key to the proof is the following observation: if $\beta: V \times V \to F$ is an alternate bilinear form, then

$$\begin{aligned} 0 &= \beta(x_1 + x_2, x_1 + x_2) \\ &= \beta(x_1, x_1) + \beta(x_1, x_2) + \beta(x_2, x_1) + \beta(x_2, x_2) \\ &= 0 + \beta(x_1, x_2) + \beta(x_2, x_1) + 0. \end{aligned}$$

In the situation at hand, regard x_i, x_{i+1} as variable and the other x_j's as fixed, and define

$$\beta(x_i, x_{i+1}) = f(x_1, \ldots, x_n);$$

then β is an alternate bilinear form, so by the preceding paragraph $\beta(x_{i+1}, x_i) + \beta(x_i, x_{i+1}) = 0$, which is the desired equation $(*)$. ∎

7.1.8 Lemma *With notations as in* 7.1.5, *if f is multilinear then so is $\sigma \cdot f$; if f is alternate then so is $\sigma \cdot f$.*

Proof. Assuming f multilinear, we are to show that $\sigma \cdot f$ is multilinear. Fix an index j; let's show that $\sigma \cdot f$ is an additive function of the j'th coordinate (the proof of homogeneity is similar). Notation is 99% of the battle.

Let $x = (x_1, \ldots, x_n)$, $y = (y_1, \ldots, y_n)$, $z = (z_1, \ldots, z_n)$; assuming $z_j = x_j + y_j$ and $z_i = x_i = y_i$ for $i \neq j$, we are to show that $(\sigma \cdot f)(z) = (\sigma \cdot f)(x) + (\sigma \cdot f)(y)$. It is convenient to write

$$x_\sigma = (x_{\sigma(1)}, x_{\sigma(2)}, \ldots, x_{\sigma(n)});$$

thus $f(x_\sigma) = (\sigma \cdot f)(x)$. The effect of passing from x to x_σ is that the $\sigma(i)$'th coordinate of x becomes the i'th coordinate of x_σ. The notations y_σ and z_σ have the analogous meanings. Our problem is to show that $f(z_\sigma) = f(x_\sigma) + f(y_\sigma)$.

Let k be the index such that $\sigma(k) = j$ (where j is the index fixed at the beginning). If $i \neq k$ then $\sigma(i) \neq j$, so the i'th coordinates of $z_\sigma, x_\sigma, y_\sigma$ are equal, whereas the k'th coordinates of $z_\sigma, x_\sigma, y_\sigma$ are $x_j + y_j$, x_j, y_j, respectively; it follows that $f(z_\sigma) = f(x_\sigma) + f(y_\sigma)$, by the additivity of f as a function of the k'th coordinate.

Finally, if f is alternate and $\sigma \in S_n$, we are to show that (the multilinear form) $\sigma \cdot f$ is also alternate; it will obviously suffice to show that $\sigma \cdot f = \pm f$. Since every permutation is a product of transpositions, we can suppose by 7.1.6 that σ is a transposition; but every transposition is a product of transpositions of the form $(i \;\; i+1)$—for example, $(25) = (23)(34)(45)(34)(23)$—so we are reduced to 7.1.7. ∎

The assertions of 7.1.6 and 7.1.8 are expressed by saying that the group $\mathbf{S}_n$ *acts* on $\mathcal{M}(V)$ and on $\mathcal{A}(V)$; for the action on $\mathcal{A}(V)$, we can be more precise:

7.1.9 Theorem *If $f \in \mathcal{A}(V)$ then $\sigma \cdot f = (\operatorname{sgn} \sigma) f$ for every permutation σ, where $\operatorname{sgn} \sigma = \pm 1$ according as σ is even or odd.*

Proof. As noted in the proof of 7.1.8, every permutation σ is product of transpositions of the form $(i \quad i+1)$; if the number of factors is k, then $\sigma \cdot f = (-1)^k f$ by 7.1.6 and 7.1.7, and $(-1)^k = \operatorname{sgn} \sigma$ by the definition of $\operatorname{sgn} \sigma$. ■

The indices in 7.1.3 need not be successive:

7.1.10 Corollary *If $f \in \mathcal{A}(V)$ then $f(x_1, \ldots, x_n) = 0$ whenever $x_i = x_j$ for some $i \neq j$.*

Proof. Let $(x_1, \ldots, x_n) \in V^n$ and suppose $x_i = x_j$ for some pair of distinct indices i and j. Choose any $\sigma \in \mathbf{S}_n$ such that $\sigma(1) = i$ and $\sigma(2) = j$. In the notation of the proof of Lemma 7.1.8, $x_\sigma = (x_{\sigma(1)}, x_{\sigma(2)}, \ldots, x_{\sigma(n)}) = (x_i, x_j, x_{\sigma(3)}, \ldots, x_{\sigma(n)})$, thus $(\sigma \cdot f)(x_1, \ldots, x_n) = f(x_\sigma) = 0$ (because the first two coordinates of x_σ are equal), that is (Theorem 7.1.9), $\pm f(x_1, \ldots, x_n) = 0$. ■

7.1.11 Theorem *Let $x_1, \ldots, x_n \in V$, let (a_{ij}) be an $n \times n$ matrix over* F, *and define*

$$y_j = \sum_{i=1}^{n} a_{ij} x_i \quad (j = 1, \ldots, n).$$

Then, for every $f \in \mathcal{A}(V)$,

$$f(y_1, \ldots, y_n) = \left(\sum_{\sigma \in \mathbf{S}_n} (\operatorname{sgn} \sigma) a_{\sigma(1)1} a_{\sigma(2)2} \ldots a_{\sigma(n)n} \right) \cdot f(x_1, \ldots, x_n).$$

Proof. Since f is multilinear,

$$\begin{aligned} f(y_1, \ldots, y_n) &= f\left(\sum_{i=1}^{n} a_{i1} x_i, \sum_{i=1}^{n} a_{i2} x_i, \ldots, \sum_{i=1}^{n} a_{in} x_i \right) \\ &= \sum a_{i_1 1} a_{i_2 2} \ldots a_{i_n n} f(x_{i_1}, x_{i_2}, \ldots, x_{i_n}), \end{aligned}$$

summed over all n-ples $(i_1, \ldots, i_n)$ of indices in $\{1, \ldots, n\}$. If $i_1, \ldots, i_n$ are not all distinct, then $f(x_{i_1}, \ldots, x_{i_n}) = 0$ by Corollary 7.1.10; such terms can be omitted from the sum. The remaining terms correspond to the choices of $(i_1, \ldots, i_n)$ for which $i_1, \ldots, i_n$ are distinct elements of $\{1, \ldots, n\}$, hence constitute a rearrangement of $1, \ldots, n$. In other words, the remaining terms correspond to the $n!$ choices

$$(\sigma(1), \ldots, \sigma(n)) \qquad (\sigma \in \mathbf{S}_n),$$

thus

$$f(y_1, \ldots, y_n) = \sum a_{\sigma(1)1} \ldots a_{\sigma(n)n} f(x_{\sigma(1)}, \ldots, x_{\sigma(n)}),$$

summed over all $\sigma \in S_n$. Finally $f(x_{\sigma(1)}, \ldots, x_{\sigma(n)}) = (\operatorname{sgn} \sigma) f(x_1, \ldots, x_n)$ by Theorem 7.1.9; factoring $f(x_1, \ldots, x_n)$ from every term of the sum, we arrive at the desired formula. ■

We remark that, so far, V could have been *any* vector space over F; from now on, the fact that V is n-dimensional will be crucial.

If alternate multilinear forms agree on a single basis n-ple then they are identical:

7.1.12 Lemma *Let $x_1, \ldots, x_n$ be a basis of* V. *If $f, g \in \mathcal{A}(V)$ and $f(x_1, \ldots, x_n) = g(x_1, \ldots, x_n)$, then $f = g$.*

Proof. Switching to the alternate multilinear form $f - g$ and changing notations, the problem is to show that if $f(x_1, \ldots, x_n) = 0$ then $f = 0$. Given any n-ple $(y_1, \ldots, y_n) \in V^n$, we have to show that $f(y_1, \ldots, y_n) = 0$. Expressing each y_j as a linear combination of the basis vectors $x_1, \ldots, x_n$, we see from the formula of Theorem 7.1.11 that $f(y_1, \ldots, y_n)$ is a multiple of $f(x_1, \ldots, x_n) = 0$. ■

7.1.13 Lemma *If $x_1, \ldots, x_n$ is a basis of* V, *there exists a unique $f \in \mathcal{A}(V)$ such that $f(x_1, \ldots, x_n) = 1$.*

Proof. Uniqueness is assured by 7.1.12.

Existence: Given any $(y_1, \ldots, y_n) \in V^n$, express each y_j as a linear combination of $x_1, \ldots, x_n$, say

$$y_j = \sum_{i=1}^{n} a_{ij} x_i,$$

and *define* $f(y_1, \ldots, y_n)$ by the formula

$$(*) \qquad f(y_1, \ldots, y_n) = \sum_{\sigma \in S_n} (\operatorname{sgn} \sigma) a_{\sigma(1)1} a_{\sigma(2)2} \ldots a_{\sigma(n)n}.$$

First, let us argue that f is multilinear. Fix an index j; regard y_j as a variable vector, and the y_i for $i \neq j$ as fixed vectors. For each permutation σ, $a_{\sigma(j)j}$ is the $\sigma(j)$'th coefficient in the representation of y_j, hence is a linear function of y_j; it follows that each term of the sum on the right side of $(*)$ is a linear functions of y_j, hence so is their sum $f(y_1, \ldots, y_n)$.

Next, we observe that $f(x_1, \ldots, x_n) = 1$. For, if $y_j = x_j$ for all j, then $a_{ij} = \delta_{ij}$ for all i and j; thus, for $\sigma \in S_n$, $a_{\sigma(i)i}$ is 1 if $\sigma(i) = i$ and it is 0 if $\sigma(i) \neq i$. If σ is not the identity permutation, then there is an index i for which $\sigma(i) \neq i$, consequently $a_{\sigma(i)i} = 0$, and the term of the sum corresponding to σ is guaranteed to be 0. Thus only the term for $\sigma = \iota$ survives, therefore

$$f(x_1, \ldots, x_n) = (\operatorname{sgn} \iota) a_{11} \ldots a_{nn} = 1.$$

Finally, let us show that f is alternate. Let $(y_1, \ldots, y_n) \in V^n$ and suppose $y_i = y_{i+1}$ for some index i. Express the y's in terms of the x's, with coefficients a_{ij} as before. The strategy is to show that the terms of the sum $(*)$ cancel in pairs.

Let $\tau = (i \quad i+1)$, and let $\mathbf{A}_n$ be the alternating subgroup of $\mathbf{S}_n$, that is, the group of all *even* permutations of $1, \ldots, n$. Since the transposition τ is odd and $\mathbf{A}_n$ is a subgroup of $\mathbf{S}_n$ of index 2, we have $\mathbf{S}_n = \mathbf{A}_n \cup \mathbf{A}_n\tau$ and the terms of the union are disjoint. Given any $\sigma \in \mathbf{A}_n$ it will suffice to show that the terms of (*) corresponding to σ and $\sigma\tau$ are negatives of each other. At any rate, $\operatorname{sgn}\sigma = 1$ and $\operatorname{sgn}\sigma\tau = -1$. The term of (*) for σ is

(1) $$a_{\sigma(1)1} \ldots a_{\sigma(i)i} a_{\sigma(i+1),i+1} \ldots a_{\sigma(n)n},$$

and the term for $\rho = \sigma\tau$ is

(2) $$-a_{\rho(1)1} \ldots a_{\rho(i)i} a_{\rho(i+1),i+1} \ldots a_{\rho(n)n}.$$

Since τ interchanges i and $i+1$, and leaves all other indices fixed, we have

$$\rho(i) = \sigma(\tau(i)) = \sigma(i+1),$$
$$\rho(i+1) = \sigma(\tau(i+1)) = \sigma(i),$$

and for all other indices k,

$$\rho(k) = \sigma(\tau(k)) = \sigma(k).$$

Looking at (1) and (2), and keeping in mind that $a_{ji} = a_{j,i+1}$ for all j (because $y_i = y_{i+1}$), we see that the n factors that make up the product in (2) are a permutation of those that make up the product of (1), thus the expression in (2) is indeed the negative of the expression in (1). ■

7.1.14 Theorem $\mathcal{A}(\mathrm{V})$ *is 1-dimensional.*

Proof. Let $x_1, \ldots, x_n$ be a basis of V and let f be the alternate multilinear form on V such that $f(x_1, \ldots, x_n) = 1$ (7.1.13). Given any $g \in \mathcal{A}(\mathrm{V})$, we seek a scalar $c \in \mathrm{F}$ such that $g = cf$. Set $c = g(x_1, \ldots, x_n)$; then cf is an alternate multilinear form such that

$$(cf)(x_1, \ldots, x_n) = c \cdot 1 = g(x_1, \ldots, x_n),$$

so $cf = g$ by Lemma 7.1.12. ■

It follows that each linear mapping in $\mathcal{A}(\mathrm{V})$ is multiplication by a fixed scalar.

▶ Exercises

1. If $f \in \mathcal{A}(\mathrm{V})$ then $f(x_1, \ldots, x_n) = 0$ whenever $x_1, \ldots, x_n$ are linearly dependent.

2. (i) If $f_1, \ldots, f_n$ are linear forms on V, then the function $f: \mathrm{V}^n \to \mathrm{F}$ defined by

$$f(x_1, \ldots, x_n) = f_1(x_1) \ldots f_n(x_n)$$

is multilinear.

(ii) If $n = 2$ and f_1, f_2 are nonzero linear forms on V, then the bilinear form f of (i) is not alternate. {Hint: §1.6, Exercise 18.}

(iii) If $n = 2$ and f_1, f_2 are linearly independent linear forms on V, then the bilinear form f defined by

$$f(x_1, x_2) = f_1(x_1)f_2(x_2) - f_1(x_2)f_2(x_1)$$

is alternate and not identically zero. {Hint: $\operatorname{Ker} f_1$ is not contained in $\operatorname{Ker} f_2$.}

(iv) If $x_1, \ldots, x_n$ is a basis of V, and $f_1, \ldots, f_n$ is the dual basis of V′, then the function f defined by the formula

$$(*) \qquad f(y_1, \ldots, y_n) = \sum_{\sigma \in S_n} (\operatorname{sgn} \sigma) f_{\sigma(1)}(y_1) \ldots f_{\sigma(n)}(y_n)$$

is the nonzero alternate multilinear form constructed in Lemma 7.1.13.

3. Let V_1, V_2, W be vector spaces over F.

(i) Extend the concept of bilinearity to mappings $f: V_1 \times V_2 \to W$ and prove that the set $\mathcal{M} = \mathcal{M}(V_1, V_2; W)$ of all such f is a vector space over F for the pointwise linear operations.

(ii) Suppose V_1, V_2, W are finite-dimensional; choose bases of the three spaces. A bilinear mapping $f: V_1 \times V_2 \to W$ is determined by the vectors $f(x, y)$, where x and y are members of the chosen bases of V_1 and V_2; in turn, $f(x, y)$ is determined by its coordinates relative to the chosen basis of W. Determine the dimension of $\mathcal{M} = \mathcal{M}(V_1, V_2; W)$.

(iii) Extend the preceding discussion to multilinear mappings $V_1 \times \ldots \times V_r \to W$.

4. As assumed at the outset of the chapter, $\dim V = n \geq 2$. Let r be any positive integer ≥ 2 and consider the vector space of multilinear forms $f: V^r \to F$ (cf. Exercise 3). Call such an f *alternate* if $f(x) = 0$ whenever $x = (x_1, \ldots, x_r)$ with $x_i = x_{i+1}$ for some i. The set $\mathcal{A}_r(V)$ of all alternate multilinear forms is a vector space for the pointwise linear operations; in particular, $\mathcal{A}_n(V)$ is the space $\mathcal{A}(V)$ of Theorem 7.1.3.

(i) If $r > n$ then $\mathcal{A}_r(V) = \{0\}$.

(ii) Determine the dimension of $\mathcal{A}_r(V)$ for $r < n$.

{Hint: (ii) If $x_1, \ldots, x_n$ is a basis of V, then an alternate form f is determined by its values on the r-ples $(x_{i_1}, x_{i_2}, \ldots, x_{i_r})$ with $i_1 < i_2 < \ldots < i_r$.}

7.2 Determinant of a linear mapping

7.2.1 Definition If $f \in \mathcal{A}(V)$ and $T: V \to V$ is a linear mapping on V, then the function $V^n \to F$ defined by $(x_1, \ldots, x_n) \mapsto f(Tx_1, \ldots, Tx_n)$ is clearly an alternate multilinear form, that is, an element of $\mathcal{A}(V)$; we define the mapping $T^{\wedge}: \mathcal{A}(V) \to \mathcal{A}(V)$ by the formula

$$(T^{\wedge} f)(x_1, \ldots, x_n) = f(Tx_1, \ldots, Tx_n)$$

for all n-ples $(x_1, \ldots, x_n) \in V^n$. It is obvious that $T^{\wedge}$ is a linear mapping, so by Theorem 7.1.14, $T^\wedge$ is multiplication by a unique scalar; this scalar is called the **determinant** of T, written $\det T$ or $|T|$, thus

$$T^\wedge = (\det T)I,$$

where I is the identity mapping on $\mathcal{A}(\mathrm{V})$.

7.2.2 Example Let $x_1, \ldots, x_n$ be a basis of V, let $\sigma \in \mathrm{S}_n$ and let $T_\sigma : \mathrm{V} \to \mathrm{V}$ be the linear mapping such that $T_\sigma x_i = x_{\sigma(i)}$ for all i. (Caution: T_σ depends on a choice of basis.) Then $\det T_\sigma = \operatorname{sgn} \sigma$. In particular, $\det I_\mathrm{V} = 1$, where I_V is the identity linear mapping on V.

{Proof. Write $T = T_\sigma$. For all $f \in \mathcal{A}(\mathrm{V})$,

$$\begin{aligned}(T^\wedge f)(x_1, \ldots, x_n) &= f(Tx_1, \ldots, Tx_n)\\ &= f(x_{\sigma(1)}, \ldots, x_{\sigma(n)})\\ &= (\sigma \cdot f)(x_1, \ldots, x_n) \qquad \text{(cf. 7.1.5)}\end{aligned}$$

therefore $T^\wedge f = \sigma \cdot f = (\operatorname{sgn} \sigma) f$ by 7.1.9 and 7.1.12, thus $T^\wedge = (\operatorname{sgn} \sigma) I$.}

7.2.3 Lemma *For all* $S, T \in \mathcal{L}(\mathrm{V})$, $(ST)^\wedge = T^\wedge S^\wedge$.

Proof. Let $S, T \in \mathcal{L}(\mathrm{V})$ and let $x_1, \ldots, x_n$ be a basis of V. For all $f \in \mathcal{A}(\mathrm{V})$,

$$\begin{aligned}[(ST)^\wedge f](x_1, \ldots, x_n) &= f((ST)x_1, \ldots, (ST)x_n)\\ &= f(S(Tx_1), \ldots, S(Tx_n))\\ &= (S^\wedge f)(Tx_1, \ldots, Tx_n)\\ &= [T^\wedge(S^\wedge f)](x_1, \ldots, x_n),\end{aligned}$$

therefore $(ST)^\wedge f = T^\wedge(S^\wedge f)$ by 7.1.12, that is, $(ST)^\wedge f = (T^\wedge S^\wedge) f$. ■

7.2.4 Theorem *For all* $S, T \in \mathcal{L}(\mathrm{V})$, $\det(ST) = (\det S)(\det T)$.

Proof. If $S, T \in \mathcal{L}(\mathrm{V})$ and I is the identity mapping on $\mathcal{A}(\mathrm{V})$, then

$$\begin{aligned}[\det(ST)]I &= (ST)^\wedge && (7.2.1)\\ &= T^\wedge S^\wedge && (7.2.3)\\ &= (\det T)I \cdot (\det S)I && (7.2.1)\\ &= [(\det S)(\det T)]I,\end{aligned}$$

therefore $\det(ST) = (\det S)(\det T)$. ■

7.2.5 Remark For all $T \in \mathcal{L}(\mathrm{V})$ and $a \in \mathrm{F}$, $\det(aT) = a^n \det T$.

{**Proof.** Let $x_1, \ldots, x_n$ be a basis of V. For all $f \in \mathcal{A}(\mathrm{V})$,

$$\begin{aligned}((aT)^\wedge f)(x_1, \ldots, x_n) &= f((aT)x_1, \ldots, (aT)x_n)\\ &= f(a(Tx_1), \ldots, a(Tx_n))\\ &= a^n f(Tx_1, \ldots, Tx_n)\\ &= a^n(T^\wedge f)(x_1, \ldots, x_n),\end{aligned}$$

therefore $(aT)\hat{}f = a^n(T\hat{}f)$ by Lemma 7.1.12; thus $(aT)\hat{} = a^n T\hat{}$, that is, $[\det(aT)]I = a^n[(\det T)I]$, whence $\det(aT) = a^n(\det T)$.}

7.2.6 Theorem *For $T \in \mathcal{L}(\mathrm{V})$, T is bijective if and only if* $\det T \neq 0$.

Proof. If T is bijective, say $ST = I$, then $1 = \det I = \det(ST) = (\det S)(\det T)$, so $\det T \neq 0$.

Let's prove the converse ($\det T \neq 0 \Rightarrow T$ bijective) in contrapositive form: assuming T not bijective, we must show that $\det T = 0$. Since T is not bijective and V is finite-dimensional, $\operatorname{Ker} T \neq \{\theta\}$ (cf. Theorem 3.7.4). Choose a nonzero vector x_1 such that $Tx_1 = \theta$ and expand to a basis $x_1, \ldots, x_n$ of V (apply Corollary 3.5.16 with $\mathrm{M} = \mathrm{F}x_1$ the 1-dimensional subspace spanned by x_1). For all $f \in \mathcal{A}(\mathrm{V})$,

$$\begin{aligned}(T\hat{}f)(x_1, \ldots, x_n) &= f(Tx_1, Tx_2, \ldots, Tx_n)\\ &= f(\theta, Tx_2, \ldots, Tx_n)\\ &= 0,\end{aligned}$$

so $T\hat{}f = 0$; thus $0 = T\hat{} = (\det T)I$, whence $\det T = 0$. ■

7.2.7 Corollary *Let $f \in \mathcal{A}(\mathrm{V})$, $f \neq 0$. A list of vectors $x_1, \ldots, x_n \in \mathrm{V}$ of length n is linearly independent if and only if $f(x_1, \ldots, x_n) \neq 0$.*

Proof. Let $y_1, \ldots, y_n$ be a basis of V and let $T \in \mathcal{L}(\mathrm{V})$ be the linear mapping such that $Ty_i = x_i$ for all i. Then

$$f(x_1, \ldots, x_n) = f(Ty_1, \ldots, Ty_n) = (T\hat{}f)(y_1, \ldots, y_n),$$

so by Lemma 7.1.12 we have $T\hat{}f \neq 0 \Leftrightarrow f(x_1, \ldots, x_n) \neq 0$; but $T\hat{}f = (\det T)f$ and f is nonzero, so $T\hat{}f \neq 0 \Leftrightarrow \det T \neq 0$. Thus $f(x_1, \ldots, x_n) \neq 0 \Leftrightarrow T$ is bijective $\Leftrightarrow$ the $x_i = Ty_i$ are linearly independent. ■

7.2.8 Theorem *Let $T \in \mathcal{L}(\mathrm{V})$, let $x_1, \ldots, x_n$ be a basis of V, and let (a_{ij}) be the matrix of T relative to the basis $x_1, \ldots, x_n$. Then*

$$\det T = \sum_{\sigma \in \mathrm{S}_n} (\operatorname{sgn} \sigma) a_{\sigma(1)1} a_{\sigma(2)2} \ldots a_{\sigma(n)n}.$$

Proof. Write c for the scalar on the right side of the equation and let $y_j = Tx_j$ $(j = 1, \ldots, n)$. For all $f \in \mathcal{A}(\mathrm{V})$,

$$\begin{aligned}(T\hat{}f)(x_1, \ldots, x_n) &= f(Tx_1, \ldots, Tx_n)\\ &= f(y_1, \ldots, y_n)\\ &= cf(x_1, \ldots, x_n)\end{aligned}$$

by the formula of 7.1.11, therefore $T\hat{}f = cf$ by 7.1.12; thus $\det T = c$ by the definition of determinant (7.2.1). ■

▶ ***Exercises***

1. (i) For every $g \in \mathrm{V}'$ and $y \in \mathrm{V}$, the linear mapping $g \otimes y \in \mathcal{L}(\mathrm{V})$ defined in Example 2.3.4 has determinant 0.

(ii) Every $T \in \mathcal{L}(V)$ is a sum $T = T_1 + \ldots + T_n$ with $\det T_1 = \ldots = \det T_n = 0$, where $n = \dim V$. {Hint: §3.8, Exercise 9.}

(iii) Express the identity mapping $I \in \mathcal{L}(V)$ as a sum $I = T_1 + \ldots + T_n$ as in (ii), and describe explicitly the mappings T_i. {Hint: Dual basis.}

2. Let $x_1, \ldots, x_n$ be a basis of V, $T \in \mathcal{L}(V)$ a linear mapping such that $Tx_k \in [\{x_1, \ldots, x_k\}]$ for $k = 1, \ldots, n$. Show that if $A = (a_{ij})$ is the matrix of T relative to this basis, then $\det T = a_{11}a_{22} \ldots a_{nn}$.

3. Let V be the (n-dimensional) vector space of real polynomials of degree $\leq n-1$ (1.3.6) and let $D \in \mathcal{L}(V)$ be the differentiation mapping $Dp = p'$ (cf. 2.1.4). Show that $\det D = 0$ and $\det(I + D) = 1$.
{Hint: Look at Exercise 2 in the light of the basis $1, t, t^2, \ldots, t^{n-1}$ of V.}

4. Let $V \subset \mathcal{F}(\mathbf{R})$ be the 2-dimensional vector space spanned by the functions sin, cos and let $T \in \mathcal{L}(V)$ be the differentiation mapping. Show that $\det T = 1$.

7.3 Determinant of a square matrix

7.3.1 Definition Let A be an $n \times n$ matrix over F. Let $T \in \mathcal{L}(F^n)$ be the linear mapping whose matrix relative to the canonical basis of F^n is A. With $V = F^n$ in Definition 7.2.1, $\det T$ is called the **determinant** of A, written $\det A$ or $|A|$.

7.3.2 Theorem *If $A = (a_{ij})$ is an $n \times n$ matrix over* F, *then*

$$\det A = \sum_{\sigma \in \mathbf{S}_n} (\operatorname{sgn} \sigma) a_{\sigma(1)1} a_{\sigma(2)2} \ldots a_{\sigma(n)n}.$$

Proof. Immediate from 7.2.8 and the definition of $\det A$. ■

We can think of the computation of $\det A$ in the following way. For each $\sigma \in \mathbf{S}_n$, let $\sigma \cdot A$ be the matrix whose i'th row is the $\sigma(i)$'th row of A $(i = 1, \ldots, n)$; thus $\sigma \cdot A = (b_{ij})$, where $b_{ij} = a_{\sigma(i)j}$. {In effect, $\sigma \cdot A$ is obtained by applying the permutation σ^{-1} to the rows of A.} The diagonal elements of $\sigma \cdot A$ are the $a_{\sigma(i)i}$; form the product of these diagonal elements and prefix it with the sign $\operatorname{sgn} \sigma$:

$$(\operatorname{sgn} \sigma) \prod_{i=1}^{n} a_{\sigma(i)i}.$$

The sum of these expressions, as σ varies over $\mathbf{S}_n$, is $|A|$. Since $\operatorname{sgn} \sigma^{-1} = \operatorname{sgn} \sigma$, $|A|$ is also the sum of the expressions

$$(\operatorname{sgn} \sigma) \prod_{i=1}^{n} a_{\sigma^{-1}(i)i}.$$

over all σ. This is easier to verbalize: For each $\sigma \in \mathbf{S}_n$, apply σ to the rows of A, take the product of the diagonal elements of the resulting matrix and prefix it with the sign $\operatorname{sgn} \sigma$; the sum of all these expressions, as σ varies over $\mathbf{S}_n$, is $|A|$.

The following theorem liberates $\det A$ from the particular space F^n and a particular choice of basis:

7.3.3 Theorem *If $x_1, \ldots, x_n$ is a basis of* V, *if $S \in \mathcal{L}(V)$ and if A is the matrix of S relative to the basis $x_1, \ldots, x_n$, then* $\det A = \det S$.

Proof. The formula for $\det A$ in 7.3.2 is also a formula for $\det S$ (7.2.8). ■

7.3.4 Theorem *If A and B are the matrices of the same linear mapping $T \in \mathcal{L}(V)$, relative to two bases of* V, *then* $\det A = \det B$.

Proof. Citing 7.3.3 twice, $\det A = \det T = \det B$. ■

7.3.5 Theorem $\det A' = \det A$ *for every $n \times n$ matrix A over F.*

Proof. Assuming $A = (a_{ij})$, write $B = A' = (b_{ij})$, where $b_{ij} = a_{ji}$. By Theorem 7.3.2,

$$\det A = \sum_{\sigma \in S_n} (\operatorname{sgn} \sigma) \prod_{i=1}^{n} a_{\sigma(i)i}.$$

Since $\sigma \mapsto \sigma^{-1}$ is a bijection of S_n,

$$(*) \qquad \det A = \sum_{\sigma \in S_n} (\operatorname{sgn} \sigma^{-1}) \prod_{i=1}^{n} a_{\sigma^{-1}(i)i}.$$

For each σ, the substitution $i \mapsto \sigma(i)$ yields

$$\prod_{i=1}^{n} a_{\sigma^{-1}(i)i} = \prod_{i=1}^{n} a_{i\sigma(i)} = \prod_{i=1}^{n} b_{\sigma(i)i};$$

since also $\operatorname{sgn} \sigma^{-1} = \operatorname{sgn} \sigma$, $(*)$ may be rewritten as

$$\det A = \sum_{\sigma \in S_n} (\operatorname{sgn} \sigma) \prod_{i=1}^{n} b_{\sigma(i)i} = \det B,$$

in other words, $\det A = \det A'$. ■

7.3.6 Theorem *If A and B are $n \times n$ matrices over* F, *then* $\det(AB) = (\det A)(\det B)$.

Proof. Choose a basis $x_1, \ldots, x_n$ of V and let S, T be the linear mappings with matrices A, B, respectively, relative to the basis $x_1, \ldots, x_n$. Then ST has matrix AB relative to this basis (4.3.3), thus

$$\det(AB) = \det(ST) = (\det S)(\det T) = (\det A)(\det B)$$

by 7.3.3 and 7.2.4. ■

In particular, $\det(AB) = \det(BA)$ for every pair of $n \times n$ matrices A and B. Another important function with this property is the 'trace':

7.3.7 Definition If $A = (a_{ij})$ is an $n \times n$ matrix over F, the **trace** of A, denoted $\operatorname{tr} A$, is the sum of the diagonal elements of A:

$$\operatorname{tr} A = \sum_{i=1}^{n} a_{ii}.$$

Writing $M_n(F)$ for the algebra of all $n \times n$ matrices over F, one verifies easily that $\mathrm{tr}: M_n(F) \to F$ is a linear form on $M_n(F)$.

7.3.8 Theorem $\mathrm{tr}(AB) = \mathrm{tr}(BA)$ *for every pair of* $n \times n$ *matrices* A *and* B *over* F.

Proof. If $AB = (c_{ij})$ then

$$\mathrm{tr}\, AB = \sum_{i=1}^{n} c_{ii} = \sum_{i=1}^{n}\sum_{j=1}^{n} a_{ij}b_{ji} = \sum_{i,j} a_{ij}b_{ji},$$

the last sum being extended over all ordered pairs (i, j) of indices; similarly

$$\mathrm{tr}\, BA = \sum_{i,j} b_{ij}a_{ji} = \sum_{i,j} a_{ji}b_{ij} = \sum_{i,j} a_{ij}b_{ji},$$

(the last equality by the substitution $(i, j) \mapsto (j, i)$), thus $\mathrm{tr}\, BA = \mathrm{tr}\, AB$. ■

Putting together the definitions of this section with the results of the preceding two, we have the following characterization of the determinant function on matrices:

7.3.9 Theorem *Identify the set* $M_n(F)$ *with the product* $F^n \times \ldots \times F^n$ *of* n *copies of* F^n, *by identifying an* $n \times n$ *matrix with its n-ple of column vectors. The determinant function* $A \mapsto \det A$ *is the unique function* $M_n(F) \to F$ *that is an alternate multilinear function of column vectors and has the value* 1 *for the identity matrix* I.

Proof. Let $e_1, \ldots, e_n$ be the canonical basis of F^n and let f be the unique alternate multilinear form on $F^n \times \ldots \times F^n$ (n factors) such that $f(e_1, \ldots, e_n) = 1$ (7.1.13). If $A \in M_n(F)$ and $y_1, \ldots, y_n$ are the column vectors of A (in order), the problem is to show that $\det A = f(y_1, \ldots, y_n)$.

Let $T: F^n \to F^n$ be the unique linear mapping whose matrix relative to the basis $e_1, \ldots, e_n$ is A; by definition, $\det A = \det T$ (7.3.1). As shown in the proof of 4.2.8, $Te_j = y_j$ (the j'th column vector of A); in the notations of 7.2.1, $T^{\wedge} = (\det T)I$, thus

$$\begin{aligned} f(y_1, \ldots, y_n) &= f(Te_1, \ldots, Te_n) = (T^{\wedge}f)(e_1, \ldots, e_n) \\ &= (\det T)f(e_1, \ldots, e_n) = \det A. \ ■ \end{aligned}$$

▶ Exercises

1. If $A \in M_n(F)$ is a triangular matrix (say with 0's below the main diagonal), then $|A|$ is the product of the diagonal elements. {Hint: With notations as in Theorem 7.3.2, if $\sigma(i) > i$ then $a_{\sigma(i)i} = 0$. Alternatively, cite §7.2, Exercise 2.}

2. The theory of determinants can be extended to matrices $A \in M_n(S)$, where S is any commutative ring.[1] Here is an informal way of seeing why this is so.

(i) Construct an integral domain R by adjoining n^2 indeterminates t_{ij} to the ring $\mathbf{Z}$ of integers. {For example, let $R_1 = \mathbf{Z}[t_{11}]$ be the ring of

[1] Cf. S. MacLane and G. Birkhoff, *Algebra* (3rd edn), Chelsea,, 1988, Chapter IX.

polynomials in t_{11} with integer coefficients, $R_2 = R_1[t_{12}] = \mathbf{Z}[t_{11}, t_{12}]$ the ring of polynomials in t_{11} and t_{12} with integral coefficients, etc.; in n^2 steps, we reach the desired ring R.}

(ii) Let F be the field of fractions of R (Appendix B.1.2) and let T be the $n \times n$ matrix whose (i, j) entry is t_{ij}; thus $T = (t_{ij}) \in M_n(R) \subset M_n(F)$. Then $|T|$ makes sense and is an element of R (7.3.1, 7.3.2), that is, a polynomial in the t_{ij} with integral coefficients.

(iii) With $A \in M_n(S)$ as above, define $|A|$ to be the result of substituting a_{ij} for t_{ij} (for all i and j) in the formula for $|T|$. (Ignore all properties of $\mathbf{Z}$ except those that it shares with every commutative ring.)

(iv) To see that $|AB| = |A| \cdot |B|$ for all A, $B \in M_n(S)$, adjoin instead $2n^2$ indeterminates t_{ij}, u_{ij} to $\mathbf{Z}$, let $T = (t_{ij})$, $U = (u_{ij})$ and make the appropriate substitutions in the equation $|TU| = |T| \cdot |U|$.

3. State and prove the analogue of Theorem 7.3.9 for row vectors. {Hint: Theorem 7.3.5.}

7.4 Cofactors

Cofactors provide a recursive technique for calculating determinants ('recursive' means that determinants of order n are expressed in terms of determinants of order $n - 1$).

7.4.1 Definition Let $A = (a_{ij})$ be an $n \times n$ matrix over F. For each ordered pair of indices (i, j), let $A^{i,j}$ be the $(n-1) \times (n-1)$ matrix obtained from A by deleting row i and column j; one calls $A^{i,j}$ the (i, j)-th **minor** of A, and one defines

$$A_{ij} = (-1)^{i+j}|A^{i,j}|.$$

Thus, the scalar A_{ij} is obtained from A by striking out the i'th row and j'th column, taking the determinant of what's left, and prefixing it with the sign $(-1)^{i+j}$; A_{ij} is called the **cofactor** of a_{ij} in A. {Observe that, as a function of (i, j), the sign $(-1)^{i+j}$ follows the 'checkerboard' pattern: it is 1 when $i = j = 1$, and changes sign when either i or j changes by 1.}

7.4.2 Theorem **(Expansion of a determinant by cofactors)** *With notations as in* 7.4.1, *for each fixed index i we have*

$$|A| = \sum_{j=1}^{n} a_{ij}A_{ij}.$$

Proof. We consider first the case that $i = n$ ('expansion by the last row'); at the end of the proof we show how to deduce the general case from this special case. Our point of departure is the formula

(1) $$|A| = \sum_{\sigma \in S_n} (\operatorname{sgn} \sigma) a_{\sigma(1)1} a_{\sigma(2)2} \ldots a_{\sigma(n)n}.$$

Let $T = \{1, \ldots, n\}$; S_n is the group of bijections of the set T.

Fix an index $j \in T$; we are going to study the terms of the sum in (1) in which a_{nj} appears. {As j varies over T, all terms of (1) will be accounted for.} Let

$$H = \{\rho \in S_n : \rho(n) = n\},$$

the subgroup of S_n consisting of those $\rho \in S_n$ that leave n fixed (hence permute $1, \ldots, n-1$); it is clear that $H \cong S_{n-1}$ via the restriction mapping $\rho \mapsto \rho | T - \{n\}$.

The terms of (1) that contain a_{nj} are

$$\sum_{\sigma(j)=n} (\operatorname{sgn} \sigma) a_{\sigma(1)1} a_{\sigma(2)2} \ldots a_{\sigma(n)n} = \sum_{\sigma(j)=n} (\operatorname{sgn} \sigma) a_{nj} \prod_{k \neq j} a_{\sigma(k)k}.$$

Let's explore the set of indices for this sum. Write τ for the element of S_n that interchanges j and n and leaves all other indices fixed. {If $j \neq n$ then τ is the transposition $(j\,n)$; if $j = n$ then τ is the identity permutation ι. In either case, $\tau = \tau^{-1}$.} Then

$$\begin{aligned} \{\sigma \in S_n : \sigma(j) = n\} &= \{\sigma : \sigma(\tau(n)) = n\} \\ &= \{\sigma : \sigma\tau \in H\} = \{\sigma : \sigma \in H\tau^{-1}\} \\ &= H\tau^{-1} = H\tau = \{\rho\tau : \rho \in H\}. \end{aligned}$$

Thus the terms of (1) containing a_{nj} are

(2) $$\sum_{\rho \in H} (\operatorname{sgn} \rho\tau) a_{nj} \prod_{k \neq j} a_{(\rho\tau)(k)k} = a_{nj} (\operatorname{sgn} \tau) \sum_{\rho \in H} (\operatorname{sgn} \rho) \prod_{k \neq j} a_{\rho(\tau(k))k}.$$

Consider the sum that appears on the right side of (2):

(3) $$\sum_{\rho \in H} (\operatorname{sgn} \rho) \prod_{k \neq j} a_{\rho(\tau(k))k};$$

we will show that this sum is equal to $(\operatorname{sgn} \tau) A_{nj}$.

The condition $k \neq j$ means that $k \in T - \{j\}$; since $\tau(\{j\}) = \{n\}$, the correspondence $k \mapsto \tau(k)$ maps $T - \{j\}$ bijectively onto $T - \{n\} = \{1, \ldots, n-1\}$, with inverse mapping $i \mapsto \tau^{-1}(i) = \tau(i)$. Writing $i = \tau(k)$ for $k \in T - \{j\}$, we have $k = \tau^{-1}(i) = \tau(i)$ and

$$a_{\rho(\tau(k))k} = a_{\rho(i)\tau(i)},$$

thus (3) can be written

(4) $$\sum_{\rho \in H} (\operatorname{sgn} \rho) \prod_{i=1}^{n-1} a_{\rho(i)\tau(i)}.$$

case 1: $j = n$

Then $\tau = \iota$ and (4) simplifies to

$$\sum_{\rho \in H} (\operatorname{sgn} \rho) \prod_{i=1}^{n-1} a_{\rho(i)i};$$

identifying H with S_{n-1} in the way indicated earlier, this is just $|A^{n,n}| = A_{nn} = (\operatorname{sgn}\tau)A_{nn}$, as we wished to show.

case 2: $j < n$

Then $\tau = (j\ n)$ and $\operatorname{sgn}\tau = -1$. As i increases from 1 to $n-1$, $\tau(i)$ runs through the values

$$1, \ldots, j-1, n, j+1, \ldots, n-1$$

in the indicated order, and the row vector of length $n-1$

$$(a_{i\tau(1)} \ldots a_{i\tau(n-1)})$$

runs through the rows of the matrix

$$B = \begin{pmatrix} a_{11} & \ldots & a_{1,j-1} & a_{1n} & a_{1,j+1} & \ldots & a_{1,n-1} \\ a_{21} & \ldots & a_{2,j-1} & a_{2n} & a_{2,j+1} & \ldots & a_{2,n-1} \\ \vdots & & & & & & \\ a_{n-1,1} & \ldots & a_{n-1,j-1} & a_{n-1,n} & a_{n-1,j+1} & \ldots & a_{n-1,n-1} \end{pmatrix}$$

from top to bottom. We have $B = (b_{ih})$, where

$$b_{ih} = a_{i\tau(h)} \quad \text{for} \quad i, h = 1, \ldots, n-1.$$

If the j'th column of B is moved to the right end, one obtains the minor $A^{n,j}$ of A; this can be accomplished by successive transpositions to the right past

$$(n-1) - (j+1) + 1 = n - j - 1$$

columns, therefore

$$|B| = (-1)^{n-j-1}|A^{n,j}| = (-1)^{n-j-1} \cdot (-1)^{n+j} A_{nj} = -A_{nj},$$

that is,

(5) $$|B| = (\operatorname{sgn}\tau)A_{nj}.$$

On the other hand, the expression in (4) is

$$\sum_{\rho \in H} (\operatorname{sgn}\rho) \prod_{i=1}^{n-1} b_{\rho(i)i},$$

which is the formula for $|B|$ (7.3.2); in view of (5), we have again shown that the expression in (4) is equal to $(\operatorname{sgn}\tau)A_{nj}$.

Summarizing, the expression in (4) (and (3)) is equal to $(\operatorname{sgn}\tau)A_{nj}$ in both cases ($j = n$ and $j < n$), therefore the expression in (2) simplifies to

$$a_{nj}(\operatorname{sgn}\tau) \cdot (\operatorname{sgn}\tau)A_{nj} = a_{nj}A_{nj};$$

thus, for each $j = 1, \ldots, n$, the terms of (1) in which a_{nj} appears have sum $a_{nj}A_{nj}$, consequently

$$|A| = \sum_{j=1}^{n} a_{nj}A_{nj}.$$

This completes the verification of the formula for expanding $|A|$ by its n'th row.

Consider now the case that $i < n$. Let C be the matrix obtained from A by moving its i'th row to the bottom (i.e., under the original n'th row); this entails moving the i'th row past rows $i+1, \dots, n$, and each of the $n-(i+1)+1 = n-i$ moves changes the sign of the determinant, therefore

$$|C| = (-1)^{n-i}|A|.$$

Expand $|C|$ by its last row (the case already considered); clearly $C^{n,j} = A^{i,j}$, and the last row of C is $a_{i1}, \dots, a_{in}$, consequently

$$|C| = \sum_{j=1}^{n} a_{ij}(-1)^{n+j}|C^{n,j}| = \sum_{j=1}^{n} a_{ij}(-1)^{n+j}|A^{i,j}|,$$

whence

$$|A| = (-1)^{i-n}|C| = \sum_{j=1}^{n} a_{ij}(-1)^{i+j}|A^{i,j}| = \sum_{j=1}^{n} a_{ij}A_{ij}$$

as claimed. ■

The formula of 7.4.2 is, so to speak, the expansion of $|A|$ by the i'th row; the same technique works for columns (consider row expansions for the transposed matrix A').

7.4.3 Definition With notations as in 7.4.1, the **adjoint matrix** of A, denoted adj A is the $n \times n$ matrix

$$\operatorname{adj} A = (A_{ij})'$$

(the transposed matrix of cofactors).

7.4.4 Theorem *For every $n \times n$ matrix A over* F, $A(\operatorname{adj} A) = (\operatorname{adj} A)A = |A| \cdot I$.

Proof. With notations as in 7.4.3, let $b_{ij} = A_{ji}$ and write $B = (b_{ij})$; thus $B = (A_{ij})' = \operatorname{adj} A$. Write $AB = (c_{ij})$; the problem is to show that $c_{ij} = 0$ when $i \neq j$ and $c_{ii} = |A|$ for all i (and similarly for the product BA). We have

$$(*) \qquad c_{ij} = \sum_{k=1}^{n} a_{ik}b_{kj} = \sum_{k=1}^{n} a_{ik}A_{jk}.$$

When $j = i$,

$$c_{ii} = \sum_{k=1}^{n} a_{ik}A_{ik} = |A|$$

by 7.4.2 (expansion of $|A|$ by row i).

Suppose $j \neq i$. Let $D = (d_{uv})$ be the matrix obtained from A by replacing the j'th row of A by a copy of the i'th row of A. Then $|D| = 0$ (two equal rows) and it is clear that $D_{jk} = A_{jk}$ for all k (when the j'th rows of D and A are deleted, the matrices that remain are identical). Also, $d_{jk} = a_{ik}$ for all k (by the way D was constructed). Thus, for the particular pair (i, j) that has been fixed, the equation $(*)$ can be written

$$c_{ij} = \sum_{k=1}^{n} d_{jk}D_{jk},$$

that is (7.4.2), $c_{ij} = |D|$ (expansion by row j); in other words $c_{ij} = 0$, as we wished to show. Thus $AB = |A|I$.

The proof is easily modified to show that $BA = |A|I$. Alternatively, observe that $(A')^{i,j} = (A^{j,i})'$ for all i and j, therefore $(A')_{ij} = A_{ji}$ by Theorem 7.3.5, consequently $\operatorname{adj} A' = (\operatorname{adj} A)'$. Since $A'(\operatorname{adj} A') = |A'|I$ by the earlier argument, that is, $A'(\operatorname{adj} A)' = |A|I$, in other words $A'B' = |A|I$, the desired equation $BA = |A|I$ results on taking transpose. ■

7.4.5 Corollary *An $n \times n$ matrix A over F is invertible if and only if $|A| \neq 0$, and then $A^{-1} = |A|^{-1}(\operatorname{adj} A)$.*

Proof. If A is invertible, say $AB = I$, then $1 = |I| = |AB| = |A||B|$, therefore $|A| \neq 0$. Conversely, if $|A| \neq 0$ then it is clear from Theorem 7.4.4 that $|A|^{-1}(\operatorname{adj} A)$ serves as inverse for A. ■

The first statement of 7.4.5 also follows from 7.2.6; what is new here is the explicit formula for the inverse. This formula is a classical determinant formula for solving linear systems, in disguise (Exercises 3 and 4).

▶ Exercises

1. Let $A \in M_n(\mathbf{Z})$. In order that A have an inverse in $M_n(\mathbf{Z})$, it is necessary and sufficient that $|A| = \pm 1$.

2. Let F be any field, R a subring of F containing the unity element 1. Let $A \in M_n(R)$. Prove that A has an inverse in $M_n(R)$ if and only if $|A|$ has an inverse in R.

3. Let $A = (a_{ij}) \in M_n(F)$ be an invertible matrix, $\operatorname{adj} A = (A_{ij})'$ the adjoint matrix, and

$$\gamma = \begin{pmatrix} c_1 \\ c_2 \\ \vdots \\ c_n \end{pmatrix}$$

a column vector with entries in F. Show that for each $j = 1, \ldots, n$, the j'th coordinate of the column vector $(\operatorname{adj} A)\gamma$ is

$$\sum_{i=1}^{n} c_i A_{ij};$$

this expression is the expansion, by the j'th column, of the determinant of the matrix obtained by replacing the j'th column of A by the column vector γ.

4. Given a system of linear equations

$$(*) \quad \begin{cases} a_{11}x_1 + a_{12}x_2 + \ldots + a_{1n}x_n = c_1 \\ a_{21}x_1 + a_{22}x_2 + \ldots + a_{2n}x_n = c_2 \\ \ldots \\ a_{n1}x_1 + a_{n2}x_2 + \ldots + a_{nn}x_n = c_n \end{cases}$$

whose coefficient matrix $A = (a_{ij}) \in M_n(F)$ is invertible. Prove that the system (*) has a unique solution vector $(x_1, \ldots, x_n)$, where $x_j = |A|^{-1}|B_j|$ and B_j is the matrix obtained by replacing the j'th column of A by the column of constants $c_1, \ldots, c_n$. (These formulas for the solution are known as *Cramer's rule*.)

{Hint: Solve the matrix equation

$$A\begin{pmatrix} x_1 \\ x_2 \\ \vdots \\ x_n \end{pmatrix} = \begin{pmatrix} c_1 \\ c_2 \\ \vdots \\ c_n \end{pmatrix}$$

for the column vector on the left, and cite Exercise 3.}

5. Let $A \in M_n(F)$. Prove:

(i) A is invertible if and only if $\operatorname{adj} A$ is invertible, in which case $(\operatorname{adj} A)^{-1} = |A|^{-1}A$.

(ii) $|\operatorname{adj} A| = |A|^{n-1}$.

(iii) If $|A| = 1$ then $\operatorname{adj}(\operatorname{adj} A) = A$.

6. Write $\rho(A)$ for the rank of a matrix $A \in M_n(F)$. We are assuming that $n \geq 2$.

(i) If $|A| \neq 0$ then $\rho(A) = \rho(\operatorname{adj} A) = n$.

(ii) If $|A| = 0$ then $\rho(A) + \rho(\operatorname{adj} A) \leq n$.

(iii) If $\rho(A) = n - 1$ then $\rho(\operatorname{adj} A) = 1$.

(iv) If $\rho(A) \leq n - 2$ then $\operatorname{adj} A = 0$.

(v) If $n = 2$ then $\operatorname{adj}(\operatorname{adj} A) = A$.

(vi) If $n \geq 3$ and $\rho(A) = n - 1$, then $\operatorname{adj}(\operatorname{adj} A) = 0$.

(vii) $\operatorname{adj}(\operatorname{adj} A) = |A|^{n-2}A$ (with the convention that $0^{n-2} = 1$ when $n = 2$).

{Hint: (ii) Call $\nu(A) = n - \rho(A)$ the *nullity* of A. From $A(\operatorname{adj} A) = 0$ infer than $\rho(\operatorname{adj} A) \leq \nu(A)$. The inequality of (ii) can be strict for a nonzero matrix (§11.4, Exercise 3).

(iii) By §10.4, Exercise 6, $\operatorname{adj} A \neq 0$.

(iv) By §10.4, Exercise 6, all $(n - 1) \times (n - 1)$ subdeterminants of A are 0.

(vi) Cite (iii) and (iv).

(vii) For $n = 2$, cite (v). For $n \geq 3$, cite (i), (iv) or (vi).}

7. Let $A, B \in M_n(F)$.

(i) $\operatorname{adj}(AB) = (\operatorname{adj} B)(\operatorname{adj} A)$.

(ii) If B is invertible, then $\operatorname{adj}(B^{-1}AB) = B^{-1}(\operatorname{adj} A)B$.

{Hint: (i) If A and B are both invertible, the formula is easily deduced from $(AB)^{-1} = B^{-1}A^{-1}$. In the general case, apply the preceding argument to the matrices $A - tI$, $B - tI$ in the context of the field of rational forms $\mathrm{F}(t)$, then substitute $t = 0$.}

8

Similarity (Act I)

The concept of similarity was observed passively in Chapter 4 (§4.10), as a byproduct of change of basis. In the present chapter we take an active view of similarity as a means of transforming matrices into simpler forms. Eventually (Ch. 11) the theory of similarity will be seen to be a powerful tool for classifying linear mappings: understanding what a linear mapping 'looks like' is a matter of looking at its matrix relative to the 'right' basis.

Reserved notations for the chapter:

F a field

n a positive integer

V an n-dimensional vector space over F

$A, B, C, P, Q, \ldots$ $n \times n$ matrices over F.

8.1 Similarity

We recall a definition from Chapter 4 (cf. 4.10.6):

8.1.1 Definition Matrices $A, B \in \mathrm{M}_n(\mathrm{F})$ are said to be **similar** (over F), written $A \sim B$, if there exists an invertible matrix $P \in \mathrm{M}_n(\mathrm{F})$ such that $P^{-1}AP = B$.

8.1.2 Theorem *Similarity is an equivalence relation in the set* $\mathrm{M}_n(\mathrm{F})$ *of all* $n \times n$ *matrices:* (1) $A \sim A$, (2) $A \sim B \Rightarrow B \sim A$, *and* (3) $(A \sim B) \,\&\, (B \sim C) \Rightarrow A \sim C$.

Proof.
(1) $A = I^{-1}AI$; (2) if $P^{-1}AP = B$ then $(P^{-1})^{-1}BP^{-1} = A$; (3) if $P^{-1}AP = B$ and $Q^{-1}BQ = C$ then $(PQ)^{-1}A(PQ) = C$. ■

8.1.3 Theorem *Matrices A, B are similar if and only if there exists a linear mapping T:V → V such that A and B are the matrices of T relative to two bases of* V *(as in* 4.10.5).

Proof. 'Only if': Suppose A and B are similar matrices over F of order n, say $C^{-1}AC = B$, $C = (c_{ij})$. Let V be an n-dimensional vector space over F and $x_1, \ldots, x_n$ a basis of V; let T be the linear mapping with matrix A relative to the basis $x_1, \ldots, x_n$ and S the (bijective) linear mapping with matrix C relative to this basis. Let $y_j = Sx_j$ for all j; thus

$$y_j = Sx_j = \sum_{i=1}^{n} c_{ij}x_i \qquad (j = 1, \ldots, n)$$

and since S is invertible, $y_1, \ldots, y_n$ is a basis of V. By 4.10.5, the matrix of T relative to the basis $y_1, \ldots, y_n$ is $C^{-1}AC$, in other words B.

'If': Immediate from 4.10.5. ■

8.1.4 Theorem *If $A \sim B$ then $|A| = |B|$ and* $\operatorname{tr} A = \operatorname{tr} B$.

Proof. Suppose $B = C^{-1}AC$. By 7.3.8, $\operatorname{tr}(C^{-1} \cdot AC) = \operatorname{tr}(AC \cdot C^{-1})$, in other words $\operatorname{tr} B = \operatorname{tr} A$; similarly for the determinant (cf. 7.3.6 or 7.3.3). ■

8.1.5 Example There exist 2×2 matrices A and B such that $|A| = |B|$ and $\operatorname{tr} A = \operatorname{tr} B$, but A and B are not similar.

{For example, if

$$A = I = \begin{pmatrix} 1 & 0 \\ 0 & 1 \end{pmatrix}, \quad B = \begin{pmatrix} 1 & 1 \\ 0 & 1 \end{pmatrix}$$

then $\operatorname{tr} A = \operatorname{tr} B = 2$ and $|A| = |B| = 1$; for every invertible matrix C, $C^{-1}AC = I \neq B$.}

A scalar matrix aI commutes with all matrices; conversely:

8.1.6 Theorem *If $C^{-1}AC = A$ for all invertible matrices C, then $A = aI$ for a suitable scalar a.*

Proof. The theorem is trivial if $n = 1$; suppose $n \geq 2$. Assuming A is a matrix of order n, let V be an n-dimensional vector space, $x_1, \ldots, x_n$ a basis of V, and $T \in \mathcal{L}(\mathrm{V})$ the linear mapping whose matrix is A relative to this basis. In view of 4.10.5, our hypothesis is that A is the matrix of T relative to *every* basis of V; our problem is to show that T is a scalar multiple of the identity.

The first step is to show that if x is any nonzero vector, then Tx is a scalar multiple of x. Assume to the contrary; then x, Tx are linearly independent (3.3.5) hence can be expanded to a basis x, Tx, $y_3, \ldots, y_n$ (3.5.15). Write $y_1 = x$, $y_2 = Tx$ to round out the notation. Since

$Ty_1 = y_2 = 0y_1 + 1y_2 + 0y_3 + \ldots + 0y_n$, the matrix of T relative to the basis $y_1, \ldots, y_n$ has first column

$$\begin{pmatrix} 0 \\ 1 \\ 0 \\ \vdots \\ 0 \\ 0 \end{pmatrix};$$

by our hypothesis, this is the first column of A, and it follows that if $z_1, \ldots, z_n$ is *any* basis of V then $Tz_1 = z_2$. Any two independent vectors y, z can be enlarged to a basis, so $Ty = z$ by the preceding argument; applying this to the independent vectors x, $x + Tx$, we have $Tx = x + Tx$, whence the contradiction $x = \theta$.

Thus, for every nonzero vector x, there exists a (unique) scalar $c(x)$ such that $Tx = c(x)x$; it will suffice to show that $c(x) = c(y)$ for all nonzero vectors x and y. If y is proportional to x, say $y = ax$ $(a \neq 0)$, then $a(Tx) = Ty$, so $a \cdot c(x)x = c(y)y = c(y) \cdot ax$, whence $c(x) = c(y)$. If y is not proportional to x, then x, y are independent and in particular $x + y \neq \theta$, therefore

$$c(x + y) \cdot (x + y) = T(x + y) = Tx + Ty = c(x)x + c(y)y,$$

whence $c(x) = c(x + y) = c(y)$ on comparing coefficients of x and y. ■

Thus, the equivalence class of A for $\sim$ consists of A alone if and only if A is a scalar matrix. In particular, A commutes with all other $n \times n$ matrices if and only if it is a scalar multiple of the identity. (This is also an easy consequence of 3.8.3, from whose proof 8.1.6 borrows abundantly.)

▶ **Exercises**

1. If n is an even integer and $A \in \mathrm{M}_n(\mathbf{R})$ is a matrix such that $AB = BA$ for all $B \in \mathrm{M}_n(\mathbf{R})$, then $|A| \geq 0$.

2. Let $A, B \in \mathrm{M}_n(\mathrm{F})$.

(i) If either A or B is invertible, then AB and BA are similar.

(ii) Can the invertibility assumption be dropped? {Hint: Suppose $AB = 0$.}

3. (i) If $A, B \in \mathrm{M}_n(\mathrm{F})$ are similar, then $p(A)$ and $p(B)$ are similar for every polynomial $p \in \mathrm{F}[t]$.

(ii) If A^2 and B^2 are similar, does it follow that A and B are similar?

4. If A and B are similar, then so are A' and B' (easy). {In fact, A is similar to A' (hard); cf. Corollary 11.1.10.}

5. Define the *trace* (denoted $\operatorname{tr} T$) of a linear mapping $T \in \mathscr{L}(\mathrm{V})$. {Hint: Theorem 8.1.4.}

8.2 Eigenvalues and eigenvectors

A central problem: Given a linear mapping $T \in \mathcal{L}(\mathrm{V})$, find a basis of V relative to which the matrix of T is 'as simple as possible'. An ideal of 'simplicity' (not always attainable): a diagonal matrix

$$\operatorname{diag}(c_1, c_2, \ldots, c_n) = \begin{pmatrix} c_1 & & & \\ & c_2 & & \\ & & \ddots & \\ & & & c_n \end{pmatrix}.$$

This amounts to finding a basis $x_1, \ldots, x_n$ of V such that $Tx_j = c_j x_j$ for all j. The equivalent matricial problem: Given an $n \times n$ matrix A, find an invertible matrix C such that $C^{-1}AC$ is diagonal. This is not always possible, for example the matrix

$$\begin{pmatrix} 1 & 1 \\ 0 & 1 \end{pmatrix}$$

is not similar to a diagonal matrix.

'Next best' would be a triangular matrix (say with 0's below the main diagonal), but this too is not always possible: Assuming F = **R** (the real field) and

$$A = \begin{pmatrix} 0 & 1 \\ -1 & 0 \end{pmatrix},$$

there exists no matrix C over **R** for which $C^{-1}AC$ is triangular. {However, when F = **C** triangular form is always attainable, essentially by the 'Fundamental Theorem of Algebra' (more about this is the next section).}

The crux of the above-mentioned constraints is that, for a linear mapping T, it may not be possible to find a nonzero vector x such that $Tx = cx$ for some scalar c.

8.2.1 Definition Let $T \in \mathcal{L}(\mathrm{V})$. If $Tx = cx$ with $x \neq \theta$ and $c \in \mathrm{F}$, then c is called an **eigenvalue** of T, and x an **eigenvector** of T.

Thus, an eigenvector of T is a nonzero vector x such that Tx is a scalar multiple of x; an eigenvalue of T is a scalar c such that $Tx = cx$ for some nonzero vector x, that is, $\operatorname{Ker}(T - cI) \neq \{\theta\}$.

The assumed finite-dimensionality of V permits the following useful characterization of eigenvalues:

8.2.2 Theorem *If $T \in \mathcal{L}(\mathrm{V})$ and $c \in \mathrm{F}$ then the following conditions are equivalent:* (a) *c is an eigenvalue of T;* (b) *$T - cI$ is not bijective;* (c) $|T - cI| = 0$.

Proof. As remarked above, c is an eigenvalue of $T \Leftrightarrow \operatorname{Ker}(T - cI) \neq \{\theta\}$. It follows from the finite-dimensionality of V that, for $S \in \mathcal{L}(\mathrm{V})$, injectivity, surjectivity and bijectivity are equivalent conditions on S (3.7.4); apply this remark to $S = T - cI$:

$$\begin{aligned} \operatorname{Ker} S \neq \{\theta\} &\Leftrightarrow S \text{ not injective} \\ &\Leftrightarrow S \text{ not bijective} \\ &\Leftrightarrow |S| = 0 \qquad (7.2.6). \ \blacksquare \end{aligned}$$

It is the same to say that $cI - T$ is not bijective, or that $|cI - T| = 0$.

Eigenvectors corresponding to *distinct* eigenvalues are linearly independent:

8.2.3 Theorem *Let $T \in \mathcal{L}(V)$ and let $x_1, \ldots, x_r$ be eigenvectors of T, say $Tx_i = c_i x_i$ for $i = 1, \ldots, r$. If the eigenvalues $c_1, \ldots, c_r$ are distinct, then $x_1, \ldots, x_r$ are linearly independent.*

Proof. Assume to the contrary that there exists a relation

$$a_1x_1 + \ldots + a_mx_m = \theta$$

with not every coefficient equal to 0 (and $m \leq r$); we can suppose that m is as small as possible. Note that $m \geq 2$ (for, $a_1x_1 = \theta$, $a_1 \neq 0$, would imply $x_1 = \theta$, contrary to hypothesis). Also, $a_m \neq 0$ (otherwise $a_1x_1 + \ldots + a_{m-1}x_{m-1} = \theta$ would contradict the minimality of m), so we can solve for x_m:

$$(*) \qquad x_m = b_1x_1 + \ldots + b_{m-1}x_{m-1}$$

(where $b_i = -a_i/a_m$). Then

$$Tx_m = b_1Tx_1 + \ldots + b_{m-1}Tx_{m-1},$$

that is,

$$c_mx_m = b_1c_1x_1 + \ldots + b_{m-1}c_{m-1}x_{m-1};$$

substituting the formula $(*)$ for x_m into the preceding equation,

$$c_m(b_1x_1 + \ldots + b_{m-1}x_{m-1}) = b_1c_1x_1 + \ldots + b_{m-1}c_{m-1}x_{m-1},$$

thus

$$b_1(c_m - c_1)x_1 + \ldots + b_{m-1}(c_m - c_{m-1})x_{m-1} = \theta.$$

By the minimality of m, all coefficients in the preceding equation are 0,

$$b_i(c_m - c_i) = 0 \quad \text{for} \quad i = 1, \ldots, m-1,$$

therefore $b_i = 0$ for $i = 1, \ldots, m-1$ (because $c_m - c_i \neq 0$ for these i) and the formula $(*)$ then yields the contradiction $x_m = \theta$. ■

► Exercises

1. Let c be an eigenvalue of $T \in \mathcal{L}(V)$ and let $M = \operatorname{Ker}(T - cI)$. If $S \in \mathcal{L}(V)$ and $ST = TS$, then $S(M) \subset M$.

2. Let $T \in \mathcal{L}(V)$, $p \in F[t]$ a nonconstant polynomial.

(i) If c is an eigenvalue of T, then $p(c)$ is an eigenvalue of $p(T)$.

(ii) If F is algebraically closed (Definition 8.3.10) and d is an eigenvalue of $p(T)$, then $d = p(c)$ for some eigenvalue c of T.

{Hint:

(i) If $Tx = cx$ calculate $T^k x$ for $k = 2, 3, 4, \ldots$

(ii) Factor $p - d1$ into linear factors and note that $p(T) - dI$ is not injective.}

3. Suppose $V = M_1 \oplus \ldots \oplus M_r$ (in the sense of §3.3, Exercise 20). For $i = 1, \ldots, r$ let $P_i \in \mathscr{L}(V)$ be the linear mapping such that $P_i = I$ on M_i and $P_i = 0$ on M_j for $j \neq i$.

(i) $P_i^2 = P_i$, $P_i P_j = 0$ when $i \neq j$.

(ii) $P_i + \ldots + P_n = I$.

(iii) For $T \in \mathscr{L}(V)$, $TP_i = P_i T$ for all i if and only if $T(M_i) \subset M_i$ for all i.

4. Let $T \in \mathscr{L}(V)$. The set of all $S \in \mathscr{L}(V)$ such that $ST = TS$ is called the *commutant* of T, written $\{T\}'$; thus,

$$\{T\}' = \{S \in \mathscr{L}(V)\colon ST = TS\}.$$

The set of all $R \in \mathscr{L}(V)$ such that $RS = SR$ for every $S \in \{T\}'$ is called the *bicommutant* of T, written $\{T\}''$; thus

$$\{T\}'' = \{R \in \mathscr{L}(V)\colon ST = TS \Rightarrow RS = SR\}.$$

With notations as in Exercise 3, suppose $T(M_i) \subset M_i$ for all i. Prove that $R(M_i) \subset M_i$ for every $R \in \{T\}''$.

{Hint: $P_i \in \{T\}'$ for all i.}

5. Let $T \in \mathscr{L}(V)$ be a linear mapping such that V has a basis consisting of eigenvectors for T (it is the same to say that the matrix of T relative to any basis of V is similar to a diagonal matrix); such mappings are called *diagonalizable*. Let $c_1, \ldots, c_r$ be the distinct eigenvalues of T and let $M_i = \operatorname{Ker}(T - c_i I)$ for $i = 1, \ldots, r$.

(i) $V = M_1 \oplus \ldots \oplus M_r$ (in the sense of §3.3, Exercise 20).

(ii) For $S \in \mathscr{L}(V)$, $ST = TS$ if and only if $S(M_i) \subset M_i$ for all i.

(iii) For $R \in \mathscr{L}(V)$, $R \in \{T\}''$ if and only if, for every i, $R(M_i) \subset M_i$ and $R|M_i$ is a scalar multiple of the identity.

(iv) For every polynomial $p \in F[t]$, $p(T) \in \{T\}''$ (this does not require that T be diagonalizable).

(v) For any scalars $d_1, \ldots, d_r \in F$, there exists a polynomial $p \in F[t]$ such that $p(c_i) = d_i$ for all i.

(vi) Every $R \in \{T\}''$ has the form $R = p(T)$ for a suitable polynomial $p \in F[t]$. (This is true without assuming T diagonalizable, but the proof is much harder.[1])

{Hints:

(i) Cf. Theorem 8.2.3.

(ii) Exercise 1.

[1] W. H. Greub, *Linear algebra*, Springer (3rd edn), 1967, p. 403.

(iii) For the 'if' part, see part (ii); for the 'only if' part, see also Exercise 4 and Theorem 8.1.6.

(v) For each $i = 1, \ldots, r$, the polynomial

$$p_i = d_i \prod_{j \neq i} (c_i - c_j)^{-1}(t - c_j)$$

takes the value d_i at c_i, and 0 at the c_j for $j \neq i$.

(vi) Use (iii) and (v). Note that $p(T)|M_i = p(T|M_i)$ for every polynomial p and for all i.}

6. Let $T \in \mathcal{L}(V)$ be a linear mapping such that V has a basis $x_1, \ldots, x_n$ consisting of eigenvectors for T, say $Tx_i = c_i x_i$ for all i. Prove that every eigenvalue of T is equal to one of the c_i. {Hint: If $Tx = cx$, express x as a linear combination of the x_i, then apply T.}

8.3 Characteristic polynomial

The polynomial ring $F[t]$, t an indeterminate, is an integral domain (in fact, a Euclidean domain, therefore a principal ideal ring; cf. Appendix B.4.3); its field of fractions is the ring $F(t)$ of rational forms p/q with $p, q \in F[t]$ (Appendix B.1.3). The theory of determinants developed in Chapter 7 is applicable to the ring $M_n(F(t))$ of $n \times n$ matrices over the field $F(t)$; in particular, the determinant function is applicable to the ring $M_n(F[t])$ of matrices with polynomial entries, and it is obvious from the formula for determinants (7.3.2) that the determinant of such a matrix is itself a polynomial.

8.3.1 Definition If $A \in M_n(F)$, the **characteristic polynomial** of A, denoted p_A, is defined by the formula $p_A = |tI - A|$; thus $p_A \in F[t]$ by the preceding remarks; the matrix $tI - A$ is called the **characteristic matrix** of A.

8.3.2 Lemma *If $A \in M_n(F)$ then $p_A(c) = |cI - A|$ for every scalar $c \in F$.*

Proof. The key algebraic observation is that if $P = (p_{ij})$ is any matrix whose entries are polynomials $p_{ij} \in F[t]$ then, by the formula for determinant (7.3.2), $d = |P|$ is a polynomial in the p_{ij}, hence $d \in F[t]$ and $d(c) = |(p_{ij})(c)|$ for all $c \in F$. In gory detail (it has to be seen once),

$$\begin{aligned} d(c) = |P|(c) &= \left(\sum_{\sigma \in S_n} (\operatorname{sgn} \sigma) \prod_{i=1}^{n} p_{\sigma(i)i} \right)(c) \\ &= \sum_{\sigma \in S_n} (\operatorname{sgn} \sigma) \prod_{i=1}^{n} p_{\sigma(i)i}(c) \\ &= |(p_{ij}(c))|; \end{aligned}$$

writing $P(c) = (p_{ij}(c)) \in M_n(F)$, we thus have $|P|(c) = |P(c)|$ for all $c \in F$.

In the situation at hand, $p_{ij} = \delta_{ij} t - a_{ij}$, $P = tI - A$, thus

$$p_A(c) = |P|(c) = |P(c)| = |(\delta_{ij} c - a_{ij})| = |cI - A|. \blacksquare$$

If A is a 2×2 matrix then $p_A = t^2 - (\mathrm{tr}\, A)t + \det A$. More generally:

8.3.3 Theorem *If* $A \in \mathrm{M}_n(\mathrm{F})$ *then* $\deg p_A = n$. *More precisely,* $p_A = t^n - (\mathrm{tr}\, A)t^{n-1} + \ldots + (-1)^n(\det A)$.

Proof. By the lemma, the constant term of p_A is $p_A(0) = |0I - A| = |-A| = (-1)^n|A|$.

Next, we show that the term of highest degree is t^n. Writing $p_A = |P| = |(p_{ij})|$ as in the proof of 8.3.2, we have

$$(*) \qquad p_A = \sum_{\sigma \in S_n} (\mathrm{sgn}\, \sigma) \prod_{i=1}^{n} p_{\sigma(i)i},$$

where

$$p_{ii} = t - a_{ii} \text{ for all } i,$$
$$p_{ij} = -a_{ij} \text{ when } i \neq j.$$

If $\sigma \in S_n$ then, in the term

$$(\mathrm{sgn}\, \sigma) \prod_{i=1}^{n} p_{\sigma(i)i},$$

the number of factors that contain t is equal to the number of indices i for which $\sigma(i) = i$; there is exactly one choice of σ for which this number takes on the maximum value n, namely $\sigma = \iota$ (the identity permutation), therefore

$$(\mathrm{sgn}\, \iota) \prod_{i=1}^{n} p_{ii} = \prod_{i=1}^{n} (t - a_{ii})$$

is the unique term containing the highest power of t (namely t^n). In short: the leading term of p_A is t^n.

It remains to show that t^{n-1} occurs in p_A with coefficient $-\mathrm{tr}\, A$. The term of $(*)$ for $\sigma = \iota$ described in the preceding paragraph may be written

$$t^n - (a_{11} + \ldots + a_{nn})t^{n-1} + \text{lower degree terms}$$
$$= t^n - (\mathrm{tr}\, A)t^{n-1} + \text{lower degree terms},$$

so we need only show that t^{n-1} does not occur in the term for any σ other than ι. Indeed, if $\sigma \neq \iota$, then σ moves at least two indices i, consequently the term of $(*)$ corresponding to σ (is either 0 or) has degree at most $n - 2$. ■

8.3.4 Theorem *If* $A, B \in \mathrm{M}_n(\mathrm{F})$ *are similar, then* $p_A = p_B$.

Proof. Let $C \in \mathrm{M}_n(\mathrm{F})$ be an invertible matrix such that $C^{-1}AC = B$. Regarding C as an element of $\mathrm{M}_n(\mathrm{F}(t))$,

$$C^{-1}(tI - A)C = tC^{-1}IC - C^{-1}AC = tI - B,$$

whence $|tI - B| = |tI - A|$ by 8.1.4, that is, $p_B = p_A$. ■

The converse of 8.3.4 is false even for 2×2 matrices (8.1.5).

8.3.5 Theorem *If* $A \in \mathrm{M}_n(\mathrm{F})$ *and* $c \in \mathrm{F}$, *then* $p_A(c) = 0 \Leftrightarrow cI - A$ *is not invertible.*

Proof. Immediate from 8.3.2 and 7.4.5 (or 8.2.2). ■

8.3.6 Definition Let $A \in \mathrm{M}_n(\mathrm{F})$, let K be an overfield of F, and let $c \in \mathrm{K}$. If $p_A(c) = 0$ we say that c is a **characteristic root** of A.

Note that the characteristic roots of $A \in \mathrm{M}_n(\mathrm{F})$ need not belong to F (in the introduction to the preceding section, an example is given of a 2×2 matrix over **R** whose characteristic roots are both complex imaginary).

If $T \in \mathcal{L}(\mathrm{V})$ then the matrices of T relative to the various bases of V are all similar (4.10.5), hence have the same characteristic polynomial (8.3.4); the following definition is therefore unambiguous:

8.3.7 Definition If $T \in \mathcal{L}(\mathrm{V})$ and A is the matrix of T relative to some basis of V, then p_A is called the **characteristic polynomial** of T, denoted p_T; thus $p_T = p_A$ for every matricial representation A of T.

Thus, with the foregoing notations, $p_T = |tI - A|$. {Informally we also write $p_T = |tI - T|$, but to make rigorous sense of this formula would take us too far afield (for our purposes, an unnecessary trip).}

8.3.8 Definition With notations as in 8.3.7, the roots of p_T (in any overfield $\mathrm{K} \supset \mathrm{F}$) are called the **characteristic roots** of T; these are precisely the characteristic roots of any matricial representation A of T (cf. 8.3.6).

8.3.9 Theorem *If* $T \in \mathcal{L}(\mathrm{V})$ *and* $c \in \mathrm{F}$, *then* c *is an eigenvalue of* T *if and only if* $p_T(c) = 0$.

Proof. Let $c \in \mathrm{F}$. It follows from 8.3.2 and the definitions that $p_T(c) = |cI - T|$. By 8.2.2, c is an eigenvalue of T if and only if $|cI - T| = 0$, in other words, $p_T(c) = 0$. ■

Thus, the eigenvalues of T are those of its characteristic roots (if any) *that belong to* F. There is a situation in which belonging to F is guaranteed:

8.3.10 Definition A field F is said to be **algebraically closed** if every nonconstant polynomial $p \in \mathrm{F}[t]$ is the product of polynomials in $\mathrm{F}[t]$ of degree 1 (that is, 'linear' polynomials over F), equivalently, the irreducible (= 'prime') elements of the principal ideal ring $\mathrm{F}[t]$ are the polynomials of degree 1. In view of the 'factor theorem', it is the same to say that every nonconstant polynomial $p \in \mathrm{F}[t]$ has a root in F.

8.3.11 Examples The complex field **C** is algebraically closed ('Fundamental Theorem of Algebra').[1] More generally, it can be shown[2] that for every field F there exists an (essentially unique) algebraically closed overfield $\mathrm{K} \supset \mathrm{F}$ such that

[1] For a proof based on functions of two real variables, see E. Landau's *Differential and integral calculus*, Chelsea, 1951.

[2] S. Lang, *Algebra*, Addison-Wesley, 1965, Chapter 7, §2.

every element of K is a root of some polynomial $p \in F[t]$; one calls K the *algebraic closure* of F. For example, **C** is the algebraic closure of **R**; the algebraic closure of the rational field **Q** is called the field of *algebraic numbers*.

We need not enter into algebraic closures here. Most of our theorems are true for arbitrary fields F; when algebraic closedness is needed, we will *assume* it. Often the following simpler fact suffices: for each polynomial $p \in F[t]$ there is an overfield $K \supset F$ in which p factors completely into linear factors (Appendix B.5.3). Just one piece of 'hard news'—that **C** is algebraically closed—is cited in the last chapter in the proof of the Spectral Theorem (§12.7), but even this can be avoided (§12.8).

8.3.12 Theorem *If the field* F *is algebraically closed, then every matrix* $A \in M_n(F)$ *is similar (over* F*) to a triangular matrix* B.

Proof. Let V be an n-dimensional vector space over F, choose a basis of V and let $T \in \mathcal{L}(V)$ be the linear mapping whose matrix relative to the chosen basis is A. The problem is to find a basis $x_1, \ldots, x_n$ such that, for every index j, Tx_j is a linear combination of $x_1, \ldots, x_j$; the matrix $B = (b_{ij})$ of T relative to such a basis will (a) be similar to A (8.1.3), and (b) have j'th column

$$\begin{pmatrix} b_{1j} \\ \vdots \\ b_{jj} \\ 0 \\ \vdots \\ 0 \end{pmatrix}$$

for $j = 1, \ldots, n$, hence will be a triangular matrix.

The existence of such a basis is proved by induction on n. For $n = 1$, there is nothing to prove. Let $n \geq 2$ and assume that all's well with $n - 1$. Since F is algebraically closed, p_T has a root c_1 in F; by 8.3.9, c_1 is an eigenvalue of T. Choose a nonzero vector x_1 such that $Tx_1 = c_1x_1$ and let $M = Fx_1$ be the 1-dimensional linear subspace spanned by x_1. The quotient space V/M is $(n-1)$-dimensional (3.6.1); let $Q: V \to V/M$ be the quotient mapping.

Since $T(M) \subset M$ we can define a mapping $S: V/M \to V/M$ as follows. Let $u \in V/M$, say $u = Qx$. We propose to define $Su = Q(Tx)$; if also $u = Qy$ then $Qx = Qy$, $x - y \in \operatorname{Ker} Q = M$, $Tx - Ty = T(x - y) \in M$, $Q(Tx) = Q(Ty)$, thus the proposed formula yields a well-defined mapping S. Since $S(Qx) = Q(Tx)$ for all $x \in V$, it follows easily that S is linear. Thus $S \in \mathcal{L}(V/M)$, where V/M is $(n-1)$-dimensional; by the induction hypothesis, V/M has a basis $u_2, \ldots, u_n$ such that, for each $j \geq 2$, Su_j is a linear combination of $u_2, \ldots, u_j$, say

$$Su_j = b_{2j}u_2 + \ldots + b_{jj}u_j \qquad (j = 2, \ldots, n).$$

Choose $x_j \in V$ with $Qx_j = u_j$ $(j = 2, \ldots, n)$; then

$$Q(Tx_j) = b_{2j}Qx_2 + \ldots + b_{jj}Qx_j \qquad (j = 2, \ldots, n)$$
$$= Q(b_{2j}x_2 + \ldots + b_{jj}x_j),$$

therefore $Tx_j - (b_{2j}x_2 + \ldots + b_{jj}x_j) \in M$, say

$$Tx_j - (b_{2j}x_2 + \ldots + b_{jj}x_j) = b_{1j}x_1,$$

thus

$$Tx_j = b_{1j}x_1 + \ldots + b_{jj}x_j \qquad (j = 2, \ldots, n).$$

Writing $b_{11} = c_1$, we have also $Tx_1 = b_{11}x_1$. Finally, $x_1, x_2, \ldots, x_n$ generate V (3.4.8) hence are a basis of V (3.5.13). {See also 11.4.14.} ■

8.3.13 Remark With notations as in Theorem 8.3.12, if $B = (b_{ij})$ then $p_A = (t - b_{11})(t - b_{22}) \ldots (t - b_{nn})$, thus the diagonal entries of B are the characteristic roots of A.

{**Proof.** This is trivial for $n = 1$, and an easy induction completes the proof (expand $|tI - B|$ by row n).}

8.3.14 Theorem *If $A \in M_n(F)$ and p_A has n distinct roots in* F, *then A is similar (over* F*) to a diagonal matrix.*

Proof. As in the proof of 8.3.12, we can suppose that A is the matrix of $T \in \mathcal{L}(V)$ relative to some basis of V. Let $c_1, \ldots, c_n$ be the (assumed) distinct roots of p_T in F. Each c_i is an eigenvalue of T (8.3.9), so we can choose a nonzero vector x_i such that $Tx_i = c_ix_i$. The vectors $x_1, \ldots, x_n$ are linearly independent by 8.2.3, hence are a basis of V (3.5.13); the matrix of T relative to the basis $x_1, \ldots, x_n$ is a diagonal matrix B to which A is similar (8.1.3). ■

The preceding theorem heightens our interest in being able to decide when a polynomial has repeated roots (short of finding them all!); there is a simple test based on the division algorithm (Appendix B.5.4).

► Exercises

1. Suppose $A \in M_5(\mathbf{C})$ has characteristic polynomial $p = t^5 + 5t - 7$. Prove that A is similar (over **C**) to a diagonal matrix.

2. Let $A \in M_n(F)$.

(i) If $a_{1j} = 0$ for $j = 2, \ldots, n$, so that

$$A = \left(\begin{array}{c|ccc} a_{11} & a_{12} & \ldots & a_{1n} \\ \hline 0 & & & \\ \vdots & & C & \\ 0 & & & \end{array}\right)$$

for a suitable matrix $C \in M_{n-1}(F)$, then $|A| = a_{11}|C|$ and $p_A = (t - a_{11})p_C$.

(ii) Prove Theorem 8.3.12 assuming only that the characteristic polynomial of A factors completely over F (into linear factors).

(iii) Infer from (ii) that if $A \in \mathrm{M}_n(\mathrm{F})$ and

$$A = \left(\begin{array}{c|c} B & D \\ \hline 0 & C \end{array}\right),$$

where B and C are square matrices and 0 represents a zero matrix of the appropriate size, then $|A| = |B| \cdot |C|$ and $p_{\mathrm{A}} = p_{\mathrm{B}} \cdot p_{\mathrm{C}}$.

{Hint:

(ii) Use (i) in the induction step of the proof.

(iii) The assertion is obvious when B and C are triangular. In the general case, let K be an overfield of F in which the characteristic polynomials of B and C factor completely (Appendix B.5.3), and apply (ii) to each of B and C; the similarity transformations that send B and C to triangular form can be combined into a similarity transformation that sends A to triangular form.}

3. If $T \in \mathscr{L}(\mathrm{V})$, M is a nonzero linear subspace of V such that $T(\mathrm{M}) \subset \mathrm{M}$, and $R = T|\mathrm{M}$ is the restriction of T to M, then p_{R} divides p_{T}, that is, $p_{\mathrm{T}} = q \cdot p_{\mathrm{R}}$ for a suitable polynomial q.
{Hint: Let $x_1, \ldots, x_m$ be a basis of M and consider the matrix of T relative to a basis $x_1, \ldots, x_m, x_{m+1}, \ldots, x_n$ of V.}

4. With assumptions as in Exercise 3, suppose in addition that $\mathrm{M} \neq \mathrm{V}$ and let $S \in \mathscr{L}(\mathrm{V/M})$ be the linear mapping induced by T in the quotient space V/M (see the proof of Theorem 8.3.12). Prove that $p_{\mathrm{T}} = p_{\mathrm{R}} \cdot p_{\mathrm{S}}$.
{Hint: With notations as in the hint for Exercise 3, the cosets $u_i = x_i + \mathrm{M}$ $(i = m+1, \ldots, n)$ are a basis of V/M; consider the matrix of S relative to this basis.}

5. Let $A \in \mathrm{M}_n(\mathrm{F})$.

(i) There exist nonconstant polynomials $p \in \mathrm{F}[t]$ such that $p(A) = 0$.

(ii) The set of all $p \in \mathrm{F}[t]$ such that $p(A) = 0$ is a nonzero ideal of $\mathrm{F}[t]$, generated by a unique nonconstant monic polynomial (Appendix B.4.3); it is called the *minimal polynomial* of A, denoted m_{A}. (It is shown in Corollary 11.3.6 that m_{A} is a divisor of the characteristic polynomial p_{A}.)

(iii) For a polynomial $p \in \mathrm{F}[t]$ and an invertible matrix $C \in \mathrm{M}_n(\mathrm{F})$, $p(C^{-1}AC) = C^{-1}p(A)C$. It follows that the minimal polynomial m_{T} of a linear mapping T on a finite-dimensional vector space can be defined as the minimal polynomial of any matricial representation of T.

(iv) If $T \in \mathscr{L}(\mathrm{V})$, M and N are nonzero linear subspaces of V such that $\mathrm{V} = \mathrm{M} \oplus \mathrm{N}$ (§1.6, Exercise 12) and M, N are both invariant under T, then the minimal polynomial of T is divisible by the minimal polynomials of $T|\mathrm{M}$ and $T|\mathrm{N}$.

(v) With notations as in (iv), if the minimal polynomials of $T|\mathrm{M}$ and $T|\mathrm{N}$ are relatively prime, then their product is the minimal polynomial of T.

{Hints:

(i) §4.4, Exercise 2.

(iv) $p(T|\mathrm{M}) = p(T)|\mathrm{M}$.

(v) $p(T) = 0$ if and only if its restrictions to both M and N are 0.}

6. Suppose $AB = BA$, where A, $B \in \mathrm{M}_n(\mathrm{F})$ and the characteristic polynomials of A and B factor completely over F. Prove: A and B can be simultaneously triangularized, that is, there exists an invertible matrix $C \in \mathrm{M}_n(\mathrm{F})$ such that both $C^{-1}AC$ and $C^{-1}BC$ have only zeros below the main diagonal.

{Hint: Viewing A and B as the matrices of a pair of commuting linear mappings S, $T \in \mathcal{L}(\mathrm{V})$, the crux of the matter is to find a vector $x \in \mathrm{V}$ that is an eigenvector for both S and T. Let $c \in \mathrm{F}$ be an eigenvector of T, let $\mathrm{N} = \mathrm{Ker}\,(T - cI)$, and observe that $S(\mathrm{N}) \subset \mathrm{N}$. After finding the desired vector x, let $\mathrm{M} = \mathrm{F}x$ and proceed as in Theorem 8.3.12 and Exercise 2.}

7. Assume F is algebraically closed, and let $\mathcal{S}$ be a set of linear mappings in the finite-dimensional vector space V such that $ST = TS$ for all S, T in $\mathcal{S}$.

(i) Prove: There exists a vector $x \in \mathrm{V}$ that is an eigenvector for every $S \in \mathcal{S}$.

(ii) Extend Exercise 6 to an arbitary set of commuting matrices.

{Hint: (i) Fix $T \in \mathcal{S}$, let c be an eigenvalue of T, and let $\mathrm{N} = \mathrm{Ker}\,(T - cI)$; as noted in the hint for Exercise 6, $S(\mathrm{N}) \subset \mathrm{N}$ for all $S \in \mathcal{S}$. Among all nonzero subspaces M of N such that $S(\mathrm{M}) \subset \mathrm{M}$ for all $S \in \mathcal{S}$, choose one (call it M) with dimension as small as possible. Argue that every $S|\mathrm{M}$ ($S \in \mathcal{S}$) is a scalar multiple of the identity (let $S|\mathrm{M}$ play the role of T in the preceding argument and cite the minimality of M), and infer that M is 1-dimensional.}

8. If F is algebraically closed, then every $T \in \mathcal{L}(\mathrm{V})$ has an eigenvalue (proof of Theorem 8.3.12). Prove this without appealing to the characteristic polynomial, then infer a 'determinant-free' proof of Theorem 8.3.12.

{Hint: Assuming $T \neq 0$, let $p \in \mathrm{F}[t]$ be the minimal polynomial of T (Exercise 5), let $c \in \mathrm{F}$ be a root of p and write $p = (t - c)q$; argue that $T - cI$ can't be bijective.}

9. Let F be an algebraically closed field, V a finite-dimensional vector space over F, and $T \in \mathcal{L}(\mathrm{V})$. The *spectrum* of T, denoted $\sigma(T)$, is defined to be the set of all scalars $c \in \mathrm{F}$ such that $T - cI$ is *not* bijective. Prove:

(i) $\sigma(T)$ is the set of all eigenvalues of T.

(ii) $\sigma(T)$ is the set of all characteristic roots of T.

(iii) For all $a \in \text{F}$, $\sigma(aT) = \{ac: c \in \sigma(T)\}$ and $\sigma(aI + T) = \{a + c: c \in \sigma(T)\}$.

(iv) If T is bijective (equivalently, $0 \notin \sigma(T)$) then $\sigma(T^{-1}) = \{c^{-1}: c \in \sigma(T)\}$.

(v) For every polynomial $p \in \text{F}[t]$, $\sigma(p(T)) = p(\sigma(T)) = \{p(c): c \in \sigma(T)\}$.

(vi) $\sigma(ST) = \sigma(TS)$ for all $S, T \in \mathcal{L}(\text{V})$.

(vii) If a basis of V is chosen relative to which the matrix $B = (b_{ij})$ of T is triangular (cf. 8.3.12), then $\sigma(T)$ is the set $\{b_{11}, \ldots, b_{nn}\}$ of diagonal elements of B.

{Hints:

(i) See Theorem 3.7.4.

(v) See §8.2, Exercise 2.

(vi) If $c \in \text{F} - \{0\}$, infer from §2.3, Exercise 26 that $c \in \sigma(ST) \Leftrightarrow c \in \sigma(TS)$. For the equivalence $0 \in \sigma(ST) \Leftrightarrow 0 \in \sigma(TS)$, see §3.7, Exercise 9.}

9

Euclidean spaces (Spectral theory)

The study of linear mappings in Euclidean spaces begun in Chapter 5 is pursued in the present chapter to obtain the structure theorem for self-adjoint linear mappings (known as the Spectral Theorem). This theorem is foreshadowed in Chapter 6 for dimensions 2 and 3, but we make here a fresh start in dimension n; nothing from Chapter 6 will be presupposed (a few of the concepts introduced there will be repeated here). As contrasted with Chapter 6, the discussion in dimension n will be 'determinant-free'; the crux of the matter is that we arrive at eigenvalues by a method other than the characteristic polynomial.

9.1 Invariant and reducing subspaces

Let's review some key facts from Chapter 5. If E is a Euclidean space (5.1.1), M is a linear subspace of E, and $\mathrm{M}^\perp$ is the orthogonal complement of M, then $\mathrm{E} = \mathrm{M} + \mathrm{M}^\perp$ and $\mathrm{M} \cap \mathrm{M}^\perp = \{\theta\}$ (5.2.7, 5.2.9). These formulas are expressed by writing $\mathrm{E} = \mathrm{M} \oplus \mathrm{M}^\perp$. This is a special case of a concept that has been mentioned in the exercises (cf. §1.6, Exercise 12):

9.1.1 Definition If M and N are linear subspaces of a vector space V, such that $\mathrm{V} = \mathrm{M} + \mathrm{N}$ and $\mathrm{M} \cap \mathrm{N} = \{\theta\}$, we say that V is the **direct sum** of M and N and we write $\mathrm{V} = \mathrm{M} \oplus \mathrm{N}$.

9.1.2 Theorem *With notations as in 9.1.1, every vector* $x \in \mathrm{V}$ *has a unique representation* $x = y + z$ *with* $y \in \mathrm{M}$ *and* $z \in \mathrm{N}$.

Proof. The existence of such a representation is immediate from $\mathrm{V} = \mathrm{M} + \mathrm{N}$. If $x = y_1 + z_1 = y_2 + z_2$ with $y_i \in \mathrm{M}$ and $z_i \in \mathrm{N}$ $(i = 1, 2)$, then $y_1 - y_2 = z_2 - z_1 \in \mathrm{M} \cap \mathrm{N} = \{\theta\}$, therefore $y_1 = y_2$ and $z_1 = z_2$. ■

In a sense, the formula $\mathrm{V} = \mathrm{M} \oplus \mathrm{N}$ decomposes V into its subspaces M and N. One of the important uses of direct sums is in the study of a linear mapping $T \in \mathcal{L}(\mathrm{V})$: one is interested in decomposing V into subspaces on which the behaviour of T is easier to describe.

9.1.3 Definition Let V be a vector space, $T \in \mathcal{L}(\mathrm{V})$, M a linear subspace of V. If $T(\mathrm{M}) \subset \mathrm{M}$ we say that M is **invariant** under T, or that M is an invariant subspace *for* T.

A linear mapping $T \in \mathcal{L}(\mathrm{V})$ always has the *trivial* invariant subspaces $\{\theta\}$ and V; there may be no others (Exercise 11).

9.1.4 Theorem *Let* E *be a Euclidean space,* $T \in \mathcal{L}(\mathrm{E})$, M *a nontrivial linear subspace of* E *invariant under* T. *Let* B *be the matrix of* $T|\mathrm{M}$ *relative to some basis of* M. *Then, relative to the basis of* E *obtained by adjoining a basis of* $\mathrm{M}^\perp$ *to that of* M, T *has matrix*

$$\begin{pmatrix} B & D \\ 0 & C \end{pmatrix}$$

for suitable matrices C *and* D.

Proof. Let $n = \dim \mathrm{E}$, $m = \dim \mathrm{M}$, $x_1, \ldots, x_m$ the basis of M in question, $x_{m+1}, \ldots, x_n$ a basis of $\mathrm{M}^\perp$, and $A = (a_{ij})$ the matrix of T relative to the basis

$$x_1, \ldots, x_m, x_{m+1}, \ldots, x_n$$

of E. Thus,

$$Tx_j = \sum_{i=1}^{n} a_{ij} y_i \qquad (j = 1, \ldots, n).$$

Since M is invariant under T, $a_{ij} = 0$ when $j = 1, \ldots, m$ and $i = m+1, \ldots, n$, thus

$$(*) \qquad Tx_j = \sum_{i=1}^{m} a_{ij} y_i \qquad (j = 1, \ldots, m),$$

where the matrix of coefficients for the equations $(*)$ is the matrix B of the statement of the theorem; directly below B in the array A is an $(n - m) \times m$ block of zeros, and we may denote by C and D the matrices (of sizes $(n - m) \times (n - m)$ and $m \times (n - m)$, respectively) needed to fill out the array A. ■

9.1.5 Theorem *Let* E *be a Euclidean space,* $T \in \mathcal{L}(\mathrm{E})$, M *a linear subspace of* E.

(1) M *is invariant under* T *if and only if* $\mathrm{M}^\perp$ *is invariant under* T^*.

(2) *If* M *is invariant under both* T *and* T^*, *then* $(T|\mathrm{M})^* = T^*|\mathrm{M}$.

Proof.

(1) If $T(\mathrm{M}) \subset \mathrm{M}$ then $T^*(\mathrm{M}^\perp) \subset \mathrm{M}^\perp$ by Theorem 5.3.6. Conversely, if $T^*(\mathrm{M}^\perp) \subset \mathrm{M}^\perp$ then $T^{**}(\mathrm{M}^{\perp\perp}) \subset \mathrm{M}^{\perp\perp}$, in other words $T(\mathrm{M}) \subset \mathrm{M}$ (5.2.11, 5.3.3).

(2) Write $S = T|\mathrm{M} \in \mathcal{L}(\mathrm{M})$. For all $x, y \in \mathrm{M}$,

$$(Sx|y) = (Tx|y) = (x|T^*y) = (x|(T^*|\mathrm{M})y),$$

therefore $S^* = T^*|\mathrm{M}$ by 5.3.2. ■

9.1.6. Definition With notation as in 9.1.5, if M and $\mathrm{M}^\perp$ are both invariant under T (equivalently, M is invariant under both T and T^*), then M is said to **reduce** T (or to be *reducing* for T).

The trivial subspaces $\{\theta\}$ and E reduce every $T \in \mathcal{L}(\mathrm{E})$; a linear mapping may have no other reducing subspaces (Exercise 11).

When a nontrivial linear subspace M reduces T, the decomposition $\mathrm{E} = \mathrm{M} \oplus \mathrm{M}^\perp$ of E effects a decomposition of T: if $x = y + z$, where $y \in \mathrm{M}$ and $z \in \mathrm{M}^\perp$, then $Tx = (T|\mathrm{M})y + (T|\mathrm{M}^\perp)z$; in a sense, $T = (T|\mathrm{M}) \oplus (T|\mathrm{M}^\perp)$. The matricial form of this remark:

9.1.7 Theorem *Let* E *be a Euclidean space,* $T \in \mathcal{L}(\mathrm{E})$, M *a linear subspace of* E *that reduces* T. *Let* B *be the matrix of* $T|\mathrm{M}$ *relative to some basis of* M, C *the matrix of* $T|\mathrm{M}^\perp$ *relative to some basis of* $\mathrm{M}^\perp$. *Then* T *has matrix*

$$\begin{pmatrix} B & 0 \\ 0 & C \end{pmatrix}$$

relative to the basis of E *obtained by joining the chosen bases of* M *and* $\mathrm{M}^\perp$.

Proof. Adopt the notations in 9.1.4 and its proof (including the meaning of A and its blocks B, C, D). If $j \geq m+1$ then Tx_j is a linear combination of $x_{m+1}, \ldots, x_n$ (because $\mathrm{M}^\perp$ is invariant under T), therefore $a_{ij} = 0$ when $j \geq m+1$ and $i = 1, \ldots, m$; this accounts for an $m \times (n-m)$ block of zeros in the northeast corner of A, and

$$Tx_j = \sum_{i=m+1}^{n} a_{ij} y_i \qquad (j = m+1, \ldots, n)$$

means that the $(m-n) \times (m-n)$ block matrix in the southeast corner of A is the matrix of $T|\mathrm{M}^\perp$ relative to the basis $x_{m+1}, \ldots, x_n$ of $\mathrm{M}^\perp$. ■

Invariant subspaces are sometimes automatically reducing:

9.1.8 Theorem *If* E *is a Euclidean space and* $T \in \mathcal{L}(\mathrm{E})$ *is orthogonal, then every linear subspace* M *of* E *that is invariant under* T *reduces* T, *and* $T|\mathrm{M}$ *is an orthogonal linear mapping in the Euclidean space* M.

Proof. Suppose M is a linear subspace of E invariant under T; the problem is to show that $T^*(\mathrm{M}) \subset \mathrm{M}$. Since $T(\mathrm{M}) \subset \mathrm{M}$ and T is injective, it follows that $T(\mathrm{M}) = \mathrm{M}$ on grounds of dimensionality (Corollary 3.5.14). Then

$M = T^{-1}(M) = T^*(M)$, so M is also invariant under T^*. The orthogonality of $T|M$ is obvious from criterion (b) of 5.4.2. ∎

9.1.9 Definition Let E be a Eculidean space. A linear mapping $T \in \mathscr{L}(E)$ such that $T^* = T$ is said to be **self-adjoint.**

9.1.10 Theorem *If* E *is a Euclidean space,* $T \in \mathscr{L}(E)$ *is self-adjoint, and* M *is a linear subspace of* E *invariant under* T*, then* (1) M *reduces* T*, and* (2) $T|M$ *and* $T|M^{\perp}$ *are self-adjoint linear mappings on the Euclidean spaces* M *and* $M^{\perp}$*, respectively.*

Proof. Immediate from 9.1.5 and the definitions. ∎

In §9.4 we prove the following generalization of 9.1.8 and 9.1.10: If $T \in \mathscr{L}(E)$ satisfies $T^*T = TT^*$ (such linear mappings are called 'normal'), then every invariant subspace for T is reducing.

► ***Exercises***

1. If V is any vector space, S, $T \in \mathscr{L}(V)$ and $ST = TS$, then the kernel and range of T are invariant under S.

2. If E is a Euclidean space, $T \in \mathscr{L}(E)$ and $T^*T = TT^*$, then

(i) $\operatorname{Ker} T = \operatorname{Ker} T^*$ and $\operatorname{Im} T = \operatorname{Im} T^*$;

(ii) $\operatorname{Ker} T$ and $\operatorname{Im} T$ are reducing subspaces for T;

(iii) $\operatorname{Ker} T \perp \operatorname{Im} T$;

(iv) $E = \operatorname{Ker} T \oplus \operatorname{Im} T$.

{Hints: (ii)–(iv): Exercise 1 and §5.3, Exercise 7.}

3. Let V be a vector space over a field K, $T \in \mathscr{L}(V)$, and $p \in K[t]$ a polynomial with coefficients in K.

(i) The range of $p(T)$ is invariant under T.

(ii) If $p(T) = 0$, $p = qr$ is a factorization of p with q, $r \in K[t]$, M is the range of $r(T)$, and $R = T|M$ (cf. part (i)), then $q(R) = 0$. {Hint: (i) Exercise 1. (ii) $q(T)r(T) = 0$.}

4. Let E be a Euclidean space, $T \in \mathscr{L}(E)$, p_T the characteristic polynomial of T, and M a reducing subspace for T with $M \neq \{\theta\}$, $M \neq E$; write $R = T|M$ and $S = T|M^{\perp}$. Then

(i) $p_T = p_R p_S$;

(ii) the minimal polymial of T (§8.3, Exercise 5) is divisible by the minimal polynomials of R and S.

{Hint: Construct a basis of E by joining bases of M and $M^{\perp}$, and apply §8.3, Exercise 2 to the matrix of T relative to this basis.}

5. Let E be a Euclidean space, $E \neq \{\theta\}$. A linear mapping $T \in \mathscr{L}(E)$ is said to be (orthogonally) *decomposable* if there exists a reducing subspace

for T other than $\{\theta\}$ and E (cf. Theorem 9.1.7), and *indecomposable* if it is not decomposable. For every $T \in \mathcal{L}(\mathrm{E})$ there exists a direct sum decomposition $\mathrm{E} = \mathrm{M}_1 \oplus \ldots \oplus \mathrm{M}_r$ (possibly $r = 1$), where the M_i are pairwise orthogonal reducing subspaces for T, and every $T|\mathrm{M}_i$, is indecomposable.
{Hint: Induction on $\dim \mathrm{E}$.}

6. (i) Let V be a vector space, $T \in \mathcal{L}(\mathrm{V})$, M and N linear subspaces of V. If M and N are invariant under T, then so are $\mathrm{M} + \mathrm{N}$ and $\mathrm{M} \cap \mathrm{N}$.
(ii) Let E be a Euclidean space, $T \in \mathcal{L}(\mathrm{E})$, M and N linear subspaces of E. If M and N reduce T, then so do $\mathrm{M} + \mathrm{N}$ and $\mathrm{M} \cap \mathrm{N}$. {Hint: §5.2, Exercise 6.}

7. Let E be a Euclidean space, M a linear subspace of E. Define $P\colon \mathrm{E} \to \mathrm{E}$ as follows; if $x \in \mathrm{E}$ write $x = y + z$ with $y \in \mathrm{M}$, $z \in \mathrm{M}^{\perp}$ and define $Px = y$. Prove that P is linear, $\operatorname{Im} P = \mathrm{M}$, $\operatorname{Ker} P = \mathrm{M}^{\perp}$ and $P^2 = P = P^*$. (P is called the *projection* with range M; its dependence on M is indicated by writing $P = P_{\mathrm{M}}$.)

8. Let E be a Euclidean space, $P \in \mathcal{L}(\mathrm{E})$ a linear mapping such that $P^2 = P = P^*$, and let $\mathrm{M} = \operatorname{Im} P$. Show that $P = P_{\mathrm{M}}$ in the sense of Exercise 7, that is, P is the projection with range M.

9. With notations as in Exercise 7, let $T \in \mathcal{L}(\mathrm{E})$. Prove:

(i) M is invariant under T if and only if $PTP = TP$.
(ii) M reduces T if and only if $PT = TP$.

10. Let M and N be linear subspaces of a Euclidean space E and let P_{M}, P_{N} be the projections with ranges M and N, respectively (Exercise 7). Prove:

(i) $\mathrm{M} \perp \mathrm{N} \Leftrightarrow P_{\mathrm{M}} P_{\mathrm{N}} = 0$.
(ii) $\mathrm{M} \subset \mathrm{N} \Leftrightarrow P_{\mathrm{M}} = P_{\mathrm{N}} P_{\mathrm{M}}$.
(iii) If $P_{\mathrm{M}} P_{\mathrm{N}} = P_{\mathrm{N}} P_{\mathrm{M}}$, then $P_{\mathrm{M}} P_{\mathrm{N}} = P_{\mathrm{M} \cap \mathrm{N}}$.
(iv) If $\mathrm{M} \perp \mathrm{N}$ then $P_{\mathrm{M}} + P_{\mathrm{N}} = P_{\mathrm{M}+\mathrm{N}}$.

11. If u_1, u_2 is an orthonormal basis of the Euclidean space E^2 and if $T \in \mathcal{L}(\mathrm{E}^2)$ is the linear mapping such that $Tu_1 = u_1$ and $Tu_2 = u_1 + u_2$, then T has a nontrivial invariant subspace but only the trivial reducing subspaces. The linear mapping $S \in \mathcal{L}(\mathrm{E}^2)$ such that $Su_1 = u_2$ and $Su_2 = -u_1$ has only the trivial invariant subspaces.

12. Let E be a Euclidean space and suppose $T \in \mathcal{L}(\mathrm{E})$ has the property that (∗) every invariant subspace for T is reducing (i.e., is also invariant under T^*). Prove:

(i) If M is a linear subspace invariant for T, then $T|\mathrm{M}$ also has the property (∗).
(ii) If T is indecomposable (Exercise 5) then $\dim \mathrm{E} \le 2$.

(iii) In the light of Exercise 5 and (ii), describe the structure of T. {Hint: (ii) We can suppose that T is not a scalar multiple of the identity. Let $p \in \mathbf{R}[t]$ be the minimal polynomial of T (§8.3, Exercise 5); by assumption, $\deg p > 1$. Argue that p is irreducible over $\mathbf{R}$ (hence is quadratic); for, if $p = qr$ is a factorization of p in $\mathbf{R}[t]$ with q nonconstant, then $r(T) \neq 0$, so the range of $r(T)$ is E by the condition (*), whence $q(T) = 0$ and q is a multiple of p. Infer that if x is any nonzero vector in E then $[\{x, Tx\}] = \mathrm{E}$.}

9.2 Bounds of a linear mapping

Throughout this section, E and F are Euclidean spaces; we assume $\mathrm{E} \neq \{\theta\}$ to keep things interesting.

9.2.1 Theorem *For every linear mapping* $T{:}\mathrm{E} \rightarrow \mathrm{F}$, *there exists a constant* $K \geqslant 0$ *such that* $\|Tx\| \leqslant K\|x\|$ *for all* $x \in \mathrm{E}$.

Proof. Let $u_1, \ldots, u_n$ be an orthonormal basis of E. For $x \in \mathrm{E}$, write

$$x = \sum_{i=1}^{n} c_i u_i;$$

by the Pythagorean theorem (Theorem 5.1.10) and an easy induction argument, we have

$$\|x\|^2 = \sum_{i=1}^{n} \|c_i u_i\|^2 = \sum_{i=1}^{n} |c_i|^2,$$

thus $\|x\| = (\sum |c_i|^2)^{\frac{1}{2}}$. On the other hand,

$$Tx = \sum_{i=1}^{n} c_i T u_i,$$

so by the triangle inequality (5.1.9),

$$\|Tx\| \leqslant \sum_{i=1}^{n} \|c_i T u_i\| = \sum_{i=1}^{n} |c_i| \, \|Tu_i\|;$$

but

$$\sum_{i=1}^{n} |c_i| \, \|Tu_i\| \leqslant \left(\sum_{i=1}^{n} |c_i|^2\right)^{\frac{1}{2}} \left(\sum_{i=1}^{n} \|Tu_i\|^2\right)^{\frac{1}{2}}$$

by the Cauchy–Schwarz inequality in the Euclidean space E^n (5.1.8), thus $\|Tx\| \leqslant K\|x\|$ with $K = (\sum \|Tu_i\|^2)^{\frac{1}{2}}$. ■

This implies a kind of 'continuity' for linear mappings (here '$\rightarrow$' means convergence of sequences in the set of real numbers):

9.2.2 Corollary *With notations as in* 9.2.1, *if* $x_k, x \in \mathrm{E}$ *and* $\|x_k - x\| \rightarrow 0$, *then* $\|Tx_k - Tx\| \rightarrow 0$.

Proof. With K as in 9.2.1, $\|Tx_k - Tx\| = \|T(x_k - x)\| \leqslant K\|x_k - x\| \rightarrow 0$ as $\|x_k - x\| \rightarrow 0$. ■

9.2.3 Definition The set $S = \{u \in E: \|u\| = 1\}$ is called the **unit sphere** of E, and the set $B = \{x \in E: \|x\| \leq 1\}$ is called the (closed) **unit ball** of E.

Every linear image of the unit ball is bounded in the following sense:

9.2.4 Corollary *With notations as in 9.2.1, the set* $\{\|Tx\|: \|x\| \leq 1\}$ *is bounded.*

Proof. If $\|x\| \leq 1$ then $\|Tx\| \leq K\|x\| \leq K$, thus K is an upper bound for the set in question. ■

9.2.5 Definition The **norm** of a linear mapping $T \in \mathcal{L}(E, F)$ is the number $\|T\|$ defined by the formula $\|T\| = \sup\{\|Tx\|: \|x\| \leq 1\}$.

9.2.6 Theorem *For a linear mapping* $T \in \mathcal{L}(E, F)$, $\|T\|$ *is the smallest number* ≥ 0 *such that* $\|Tx\| \leq \|T\|\,\|x\|$ *for all* $x \in E$.

Proof. If $x \neq \theta$ then $u = \|x\|^{-1}x$ is a unit vector, so $\|Tu\| \leq \|T\|$, whence $\|Tx\| \leq \|T\|\,\|x\|$ (an inequality that holds trivially when $x = \theta$). On the other hand, if K is any number such that $\|Tx\| \leq K\|x\|$ for all vectors x, then $\|Tx\| \leq K$ whenever $\|x\| \leq 1$, whence $\|T\| \leq K$; thus $\|T\|$ is minimal in the appropriate sense. ■

In calculating the norm of a linear mapping, unit vectors suffice:

9.2.7 Theorem *For every linear mapping* $T \in \mathcal{L}(E, F)$, $\|T\| = \sup\{\|Tu\|: \|u\| = 1\}$.

Proof. Write s for the supremum on the right side. For all unit vectors u, $\|Tu\| \leq \|T\|$; therefore $s \leq \|T\|$. On the other hand if $x \neq \theta$ then $u = \|x\|^{-1}x$ is a unit vector, so $\|Tu\| \leq s$; thus $\|Tx\| \leq s\|x\|$ (an inequality that holds trivially when $x = \theta$), so $\|T\| \leq s$ by 9.2.6 ■

Inner products and norm are continuous in the following sense:

9.2.8 Lemma *For vectors in* E, *if* $\|x_k - x\| \to 0$ *and* $\|y_k - y\| \to 0$, *then* $(x_k|y_k) \to (x|y)$ *and* $\|x_k\| \to \|x\|$.

Proof. By the distributive law for inner products

$$(x_k|y_k) - (x|y) = (x_k - x|y_k - y) + (x|y_k - y) + (x_k - x|y),$$

so by the triangle inequality in **R** and the Cauchy–Schwarz inequality in E, we have

$$|(x_k|y_k) - (x|y)| \leq \|x_k - x\|\,\|y_k - y\| + \|x\|\,\|y_k - y\| + \|x_k - x\|\,\|y\|;$$

as $k \to \infty$ the right side of the inequality tends to 0, therefore so does the left side.

In particular (take $y_k = x_k$, $y = x$), $\|x_k\|^2 \to \|x\|^2$, whence $\|x_k\| \to \|x\|$. ■

9.2.9 Lemma *Let* $u_1, \ldots, u_n$ *be an orthonormal basis of* E. *Given vectors* $x \in E$, *and* $x_k \in E$ $(k = 1, 2, 3, \ldots)$, *suppose*

$$x_k = \sum_{i=1}^{n} a_{ik} u_i, \qquad x = \sum_{i=1}^{n} a_i u_i.$$

Then

$$\|x_k - x\| \to 0 \quad \Leftrightarrow \quad \textit{for} \quad i = 1, \ldots, n, \quad |a_{ik} - a_i| \to 0.$$

Proof. By the Pythagorean theorem (cf. the proof of 9.2.1),

$$\|x_k - x\|^2 = \sum_{i=1}^{n} |a_{ik} - a_i|^2. \ \blacksquare$$

When $E = F$, expressions $(Tx|x)$ make sense.

9.2.10 Lemma *If $T \in \mathcal{L}(E)$, and if x_k, $x \in E$ and $\|x_k - x\| \to 0$, then $(Tx_k|x_k) \to (Tx|x)$.*

Proof. Let $y_k = Tx_k$, $y = Tx$; then $\|y_k - y\| \to 0$ by Corollary 9.2.2, therefore $(y_k|x_k) \to (y|x)$ by 9.2.8. $\blacksquare$

The key proposition needed from analysis (the Weierstrass–Bolzano theorem) is cited in the proof of the following theorem:

9.2.11. Theorem *If $T \in \mathcal{L}(E)$ then the set*

$$\{(Tu|u) \colon \|u\| = 1\}$$

has a largest element; that is, there exists a unit vector $v \in E$ such that $(Tu|u) \leq (Tv|v)$ for all unit vectors u.

Proof. For all unit vectors u, $|(Tu|u)| \leq \|Tu\|\,\|u\| = \|Tu\| \leq \|T\|$; let

$$M = \sup\{(Tu|u) \colon \|u\| = 1\}.$$

The problem is to find a unit vector v such that $M = (Tv|v)$. Choose a sequence of unit vectors (u_k) such that $(Tu_k|u_k) \to M$. It would be nice if there existed a vector v such that $\|u_k - v\| \to 0$, for it would then follow that $\|v\| = \lim \|u_k\| = 1$ and $M = \lim (Tu_k|u_k) = (Tv|v)$; the strategy is to pass to a subsequence of (u_k) for which this is true.

Let $w_1, \ldots, w_n$ be an orthonormal basis of E and suppose

$$u_k = \sum_{i=1}^{n} a_{ik} w_i \qquad (k = 1, 2, 3, \ldots).$$

Then

$$1 = \|u_k\|^2 = \sum_{i=1}^{n} |a_{ik}|^2 \geq |a_{ik}|^2$$

shows that for each $i = 1, \ldots, n$, the sequence $(a_{ik})_{k \geq 1}$ is bounded.

By the Weierstrass–Bolzano theorem (Appendix C), the bounded sequence (a_{1k}) has a convergent subsequence, and the corresponding subsequence of $(Tu_k|u_k)$ necessarily converges to M; passing to a subsequence, we can suppose that $a_{1k} \to a_1$ for a suitable real number a_1. Again passing to a subsequence, we can suppose also that $a_{2k} \to a_2$ for suitable a_2. Continuing, we can suppose that for each $i = 1, \ldots, n$,

$$a_{ik} \to a_i \quad \text{as} \quad k \to \infty.$$

Let

$$v = \sum_{i=1}^{n} a_i w_i.$$

By Lemma 9.2.9, $\|u_k - v\| \to 0$, and the proof that $\|v\| = 1$ and $(Tv|v) = M$ is concluded in the way indicated earlier. ■

9.2.12 Corollary *If* $T \in \mathscr{L}(\mathrm{E})$ *and if*

$$M = \sup\{(Tu|u)\colon \|u\| = 1\},$$
$$m = \inf\{(Tu|u)\colon \|u\| = 1\},$$

then

$$\{m, M\} \subset \{(Tu|u)\colon \|u\| = 1\} \subset [m, M],$$

thus M and m are the largest and smallest elements of $\{(Tu|u)\colon \|u\| = 1\}$.[1]

Proof.
Let $\mathrm{W} = \{(Tu|u)\colon \|u\| = 1\}$; obviously $\mathrm{W} \subset [m, M]$ and by the theorem, $M \in \mathrm{W}$. Applying the theorem to $-T$, there exists a unit vector w such that $(-Tu|u) \leqslant (-Tw|w)$ for all unit vectors u; thus $(Tw|w) \leqslant (Tu|u)$ for all unit vectors u, therefore $m = (Tw|w) \in \mathrm{W}$. ■

9.2.13 Remarks With notations as in the preceding corollary and $\mathrm{S} = \{u\colon \|u\| = 1\}$ the unit sphere of E, the function $\mathrm{S} \to \mathbf{R}$ defined by $u \mapsto (Tu|u)$ is called the **quadratic form**[2] associated with T; its range $\{(Tu|u)\colon \|u\| = 1\}$ is called the **numerical range** of T, and its extremal values m, M are called the **bounds** of T.

9.2.14 Theorem *With notations as in 9.2.12,*

$$\sup\{|(Tu|u)|\colon \|u\| = 1\} = \max\{|m|, |M|\}.$$

If $K = \max\{|m|, |M|\}$, *then* $|(Tx|x)| \leqslant K\|x\|^2$ *for all* $x \in \mathrm{E}$.

Proof. Write s for the supremum on the left side. For all unit vectors u,

$$(Tu|u) \in [m, M] \subset [-K, K];$$

thus $|(Tu|u)| \leqslant K$ for all unit vectors u, whence $s \leqslant K$.

On the other hand, $M = (Tv|v)$ and $m = (Tw|w)$ for suitable unit vectors v and w, therefore

$$|m| = |(Tw|w)| \leqslant s \quad \text{and} \quad |M| = |(Tv|v)| \leqslant s,$$

whence $K \leqslant s$. Thus $s = K$.

If x is a nonzero vector, then $u = \|x\|^{-1}x$ is a unit vector, therefore $|(Tu|u)| \leqslant K$, whence $|(Tx|x)| \leqslant K\|x\|^2$ (an inequality that holds trivially when $x = \theta$). ■

[1] It can be shown that $\{(Tu|u)\colon \|u\| = 1\} = [m, M]$ (Exercise 1).

[2] See 6.7.2 for the motivation for the terminology.

9.2.15 Remarks With notations as in Corollary 9.2.12, $\max\{|m|, |M|\} \leqslant \|T\|$. If $T \in \mathscr{L}(\mathrm{E}^2)$ is rotation by 90° then $m = M = 0$ but $\|T\| = 1$.

For an arbitrary linear mapping $T \in \mathscr{L}(\mathrm{E})$, the bounds m, M are not very useful (cf. the preceding remark); we shall see in the next section that when T is self-adjoint, they are decisive for the analysis of T.

▶ Exercises

1. With notations as in Corollary 9.2.12, $\{(Tu|u): \|u\| = 1\} = [m, M]$. {Hint: If $\dim \mathrm{E} = 1$ then T is multiplication by a real number and the assertion is trivial. Suppose $\dim \mathrm{E} \geqslant 2$. Let $\mathrm{S} = \{u \in \mathrm{E}: \|u\| = 1\}$ (the 'unit sphere' of E) and let $f: \mathrm{S} \to \mathbf{R}$ be the function defined by $f(u) = (Tu|u)$. The function f is continuous in the sense that $\|u_n - u\| \to 0 \Rightarrow f(u_n) \to f(u)$ (Lemma 9.2.10); since S is 'connected', it follows that $f(\mathrm{S})$ is an interval.[3]}

2. In a Euclidean space, if (x_k) is a sequence of vectors such that $\|x_k\|$ is bounded, then there exist a vector x and a subsequence (x_{k_i}) such that $\|x_{k_i} - x\| \to 0$. {Hint: Proof of Theorem 9.2.11.}

9.3 Bounds of a self-adjoint mapping, Spectral Theorem

In this section, E is a Euclidean space and $T \in \mathscr{L}(\mathrm{E})$ is self-adjoint; our objective is to show that the structure of T can be determined by studying the associated quadratic form $u \mapsto (Tu|u)$ (9.2.13). The hypothesis $T^* = T$ is repeated in the statements to guard against the suggestion that they are true for every linear mapping.

9.3.1 Lemma *If $T^* = T$ then*

$$(Tx|y) = \tfrac{1}{4}\{(T(x+y)|x+y) - (T(x-y)|x-y)\}$$

for all $x, y \in \mathrm{E}$.

Proof. Write $[x,y] = (Tx|y)$ for all vectors x and y; this number is a linear function of x and a linear function of y (that is, $[x, y]$ is a bilinear form in the sense of Definition 7.1.1). Since T is self-adjoint,

$$(Tx|y) = (T^*x|y) = (x|Ty) = (Ty|x)$$

by symmetry of the inner product, thus $[x,y] = [y,x]$ (one calls $[x,y]$ a 'symmetric' bilinear form). It follows that the form $[x,y]$ satisfies the 'polarization identity'

$$[x, y] = \tfrac{1}{4}\{[x+y, x+y] - [x-y, x-y]\}$$

(by a proof formally the same as in Theorem 5.1.7), which is the asserted identity. ∎

9.3.2 Lemma *If $T^* = T$ and if $K = \max\{|m|, |M|\}$, then $|(Tx|y)| \leqslant K\|x\|\,\|y\|$ for all $x, y \in \mathrm{E}$.*

[3] Cf. J. Dixmier, *General topology*, Springer, 1984, p. 21, 2.5.7, and Chapter X, §10.1.

Proof. From the preceding lemma and the inequality in 9.2.14, we have

$$|(Tx|y)| \leqslant \tfrac{1}{4}\{K\|x + y\|^2 + K\|x - y\|^2\},$$

thus

$$|(Tx|y)| \leqslant \tfrac{1}{4}K\{2\|x\|^2 + 2\|y\|^2\} = \tfrac{1}{2}K\{\|x\|^2 + \|y\|^2\}$$

by the parallelogram law. In particular, if u and v are unit vectors, then

$$(*) \qquad |(Tu|v)| \leqslant \tfrac{1}{2}K\{1 + 1\} = K;$$

if x and y are any nonzero vectors, application of $(*)$ to the unit vectors $u = \|x\|^{-1}x$ and $v = \|y\|^{-1}y$ yields $|(Tx|y)| \leqslant K\|x\|\,\|y\|$ (an inequality that holds trivially when $x = \theta$ or $y = \theta$). ■

9.3.3 Theorem *If $T^* = T$ then there exists a unit vector u such that $\|T\| = |(Tu|u)|$.*

Proof. With notations as in 9.3.2, if x is any vector and $y = Tx$, we have

$$\|Tx\|^2 = (Tx|Tx) = (Tx|y) \leqslant K\|x\|\,\|y\| = K\|x\|\,\|Tx\|,$$

whence $\|Tx\| \leqslant K\|x\|$ (even if $\|Tx\| = 0$), therefore $\|T\| \leqslant K$.

By 9.2.12, there exist unit vectors v and w such that $(Tv|v) = M$ and $(Tw|w) = m$; one of $|m|$, $|M|$ is equal to K, thus one of the vectors v, w—call it u—is a unit vector such that $|(Tu|u)| = K$. Then

$$K = |(Tu|u)| \leqslant \|Tu\|\,\|u\| = \|Tu\| \leqslant \|T\|,$$

thus $\|T\| = K = |(Tu|u)|$. ■

9.3.4 Theorem. (Spectral Theorem)[1] *If $T \in \mathcal{L}(\mathrm{E})$ is self-adjoint then E has an orthonormal basis consisting of eigenvectors of T.*

Proof. The proof is by induction on $n = \dim \mathrm{E}$. If $n = 1$ then any unit vector will serve as a basis of the required sort. Suppose $n \geqslant 2$ and assume that all's well with $n - 1$.

Choose a unit vector u as in 9.3.3. Then

$$\|T\| = |(Tu|u)| \leqslant \|Tu\|\,\|u\| = \|Tu\| \leqslant \|T\|,$$

whence $|(Tu|u)| = \|Tu\|\,\|u\|$; it follows that Tu is a scalar multiple of u (5.1.8), in other words u is an eigenvector for T.

Let $\mathrm{M} = \mathbf{R}u$ be the 1-dimensional subspace of E spanned by u; M is invariant under T, therefore M reduces T and $T|\mathrm{M}^\perp$ is a self-adjoint linear mapping on the Euclidean space $\mathrm{M}^\perp$ (9.1.10). Since $\mathrm{M}^\perp$ is $(n-1)$-dimensional (5.2.8), it follows from the induction hypothesis that $\mathrm{M}^\perp$ has an orthonormal basis $u_2, \ldots, u_n$ of eigenvectors for $T|\mathrm{M}^\perp$—in other words, for T—and the definition $u_1 = u$ rounds out the sought-for orthonormal basis $u_1, u_2, \ldots, u_n$ of E. ■

[1] Also called the Principal Axis Theorem (cf. §6.7)

The proof of the Spectral Theorem makes no use of determinants. However, when the theorem is related to characteristic roots (the roots of the characteristic polynomial), determinants are unavoidable since the characteristic polynomial is *defined* by a determinant (8.3.1, 8.3.7). For example:

9.3.5 Corollary *If A is a symmetric real matrix, then there exists an orthogonal matrix P such that $P'AP = P^{-1}AP$ is diagonal; in particular, the characteristic roots of A are real.*

Proof. Assuming A is $n \times n$, let E be an n-dimensional Euclidean space (for example, $\mathrm{E} = \mathrm{E}^n$), choose an orthonormal basis of E (for example, the canonical orthonormal basis of E^n) and let $T \in \mathcal{L}(\mathrm{E})$ be the linear mapping whose matrix relative to the chosen basis is A. Since $A' = A$, T is self-adjoint (5.3.5), so by the Spectral Theorem, E has an orthonormal basis $u_1, \ldots, u_n$ consisting of eigenvectors for T; the matrix B of T relative to the basis $u_1, \ldots, u_n$ is diagonal. Express the u_j as linear combinations of the basis vectors chosen earlier, and let P be the matrix of coefficients; since both bases are orthonormal, P is an orthogonal matrix (5.4.7), and $P'AP = P^{-1}AP = B$ (4.10.5). This proves the first assertion of the theorem (without using determinants).

Finally, since $tI - B = \mathrm{diag}(t - b_{11}, \ldots, t - b_{nn})$, $p_B = |tI - B|$ is the product of the $t - b_{ii}$ (cf. 6.4.18 or §7.3, Exercise 1); thus the b_{ii} are the roots of p_B, in other words of p_A (8.3.4), and we have shown that the characteristic roots of A are the real numbers b_{ii}. ■

9.3.6 Corollary *If $T^* = T$ then the characteristic roots of T are real.*

Proof. Let A be the matrix of T relative to any orthonormal basis of E; by 5.3.5, A is symmetric. By definition, $p_T = p_A$, and the roots of p_A are real by 9.3.5. ■

▶ Exercises

1. If $x_1 \ldots, x_n$ is any finite list of vectors in an inner product space and if $a_{ij} = (x_i|x_j)$ for all i and j, then the $n \times n$ matrix $A = (a_{ij})$ has determinant $\geqslant 0$; in fact, A is orthogonally similar to a matrix $\mathrm{diag}(r_1, \ldots, r_n)$ with $r_i \geqslant 0$ for all i. (The matrix A is called the *Gram matrix*, or 'Gramian', of the list $x_1, \ldots, x_n$.)

{Hint: Note that A is symmetric; let $C = (c_{ij})$ be an orthogonal matrix such that $C'AC = \mathrm{diag}(r_1, \ldots, r_n)$. Argue that if

$$y_j = \sum_{i=1}^{n} c_{ij} x_i \qquad (j = 1, \ldots, n)$$

then $r_j = \|y_j\|^2$ for all j.}

2. If T is self-adjoint and p is a polynomial with real coefficients, then the eigenvalues of $p(T)$ are the numbers $p(r)$, where r runs over the eigenvalues of T. {Hint: Spectral Theorem and §8.2, Exercise 6.}

3. The following conditions on a linear mapping T in a Euclidean space are equivalent:

(a) $T^* = T$ and the eigenvalues of T are $\geqslant 0$;

(b) $T = S^2$ for some self-adjoint linear mapping S with eigenvalues $\geqslant 0$;

(c) $T = S^2$ for some self-adjoint linear mapping S.

{Hint: (a) $\Rightarrow$ (b): Spectral Theorem. (c) $\Rightarrow$ (a): Exercise 2.}

9.4 Normal linear mappings in Euclidean spaces

In this section (including the exercises), E is a Euclidean space.

9.4.1 Definition A linear mapping $T \in \mathcal{L}(\mathrm{E})$ is said to be **normal** if $T^*T = TT^*$.

The structure of normal mapping is worked out in Exercises 5–7; the ke to the analysis is the theorem that invariant subspaces of a normal linea mapping are automatically reducing (9.4.5). Three lemmas prepare the way.

9.4.2 Lemma *If $R \in \mathcal{L}(\mathrm{E})$ is self-adjoint, then $R = 0 \Leftrightarrow (Rx|x) = 0$ for all $x \in \mathrm{E}$.*

Proof. This is immediate from 9.3.1 (or 9.3.3). ■

9.4.3 Lemma *If $T \in \mathcal{L}(\mathrm{E})$, the following conditions are equivalent:*

(a) *T is normal;*

(b) *$\|Tx\| = \|T^*x\|$ for all $x \in \mathrm{E}$.*

Proof. For all $x \in \mathrm{E}$,

$$\begin{aligned}\|Tx\|^2 - \|T^*x\|^2 &= (Tx|Tx) - (T^*x|T^*x)\\ &= (T^*Tx|x) - (TT^*x|x)\\ &= ((T^*T - TT^*)x|x).\end{aligned}$$

Writing $R = T^*T - TT^*$, we have $R^* = R$ and $(Rx|x) = \|Tx\|^2 - \|T^*x\|$ for all vectors x. Citing 9.4.2, we have $R = 0 \Leftrightarrow \|Tx\|^2 - \|T^*x\| = 0$ for all x in other words (a) $\Leftrightarrow$ (b). ■

Better yet:

9.4.4 Lemma *If $T \in \mathcal{L}(\mathrm{E})$ and $\|T^*x\| \leqslant \|Tx\|$ for all $x \in \mathrm{E}$, then T is normal.*

Proof. $R = T^*T - TT^*$ is self-adjoint and satisfies $(Rx|x) \geqslant 0$ for all vec tors x; it follows that if r is an eigenvalue of R and u is a unit vector suc that $Ru = ru$, then $r = r(u|u) = (Ru|u) \geqslant 0$. By the spectral theorer (9.3.4), E has an orthonormal basis of eigenvectors for R. If A is th matrix of T for this basis, then $A^*A - AA^* = D$, where D is a diagona matrix whose entries are $\geqslant 0$. Since $\operatorname{tr} D = 0$ (7.3.8), it follows that $D = 0$ thus $A^*A = AA^*$, therefore $T^*T = TT^*$. ■

9.4.5 Theorem *If $T \in \mathcal{L}(\mathrm{E})$ is normal and M is an invariant subspace for T, then M reduces T.*

Proof. The restriction $S = T|\mathrm{M}$ is a linear mapping in the Euclidean space M. We assert that S is normal. By the lemma, it suffices to show that if $x \in \mathrm{M}$ then $\|S^*x\| \le \|Sx\|$. For all $y \in \mathrm{M}$ we have

$$(T^*x|y) = (x|Ty) = (x|Sy) = (S^*x|y),$$

thus

$$(T^*x - S^*x|y) = 0 \quad \text{for all} \quad y \in \mathrm{M};$$

in other words, $T^*x - S^*x \in \mathrm{M}^\perp$. But $S^*x \in \mathrm{M}$, so

$$T^*x = (T^*x - S^*x) + S^*x,$$

where the two terms of the sum on the right are in $\mathrm{M}^\perp$ and M, respectively; citing the Pythagorean formula, we have

$$\|T^*x\|^2 = \|T^*x - S^*x\|^2 + \|S^*x\|^2 \ge \|S^*x\|^2,$$

thus $\|S^*x\| \le \|T^*x\| = \|Tx\| = \|Sx\|$. This completes the proof that S is normal.

Assuming $x \in \mathrm{M}$ we have to show that $T^*x \in \mathrm{M}$. Since S is normal, we know that $\|S^*x\| = \|Sx\|$; the earlier formula

$$\|T^*x\|^2 = \|T^*x - S^*x\|^2 + \|S^*x\|^2$$

may now be written

$$\|Tx\|^2 = \|T^*x - S^*x\|^2 + \|Sx\|^2 = \|T^*x - S^*x\|^2 + \|Tx\|^2,$$

whence $\|T^*x - S^*x\| = 0$, $T^*x = S^*x \in \mathrm{M}$. ■

9.4.6 Corollary *If $A \in \mathrm{M}_n(\mathbf{R})$, $A'A = AA'$ and A is triangular, then A is diagonal.*

Proof. Let $T \in \mathcal{L}(\mathrm{E}^n)$ be the linear mapping whose matrix is A relative to the canonical orthonormal basis $e_1, \ldots, e_n$ of E^n (5.2.14). Since A commutes with A', it follows that T commutes with T^* (5.3.5), that is, T is normal. Let $\mathrm{N}_k = [\{e_1, \ldots, e_k\}]$; then $\mathrm{N}_1 \subset \mathrm{N}_2 \subset \ldots \subset \mathrm{N}_n = \mathrm{E}^n$. Suppose, for example, that all entries below the main diagonal of A are 0 (otherwise, consider A'); it follows that the N_k are invariant under T, therefore they reduce T by the theorem. This means that the matrix A' of T^* also has zeros below the main diagonal, thus A has zeros above its main diagonal. ■

▶ *Exercises*

1. Let $T \in \mathcal{L}(\mathrm{E})$ be normal. Then:
   (i) For every real number c, $T - cI$ is normal.
   (ii) For a real number c and a vector x, $Tx = cx \Leftrightarrow T^*x = cx$.
   (iii) If c and d are distinct real numbers, then $\mathrm{Ker}(T - cI) \perp \mathrm{Ker}(T - dI)$.
   (iv) If $c_1, \ldots, c_r \in \mathbf{R}$ are the distinct eigenvalues of T (if there are any!) and $\mathrm{M} = \mathrm{Ker}(T - c_1 I) + \ldots + \mathrm{Ker}(T - c_r I)$, then M reduces T, M has an orthonormal basis of eigenvectors for T, and $T|\mathrm{M}^\perp$ has no eigenvalues.

{Hints:

(i) Theorem 5.3.3.

(ii) §9.1, Exercise 2.

(iii) If $Tx = cx$ and $Ty = dy$, look at $(Tx|y) = (x|T^*y)$.}

2. If $T \in \mathcal{L}(E)$ is normal and $T^k = 0$ for some positive integer k, then $T = 0$. {Hint: We can suppose that k is a power of 2. Rewrite $0 = (T^k)^* T^k$.}

3. Let $T \in \mathcal{L}(E)$ be normal and let m_T be the minimal polynomial of T (§8.3, Exercise 5). Express m_T as a product of real-irreducible factors (Appendix B.3.13, B.4.3; by the Fundamental Theorem of Algebra, the factors are either linear, or quadratic with nonreal roots).

Prove: No factor of m_T is repeated.

{Hint: If $m_T = q^2 r$ then $0 = q(T)^2 r(T)$ implies $q(T)^2 r(T)^2 = 0$; cf. Exercise 2.}

4. Let $S, T \in \mathcal{L}(E)$.

(i) If $\operatorname{Ker} S \subset \operatorname{Ker} T$ then $T = US$ for some $U \in \mathcal{L}(E)$.

(ii) If $\|Sx\| = \|Tx\|$ for all $x \in E$, then $T = US$ with $U \in \mathcal{L}(E)$ orthogonal.

(iii) If T is normal then $T = UT^*$ with $U \in \mathcal{L}(E)$ orthogonal.

(iv) The converses of (i)–(iii) are also true.

{Hint:

(i) Define $U \in \mathcal{L}(E)$ so that $U(Sx) = Tx$ for all $x \in E$ and (for example) $Uy = \theta$ for all $y \in (\operatorname{Im} S)^\perp$.

(ii) Since $\operatorname{Ker} S = \operatorname{Ker} T$, $(\operatorname{Im} S)^\perp$ and $(\operatorname{Im} T)^\perp$ have the same dimension.}

5. If $T \in \mathcal{L}(E)$ is normal and indecomposable (§9.1, Exercise 5), then $\dim E \le 2$.

{Hint: §9.1, Exercise 12.}

6. If $T \in \mathcal{L}(E)$ is normal and T has no eigenvalues, then there exists a direct sum decomposition $E = M_1 \oplus \ldots \oplus M_r$, where the M_i are pairwise orthogonal, 2-dimensional reducing subspaces for T, and $T|M_i$ is indecomposable for all i.

{Hint: §9.1, Exercise 5, and Exercise 5 above; a reducing subspace of dimension 1 would produce an eigenvalue.}

7. If $T \in \mathcal{L}(E)$ is normal, then there exists a direct sum decomposition $E = M \oplus M_1 \oplus \ldots \oplus M_r$, where M and the M_i are pairwise orthogonal subspaces that reduce T, such that M has an orthonormal basis of eigenvectors for T and, for each $i = 1, \ldots r$, M_i is 2-dimensional and $T|M_i$ is indecomposable (and is a scalar multiple of a rotation).

{Hint: Exercises 1 and 6. The structure of each $T|M_i$ is known from §6.7, Exercise 2.}

8. A matrix $A \in M_n(\mathbf{R})$ is said to be *normal* if $A'A = AA'$.

(i) Let $T \in \mathcal{L}(E)$ and let A be the matrix of T relative to an orthonormal basis of E. Then T is normal if and only if A is normal.

(ii) If $A \in M_n(\mathbf{R})$ is normal and $P \in M_n(\mathbf{R})$ is orthogonal, then $P'AP$ is normal.

(iii) With an eye on Exercise 7, complete the following statement: If $A \in M_n(\mathbf{R})$ is normal, then there exists an orthogonal matrix P such that . . . .

(iv) If $A, B \in M_n(\mathbf{R})$, A is normal and $AB = BA$, then $AB' = B'A$.

(v) If A and B are normal and $AB = BA$, then AB and $A + B$ are normal.

{Hint:

(i) Theorem 5.3.5.

(ii) Let $C = AB' - B'A$, show that $C'C = A \cdot BA'B' - BA'B' \cdot A - A \cdot A'BB' + A'BB' \cdot A$ and infer that $\operatorname{tr}(C'C) = 0$ (cf. 7.3.8 and §4.6, Exercise 2).

(v) Use (iv).}

9. If $A \in M_n(\mathbf{R})$ is normal (orthogonal, skew-symmetric) then so is its transpose A'.

10. If $T \in \mathcal{L}(E)$ is normal then $\operatorname{Ker} T^k = \operatorname{Ker} T$ for all positive integers k. {Hint: If $T(Tx) = \theta$ then $T^*(Tx) = \theta$.} Exercise 2 is a special case.

11. Let $T \in \mathcal{L}(E)$ (E a Euclidean space).

(i) T is normal if and only if $T^* \in \{T\}''$ (the bicommutant of T; cf. §8.2, Exercise 4).

(ii) T is normal if and only if $T^* = p(T)$ for some polynomial $p \in \mathbf{R}[t]$.

{Hint:

(i) Exercise 8.

(ii) Parenthetical remark in §8.2, Exercise 5, (vi).}

12. The commutants appearing in Exercise 11 can be generalized as follows. Let V be a vector space over a field F. For a set $\mathcal{S} \subset \mathcal{L}(V)$ of linear mappings, the *commutant* of $\mathcal{S}$ is the set $\mathcal{S}'$ of all $R \in \mathcal{L}(V)$ that commute with every $S \in \mathcal{S}$:

$$\mathcal{S}' = \{R \in \mathcal{L}(V)\colon RS = SR \text{ for all } S \in \mathcal{S}\}.$$

The commutant $(\mathcal{S}')'$ of $\mathcal{S}'$ is called the *bicommutant* of $\mathcal{S}$, also written $\mathcal{S}''$. There are similar definitions and notations for sets $\mathcal{S} \subset M_n(F)$ of matrices. In the following, $\mathcal{S}$ and $\mathcal{T}$ are subsets of $\mathcal{L}(V)$.

(i) $\mathcal{S}'$ is a subalgebra of $\mathcal{L}(V)$, that is, $\mathcal{S}'$ is a linear subspace of $\mathcal{L}(V)$ that is closed under multiplication; also $I \in \mathcal{S}'$, so $\mathcal{S}'$ contains the subalgebra of scalar multiples of the identity mapping.

(ii) $\mathcal{S} \subset \mathcal{S}''$.
(iii) $\mathcal{S} \subset \mathcal{T} \Rightarrow \mathcal{S}' \supset \mathcal{T}'$.
(iv) $(\mathcal{S}'')' = (\mathcal{S}')'' = \mathcal{S}'$; so to speak, the two candidates for $\mathcal{S}'''$ coincide with $\mathcal{S}'$.
(v) $(\mathcal{S} \cup \mathcal{T})' = \mathcal{S}' \cap \mathcal{T}'$.
(vi) $\mathcal{S}$ is commutative ($ST = TS$ for all $S, T \in \mathcal{S}$) if and only if $\mathcal{S} \subset \mathcal{S}'$.
(vii) If $\mathcal{S}$ is commutative then so is $\mathcal{S}''$.
(viii) The following conditions are equivalent: (a) $\mathcal{S} = \mathcal{S}'$; (b) $\mathcal{S}$ is a commutative subalgebra of $\mathcal{L}(V)$, and if $\mathcal{T}$ is a commutative subalgebra of $\mathcal{L}(V)$ with $\mathcal{S} \subset \mathcal{T}$ then $\mathcal{S} = \mathcal{T}$. (Such a set $\mathcal{S}$ is called a **maximal abelian** subalgebra of $\mathcal{L}(V)$.)

{Hints:

(iv) Note that $(\mathcal{S}'')' = ((\mathcal{S}')')' = (\mathcal{S}')''$. By (ii), $\mathcal{S}' \subset (\mathcal{S}')''$; for the reverse inclusion, apply (iii) with $\mathcal{T} = \mathcal{S}''$.

(viii) (a) $\Rightarrow$ (b): $\mathcal{S} \subset \mathcal{T} \subset \mathcal{T}' \subset \mathcal{S}' = \mathcal{S}$.
(b) $\Rightarrow$ (a): For every $R \in \mathcal{S}'$, $(\mathcal{S} \cup \{R\})''$ is a commutative subalgebra, whence $\mathcal{S} = (\mathcal{S} \cup \{R\})''$ and $R \in \mathcal{S}$.}

13. Suppose $A \in M_2(\mathbf{R})$ is normal and not symmetric; thus (§6.7, Exercise 2)

$$A = \begin{pmatrix} a & -b \\ b & a \end{pmatrix}$$

with $a, b \in \mathbf{R}$ and $b \neq 0$. Show that every matrix $B \in M_2(\mathbf{R})$ that commutes with A is normal; in fact, $\{A\}'$ is the set $\mathcal{C}$ of all matrices B of the form

$$B = \begin{pmatrix} c & -d \\ d & c \end{pmatrix}$$

with $c, d \in \mathbf{R}$ (cf. §4.4, Exercise 5).

14. Let $\mathcal{S} \subset M_2(\mathbf{R})$ be the set of all symmetric matrices

$$\begin{pmatrix} a & b \\ b & c \end{pmatrix},$$

$\mathcal{Z}$ the set of all scalar multiples of the identity

$$\begin{pmatrix} a & 0 \\ 0 & a \end{pmatrix},$$

$\mathcal{D}$ the set of all diagonal matrices

$$\begin{pmatrix} a & 0 \\ 0 & b \end{pmatrix},$$

$\mathcal{C}$ the set of all matrices described in Exercise 13, and $\mathcal{N}$ the set of all normal matrices. Prove:

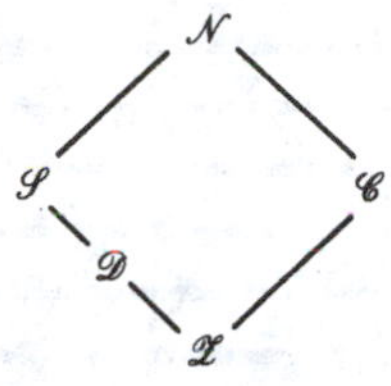

Fig. 23

(i) $\mathscr{Z} \subset \mathscr{D} \subset \mathscr{S} \subset \mathscr{N}$ and $\mathscr{Z} \subset \mathscr{C} \subset \mathscr{N}$ (cf. Fig. 23).

(ii) $\mathscr{Z}$, $\mathscr{D}$ and $\mathscr{C}$ are subalgebras of $M_2(\mathbf{R})$ (that is, linear subspaces closed under multiplication).

(iii) $\mathscr{S}$ is a linear subspace of $M_2(\mathbf{R})$ but is not closed under multiplication.

(iv) $\mathscr{N}$ is neither a linear subspace of $M_2(\mathbf{R})$ nor closed under multiplication.

(v) $\mathscr{N} = \mathscr{S} \cup \mathscr{C}$ and $\mathscr{Z} = \mathscr{S} \cap \mathscr{C}$.

(vi) $\mathscr{C} = \mathscr{C}'$ and $\mathscr{D} = \mathscr{D}'$, thus $\mathscr{C}$ and $\mathscr{D}$ are maximal abelian subalgebras of $M_2(\mathbf{R})$ (cf. Exercise 12).

(vii) $\mathscr{S}' = \mathscr{Z}$.

{Hints:

(v) §6.7, Exercise 2.

(vi) As regards $\mathscr{C}$, see Exercise 13.

(vii) First note that $\mathscr{S}' \subset \mathscr{D}' = \mathscr{D}$, then check what it means for a diagonal matrix to commute with the symmetric matrix $\begin{pmatrix} 0 & 1 \\ 1 & 0 \end{pmatrix}$.}

15. The analysis of Exercise 14 is valid with $M_2(\mathbf{R})$ replaced by $M_2(F)$, where F is any field in which $1 + 1 \neq 0$, and a matrix $A \in M_2(F)$ is called 'normal' if $A'A = AA'$. {Caution: For $F = \mathbf{C}$ this is not the usual definition of normality (12.6.2).}

16. The converse of Theorem 9.4.5 is false: there exist nonnormal linear mappings $T \in \mathscr{L}(E)$ for which every invariant subspace is reducing. {Hint: Look at dimension 2.}

10

Equivalence of matrices over a PIR

The question of whether matrices $A, B \in M_n(F)$ have the same rank can be decided by performing row and column operations on A and B (4.7.7); the gist of the present chapter is that the question of whether A and B are *similar* (over F) can be decided by performing row and column operations on their characteristic matrices $tI - A$ and $tI - B$. The analysis requires a substantial dose of commutative ring theory, especially Euclidean integral domains (reviewed in Appendix B); ultimately, the question of similarity is settled by comparing factors in a subtle factorization of the characteristic polynomials (a factorization induced by appropriate row and column operations on the characteristic matrices).

10.1 Unimodular matrices

Let R be an integral domain. Since R is a subring of a field (for example, its field of fractions), much of the theory of determinants in Chapter 7 is applicable verbatim to matrices over R. In particular, if $A \in M_n(R)$, the determinant $|A|$ is defined and is an element of R (7.3.2); the cofactors A_{ij} belong to R (7.4.1), therefore so does the adjoint matrix $\operatorname{adj} A = (A_{ij})' = (A_{ji})$ (7.4.3), and the equation

(*) $$A(\operatorname{adj} A) = (\operatorname{adj} A)A = |A| \cdot I$$

is verified (7.4.4).

10.1.1 Theorem *If* R *is an integral domain and* $A \in M_n(R)$, *the following conditions are equivalent:*

(a) A *is invertible in the ring* $M_n(R)$;
(b) $AB = I$ *for some* $B \in M_n(R)$;
(c) $CA = I$ *for some* $C \in M_n(R)$;
(d) $|A|$ *is invertible in* R.

For such a matrix A, *necessarily* $B = C = A^{-1} = |A|^{-1} \operatorname{adj} A$.

Proof.
(a) ⇒ (b) & (c): Obvious from the definition of invertibility.

(b) ⇒ (d): $1 = |I| = |AB| = |A|\,|B|$, so $|A|$ is a unit of R with inverse $|B|$.

(c) ⇒ (d): Similarly.

(d) ⇒ (a): If $r \in R$ is the inverse of $|A|$, multiplying through the equation (∗) by r yields $A \cdot r(\operatorname{adj} A) = r(\operatorname{adj} A) \cdot A = I$, therefore A is invertible in $M_n(R)$ with $A^{-1} = r(\operatorname{adj} A) = |A|^{-1}(\operatorname{adj} A)$.

The last assertion follows from the usual 'uniqueness of inverse' argument: $B = IB = (CA)B = C(AB) = CI = C$. ■

10.1.2 Definition A matrix $A \in M_n(R)$ satisfying the conditions of 10.1.1 is said to be **unimodular** (over R). Thus, the set of unimodular matrices is just the group of units of the ring $M_n(R)$.

The equivalence (a) ⇔ (d) of 10.1.1 can be expressed as follows:

10.1.3 Corollary *If* R *is an integral domain, then*

$$U_{M_n(R)} = \{A \in M_n(R) \colon |A| \in U_R\}$$

(where U *denotes group of units).*

10.1.4 Examples If $R = \mathbf{Z}$ (the ring of integers), the unimodular matrices are the matrices $A \in M_n(\mathbf{Z})$ with $|A| = \pm 1$. If $R = F[t]$, F a field, the unimodular matrices are the $A \in M_n(F[t])$ such that $|A|$ is a nonzero element of F.

▶ **Exercises**

1. Let R be an integral domain, $n \geq 2$, $A \in M_n(R)$. Prove:

(i) A is unimodular if and only if $\operatorname{adj} A$ is unimodular, in which case $(\operatorname{adj} A)^{-1} = |A|^{-1} A$.

(ii) If $|A| = 1$ then $\operatorname{adj}(\operatorname{adj} A) = A$.

{Hint: (i) 'Only if': Take determinant in the equation $A(\operatorname{adj} A) = |A| \cdot I$. 'If': Argue first that $|A|$ can't be 0.}

2. Let $F = \{0, 1\}$ be the field with two elements. The matrix ring $R = M_2(F)$ has 16 elements; find the units of R.

10.2 Preview of the theory of equivalence

Here is a preview of forthcoming events. Our eventual goal (attained in 11.1.7) is the following theorem:

10.2.1 Theorem (Fundamental theorem of similarity) *If* F *is a field and* A, $B \in \mathrm{M_n(F)}$, *then*

$$A, B \text{ are similar over } \mathrm{F} \Leftrightarrow \exists \text{ unimodular matrices } P, Q \in \mathrm{M_n(F}[t]) \text{ such that } P(tI - A)Q = tI - B.$$

The condition on the right prompts the following definition:

10.2.2 Definition Let R be an integral domain, C, $D \in \mathrm{M_n(R)}$. We say that C is **equivalent** to D over R (briefly, 'C equiv D') if there exist unimodular matrices P, $Q \in \mathrm{M_n(R)}$ such that $PCQ = D$. {The definition makes sense in any ring A with unity: call c, $d \in \mathrm{A}$ 'equivalent' if there exist units p, $q \in \mathrm{A}$ with $pcq = d$.}

Since the unimodular matrices form a group (the group of units of $\mathrm{M_n(R)}$) one sees easily that the relation «C equiv D» is an equivalence relation in $\mathrm{M_n(R)}$.

The fundamental theorem of similarity (proved in the next chapter) is relatively elementary, and leaves us with a huge question: how can we decide whether or not $tI - A$ and $tI - B$ are equivalent? That's the purpose of the present chapter, and the key result is the following (see 10.3.12 and 10.4.11):

10.2.3 Theorem (Fundamental theorem of equivalence) *Let* R *be a principal ideal ring* (Appendix B.3). *For every matrix* $A \in \mathrm{M_n(R)}$, *there exist unimodular matrices* $P, Q \in \mathrm{M_n(R)}$ *such that*

$$PAQ = \operatorname{diag}(a_1, a_2, \ldots, a_n) \quad \text{with} \quad a_i | a_{i+1} \quad (i = 1, \ldots, n-1).$$

{In other words, A is equivalent to a diagonal matrix in which each diagonal element divides the next (if there is a next).}

Moreover, the a_i are unique up to associate: if also P_1, Q_1 are unimodular matrices and

$$P_1AQ_1 = \operatorname{diag}(b_1, b_2, \ldots, b_n) \quad \text{with} \quad b_i | b_{i+1},$$

then $a_i \sim b_i$ for all i.

Recall that a, b are said to be **associate** in R, written $a \sim b$, if $a|b$ and $b|a$; it is the same to say that $b = ua$ with $u \in \mathrm{R}$ a unit (Appendix B.2.5).

Granted the Fundamental theorem of equivalence (proved in the next two sections), let's explore some of its consequences. First, some terminology:

10.2.4 Definition With notations as in 10.2.3, one calls the ordered list $a_1, a_2, \ldots, a_n$ the **invariant factors** of the matrix A (they are uniquely determined by A, up to associate). {Caution: When R is a field, the term 'invariant factor' usually means something else (11.1.1).}

10.2.5 Remark With notations as in 10.2.3,

$$|PAQ| = a_1 a_2, \dots a_n,$$

thus $a_1 a_2 \dots a_n = |P| \cdot |A| \cdot |Q|$; since $|P|$ and $|Q|$ are units of R, we have

$$|A| \sim a_1 a_2 \dots a_n \text{ and } a_i | a_{i+1}.$$

Thus, the invariant factors of A provide a (highly structured) factorization of $|A|$.

10.2.6 Remark With notations as in 10.2.3, if $a_j = 0$ then $a_j = a_{j+1} = \dots = a_n = 0$. {Reason: if $0|a$ then $a = 0$.}

10.2.7 Remark If $\mathrm{R} = \mathbf{Z}$ we can take $a_i \geq 0$ in 10.2.3; the a_i are then unique. If $\mathrm{R} = \mathrm{F}[t]$, F a field, we can take the nonzero a_i in 10.2.3 to be monic; the a_i are then unique.

10.2.8 Corollary *Let* R *be a principal ideal ring,* $A, B \in \mathrm{M_n(R)}$. *Then:*

A equiv B $\Leftrightarrow$ *A, B have the same (up to associate) invariant factors.*

Proof. $\Rightarrow$: Let P, Q be unimodular with $PAQ = B$. By 10.2.3, there exist unimodular P_1, Q_1 such that $P_1 B Q_1 = \operatorname{diag}(b_1, b_2, \dots, b_n)$ and $b_i | b_{i+1}$. Then

$$\operatorname{diag}(b_1, b_2, \dots, b_n) = P_1(PAQ)Q_1 = (P_1 P)A(QQ_1),$$

where $P_1 P$ and QQ_1 are unimodular, so the b_i also serve as the invariant factors of A.

$\Leftarrow$: If P, Q, P_1, Q_1 are unimodular with

$$PAQ = \operatorname{diag}(a_1, \dots, a_n), \quad P_1 B Q_1 = \operatorname{diag}(b_1, \dots, b_n)$$

and $a_i \sim b_i$ for all i, then A is equivalent to B. For, writing $b_i = u_i a_i$ with $u_i \in \mathrm{R}$ a unit, we have

$$\begin{aligned} P_1 B Q_1 &= \operatorname{diag}(u_1 a_1, \dots, u_n a_n) \\ &= \operatorname{diag}(u_1, \dots, u_n) \operatorname{diag}(a_1, \dots, a_n) \\ &= \operatorname{diag}(u_1, \dots, u_n) \cdot PAQ; \end{aligned}$$

the matrix $Q_2 = \operatorname{diag}(u_1, \dots, u_n)$ is obviously unimodular, and $P_1 B Q_1 = (Q_2 P)AQ$, so

$$A \text{ equiv } (Q_2 P)AQ = P_1 B Q_1 \text{ equiv } B,$$

whence A equiv B by transitivity. ∎

▶ ***Exercises***

1. If $\mathrm{R} = \mathrm{M_2(F)}$, where $\mathrm{F} = \{0,1\}$ is the field with two elements, find the set of all matrices equivalent (in the sense of Definition 10.2.2) to the matrix

$$\begin{pmatrix} 1 & 0 \\ 0 & 0 \end{pmatrix}.$$

{Hint: Use the result of §10.1, Exercise 2.}

2. In any ring R with unity, write $a \sim b$ if there exist units u, v such that $uav = b$. Check that $\sim$ is an equivalence relation in R.

10.3 Equivalence: existence of a diagonal form

Fixed notations for the section: R is a principal ideal ring (on occasion, a Euclidean integral domain), n is a positive integer; A, B, C are $n \times n$ matrices over R, and P, Q are unimodular $n \times n$ matrices over R.[1]

The general strategy for thinning down a matrix A to an equivalent diagonal matrix is to do it in little bites; the little bites are called **elementary operations** (briefly, 'elops') and come in three flavors:

10.3.1 Definition An elementary operation of **type I** on a matrix: interchange two rows (or two columns).

If B is obtained from A by an elop of type I, then $|B| = -|A|$ (7.3.9, 7.1.9).

10.3.2 Definition An elementary operation of **type II** on a matrix: add a multiple (by an element of R) of one row (or column) to another.

If B is obtained from A by an elop of type II, then $|B| = |A|$ (7.3.9).

10.3.3 Definition An elementary operation of **type III** on a matrix: multiply a row (or column) by a *unit* of R.

If B is obtained from A by multiplying a row (or column) by a unit $u \in$ R, then $|B| = u|A|$ (7.3.9).

Note that each elop is reversible (by an elop of the same type).

Suppose $n = 3$ and $A \in$ M$_3$(R). Let $r \in$ R and let u be a unit of R. Consider the matrices

$$P = \begin{pmatrix} 1 & 0 & 0 \\ 0 & 0 & 1 \\ 0 & 1 & 0 \end{pmatrix}, \quad Q = \begin{pmatrix} 1 & 0 & r \\ 0 & 1 & 0 \\ 0 & 0 & 1 \end{pmatrix}, \quad U = \begin{pmatrix} 1 & 0 & 0 \\ 0 & 1 & 0 \\ 0 & 0 & u \end{pmatrix}.$$

P is obtained from I by interchanging rows 2 and 3 (or columns 2 and 3);

Q is obtained from I by multiplying row 3 by r and adding to row 1 (or multiplying column 1 by r and adding to column 3);

U is obtained from I by multiplying row 3 (or column 3) by u.

One verifies easily that pre- and post-multiplication by P, Q or U is an elementary operation; explicitly,

PA is the result of interchanging rows 2 and 3 of A;

AP is the result of interchanging columns 2 and 3 of A.

QA is the result of multiplying row 3 of A by r, and adding to row 1;

[1] With minor modifications, the theory of diagonalization is applicable to rectangular ($m \times n$) matrices (cf. S. MacLane and G. Birkhoff, *Algebra*, 3rd. edn, Chelsea, 1988, p. 401; P. M. Cohn, *Algebra*, Vol. 1, Wiley, 1974, p. 279).

AQ is the result of multiplying column 1 of A by r, and adding to column 3;

UA is the result of multiplying row 3 of A by u;

AU is the result of multiplying column 3 of A by u.

Concisely: multiplication by P is an elop of type I; multiplication by Q is an elop of type II; multiplication by U is an elop of type III. Give or take some tedious notations, this proves:

10.3.4 Theorem *If B is obtained from A by a finite number of elementary operations, then B is equivalent to A.*

Later in the section we will see that when R is a Euclidean domain, the converse of Theorem 10.3.4 is also true: if B is equivalent to A, then B can be obtained from A by a finite number of elementary operations.

10.3.5 Definition We call a matrix $P \in \mathrm{M}_n(\mathrm{R})$ an **elementary matrix** (briefly, 'elmat') of type i ($i = \mathrm{I}, \mathrm{II}, \mathrm{III}$) if it is obtained from the identity matrix by applying an elementary operation of type i.

Concisely: an elmat is the result of applying an elop to I. Every elmat is unimodular. The message of the discussion preceding Theorem 10.3.4: every elop can be achieved by pre- or post-multiplying by a suitable elmat.

Note that interchanging rows j and k of I produces the same matrix as interchanging columns j and k; multiplying row j of I by $r \in \mathrm{R}$ and adding to row k ($k \neq j$) produces the same matrix as multiplying column k by r and adding to column j; multiplying row j of I by a unit $u \in \mathrm{R}$ produces the same matrix as multiplying column j by u. So to speak, the set of 'row elmats of type i' is the same as the set of 'column elmats of type i' ($i = \mathrm{I}, \mathrm{II}, \mathrm{III}$).

10.3.6 Remarks Let us write $[A]$ for the class of A under equivalence: $[A] = \{B : B \text{ equiv } A\}$. The statement of 10.3.4 can be expressed as the inclusion

$$\{PAQ\colon P,\ Q \text{ finite products of elmats}\} \subset [A];$$

we shall see in 10.3.10 that if R is a Euclidean integral domain, then the two sets are equal.

10.3.7 Theorem *If* R *is a Euclidean integral domain*[2] *and* $A \in \mathrm{M}_n(\mathrm{R})$, *then there exist unimodular matrices* $P, Q \in \mathrm{M}_n(\mathrm{R})$ *such that*

$$PAQ = \operatorname{diag}(a_1, \ldots, a_n) \text{ and } a_i | a_{i+1} \quad (i = 1, \ldots, n-1).$$

Moreover, P *and* Q *can be chosen to be finite products of elementary matrices of types* I *and* II.

Proof. The proof is by induction on n. For $n = 1$ there is nothing to prove; suppose $n \geq 2$ and assume that all's well for $n - 1$.

[2] Appendix B.4.1.

Let $\mathscr{C}$ be the set of all matrices that are obtainable from A by a finite number of elementary operations of types I and II; from Theorem 10.3.4 and the discussion preceding it, we see that

$$\mathscr{C} = \{PAQ\colon P, Q \text{ finite products of elmats of types I and II}\}.$$

It is clear that $\mathscr{C}$ is closed under elementary operations of types I and II; our problem is to show that $\mathscr{C}$ contains a diagonal matrix with the indicated divisibility relations.

If $A = 0$ there is nothing to prove. Assuming $A \neq 0$, let S be the set of all nonzero elements of R that occur as entries in matrices of $\mathscr{C}$:

$$S = \{c \in R - \{0\}\colon \quad c \text{ is an entry of some } B \in \mathscr{C}\};$$

by assumption $S \neq \varnothing$. If δ is the rank function[2] of R, then

$$\varnothing \neq \{\delta(c)\colon c \in S\} \subset \mathbf{N}.$$

Choose $a \in S$ with $\delta(a)$ *minimal*. Say a occurs as an entry of $B \in \mathscr{C}$. We can suppose that

$$B = \begin{pmatrix} a & \cdots \\ \vdots & \end{pmatrix};$$

for, if a is not already in the northwest corner, it can be brought there by a finite number of elops of type I.

claim: a divides every element in row 1 and column 1.

Suppose, for instance, that

$$B = \begin{pmatrix} a & b & \cdots \\ \vdots & & \\ & & \end{pmatrix};$$

let's show that $a|b$. Write $b = qa + r$, where $r = 0$ or $\delta(r) < \delta(a)$; we'd like to rule out the second alternative. Multiply column 1 of B by $-q$ and add to column 2:

$$\begin{pmatrix} a & b - qa & \cdots \\ \vdots & & \\ & & \end{pmatrix}.$$

This matrix belongs to $\mathscr{C}$ (it is obtained from B by a type II elop) and it has r in the $(1, 2)$ place; since $\delta(r) < \delta(a)$ is ruled out by the minimality of $\delta(a)$, we must have $r = 0$.

It follows from the preceding claim that a finite number of elops of type II will bring B to the form

$$C = \left(\begin{array}{c|ccc} a & 0 & \cdots & 0 \\ \hline 0 & & & \\ \vdots & & B_1 & \\ 0 & & & \end{array}\right).$$

By the induction hypothesis applied to B_1, there exist matrices P_1, Q_1 in $M_{n-1}(R)$ such that

$$P_1 B_1 Q_1 = \begin{pmatrix} a_2 & & & \\ & a_3 & & \\ & & \ddots & \\ & & & a_n \end{pmatrix}$$

with $a_i | a_{i+1}$ for $i = 2, \ldots, n-1$, and such that P_1, Q_1 are finite products of elmats of types I and II; then the n'th order matrices

$$P = \left(\begin{array}{c|c} 1 & 0 \\ \hline 0 & P_1 \end{array}\right), \quad Q = \left(\begin{array}{c|c} 1 & 0 \\ \hline 0 & Q_1 \end{array}\right)$$

are products of the same sort, therefore $\mathscr{C}$ contains the diagonal matrix

$$PCQ = \begin{pmatrix} a & & & \\ & a_2 & & \\ & & \ddots & \\ & & & a_n \end{pmatrix}.$$

It will suffice to show that $a | a_2$. Adding row 2 of this diagonal matrix to row 1, we see that $\mathscr{C}$ contains the matrix

$$\begin{pmatrix} a & a_2 & & \\ 0 & a_2 & & \\ & & \ddots & \\ & & & a_n \end{pmatrix}$$

and a repetition of the earlier argument shows that $a | a_2$. ■

The issue of *uniqueness* of the diagonal form is deferred until the next section.

10.3.8 Corollary *If* R *is a Euclidean integral domain and* $A \in M_n(R)$, *then* A *is unimodular* $\Leftrightarrow$ A *is a finite product of elementary matrices.*

Proof. $\Leftarrow$: Elementary matrices are unimodular, therefore so is any finite product of them (10.1.2).

$\Rightarrow$: If A, P and Q are unimodular then so is PAQ; by the theorem, we are reduced to the case that A is diagonal. A diagonal unimodular matrix is obviously a finite product of elmats of type III; for example,

$$\begin{pmatrix} u & 0 \\ 0 & v \end{pmatrix} = \begin{pmatrix} u & 0 \\ 0 & 1 \end{pmatrix}\begin{pmatrix} 1 & 0 \\ 0 & v \end{pmatrix}$$

(u, v units of R). ■

10.3.9 Corollary *If* F *is a field and* $A \in M_n(F)$, *then* $|A| \neq 0 \Leftrightarrow A$ *is a finite product of elementary matrices.*

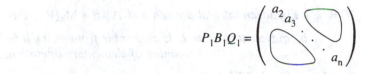

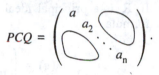

Proof. The set of units of F is $F - \{0\}$, so A is unimodular if and only if $|A| \neq 0$ (10.1.3); quote Corollary 10.3.8. ■

10.3.10 Corollary *If* R *is a Euclidean integral domain and* $A, B \in M_n(R)$, *then*

$$A \text{ is equivalent to } B \Leftrightarrow B \text{ is obtainable from } A \text{ by a finite number of elementary operations.}$$

Proof. $\Leftarrow$: This is 10.3.4.

$\Rightarrow$: By defintion, $B = PAQ$ with P, Q unimodular; by 10.3.8, P and Q are finite products of elmats, thus B is obtainable from A in the indicated way (see the discussion preceding 10.3.4). ■

To generalize the diagonalization theorem (the first statement of 10.3.7) to matrices over a principal ideal ring, we need a substitute for the rank function of a Euclidean domain to propel the inductive argument:

10.3.11 Definition If R is a principal ideal ring, we define a function $\lambda: R - \{0\} \to \mathbf{N}$ by the formula

$$\lambda(a) = \begin{cases} r \text{ if } a \text{ is the product of } r \geq 1 \text{ primes} \\ 0 \text{ if } a \text{ is a unit.} \end{cases}$$

From the (essentially) unique factorization into primes (Appendix B.3.14), it is easy to see that $\lambda(ab) = \lambda(a) + \lambda(b)$ for all nonzero a, b in R; in particular, $\lambda(ab) \geq \lambda(a)$.

In contrast with 10.3.7, what is missing in the following theorem is a statement about the composition of the unimodular matrices P and Q (see Exercise 1):

10.3.12 Theorem *If* R *is a principal ideal ring and* $A \in M_n(R)$ *then there exist unimodular matrices* P *and* Q *such that*

$$PAQ = \operatorname{diag}(a_1, \ldots, a_n) \quad \textit{and} \quad a_i | a_{i+1} \quad (i = 1, \ldots, n-1).$$

Proof. Let $\mathscr{C}$ be the set of all matrices equivalent to A, that is

$$\mathscr{C} = \{PAQ\colon P, Q \text{ unimodular}\};$$

by Theorem 10.3.4, $\mathscr{C}$ is closed under elementary operations. Our problem is to show that $\mathscr{C}$ contains a diagonal matrix with the indicated divisibility relations. The proof is by induction on n. The case $n = 1$ is trivial; let $n \geq 2$ and assume that all's well for $n - 1$.

If $A = 0$ there is nothing to prove. Assuming $A \neq 0$, let

$$S = \{c \in R - \{0\}\colon c \text{ is an entry of some } B \in \mathscr{C}\};$$

then $S \neq \varnothing$, so we can choose an element $a \in S$ with $\lambda(a)$ minimal (i.e., a has the smallest possible number of prime factors). It is conceivable that $\lambda(a) = 0$, in other words that a is a unit of R; the following proof is then even simpler (the 'claim' below being trivially verified).

As argued in the proof of 10.3.7, there exists a matrix $B \in \mathscr{C}$ such that

$$B = \begin{pmatrix} a & \cdots & \\ \vdots & & \end{pmatrix}.$$

claim: a divides every element in row 1 and column 1.

Suppose, for instance that

$$B = \begin{pmatrix} a & b & \cdots \\ \vdots & & \end{pmatrix};$$

let's show that $a|b$. Since R is a principal ideal ring, a and b have a greatest common divisor $d = (a, b)$, and $d = ra + sb$ for suitable elements $r, s \in \mathrm{R}$.[3] Write $a = da'$, $b = db'$; then $d = (ra' + sb')d$, so $ra' + sb' = 1$ and $(a', b') = 1$, that is, a' and b' are relatively prime. Also, $ab' = a'b = a'b'd$ (the least common multiple of a and b).

The relation $ra' + sb' = 1$ means that the 2×2 matrix

$$P_1 = \begin{pmatrix} r & -b' \\ s & a' \end{pmatrix}$$

is unimodular, therefore so is the $n \times n$ matrix

$$P = \left(\begin{array}{c|c} P_1 & 0 \\ \hline 0 & I \end{array}\right),$$

where I is the identity matrix of order $n - 2$.

The 2×2 northwest corner of B is of the form

$$B_1 = \begin{pmatrix} a & b \\ x & y \end{pmatrix},$$

and

$$B_1 P_1 = \begin{pmatrix} a & b \\ x & y \end{pmatrix}\begin{pmatrix} r & -b' \\ s & a' \end{pmatrix}$$

$$= \begin{pmatrix} ra + sb & -ab' + ba' \\ * & * \end{pmatrix} = \begin{pmatrix} d & 0 \\ * & * \end{pmatrix}$$

(we are not interested in the entries $*$). It follows that

$$BP = \left(\begin{array}{c|c} B_1 & * \\ \hline * & * \end{array}\right)\left(\begin{array}{c|c} P_1 & 0 \\ \hline 0 & I \end{array}\right) = \left(\begin{array}{c|c} B_1 P_1 & * \\ \hline * & * \end{array}\right)$$

$$= \begin{pmatrix} d & \cdots & \\ \vdots & & \end{pmatrix},$$

thus $\mathcal{C}$ contains a matrix with d in the northwest corner; therefore $d \in \mathrm{S}$, so $\lambda(d) \geq \lambda(a)$ by the choice of a. But $\lambda(a) = \lambda(a'd) = \lambda(a') + \lambda(d) \geq \lambda(d)$, so

[3] Appendix B.3.4.

$\lambda(d) = \lambda(a)$ and $\lambda(a') = 0$; this means that a' is a unit, so $a \sim d$. But $d|b$, therefore also $a|b$.

This establishes the claim, and the proof continues as in Theorem 10.3.7. ■

▶ *Exercises*

1. The unimodular matrix P_1 that figures in the proof of Theorem 10.3.12 comes within a whisker of being expressible as a product of elementary matrices.[4] To simplify the notation, suppose

$$P = \begin{pmatrix} r & -b \\ s & a \end{pmatrix}$$

with $|P| = ra + sb = 1$. Then

$$bP = \begin{pmatrix} b & 0 \\ 0 & 1 \end{pmatrix}\begin{pmatrix} 1 & 0 \\ -a & 1 \end{pmatrix}\begin{pmatrix} 0 & 1 \\ 1 & 0 \end{pmatrix}\begin{pmatrix} 1 & 0 \\ r & 1 \end{pmatrix}\begin{pmatrix} 1 & 0 \\ 0 & -b \end{pmatrix},$$

so if b is a unit in R then P is a product of elmats. When $b \neq 0$ (but not necessarily a unit),

$$P = \begin{pmatrix} 1 & 0 \\ 0 & b^{-1} \end{pmatrix}\begin{pmatrix} 1 & 0 \\ -a & 1 \end{pmatrix}\begin{pmatrix} 0 & 1 \\ 1 & 0 \end{pmatrix}\begin{pmatrix} 1 & 0 \\ r & 1 \end{pmatrix}\begin{pmatrix} 1 & 0 \\ 0 & -b \end{pmatrix}$$

is a product of elmats over the field of fractions of R (cf. Corollary 10.3.9 and Appendix B.1).

2. Let

$$A = \begin{pmatrix} a & b \\ c & d \end{pmatrix}$$

be a unimodular matrix with entries in a Euclidean integral domain R. Describe a procedure for factoring A as a product of elementary matrices.

{Hint: If $a = qb + r$ with $q, r \in$ R (cf. Appendix B. 4.4), then an elop of type II transforms A into the form

$$\begin{pmatrix} a - qb & b \\ c - qd & d \end{pmatrix} = \begin{pmatrix} r & b \\ c - qd & d \end{pmatrix}.$$

Applying Euclid's algorithm to a and b, a finite number of elops of type II will transform A into a matrix B having a 0 in the top row. We can suppose (interchanging columns, if necessary, that the element in the northeast corner of B is 0; the diagonal elements of B are then units.}

3. In the matrix ring $M_2(\mathbf{Z})$, express the matrix

$$A = \begin{pmatrix} 2 & 5 \\ 3 & 8 \end{pmatrix}$$

[4] The whisker is thick: P. M. Cohn has given an example of a principal ideal ring R (not Euclidean) and a unimodular matrix $A \in M_2(R)$ that is not a product of elementary matrices ('On the structure of $\mathbf{GL}_2$ of a ring', Institut des Hautes Etudes Scientifiques, *Publications mathématiques*, No. 30 (1966), 5–53).

as a product of elementary matrices.

10.4 Equivalence: uniqueness of the diagonal form

The notations fixed at the beginning of the preceding section remain in force; in particular, R is a principal ideal ring. To get at the uniqueness (up to associates) of the diagonal form (10.3.12), a new technique is needed:

10.4.1 Definition Let A be an $n \times n$ matrix over R. If $1 \leq r \leq n$, an $r \times r$ **subdeterminant** of A is the element of R obtained by striking out $n - r$ rows and $n - r$ columns of A and taking the determinant of what's left. {When $r = n$, nothing is stricken and we get $|A|$.} The greatest common divisor (GCD) of all possible $r \times r$ subdeterminants of A is denoted $\mathrm{D_r}(A)$; the elements

$$\mathrm{D_1}(A), \quad D_2(A), \ldots, \mathrm{D_n}(A)$$

of R are called the **determinantal divisors** of A.

Since GCDs are unique only up to associate, the same is true of the determinantal divisors; thus if $a, b \in \mathrm{R}$ and $a \sim b$, the statements $\mathrm{D_r}(A) = a$ and $\mathrm{D_r}(A) = b$ mean the same thing: either a or b serves as a GCD for the $r \times r$ subdeterminants of A. To be absolutely precise: $\mathrm{D_r}(A)$ stands for any generator of the (principal) ideal generated by the $r \times r$ subdeterminants of A.

For example, $\mathrm{D_n}(A) = |A|$, and if $A = (a_{ij})$ then $\mathrm{D_1}(A)$ is the GCD of all the a_{ij}.

For $r = 1, \ldots, n - 1$, each $(r + 1) \times (r + 1)$ subdeterminant of A is a linear combination of $r \times r$ subdeterminants of A (Theorem 7.4.2), therefore $\mathrm{D_r}(A) | \mathrm{D_{r+1}}(A)$.

When $\mathrm{R} = \mathbf{Z}$, the determinantal divisors can be made unique by taking them to be $\geqslant 0$; when $\mathrm{R} = \mathrm{F}[t]$, F a field, they can be made unique by taking them to be monic (when they are nonzero).

The key result to be proved in this section: $A, B \in \mathrm{M_n(R)}$ are equivalent if and only if $\mathrm{D_r}(A) \sim \mathrm{D_r}(B)$ for $r = 1, \ldots, n$. When R is a Euclidean domain—for example, when $\mathrm{R} = \mathbf{Z}$ or when $\mathrm{R} = \mathrm{F}[t]$, F a field—then the determinantal divisors of A and B can be computed in finitely many steps using Euclid's algorithm[1], thus we can decide in finitely many steps whether A and B are equivalent.

10.4.2 Lemma *If $A = \mathrm{diag}(a_1, \ldots, a_n)$ and $a_i | a_{i+1}$ for $i = 1, \ldots, n - 1$, then $\mathrm{D_r}(A) \sim a_1 \ldots a_r$ for $r = 1, \ldots, n$.*

Proof (informal). The principal is very easy to see through an example. Consider the case $n = 5$:

[1] Appendix B.4.5.

$$A = \begin{pmatrix} a_1 & 0 & 0 & 0 & 0 \\ 0 & a_2 & 0 & 0 & 0 \\ 0 & 0 & a_3 & 0 & 0 \\ 0 & 0 & 0 & a_4 & 0 \\ 0 & 0 & 0 & 0 & a_5 \end{pmatrix}.$$

Suppose we are going to calculate $D_3(A)$; we have to look at all 3×3 subdeterminants. A 3×3 subdeterminant arises from crossing out two rows and two columns. Suppose, for instance, that we cross out rows 2 and 5:

$$\begin{pmatrix} a_1 & 0 & 0 & 0 & 0 \\ 0 & a_2 & 0 & 0 & 0 \\ 0 & 0 & a_3 & 0 & 0 \\ 0 & 0 & 0 & a_4 & 0 \\ 0 & 0 & 0 & 0 & a_5 \end{pmatrix}.$$

If we do **not** cross out column 2 then we are leaving in place a column consisting entirely of 0's, and the subdeterminant we are calculating is guaranteed to be 0; similarly if we fail to cross out column 5.

Conclusion: if we cross out a row, then we should cross out the column of the same index; if we don't, we'll get 0, which will have no influence on the GCD being calculated.

In other words, the only subdeterminants that need be computed are those where we have crossed out rows and columns of the same indices; the resulting matrix is a diagonal matrix with the surviving a's down its main diagonal. If we cross out the last two rows and the last two columns,

$$\begin{pmatrix} a_1 & 0 & 0 & 0 & 0 \\ 0 & a_2 & 0 & 0 & 0 \\ 0 & 0 & a_3 & 0 & 0 \\ 0 & 0 & 0 & a_4 & 0 \\ 0 & 0 & 0 & 0 & a_5 \end{pmatrix}$$

the remaining 3×3 matrix has determinant $a_1a_2a_3$; this clearly divides all the other possibilities (because each a_i divides all later a_j's), so $D_3(A) = a_1a_2a_3$. ■

10.4.3 Lemma *If P and Q are products of elementary matrices of type* I, *then* $D_r(PAQ) \sim D_r(A)$ *for* $r = 1, \ldots, n$.

Proof. The matrix PA is the result of permuting rows of A, therefore the set of $r \times r$ subdeterminants of PA is, apart from signs, the same as the set of $r \times r$ subdeterminants of A. Similarly, $(PA)Q$ is the result of permuting columns of PA, etc. {This lemma is superseded by 10.4.7, but it smoothes the way for the proof of the next lemma.} ■

10.4.4 Lemma *If $A, B \in \mathrm{M}_n(\mathrm{R})$ and $1 \le r \le n$, then each $r \times r$ subdeterminant of AB is a linear combination of $r \times r$ subdeterminants of A (with coefficients in* R*), consequently* $\mathrm{D}_r(A) | \mathrm{D}_r(AB)$.

Proof. Say $A = (a_{ij})$, $B = (b_{ij})$, $C = AB = (c_{ij})$. If we are given an $r \times r$ subdeterminant of C then there exist permutation matrices P and Q such that PCQ has this $r \times r$ subdeterminant in its northwest corner; moreover, $PCQ = (PA)(BQ)$, where the $r \times r$ subdeterminants of PA are $\pm$ those of A. Conclusion: we need only show that the $r \times r$ subdeterminant in the northwest corner of C has the property claimed for it.

Write α_j for the j'th column vector of A, and $\alpha_j{}^{\#}$ for the column vector of length r obtained from α_j by omitting the last $n - r$ entries:

$$\alpha_j{}^{\#} = \begin{pmatrix} a_{1j} \\ \vdots \\ a_{rj} \end{pmatrix}$$

Similarly, let

$$\gamma_j{}^{\#} = \begin{pmatrix} c_{1j} \\ \vdots \\ c_{rj} \end{pmatrix}$$

(the result of omitting the last $n - r$ entries in the j'th column of C). Since

$$c_{ij} = \sum_{k=1}^{n} a_{ik} b_{kj} = \sum_{k=1}^{n} b_{kj} a_{ik}$$

for all i, in particular for $i = 1, \ldots, r$, it is clear that

$$\gamma_j{}^{\#} = \sum_{k=1}^{n} b_{kj} \alpha_k{}^{\#}.$$

It will be convenient to indicate the $r \times r$ matrix in the northwest corner of C as the r'ple $(\gamma_1{}^{\#}, \ldots, \gamma_r{}^{\#})$ of its column vectors; its determinant is (by multilinearity of the determinant function)

$$\begin{aligned} \det(\gamma_1{}^{\#}, \ldots, \gamma_r{}^{\#}) &= \det\left(\sum_{k=1}^{n} b_{k1} \alpha_k{}^{\#}, \ldots, \sum_{k=1}^{n} b_{kr} \alpha_k{}^{\#}\right) \\ &= \sum b_{k_1 1} b_{k_2 2} \ldots b_{k_r r} \det(\alpha^{\#}_{k_1}, \alpha^{\#}_{k_2}, \ldots, \alpha^{\#}_{k_r}), \end{aligned}$$

summed over all combinations of the indices $k_1, \ldots, k_r \in \{1, \ldots, n\}$. The determinant appearing in any particular term is 0 unless $k_1, \ldots, k_r$ are distinct, and then it is $\pm$ an $r \times r$ subdeterminant of A.

This proves the first assertion of the theorem. It follows that $\mathrm{D}_r(A)$ divides each $r \times r$ subdeterminant of C, therefore it divides their greatest common divisor $\mathrm{D}_r(C)$. ■

10.4.5 Lemma $\mathrm{D}_r(A) \sim \mathrm{D}_r(A')$ *for* $r = 1, \ldots, n$.

Proof. The set of $r \times r$ subdeterminants of A' is equal to the set of $r \times r$ subdeterminants of A (7.3.5). ■

10.4.6 Lemma $D_r(A)|D_r(BA)$ *for* $r = 1, \ldots, n$.

Proof. Immediate from 10.4.4, 10.4.5 and the fact that $(BA)' = A'B'$. ■

10.4.7 Lemma *If* P *and* Q *are unimodular, then* $D_r(PAQ) \sim D_r(A)$ *for* $r = 1, \ldots, n$.

Proof. By 10.4.4 and 10.4.6,

$$D_r(A)|D_r(AQ) \quad \text{and} \quad D_r(AQ)|D_r(PAQ),$$

thus $D_r(A)|D_r(PAQ)$; also $A = P^{-1}(PAQ)Q^{-1}$, so $D_r(PAQ)|D_r(A)$. ■

10.4.8 Lemma *If* P, Q *are unimodular and* $PAQ = \text{diag}(a_1, \ldots, a_n)$ *with* $a_i|a_{i+1}$ *for* $i = 1, \ldots, n-1$, *then* $D_r(A) \sim a_1 \ldots a_r$ *for* $r = 1, \ldots, n$.

Proof. Immediate from 10.4.7 and 10.4.2. ■

10.4.9 Theorem *If* R *is a principal ideal ring and* $A, B \in M_n(R)$, *then*

$$A \text{ is equivalent to } B \text{ (over R)} \Leftrightarrow D_r(A) \sim D_r(B) \text{ for } r = 1, \ldots, n.$$

Proof. $\Rightarrow$: This is 10.4.7.

$\Leftarrow$: By Theorem 10.3.12 there exist unimodular matrices P, Q, P_1, Q_1 such that

$$PAQ = \text{diag}(a_1, \ldots, a_n), \quad a_i|a_{i+1},$$
$$P_1BQ_1 = \text{diag}(b_1, \ldots, b_n), \quad b_i|b_{i+1}.$$

By 10.4.8 and the hypothesis,

$$(*) \qquad a_1, \ldots, a_r \sim D_r(A) \sim D_r(B) \sim b_1 \ldots b_r$$

for $r = 1, \ldots, n$.

Note that $a_1 = 0 \Leftrightarrow A = 0$; for, $a_1 \sim D_1(A)$ is the GCD of all the entries a_{ij} of A. By $(*)$, $a_1 \sim b_1$, so $A = 0 \Leftrightarrow B = 0$, in which case A, B are trivially equivalent.

Assume $a_1 \neq 0$ (therefore $b_1 \neq 0$). Let m be the last index with $a_m \neq 0$ (conceivably $m = n$, in which case the following argument can be simplified). Then $D_m(A) = a_1 \ldots a_m \neq 0$ (10.2.6), therefore $D_m(B) \neq 0$, whence $b_m \neq 0$. By symmetry, it is clear that m is also the last index with $b_m \neq 0$. Then

$$a_{m+1} = \ldots = a_n = 0, \quad b_{m+1} = \ldots = b_n = 0$$

by 10.2.6.

claim: $a_i \sim b_i$ for all i.

This is trivial for $i > m$. Already noted is $a_1 \sim b_1$. For $1 < r \leq m$, by $(*)$ we have

$$a_1 \ldots a_{r-1} \sim b_1 \ldots b_{r-1} \quad \text{and} \quad (a_1 \ldots a_{r-1})a_r \sim (b_1 \ldots b_{r-1})b_r,$$

whence $a_r \sim b_r$.

For $i = 1, \ldots, n$ write $b_i = u_i a_i$ with u_i a unit of R (if $b_i = 0$ take $u_i = 1$). Then

$$\operatorname{diag}(b_1, \ldots, b_n) = \operatorname{diag}(u_1, \ldots, u_n) \operatorname{diag}(a_1, \ldots, a_n).$$

The matrix $P_2 = \operatorname{diag}(u_1, \ldots, u_n)$ is unimodular, and the preceding equation says that $P_1 B Q_1 = P_2(PAQ)$; then

$$A \text{ equiv } (P_2 P)AQ = P_1 B Q_1 \text{ equiv } B. \blacksquare$$

So to speak, A and B are equivalent if and only if they have the 'same' determinantal divisors. The proof of 10.4.9 shows:

10.4.10 Theorem *Let* R *be a principal ideal ring,* $A, B \in M_n(R)$. *Say*

$$PAQ = \operatorname{diag}(a_1, \ldots, a_n) \text{ with } a_i | a_{i+1},$$
$$P_1 B Q_1 = \operatorname{diag}(b_1, \ldots, b_n) \text{ with } b_i | b_{i+1},$$

where the P's and Q's are unimodular. The following conditions imply one another:

(a) A *is equivalent to* B *(over* R*);*

(b) $D_r(A) \sim D_r(B)$ *for* $r = 1, \ldots, n$;

(c) $a_i \sim b_i$ *for* $i = 1, \ldots, n$.

10.4.11 Remark In particular, when $B = A$ (so that (a) is trivially verified), condition (c) shows that the a_i provided by Theorem 10.3.12 are unique up to associates. This completes the proof of the *Fundamental theorem of equivalence* announced in 10.2.3.

▶ Exercises

1. Let $A \in M_n(R)$ and let $p_1, \ldots, p_n$ be the invariant factors of A, with $p_i | p_{i+1}$ for all i. Prove: $D_i(A) = 1 \Leftrightarrow p_i$ is a unit.

2. Let

$$A = \left(\begin{array}{c|c} B & 0 \\ \hline 0 & C \end{array}\right),$$

where $B \in M_m(R)$, $C \in M_n(R)$, and suppose that $(|B|, |C|) = 1$. If $r = \min\{m, n\}$ then $D_r(A) = 1$.

{Hint: The invariant factors of B are relatively prime to those of C. If $a, b \in R$ and $(a, b) = 1$, then $\operatorname{diag}(a, b)$ is equivalent to $\operatorname{diag}(1, ab)$ over R.}

3. Let $R = F[t]$, F a field, and let $A \in M_n(F)$. Suppose $tI - A$ is equivalent to

$$\begin{pmatrix} p_1 & 0 \\ 0 & p_2 \end{pmatrix},$$

over R. True or false (explain): $p_1 | p_2$ or $p_2 | p_1$.

4. Let $R = \mathbf{Z}$ be the ring of integers and let $A \in M_3(\mathbf{Z})$ be the matrix

$$A = \begin{pmatrix} 3 & 2 & 3 \\ 1 & 2 & 7 \\ 5 & 4 & 5 \end{pmatrix}.$$

Find the determinantal divisors and invariant factors of A. Then do it another way.

5. Let $R = F[t]$, F a field, and let $A \in M_3(R)$ be the matrix

$$A = \begin{pmatrix} t & 2t & 0 \\ t^2 & 2t^2 + t - 1 & 0 \\ 0 & 0 & t^2 \end{pmatrix}.$$

Find the determinantal divisors and invariant factors of A.

6. If F is a field and $A \in M_n(F)$, $A \neq 0$, then the rank of A is the largest positive integer r such that A has a nonzero $r \times r$ subdeterminant.

7. Let $A \in M_n(R)$, R an integral domain, $n \geq 2$.

(i) $|\operatorname{adj} A| = |A|^{n-1}$.

(ii) $\operatorname{adj}(\operatorname{adj} A) = |A|^{n-2} A$.

{Hint: §7.4, Exercises 5, 6.}

11

Similarity (Act II)

The program sketched in §10.2 is carried to its completion in the present chapter. The highlights:

Matrices $A, B \in M_n(F)$ are similar over F if and only if $tI - A$, $tI - B$ are equivalent over $F[t]$ (11.1.7). Every square matrix is a 'root' of its characteristic polynomial (11.3.6). The rational canonical form of a matrix $A \in M_n(F)$ is constructed from a certain factorization of its characteristic polynomial p_A (into the 'invariant factors' of A); two matrices are similar if and only if they have the same rational canonical form (11.2.6). Assuming the field F is algebraically closed, the analogous theorem is proved (11.4.13) for the Jordan canonical form (a refinement of the rational canonical form obtained from factoring the invariant factors into powers of linear polynomials).

Reserved notations for the chapter:

F a field
n a positive integer
$A, B \in M_n(F)$
$P, Q, R, S \in M_n(F[t])$

11.1 Invariant factors, Fundamental theorem of similarity

If $A \in M_n(F)$ then $tI - A \in M_n(F[t])$. According to Theorem 10.3.7, there exist unimodular matrices $P, Q \in M_n(F[t])$ such that

$$P(tI - A)Q = \text{diag}(p_1, \ldots, p_n) \quad \text{with} \quad p_i | p_{i+1};$$

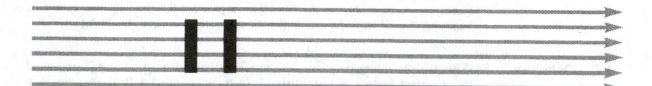

since $p_1 \cdots p_n = |P(tI - A)Q| \sim |tI - A| = p_A \neq 0$, the p_i are nonzero, hence can be taken to be *monic* (hence completely unique, by 10.4.11).

11.1.1 Definition With the preceding notations, the monic polynomials $p_1, \ldots, p_n$ are called the **invariant factors** of the matrix A (a minor conflict with the terminology in 10.2.4, according to which they are the 'invariant factors' of $tI - A$).

Since $|tI - A| = p_A$ and $p_1 \ldots p_n$ are both monic, it follows that $p_A = p_1 \ldots p_n$. Moreover,

$$n = \deg p_A = \sum_{i=1}^{n} \deg p_i;$$

since $\deg p_i \geqslant 0$, the chances are good that the first few invariant factors are 1. More precisely:

11.1.2 Remark 1 is not an invariant factor of $A \Leftrightarrow A = cI$ for some $c \in F$.

{**Proof.** Let $p_1, \ldots, p_n$ be the (monic) invariant factors of A, $p_i | p_{i+1}$.

$\Leftarrow$: If $A = cI = \operatorname{diag}(c, \ldots, c)$, then $tI - A = \operatorname{diag}(t - c, \ldots, t - c)$ and it is clear that the invariant factors are $p_i = t - c$ $(i = 1, \ldots, n)$.

$\Rightarrow$: By assumption, $\deg p_i \geq 1$ for all i; but

$$\sum_{i=1}^{n} \deg p_i = n,$$

so necessarily $\deg p_i = 1$ for all i, that is, the p_i are all linear. Say $p_1 = t - c$; since the p_i are linear and $p_1 | p_i$ for all i, necessarily $p_i = t - c$ for all i.

By Theorem 10.3.7 there exist unimodular matrices P, Q in $M_n(F[t])$ such that

$$P(tI - A)Q = \operatorname{diag}(t - c, \ldots, t - c) = (t - c)I,$$

whence $tI - A = P^{-1}((t - c)I)Q^{-1} = (t - c)I \cdot P^{-1}Q^{-1}$. Say $P^{-1}Q^{-1} = (r_{ij})$, $r_{ij} \in F[t]$; then

$$(*) \qquad (t\delta_{ij} - a_{ij}) = tI - A = ((t - c)r_{ij}).$$

The off-diagonal elements in this matrix equation are scalars (look at the left member), hence $r_{ij} = 0$ when $i \neq j$ (look at the right member), hence $a_{ij} = 0$ when $i \neq j$ (look at the left member again). Thus A is a diagonal matrix, and

$$tI - A = \operatorname{diag}(t - a_{11}, \ldots, t - a_{nn}).$$

Looking again at $(*)$ we see that $t - a_{ii} = (t - c)r_{ii}$, whence $r_{ii} = 1$ and $a_{ii} = c$ for all i, thus $A = cI$.}

The main result in this section is the *Fundamental theorem of similarity* (11.1.7); the proof follows the exposition of A.A.Albert (*Introduction to algebraic theories*, University of Chicago Press, 1941).

11.1.3 Lemma $M_n(F[t]) = M_n(F)[t]$, *that is, the matrices with entries in the polynomial ring* $F[t]$ *can be written as polynomials with coefficients in the ring* $M_n(F)$, *and vice versa*.

Proof (informal). Innocent as the lemma appears, there's a lot to talk about here (the instant you feel persuaded, read no more). On a superficial level, the lemma is 'obvious'; for example, starting with a matrix in $M_2(F[t])$ we can rewrite it as a 'polynomial' with coefficients in $M_2(F)$:

$$\begin{pmatrix} t^2+5t-3 & 2t^2-4t \\ t^3-5t & 2 \end{pmatrix}$$

$$= \begin{pmatrix} -3 & 0 \\ 0 & 2 \end{pmatrix} + \begin{pmatrix} 5t & -4t \\ -5t & 0 \end{pmatrix} + \begin{pmatrix} t^2 & 2t^2 \\ 0 & 0 \end{pmatrix} + \begin{pmatrix} 0 & 0 \\ t^3 & 0 \end{pmatrix}$$

$$= \begin{pmatrix} -3 & 0 \\ 0 & 2 \end{pmatrix} + \begin{pmatrix} 5 & -4 \\ -5 & 0 \end{pmatrix}(tI) + \begin{pmatrix} 1 & 2 \\ 0 & 0 \end{pmatrix}(t^2 I) + \begin{pmatrix} 0 & 0 \\ 1 & 0 \end{pmatrix}(t^3 I)$$

$$= A_0 + A_1(tI) + A_2(tI)^2 + A_3(tI)^3,$$

where the A_i belongs to $M_2(F)$.

Since $F[t]$ is contained in its field of fractions $F(t)$ (Appendix B.1.3), and since $M_n(F(t))$ is a vector space over $F(t)$, the above matrix can also be written

$$\begin{pmatrix} -3 & 0 \\ 0 & 2 \end{pmatrix} + t\begin{pmatrix} 5 & -4 \\ -5 & 0 \end{pmatrix} + t^2\begin{pmatrix} 1 & 2 \\ 0 & 0 \end{pmatrix} + t^3\begin{pmatrix} 0 & 0 \\ 1 & 0 \end{pmatrix}$$

or (what's the harm, since tI commutes with every element of $M_2(F[t])$?)

$$\begin{pmatrix} -3 & 0 \\ 0 & 2 \end{pmatrix} + \begin{pmatrix} 5 & -4 \\ -5 & 0 \end{pmatrix}t + \begin{pmatrix} 1 & 2 \\ 0 & 0 \end{pmatrix}t^2 + \begin{pmatrix} 0 & 0 \\ 1 & 0 \end{pmatrix}t^3.$$

{Technically, we can view $M_n(F[t])$ as a right module[1] over $F[t]$: $(p_{ij})p = (p_{ij}p)$. Since $F[t]$ is commutative, $p(p_{ij}) = (p_{ij})p$.}

More generally, if $P \in M_n(F[t])$ and m is the highest power of t that occurs in some entry of P, we can write

$$(*) \qquad P = A_0 + tA_1 + \ldots + t^m A_m$$

for suitable $A_0, A_1, \ldots, A_m \in M_n(F)$, $A_m \neq 0$ (these matrix coefficients are clearly unique).

Conversely, the right side of $(*)$ is a typical element of $M_n(F)[t]$; it determines an element of $M_n(F[t])$ on carrying out the operations indicated on the right side of $(*)$.

Well, that's *almost* a proof of the lemma. However, we should pause to worry about what $M_n(F)[t]$ means; $M_n(F)$ is a *noncommutative* ring (when $n \geq 2$), so we should review the construction of polynomial rings to make sure we can deal with a noncommutative coefficient ring. The details are thrashed out in an Appendix at the end of the chapter. ■

[1] P. M. Cohn, *Algebra*, Vol. 1, Wiley, 1974, p. 226.

11.1.4 Lemma (Noncommutative division algorithm, jr. grade) *If $P \in M_n(F[t])$ and $A \in M_n(F)$, then*
$$P = (tI - A)Q + B$$
for suitable $Q \in M_n(F[t])$ and $B \in M_n(F)$.

Proof. If $P = 0$ take $Q = B = 0$. Assuming $P \neq 0$, by 11.1.3 we can write
$$P = t^m A_m + \ldots + tA_1 + A_0$$
with $A_m, \ldots, A_0 \in M_n(F)$ and $A_m \neq 0$. The proof proceeds by inducation on m (call it the *degree* of P).

If $m = 0$ then $P = A_0 \in M_n(F)$ and the choices $Q = 0$, $B = P$ meet the requirements. If $m = 1$ then
$$P = tA_1 + A_0 = (tI)A_1 + A_0 = (tI - A)A_1 + (AA_1 + A_0),$$
so the choices $Q = A_1$, $B = AA_1 + A_0$ meet the requirements. Finally, let $m \geq 2$ and assume that all's well for degree $m - 1$. The first step is to reduce to degree $m - 1$ by sawing off the term of highest degree: define
$$\begin{aligned} R &= P - (tI - A)(t^{m-1}A_m) \\ &= (P - t^m A_m) + t^{m-1}AA_m \\ &= (t^{m-1}A_{m-1} + \ldots + tA_1 + A_0) + t^{m-1}AA_m \\ &= t^{m-1}(A_{m-1} + AA_m) + t^{m-2}A_{m-2} + \ldots + tA_1 + A_0. \end{aligned}$$
Applying the induction hypothesis to R, we can write
$$R = (tI - A)S + B \quad \text{with} \quad S \in M_n(F[t]) \quad \text{and} \quad B \in M_n(F).$$
Then
$$\begin{aligned} P &= (tI - A)(t^{m-1}A_m) + R \\ &= (tI - A)(t^{m-1}A_m) + (tI - A)S + B \\ &= (tI - A)(t^{m-1}A_m + S) + B \end{aligned}$$
and the choice $Q = t^{m-1}A_m + S$ meets the requirements. (In fact, Q and B are uniquely determined by P and A; the proof of Lemma 11.1.6 contains the seeds of the argument.) ■

With notations as in 11.1.4, we can think of P as dividend, $tI - A$ as 'left divisor', with Q and B as quotient and remainder. Application of 11.1.4 to P' and A' yields a dual version:

11.1.5 Lemma *If $P \in M_n(F[t])$ and $A \in M_n(F)$, then*
$$P = R(tI - A) + C$$
for suitable $R \in M_n(F[t])$ and $C \in M_n(F)$.

11.1.6 Lemma *If*
$$(tI - A)P(tI - B) = tC + D,$$

where $P \in M_n(F[t])$ *and* $A, B, C, D \in M_n(F)$, *then* $P = 0$ (*hence also* $C = D = 0$).

Proof. (This is reasonable: if P were nonzero, we would expect the matrix on the left side to contain entries of degree at least 2, which is not consistent with the right side.)

Assume to the contrary that $P \neq 0$. By Lemma 11.1.3 we can write

$$p = t^m A_m + \ldots + tA_1 + A_0$$

with $m \geq 0$ and $A_m \neq 0$. Then

$$\begin{aligned} tC + D &= (tI - A)(t^m A_m + \ldots + tA_1 + A_0)(tI - B) \\ &= (t^{m+1} A_m + \ldots - AA_0)(tI - B) \\ &= t^{m+2} A_m + \ldots + AA_0 B, \end{aligned}$$

which is absurd since the highest power of t occurring in the left side is the first power. ■

11.1.7 Theorem (Fundamental theorem of similarity) *If* A *and* B *are matrices over a field* F, *then*

$$A, B \text{ are similar over } F \Leftrightarrow tI - A, tI - B \text{ are equivalent over } F[t].$$

Proof. $\Rightarrow$: If $B = CAC^{-1}$, $C \in M_n(F)$, then $|C|, |C^{-1}| \in F - \{0\}$ are units of $F[t]$; thus C, C^{-1} are unimodular over $F[t]$ and

$$C(tI - A)C^{-1} = tCC^{-1} - CAC^{-1} = tI - B.$$

$\Leftarrow$: Let's begin with an elegant but **false** proof so that we can appreciate the difficulty overcome by Albert's ingenious argument. Suppose

$$tI - B = P(tI - A)Q,$$

with $P, Q \in M_n(F[t])$ unimodular. Then

$$(*) \qquad tI - B = tPQ - PAQ.$$

Let r be the largest degree among the entries of P, and s the largest degree among the entries of Q. If degree r occurs in row i of P and degree s occurs in column j of Q, then degree $r + s$ will occur in the (i, j) entry of PQ (won't it?. . .). Thus the first term on the right side of $(*)$ will have an entry of degree $r + s + 1$; the entries of the second term on the right side have degree at most $r + s$, so the right side will have an entry of degrees $r + s + 1$. Looking at the left side, we conclude that $r = s = 0$. Then $P, Q \in M_n(F)$ and comparison of coefficients in $(*)$ yields $PQ = I$ and $B = PAQ$. {Did you spot the flaw? When we take the dot product of row i and column j, the entries of highest degree might be out of step and miss each other; or they might occur in several terms and cancel out. An extreme example with $n = 2$: $(t^2, 0) \cdot (0, t^3) = (0, 0)$.}

Let's start over. Let $P, Q \in M_n(F[t])$ be unimodular matrices such that

$$(1) \qquad tI - B = P(tI - A)Q.$$

Apply 11.1.4 to P and 11.1.5 to Q:

$$(2) \qquad P = (tI - B)R + C$$

(3) $$Q = S(tI - B) + D$$

with $R, S \in M_n(F[t])$ and $C, D \in M_n(F)$; we are going to show that $CD = I$ and $B = CAC^{-1}$.

We assert that there exists a matrix $T \in M_n(F[t])$ such that

(4) $$(tI - A)D = T(tI - B).$$

For, by (1) we have $(tI - A)Q = P^{-1}(tI - B)$; substituting the formula (3) for Q,

$$(tI - A)[S(tI - B) + D] = P^{-1}(tI - B),$$

thus

$$\begin{aligned}(tI - A)D &= P^{-1}(tI - B) - (tI - A)S(tI - B)\\ &= [P^{-1} - (tI - A)S](tI - B)\end{aligned}$$

and the choice $T = P^{-1} - (tI - A)S$ validates (4).

Similarly, there exists a matrix $U \in M_n(F[t])$ such that

(5) $$C(tI - A) = (tI - B)U.$$

For, by (1) we have $P(tI - A) = (tI - B)Q^{-1}$; substituting the formula (2) for P,

$$[(tI - B)R + C](tI - A) = (tI - B)Q^{-1},$$

thus

$$\begin{aligned}C(tI - A) &= (tI - B)Q^{-1} - (tI - B)R(tI - A)\\ &= (tI - B)[Q^{-1} - R(tI - A)]\end{aligned}$$

and $U = Q^{-1} - R(tI - A)$ validates (5).

Substituting (2) and (3) into (1), we have

$$\begin{aligned}tI - B &= [(tI - B)R + C](tI - A)[S(tI - B) + D]\\ &= (tI - B)\cdot R(tI - A)S\cdot(tI - B) &\text{(a)}\\ &\quad + (tI - B)R\cdot(tI - A)D &\text{(b)}\\ &\quad + C(tI - A)\cdot S(tI - B) &\text{(c)}\\ &\quad + C(tI - A)D. &\text{(d)}\end{aligned}$$

Writing $V = R(tI - A)S$ and substituting (4) and (5) in the indicated terms (b) and (c), we have

$$\begin{aligned}tI - B &= (tI - B)V(tI - B)\\ &\quad + (tI - B)R\cdot T(tI - B)\\ &\quad + (tI - B)U\cdot S(tI - B)\\ &\quad + C(tI - A)D\\ &= (tI - B)[V + RT + US](tI - B) + C(tI - A)D\\ &= (tI - B)W(tI - B) + C(tI - A)D,\end{aligned}$$

where $W = V + RT + US$. Thus

$$(tI - B)W(tI - B) = (tI - B) - C(tI - A)D$$
$$= t(I - CD) + (CAD - B);$$

by Lemma 11.1.6, $W = 0$, $I - CD = 0$, $CAD - B = 0$, thus C is invertible and $B = CAC^{-1}$. ■

11.1.8 Corollary *For matrices $A, B \in \mathrm{M_n(F)}$, the following conditions are equivalent:*

(a) *A and B are similar over F;*

(b) *A and B have the same invariant factors;*

(c) *$\mathrm{D_r}(tI - A) = \mathrm{D_r}(tI - B)$ for $r = 1, \ldots, n$.*

Proof. Immediate from 11.1.7 and 10.4.10 (applied with F[t] in the role of the principal ideal ring R, and $tI - A$, $tI - B$ in the role of A, B). ■

11.1.9 Corollary *Let K be an overfield of F, and let $A, B \in \mathrm{M_n(F)}$. Then A and B are similar over F if and only if they are similar over K.*

Proof. The determinantal divisors can be calculated in F[t] using Euclid's algorithm; if we view the calculation as taking place in K[t], the outcome does not change. Thus the corollary is immediate form 11.1.8. {Put another way, the issue of the similarity of A and B over K can be decided in the context of any subfield of K that contains all of the entries of A and B.} ■

11.1.10 Corollary *If $A \in \mathrm{M_n(F)}$, then A and A' are similar.*

Proof. The set of determinantal divisors of $tI - A$ is equal to the set of determinantal divisors of $tI - A' = (tI - A)'$ (10.4.5), so the corollary is immediate from 11.1.8. ■

A curious corollary of 11.1.7:

11.1.11 Corollary *If $A, B \in \mathrm{M_n(F)}$ and if there exist unimodular matrices $P, Q \in \mathrm{M_n(F[}t\mathrm{])}$ such that $P(tI - A)Q = tI - B$, then P and Q can be chosen to be in $\mathrm{M_n(F)}$ with $PQ = 1$.*

11.1.12 Example For the matrices

$$A = \begin{pmatrix} 2 & 0 & 0 \\ 0 & 2 & 1 \\ 0 & 0 & 2 \end{pmatrix}, \quad B = \begin{pmatrix} 2 & 1 & 0 \\ 0 & 2 & 1 \\ 0 & 0 & 2 \end{pmatrix},$$

$\mathrm{D_3}(tI - A) = \mathrm{D_3}(tI - B) = (t - 2)^3$ and $\mathrm{D_1}(tI - A) = \mathrm{D_1}(tI - B) = 1$; but $\mathrm{D_2}(tI - A) = t - 2$ and $\mathrm{D_2}(tI - B) = 1$, consequently A and B are not similar. The determinantal divisors of A are 1, $t - 2$, $(t - 2)^3$, and those of B are 1, $1, (t - 2)^3$. With an eye on Lemma 10.4.2, we see that '$p_{i+1} = \mathrm{D_{i+1}}/\mathrm{D_i}$', consequently

A has invariant factors $1,\ t-2,\ (t-2)^2$,
B has invariant factors $1,\ 1,\ (t-2)^3$.

11.1.13 Definition Let $A \in M_n(F)$ have invariant factors $p_1, \ldots, p_n$, where $p_i | p_{i+1}$. Let r be the number of invariant factors (if any) that are equal to 1, so that A has invariant factors

$$1, \ldots, 1, p_{r+1}, \ldots, p_n,$$

where $p_{r+1}, \ldots, p_n$ are nonconstant (conceivably $r = 0$). The first r invariant factors are called **trivial** (the rest, nontrivial). The last invariant factor p_n is called the **minimal polynomial** of A (for reasons explained in §11.3).

Note that the characteristic polynomial p_A of A is the product of the nontrivial invariant factors, and the minimal polynomial of A is an invariant factor of *maximal* degree.

11.1.14 Example Suppose the nontrivial invariant factors of $A \in M_n(F)$ are

$$t^2 - 2t, \quad t, \quad t^2(t-2)^2.$$

It is implicit that, after reordering according to increasing degree, each must divide the next:

$$t, \quad t(t-2), \quad t^2(t-2)^2.$$

The characteristic polynomial of A is $p_A = t^4(t-2)^3$, consequently $n = 7$; the invariant factors are $p_1 = \ldots = p_4 = 1$, $p_5 = t$, $p_6 = t(t-2)$, $p_7 = t^2(t-2)^2$, in particular the minimal polynomial is $t^2(t-2)^2$. The 'prime powers' that make up the nontrivial invariant factors are t, t, t^2, $t-2$, $(t-2)^2$; the theory developed in section §11.4 will allow us to conclude that A is similar to the matrix

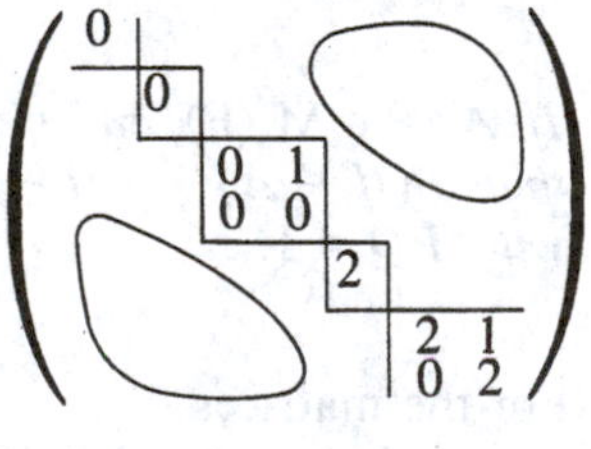

▶ Exercises

1. Similar matrices $A, B \in M_n(F)$ have the same minimal polynomial (cf. Corollary 11.1.8). Define the *minimal polynomial* m_T of a linear mapping $T \in \mathcal{L}(V)$, where V is a finite-dimensional vector space over F.

2. Let $F(t)$ be the field of rational forms over F (Appendix B.1.3) and let $A, B \in M_n(F)$. Prove that A and B are similar over F if and only if $tI - A$ and $tI - B$ are similar over $F(t)$.
{Hint: $K = F(t)$ is an overfield of F.}

11.2 Companion matrix, Rational canonical form

Let F be a field, n a positive integer, $A \in M_n(F)$. The invariant factors of A are in a sense the building blocks of its characteristic matrix $tI - A$. They arise from a 'decomposition' of the characteristic matrix as a sort of 'direct sum' of 1×1 blocks; to arrive at the decomposition, we have to use equivalence transformations over the polynomial ring $F[t]$.

There ought to be a corresponding reduction of A into simpler blocks—not necessarily 1×1—arrived at by similarity over F. Such a reduction does indeed exist; it is called the 'rational canonical form of A' and it is our next goal.

The first step is a trick for creating a matrix with specified characteristic polynomial and having just *one* nontrivial invariant factor—in other words, with minimal polynomial *equal* to the characteristic polynomial:

11.2.1 Definition Let $p \in F[t]$ be monic, of degree $m \geq 1$, say $p = t^m + c_1 t^{m-1} + \ldots + c_{m-1}t + c_m$. The $m \times m$ matrix $C_p \in M_m(F)$ defined by

$$C_p = \begin{pmatrix} 0 & 1 & & & \\ & 0 & 1 & & \\ & & \ddots & \ddots & \\ & & & 0 & 1 \\ -c_m & \cdot & \cdot & -c_2 & -c_1 \end{pmatrix}$$

is called the **companion matrix** of the polynomial p.

Convention: The companion matrix of $t + c_1$ is the 1×1 matrix $(-c_1)$.

11.2.2 Example If $p = t^4 + c_1 t^3 + c_2 t^2 + c_3 t + c_4$, the companion matrix of p is the matrix

$$C = \begin{pmatrix} 0 & 1 & 0 & 0 \\ 0 & 0 & 1 & 0 \\ 0 & 0 & 0 & 1 \\ -c_4 & -c_3 & -c_2 & -c_1 \end{pmatrix}.$$

The characteristic matrix of C is

$$tI - C = \begin{pmatrix} t & -1 & 0 & 0 \\ 0 & t & -1 & 0 \\ 0 & 0 & t & -1 \\ c_4 & c_3 & c_2 & t+c_1 \end{pmatrix};$$

its 'northeast' minor of order 3 (strike out the first column and the last row), namely

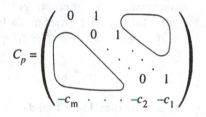

$$\begin{pmatrix} -1 & 0 & 0 \\ t & -1 & 0 \\ 0 & t & -1 \end{pmatrix},$$

has determinant -1, consequently $D_3(tI - C) = 1$. Since $D_1|D_2|D_3$ and $D_4 = p_C$, it follows that the invariant factors of C are

$$1, 1, 1, p_C$$

and by direct calculation we see that $p_C = p$. Thus, the companion matrix of p has invariant factors 1, 1, 1, p. This works for every monic polynomial:

11.2.3 Theorem (Companion matrix) *For every monic polynomial $p \in F[t]$ of degree $m \geq 1$, the companion matrix C_p has invariant factors $1, \ldots, 1, p$; that is, the first $m - 1$ invariant factors are trivial, and the minimal polynomial and characteristic polynomial coincide with p.*

Proof. Write $D_r = D_r(tI - C_p)$ for the r'th determinantal divisor of $tI - C_p$. The proof is by induction on $m = \deg p$.

If $m = 1$, say $p = t + c_1$, then C_p is the 1×1 matrix $(-c_1)$, $tI - C_p = (t + c_1)$ and $t + c_1 = p$ is the only invariant factor; this is how the statement of the theorem is to be interpreted when $m = 1$ (there are indeed $m - 1 = 0$ trivial invariant factors).

If $m = 2$, say $p = t^2 + c_1 t + c_2$, then

$$C_p = \begin{pmatrix} 0 & 1 \\ -c_2 & -c_1 \end{pmatrix}, \quad tI - C_p = \begin{pmatrix} t & -1 \\ c_2 & t + c_1 \end{pmatrix}.$$

It is clear that $D_1 = 1$ and

$$D_2 = |tI - C_p| = t(t + c_1) + c_2 = p,$$

thus the invariant factors are 1, p and the assertion of the theorem is verified.

Let $m \geq 3$ and assume that all's well with $m - 1$. {Incidentally, $m = 4$ is covered by Example 11.2.2} Say $p = t^m + c_1 t^{m-1} + \ldots + c_{m-1} t + c_m$, so that C_p has the form given in Definition 11.2.1. Cross out column 1 and row m of $tI - C_p$; the remaining 'northeast minor' of order $m - 1$ is a triangular matrix with -1's down the diagonal, so its determinant is $(-1)^{m-1} = \pm 1$; it follows that $D_{m-1} = 1$, consequently $D_1 = D_2 = \ldots = D_{m-1} = 1$. It remains to show that $D_m = p$, that is, $|tI - C_p| = p$.

The first column of $tI - C_p$ is

$$\begin{pmatrix} t \\ 0 \\ \vdots \\ 0 \\ c_m \end{pmatrix};$$

if we expand $|tI - C_p|$ by the first column, only two terms survive:

$$t\begin{vmatrix} t & -1 & & \\ & \ddots & \ddots & \\ & & t & -1 \\ c_{m-1} & \cdots & c_2 & t+c_1 \end{vmatrix} + (-1)^{m+1}c_m \begin{vmatrix} -1 & & & \\ t & -1 & & \\ & \ddots & \ddots & \\ & & t & -1 \end{vmatrix}.$$

The determinant in the first term is $|tI - C_q|$, where $q = t^{m-1} + c_1t^{m-2} + \ldots + c_{m-2}t + c_{m-1}$; this determinant is q by the induction hypothesis, so the first term is tq. The determinant in the second term is $(-1)^{m-1}$, so the second term is c_m. Conclusion:

$$|tI - C_p| = tq + c_m = p. \blacksquare$$

Companion matrices are the building blocks (under similarity) alluded to in the remarks at the beginning of the section:

11.2.4 Theorem (Rational canonical form) *If $A \in M_n(F)$ has nontrivial invariant factors $p_1, \ldots, p_r$, with $p_i | p_{i+1}$, and if C_{p_i} is the companion matrix of p_i, then A is similar to the matrix*

$$C = \begin{pmatrix} C_{p_1} & & & \\ & C_{p_2} & & \\ & & \ddots & \\ & & & C_{p_r} \end{pmatrix}.$$

Proof. We know that

$$tI - A \quad \text{equiv} \quad \operatorname{diag}(1, \ldots, 1, p_1, p_2, \ldots, p_r)$$

(see Definition 11.1.1 and the discussion preceding it); by the Fundamental theorem of similarity (11.1.7) we need only show that $tI - C$ is equivalent to the same diagonal matrix.

Write C_i for the companion matrix of p_i. By 11.2.3, C_i has invariant factors $1, \ldots, 1, p_i$ (the number of 1's is $n_i - 1$, where n_i is the order of C_i), thus

$$tI - C_i \quad \text{equiv} \quad \operatorname{diag}(1, \ldots, 1, p_i);$$

let P_i, Q_i be unimodular matrices over $F[t]$ such that

$$P_i(tI - C_i)Q_i = \operatorname{diag}(1, \ldots, 1, p_i).$$

The matrices

$$P = \begin{pmatrix} P_1 & & & \\ & P_2 & & \\ & & \ddots & \\ & & & P_r \end{pmatrix}, \quad Q = \begin{pmatrix} Q_1 & & & \\ & Q_2 & & \\ & & \ddots & \\ & & & Q_r \end{pmatrix}$$

are also unimodular, with

$$P(tI - C)Q = \operatorname{diag}(1, \ldots, 1, p_1, 1, \ldots, 1, p_2, \ldots, 1, \ldots, 1, p_r),$$

and the diagonal matrix on the right is equivalent to

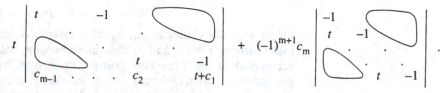

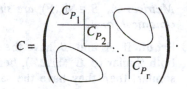

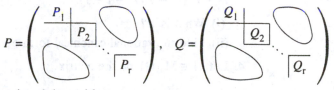

$$\operatorname{diag}(1, \ldots, 1, p_1, p_2, \ldots, p_r).$$

{Note that the elements b_{ii}, b_{jj} of a matrix B can be interchanged by first interchanging rows i and j, then interchanging columns i and j; this can be accomplished by pre- and post-multiplication by the same (self-inverse) permutation matrix, hence by a similarity transformation. When B is diagonal, the matrix that results is identical to B except that b_{ii} and b_{jj} have changed places with each other. Incidentally, blocks of a block-diagonal matrix such as C can be similarly permuted; the argument is slightly subtler but the principle is the same. It follows that the hypothesis $p_i | p_{i+1}$ can be dropped. (A good exercise: consider an 8×8 matrix with a 3×3 block in the northwest corner, a 5×5 block in the southeast corner, and 0's everywhere else; write down what has to be done to interchange the two blocks.)} ■

11.2.5 Definition With notations as in 11.2.4, the matrix C is called the **rational canonical form** of A. (Note that p_r is the minimal polynomial of A.)

11.2.6 Theorem *Matrices $A, B \in M_n(F)$ are similar if and only if they have the same rational canonical form.*

Proof. If A and B have the same rational canonical form C, then they are both similar to C (11.2.4), hence to each other. Conversely, if A and B are similar, then they have the same invariant factors (11.1.18) hence the same rational canonical form (11.2.5). ■

If there are 1×1 blocks in the matrix C of Theorem 11.2.4, they are identical (because $p_i | p_{i+1}$). The blocks of order ≥ 2 are discernible from the fact that the 1's above the main diagonal are punctuated by a 0 at the boundary of two such blocks.

► Exercises

1. Find the companion matrix of the polynomial $p = (t-1)^4$.
2. If p is the characteristic polynomial of the matrix

$$A = \begin{pmatrix} 0 & 1 \\ 3 & 2 \end{pmatrix},$$

find the companion matrix of p.

3. Let $A \in M_n(F)$ be the matrix whose nontrivial invariant factors are $t-1$, $t^2 - t$, $(t^2 - t)(t^2 + t + 1)$.

   (i) What is n?

   (ii) Find the rational canonical form of A.

4. Let $A \in M_4(Q)$ be the matrix

$$A = \left(\begin{array}{cc|cc} 2 & 1 & 0 & 0 \\ 0 & 2 & 0 & 0 \\ \hline 0 & 0 & 3 & 0 \\ 0 & 0 & 0 & 3 \end{array}\right).$$

(i) Find the invariant factors of A.

(ii) Find the rational canonical form of A.

11.3 Hamilton–Cayley theorem, minimal polynomial

Our next objective is to show that every square matrix is a 'root' of its characteristic polynomial: if $A \in \mathrm{M}_n(\mathrm{F})$, F a field, then $p_A(A) = 0$. The key to the proof is to look first at companion matrices:

11.3.1 Lemma *If* $p \in \mathrm{F}[t]$ *is a nonconstant monic polynomial and* C_p *is its companion matrix* (11.2.1), *then* $p(C_p) = 0$.

Proof. The claim is, so to speak, that the companion matrix of p is a 'root' of p. Let $D = C_p'$ be the transpose of C_p; since $(p(C_p))' = p(C_p') = p(D)$, it will suffice to show that $p(D) = 0$. {It is more convenient to work with the transpose because we are writing linear mappings on the *left* of vectors.}

Say $p = t^m + c_1 t^{m-1} + \ldots + c_{m-1}t + c_m$. Then

$$D = \begin{pmatrix} 0 & & & & -c_m \\ 1 & 0 & & & \vdots \\ & 1 & \ddots & & \vdots \\ & & \ddots & 0 & -c_2 \\ & & & 1 & -c_1 \end{pmatrix}.$$

Let V be an m-dimensional vector space over F, $x_1, \ldots, x_m$ a basis of V, and $T \in \mathcal{L}(\mathrm{V})$ the linear mapping whose matrix is D relative to the basis $x_1, \ldots, x_m$. The matrix of $p(T)$ relative to this basis is $p(\mathrm{mat}\, T) = p(D)$, so it will suffice to show that $p(T) = 0$.

By inspection of the matrix D,

$$Tx_i = x_{i+1} \quad \text{for} \quad i = 1, \ldots, m-1,$$
$$Tx_m = -c_m x_1 - c_{m-1}x_2 - \ldots - c_1 x_m.$$

It follows that

$$T^i x_1 = x_{i+1} \quad \text{for} \quad i = 0, 1, \ldots, m-1,$$

whereas

$$\begin{aligned} T^m x_1 = T(T^{m-1}x_1) &= Tx_m \\ &= -c_m x_1 - c_{m-1}x_2 - \ldots - c_1 x_m \\ &= -c_m I x_1 - c_{m-1}Tx_1 - \ldots - c_1 T^{m-1}x_1 \\ &= -(c_m I + c_{m-1}T + \ldots + c_1 T^{m-1})x_1, \end{aligned}$$

whence

$$(T^m + c_1 T^{m-1} + \ldots + c_{m-1}T + c_m I)x_1 = \theta,$$

in other words $p(T)x_1 = \theta$.

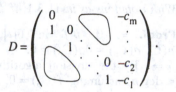

To show that $p(T)=0$ it is enough to show that $p(T)x_i=\theta$ for $i=1,\ldots,m$. For $i=1$ this is already verified. For $1<i\leq m$,

$$p(T)x_i = p(T)T^{i-1}x_1 = T^{i-1}p(T)x_1 = T^{i-1}\theta = \theta. \blacksquare$$

11.3.2 Lemma *With notations as in* 11.3.1, *if* $f\in \mathrm{F}[t]$ *is a nonzero polynomial with* $\deg f<\deg p$, *then* $f(C_p)\neq 0$.

Proof. We can suppose f is monic, say

$$f = t^k + a_1t^{k-1} + \ldots + a_{k-1}t + a_k,$$

where $0\leq k<m$. With notations as in the proof of 11.3.1, we have

$$\begin{aligned} f(T)x_1 &= T^kx_1 + a_1T^{k-1}x_1 + \ldots + a_{k-1}Tx_1 + a_kIx_1 \\ &= x_{k+1} + a_1x_k + \ldots + a_{k-1}x_2 + a_kx_1, \end{aligned}$$

so $f(T)x_1\neq\theta$ by the linear independence of $x_1,\ldots,x_{k+1}$. In particular, $f(T)\neq 0$; since $f(T)$ has matrix $f(D)$ relative to this basis, we conclude that $f(D)\neq 0$, therefore $f(C_p)\neq 0$. ■

11.3.3 Lemma *With notations as in* 11.3.1, *if* $f\in\mathrm{F}[t]$ *then* $f(C_p)=0 \Leftrightarrow p|f$.

Proof. $\Leftarrow$: If $p|f$ in $\mathrm{F}[t]$, say $f=pq$, then (by 11.3.1) $f(C_p)=p(C_p)q(C_p)=0q(C_p)=0$.

$\Rightarrow$: By the division algorithm, write $f=qp+r$, where $r=0$ or $\deg r<\deg p$; assuming $f(C_p)=0$, we are to show that $r=0$. We have

$$r(C_p) = f(C_p) - q(C_p)p(C_p) = 0 - q(C_p)0 = 0;$$

since $\deg r<\deg p$ is excluded by 11.3.2, the only way out is $r=0$. ■

11.3.4 Definition If $A\in\mathrm{M}_n(\mathrm{F})$ we write m_A for the minimal polynomial of A (11.1.13).

11.3.5 Theorem (Minimal polynomial) *Let* $A\in\mathrm{M}_n(\mathrm{F})$ *and let* m_A *be the minimal polynomial of* A. *For* $f\in\mathrm{F}[t]$,

$$f(A)=0 \Leftrightarrow m_A|f.$$

Thus, the set of all $f\in\mathrm{F}[t]$ *such that* $f(A)=0$ *is the principal ideal of* $\mathrm{F}[t]$ *generated by the minimal polynomial of* A.

Proof. Let $p_1,\ldots,p_r$ be the nontrivial invariant factors of A, where $p_i|p_{i+1}$ for $i=1,\ldots,r-1$; by definition, $m_A=p_r$. Let C be the rational canonical form of A (11.2.5), C_i the companion matrix of p_i; thus $C=\mathrm{diag}(C_1,\ldots,C_r)$ in the appropriate sense. Since $p_i|m_A$, by 11.3.3 we have $m_A(C_i)=0$ for $i=1,\ldots,r$, therefore $m_A(C)=0$.

By Theorem 11.2.4, A is similar to C, say $A=Q^{-1}CQ$. The mapping $B\mapsto Q^{-1}BQ$ is an automorphism of the ring $\mathrm{M}_n(\mathrm{F})$, in particular $Q^{-1}C^kQ=(Q^{-1}CQ)^k$ for $k=0,1,2,\ldots$, whence

$$m_A(A) = m_A(Q^{-1}CQ) = Q^{-1}m_A(C)Q = Q^{-1}0Q = 0;$$

thus A is a 'root' of its minimal polynomial.

$\Leftarrow$: If $m_A | f$ then $f(A) = 0$ is obvious from the preceding.
$\Rightarrow$: If $f(A) = 0$ then also $f(C) = f(QAQ^{-1}) = Qf(A)Q^{-1} = 0$, therefore $f(C_i) = 0$ for $i = 1, \ldots, r$; in particular, $f(C_r) = 0$, so $m_A | f$ by 11.3.3 ■

11.3.6 Corollary (Hamilton–Cayley theorem) *If $A \in M_n(F)$ then $p_A(A) = 0$.*

Proof. $m_A | p_A$. ■

When the field F is algebraically closed (8.3.10), the theory of 'diagonalizable' matrices fits nicely into the present circle of ideas. Suppose $A \in M_n(F)$, where F is algebraically closed. We know that if the characteristic polynomial p_A has distinct roots, then A is similar (over F) to a diagonal matrix (8.3.14). The converse is false: $A = cI$ is diagonal but the roots of $p_A = (t - c)^n$ are all equal. Thus, the implication

$$p_A \text{ has distinct roots} \Rightarrow A \text{ is similar to a diagonal matrix}$$

is true, but the characteristic polynomial is in general 'too big' for the reverse implication to hold. The minimal polynomial is just the right size:

11.3.7 Theorem *Let F be an algebraically closed field, $A \in M_n(F)$, m_A the minimal polynomial of A. The following conditions are equivalent:*

(a) *A is similar (over F) to a diagonal matrix;*

(b) *the roots of m_A are distinct.*

Proof. (a) $\Rightarrow$ (b): Some similar matrices have the same invariant factors, in particular the same minimal polynomial, we can assume that A is already diagonal. The diagonal entries of A are the roots of p_A, that is, the characteristic roots of A. Suppose $\lambda_1, \ldots, \lambda_s$ are the distinct characteristic roots of A; by another similarity transformation, we can suppose that equal entries on the diagonal are bunched together (cf. the proof of 11.2.4), say

$$A = \begin{pmatrix} \lambda_1 I_1 & & & \\ & \lambda_2 I_2 & & \\ & & \ddots & \\ & & & \lambda_s I_s \end{pmatrix},$$

where I_k is the identity matrix of order equal to the multiplicity of λ_k in p_A.

Let $q = (t - \lambda_1) \ldots (t - \lambda_s)$. It is clear that $q(A) = 0$, therefore $m_A | q$ by 11.3.5; since the roots of q are distinct, the same is true of m_A. {In fact, since all invariant factors of A divide m_A and since their product is p_A, m_A can omit no root of p_A; thus $m_A = q$.}

(b) $\Rightarrow$ (a): Every invariant factor of A is a divisor of m_A, hence has distinct roots. Let $p_1, \ldots, p_r = m_A$ be the nontrivial invariant factors of A, C_i the companion matrix of p_i, and $C = \operatorname{diag}(C_1, \ldots, C_r)$ the rational canonical form of A (cf. 11.2.4). Since C_i has characteristic polynomial p_i (11.2.3) and the roots of p_i are distinct, C_i is similar to a diagonal matrix by

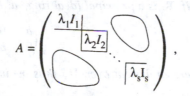

8.3.14. Say $P_i^{-1}C_iP_i$ is diagonal $(i = 1, \ldots, r)$; then $P = \operatorname{diag}(P_1, \ldots, P_r)$ is invertible and C is similar to the diagonal matrix $P^{-1}CP$, therefore so is A (11.2.4). ■

Note that the question of whether A is similar to a diagonal matrix is decidable in a finite number of steps: calculate $\mathrm{D_n}(tI - A) = |tI - A|$ and calculate $\mathrm{D_{n-1}}(tI - A)$ as in §10.4 (using Euclid's algorithm); the minimal polynomial is $m_A = |tI - A|/\mathrm{D_{n-1}}(tI - A)$. Then calculate the GCD of m_A and its derivative to see whether m_A has any repeated roots (Appendix B.5.4).

The rational canonical form shows that, under very special circumstances, the invariant factors of a block-diagonal matrix $\operatorname{diag}(C_1, \ldots, C_r)$ are the invariant factors of the diagonal blocks C_i. At the end of the section we give a very different kind of example where this is true, but let's observe that it isn't *always* true:

11.3.8 Example (Caution) Suppose $B \in \mathrm{M_m(F)}$, $C \in \mathrm{M_n(F)}$ and $A \in \mathrm{M_{m+n}(F)}$ is the matrix

$$A = \left(\begin{array}{c|c} B & 0 \\ \hline 0 & C \end{array}\right).$$

Let $f_1, \ldots, f_m$ be the invariant factors of B, and $g_1, \ldots, g_n$ the invariant factors of C. In general, $f_1, \ldots, f_m, g_1, \ldots, g_n$ are **not** the invariant factors of A. For example, if $B = (1)$ and $C = (2)$ are 1×1 matrices, and

$$A = \begin{pmatrix} 1 & 0 \\ 0 & 2 \end{pmatrix},$$

then $p_A = (t-1)(t-2)$ has distinct roots, so p_A is also the minimal polynomial, and the invariant factors of A are $1, p_A$; on the other hand, the unique invariant factor of B is $p_B = t - 1$ and that of C is $p_C = t - 2$.

Here's what's going on in the preceding example:

11.3.9 Theorem *If* R *is a principal ideal ring,* $a, b \in \mathrm{R}$ *and* $(a, b) = 1$, *then the matrices*

$$\begin{pmatrix} a & 0 \\ 0 & b \end{pmatrix}, \quad \begin{pmatrix} 1 & 0 \\ 0 & ab \end{pmatrix}$$

are equivalent over R, *thus the invariant factors*[1] *of*

$$\begin{pmatrix} a & 0 \\ 0 & b \end{pmatrix}$$

are $1, ab$.

Proof. The assumption is that a and b are relatively prime (Appendix B.3.6). Write

$$A = \begin{pmatrix} a & 0 \\ 0 & b \end{pmatrix}$$

[1] In the sense of Definition 10.2.4.

and let $D_i(A)$ $(i = 1, 2)$ be the determinantal divisors of A (10.4.1). Thus $D_2(A) = |A| = ab$ and since $(a, b) = 1$ it is clear that $D_1(A) = 1$. These are also the determinantal divisors of

$$B = \begin{pmatrix} 1 & 0 \\ 0 & ab \end{pmatrix},$$

therefore A and B are equivalent over R (10.4.9) and 1, ab are the invariant factors of A in the sense of 10.2.4. ■

11.3.10 Example Let R be a principal ideal ring, $P \in M_m(R)$, $Q \in M_n(R)$ matrices such that $(|P|, |Q|) = 1$. Let $S \in M_{m+n}(R)$ be the matrix

$$S = \left(\begin{array}{c|c} P & 0 \\ \hline 0 & Q \end{array}\right).$$

We can suppose $m \leq n$ (the positions of P and Q could be interchanged by suitable row and column permutations). Say

$$\begin{aligned} &P \text{ has invariant factors } p_1, \ldots, p_m, \quad p_i | p_{i+1}, \\ &Q \text{ has invariant factors } q_1, \ldots, q_n, \quad q_j | q_{j+i}, \end{aligned}$$

so that

$$\begin{aligned} P &\text{ equiv } \operatorname{diag}(p_1, \ldots, p_m), \\ Q &\text{ equiv } \operatorname{diag}(q_1, \ldots, q_n); \end{aligned}$$

then

$$\begin{aligned} S &\text{ equiv } \operatorname{diag}(p_1, \ldots, p_m, q_1, \ldots, q_n) \\ &\text{ equiv } \operatorname{diag}(p_1, \ldots, p_{m-1}, q_1, \ldots, q_{n-1}, p_m, q_n). \end{aligned}$$

Note that $(p_i, q_j) = 1$ for all i and j. Since $(p_m, q_n) = 1$, we have

$$\operatorname{diag}(p_m, q_n) \text{ equiv } \operatorname{diag}(1, p_m q_n)$$

by 11.3.9, therefore

$$\begin{aligned} S &\text{ equiv } \operatorname{diag}(p_1, \ldots, p_{m-1}, q_1, \ldots, q_{n-1}, 1, p_m q_n) \\ &\text{ equiv } \operatorname{diag}(1, p_1, \ldots, p_{m-1}, q_1, \ldots, q_{n-1}, p_m q_n). \end{aligned}$$

If $m = 1$ we are through:

$$S \text{ equiv } \operatorname{diag}(1, q_1, \ldots, q_{n-1}, p_1 q_n),$$

where each diagonal entry of the latter matrix divides the next, thus the diagonal entries are precisely the invariant factors of S.

If $m > 1$ we continue:

$$S \text{ equiv } \operatorname{diag}(1, p_1, \ldots, p_{m-2}, q_1, \ldots, q_{n-2}, p_{m-1}, q_{n-1}, p_m q_n)$$

and since $(p_{m-1}, q_{n-1}) = 1$ we have

$$\begin{aligned} S &\text{ equiv } \operatorname{diag}(1, p_1, \ldots, p_{m-2}, q_1, \ldots, q_{n-2}, 1, p_{m-1} q_{n-1}, p_m q_n) \\ &\text{ equiv } \operatorname{diag}(1, 1, p_1, \ldots, p_{m-2}, q_1, \ldots, q_{n-2}, p_{m-1} q_{n-1}, p_m q_n). \end{aligned}$$

This game can be played m times before we run out of p's (because $m \le n$ by assumption), resulting in

$$S \ \text{equiv} \ \operatorname{diag}(\underbrace{1, \ldots, 1}_{m}, q_1, \ldots, q_{n-m}, p_1 q_{n-m+1}, \ldots, p_m q_n);$$

since each diagonal entry of the latter matrix divides the next, the diagonal entries are the invariant factors of S.

For application in the next section, we specialize to the case that R is a polynomial ring F[t]:

11.3.11 Example Let $B \in \mathrm{M_m(F)}$, $C \in \mathrm{M_n(F)}$ and suppose $(p_B, p_C) = 1$. Let $A \in \mathrm{M_{m+n}(F)}$ be the matrix

$$A = \left(\begin{array}{c|c} B & 0 \\ \hline 0 & C \end{array}\right).$$

Assume $m \le n$ and apply the discussion of 11.3.10 to the characteristic matrix $tI - A \in \mathrm{M_{m+n}(F[}t\mathrm{])}$: If

$$B \text{ has invariant factors } p_1, \ldots, p_m \in \mathrm{F}[t], \qquad p_i | p_{i+1},$$

and

$$C \text{ has invariant factors } q_1, \ldots, q_n \in \mathrm{F}[t], \qquad q_j | q_{j+1},$$

then A has invariant factors

$$\underbrace{1, \ldots, 1}_{m}, q_1, \ldots, q_{n-m}, p_1 q_{n-m+1}, \ldots, p_m q_n.$$

▶ ***Exercises***

1. As in §8.3, Exercise 4, let $T \in \mathscr{L}(\mathrm{V})$, M a nontrivial invariant subspace for T, $R = T|\mathrm{M}$, and $S \in \mathscr{L}(\mathrm{V/M})$ the linear mapping induced by T in the quotient space V/M. Let m_T be the minimal polynomial of T (§11.1, Exercise 1). Then:

(i) m_R and m_S are divisors of m_T.

(ii) It can happen that $m_T \ne m_R m_S$.

(iii) It can happen that $m_T = m_R m_S$.

(iv) If p_R and p_S are relatively prime, then $m_T = m_R m_S$.

(v) It can happen that $m_T = m_R m_S$ and $m_R = m_S$.

{Hints:

(ii) Consider $\mathrm{V} = \mathrm{F}^2$, $T = I$, and M any one-dimensional linear subspace of V.

(iii) Consider $\mathrm{V} = \mathrm{F}^2$, $T \in \mathscr{L}(\mathrm{V})$ the linear mapping such that $Te_1 = e_1$, $Te_2 = \theta$.

(v) Consider $\mathrm{V} = \mathrm{F}^2$ and $Te_1 = e_1$, $Te_2 = e_1 + e_2$.}

2. Let F be any field, $A \in \mathrm{M_n(F)}$. The following conditions are equivalent:

(a) A is similar over F to a diagonal matrix; (b) $m_A = (t-c_1)\dots(t-c_r)$ with $c_1,\dots,c_r \in \mathrm{F}$ distinct.

{Hint: Re-examine the proof of Theorem 11.3.7. The crux of the matter is that Theorem 8.3.14 does not require that F be algebraically closed.}

3. A linear mapping $T \in \mathcal{L}(\mathrm{V})$ is diagonalizable (§8.2, Exercise 5) if and only if $m_T = (t-c_1)\dots(t-c_r)$ with $c_1,\dots,c_r \in \mathrm{F}$ distinct.

11.4 Elementary divisors, Jordan canonical form

We assume in this section (including the exercises) that the field F is *algebraically closed* (8.3.10; cf. the remarks in 8.3.11); thus every nonconstant polynomial $p \in \mathrm{F}[t]$ has a root in F, and p is the product of linear polynomials in $\mathrm{F}[t]$.

Suppose $A \in \mathrm{M}_n(\mathrm{F})$ has nontrivial invariant factors

$$p_1, \dots, p_r, \qquad p_i | p_{i+1};$$

the minimal polynomial of A is $m_A = p_r$. Suppose the characteristic polynomial p_A has s distinct roots (possibly multiple) $c_1, \dots, c_s$. Since $p_A = p_1 \dots p_r$, the roots of each p_i are among the c_j, and since the p_i are monic we have

$$\begin{aligned} p_1 &= (t-c_1)^{\alpha_{11}}(t-c_2)^{\alpha_{12}}\dots(t-c_s)^{\alpha_{1s}} \\ p_2 &= (t-c_1)^{\alpha_{21}}(t-c_2)^{\alpha_{22}}\dots(t-c_s)^{\alpha_{2s}} \\ &\dots \\ p_r &= (t-c_1)^{\alpha_{r1}}(t-c_2)^{\alpha_{r2}}\dots(t-c_s)^{\alpha_{rs}} \end{aligned}$$

for suitable nonnegative integers $\alpha_{ij} \in \mathbf{N}$. These integers have the following properties:

(1) For each $i = 1, \dots, r$,

$$\sum_{j=1}^{s} \alpha_{ij} = \deg p_i \geqslant 1 \quad \text{(the } p_i \text{ are } \textit{nontrivial}\text{)}.$$

(2) $$\sum_{i=1}^{r}\sum_{j=1}^{s} \alpha_{ij} = \sum_{i=1}^{r} \deg p_i = \deg p_A = n.$$

(3) For each $j = 1, \dots, s$,

$$\alpha_{1j} \leqslant \alpha_{2j} \leqslant \dots \leqslant \alpha_{rj} \quad (\text{because } p_i | p_{i+1}).$$

(4) For each $j = 1, \dots, s$, $\alpha_{rj} > 0$ (because c_j *is* a root of p_A, hence of m_A).

11.4.1 Definition With the preceding notations, the polynomials $(t-c_j)^{\alpha_{ij}}$ for which $\alpha_{ij} > 0$ are called the **elementary divisors** of A.

Thus, each nontrivial invariant factor is the product of certain of the elementary divisors, and the product of all the elementary divisors is the characteristic polynomial.

11.4.2 Example If A has nontrivial invariant factors

$$t^2 - 1,\ (t^2 - 1)(t - 2)^2$$

then $p_A = (t^2 - 1)^2(t - 2)^2$ has degree 6 (so A is a 6×6 matrix), the invariant factors of A are

$$p_1 = p_2 = p_3 = p_4 = 1, \quad p_5 = t^2 - 1, \quad p_6 = (t^2 - 1)(t - 2)^2,$$

and the elementary divisors are

$$t + 1, \quad t + 1, \quad t - 1, \quad t - 1, \quad (t - 2)^2.$$

11.4.3 Example If A has elementary divisors

$$(t + 5)^2, \quad t - 1, \quad (t - 1)^2, \quad (t - 1)^3, \quad t - 2, \quad (t - 2)^2,$$

their product p_A has degree 11 (thus A is 11×11). The invariant factors, in order of descending degree, are

$$\begin{aligned} p_{11} &= (t + 5)^2(t - 1)^3(t - 2)^2 \quad \text{(the minimal polynomial)} \\ p_{10} &= (t - 1)^2(t - 2) \\ p_9 &= t - 1 \\ p_8 &= \ldots = p_1 = 1. \end{aligned}$$

The invariant factors determine the elementary divisors (11.4.1), and it is clear from the preceding example that the elementary divisors determine the invariant factors (by 'reconsituting' the invariant factors, starting with the highest occurring powers of the distinct linear factors). The following theorem is therefore immediate from 11.1.8:

11.4.4 Theorem *Let* F *be an algebraically closed field. Matrices* A, $B \in \mathrm{M}_n(\mathrm{F})$ *are similar (over* F*) if and only if they have the same elementary divisors.*

Every matrix $A \in \mathrm{M}_n(\mathrm{F})$ is similar to its *rational canonical form* (11.2.4); the building blocks of the rational canonical form are the companion matrices of the nontrivial invariant factors, which arise from a special factorization of the characteristic polynomial.

The elementary divisors also constitute a special factorization of the characteristic polynomial. Again there is a special matrix to which A is similar, called the 'Jordan canonical form', in which the building block corresponding to an elementary divisor $(t - c)^m$ is the $m \times m$ matrix $J_m(c)$ defined as follows:

11.4.5 Definition If $c \in \mathrm{F}$ and m is a positive integer, we write

$$J_m(c) = \begin{pmatrix} c & 1 & & & \\ & c & 1 & & \\ & & \ddots & \ddots & \\ & & & c & 1 \\ & & & & c \end{pmatrix}$$

for the matrix obtained from the $m \times m$ scalar matrix $\operatorname{diag}(c, \ldots, c)$ by placing $m-1$ 1's just above the main diagonal. Matrices of this form are called *Jordan block* matrices. Convention: $J_1(c)$ is the 1×1 matrix (c).

11.4.6 Example If $c \in \mathrm{F}$ and

$$J_4(c) = \begin{pmatrix} c & 1 & 0 & 0 \\ 0 & c & 1 & 0 \\ 0 & 0 & c & 1 \\ 0 & 0 & 0 & c \end{pmatrix},$$

the determinant of the northeast minor of order 3 of $tI - J_4(c)$ is $(-1)^3$ (cf. 11.2.2), consequently $\mathrm{D}_3(tI - J_4(c)) = 1$. It follows that $tI - J_4(c)$ is equivalent over $\mathrm{F}[t]$ to

$$\operatorname{diag}(1, 1, 1, (t-c)^4),$$

so that $(t-c)^4$ serves as characteristic polynomial, minimal polynomial and only elementary divisor of $J_4(c)$. The argument is general:

11.4.7 Lemma *For a Jordan block matrix $J_{\mathrm{m}}(c)$, $(t-c)^{\mathrm{m}}$ is the characteristic polynomial, minimal polynomial and only elementary divisor, and $tI - J_{\mathrm{m}}(c)$ is equivalent over $\mathrm{F}[t]$ to $\operatorname{diag}(1, \ldots, 1, (t-c)^{\mathrm{m}})$.*

11.4.8 Example If

$$B = \left(\begin{array}{c|c} J_3(c) & 0 \\ \hline 0 & J_4(c) \end{array}\right),$$

then $tI - B$ is equivalent over $\mathrm{F}[t]$ to

$$\operatorname{diag}(1, 1, (t-c)^3, 1, 1, 1, (t-c)^4),$$

hence to

$$\operatorname{diag}(1, 1, 1, 1, 1, (t-c)^3, (t-c)^4).$$

Thus $(t-c)^3$, $(t-c)^4$ are the nontrivial invariant factors, as well as the elementary divisors, of B. More generally:

11.4.9 Lemma *If $\alpha_1 \le \alpha_2 \le \ldots \le \alpha_s$ are positive integers and $c \in \mathrm{F}$, and if*

$$B = \begin{pmatrix} J_{\alpha_1}(c) & & & \\ & J_{\alpha_2}(c) & & \\ & & \ddots & \\ & & & J_{\alpha_s}(c) \end{pmatrix},$$

then $tI - B$ is equivalent over $\mathrm{F}[t]$ to

$$\operatorname{diag}(1, \ldots, 1, (t-c)^{\alpha_1}, \ldots, (t-c)^{\alpha_s})$$

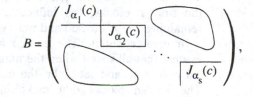

(the number of 1*'s is* $\alpha_1 + \ldots + \alpha_s - s$*), thus* $(t - c)^{\alpha_1}, \ldots, (t - c)^{\alpha_s}$ *are the nontrivial invariant factors, as well as the elementary divisors, of* B*.*

Proof. A trivial modification of the argument in the preceding example. ■

11.4.10 Example If $c, d \in \mathrm{F}$, $c \neq d$, and if

$$B = \left(\begin{array}{c|c} J_2(c) & 0 \\ \hline 0 & J_3(d) \end{array}\right),$$

then, by 11.4.7 and 11.3.11, the invariant factors of B are

$$1, 1, 1, 1, (t - c)^2(t - d)^3,$$

thus the elementary divisors of B are $(t - c)^2, (t - d)^3$.

The message of 11.4.6–11.4.10: when we form a 'direct sum' of Jordan blocks (with equal or distinct characteristic roots) and calculate the elementary divisors of the direct sum, there are no surprises. The general theorem:

11.4.11 Theorem (Jordan canonical form) *Let* F *be algebraically closed,* $A \in \mathrm{M}_n(\mathrm{F})$*, and suppose* A *has elementary divisors*

$$(t - c_1)^{k_1}, (t - c_2)^{k_2}, \ldots, (t - c_r)^{k_r}$$

(the c_i *need not be distinct and the positive integers* k_i *need not be in any special order). Then* A *is similar over* F *to the matrix*

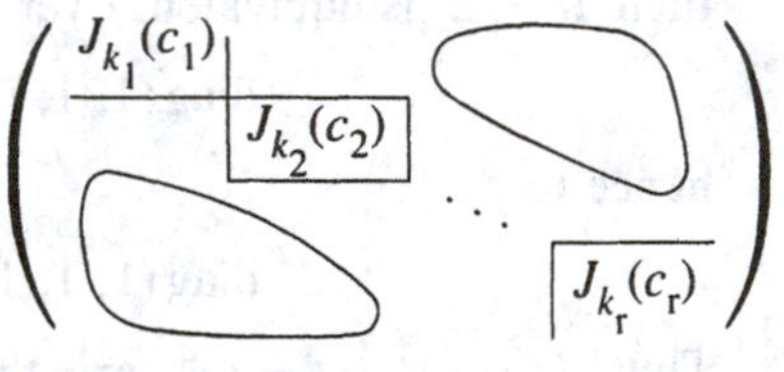

Proof. Write T for this matrix; in view of Theorem 11.4.4, we need only show that T has the same elementary divisors as A. The idea of the proof (by complete induction on r): reordering the c_i, we can suppose that $c_1 = \ldots = c_s = c$ and $c_i \neq c$ for $i > s$. Apply 11.4.9 to the direct sum B of the first s blocks, the induction hypothesis to the direct sum C of the remaining $s - r$ blocks, and put everything together with 11.3.11.

In greater detail: The case $r = 1$ is covered by 11.4.7; let $r \geq 2$ and assume that all's well when the number of blocks is $< r$.

Let $c = c_1$ and let s be the number of occurrences of c among the c_i. The Jordan blocks that make up T can be permuted (by a similarity transformation, as in the proof of 11.2.4) so that these s blocks come first, in order of increasing multiplicities of c. The matrix T can then be organized as

$$T = \begin{pmatrix} B & 0 \\ 0 & C \end{pmatrix},$$

where B contains the first s blocks and C contains the rest. By 11.4.9, B has elementary divisors

$$(t - c)^{k_1}, (t - c)^{k_2}, \ldots, (t - c)^{k_s}$$

and by the induction hypothesis the elementary divisors of C are the remaining elementary divisors of A, namely

$$(t - c_i)^{k_i} \quad (i > s).$$

Clearly $(p_B, p_C) = 1$, so the discussion of 11.3.11 is applicable: factorization of the invariant factors shows that the elementary divisors of T are those of B together with those of C, in other words, T and A have the same elementary divisors. ■

11.4.12 Definition With notations as in 11.4.11 and its proof, the matrix T is called the **Jordan canonical form** of A.

In general there is no preferred ordering of the characteristic roots, so 'the' Jordan canonical form is not 100% unique: it is defined only up to a permutation of the constitutent blocks. {The 1×1 blocks can be organized into a diagonal matrix in the northwest corner. The remark following 11.2.6 about discerning the blocks of order ≥ 2 is applicable here as well.} Thus, when we say that matrices A and B have the 'same' Jordan canonical form, we mean that if Theorem 11.4.11 is applied to each of A and B, the corresponding matrices T can be obtained from one another by a permutation of the blocks along the diagonal. (Note that such a permutation of blocks is itself a similarity transformation.)

11.4.13 Theorem *Assume* F *is algebraically closed. Matrices* $A, B \in \mathrm{M_n(F)}$ *are similar if and only if they have the 'same' Jordan canonical form.*

Proof. If A and B have the same Jordan canonical form T (see the preceding remarks), then they are both similar to T (11.4.11) hence to each other. Conversely, if A and B are similar then they have the same elementary divisors (11.4.4) hence the same Jordan canonical form (11.4.12). ■

11.4.14 Remark The Jordan canonical form gives an alternative (and excessively difficult!) proof of Theorem 8.3.12: If F is algebraically closed, then every matrix $A \in \mathrm{M_n(F)}$ is similar over F to a triangular matrix.

▶ *Exercises*

1. If $T \in \mathcal{L}(\mathrm{V})$ and A is the matrix of T relative to any basis of V, the *elementary divisors* of T are defined to be those of A (we are assuming that the field F of scalars is algebraically closed). Then T is diagonalizable (§8.2, Exercise 5) if and only if its elementary divisors are linear.
   {Hint: §11.3, Exercise 3.}

2. (i) With notations as in Definition 11.4.5, $J_m(c) = cI + N$, where $N^m = 0$. (A square matrix is said to be *nilpotent* if some positive integral power of it is 0.)

(ii) Every $A \in M_n(F)$ is similar to a matrix of the form $D + N$, where D is diagonal, N is nilpotent and $DN = ND$.

3. Let $A = J_m(0)$, $m \geq 2$, and let $\operatorname{adj} A$ be the adjoint matrix of A (Definition 7.4.3).

(i) A^i has rank $m - i$ for $i = 1, \ldots, m$; in particular, $A^m = 0$ and A has rank $m - 1$.

(ii) $\operatorname{adj} A$ has rank 1.

(iii) $\operatorname{adj}(A^2) = 0$.

(iv) If $m \geq 3$ then, in the notation of §7.4, Exercise 6, $\rho(A^2) + \rho(\operatorname{adj} A^2) = m - 2 < m$, thus the inequality in (ii) of the cited exercise can be strict for a nonzero matrix.

11.5 Appendix: proof that $M_n(F)[t] = M_n(F[t])$

Let S be a ring with unity, not necessarily commutative; the ring of polynomials with coefficients in S can be defined as follows. Consider the set X of all '∞-ples' (sequences)

$$x = (a_0, a_1, \ldots, a_m, 0, 0, \ldots)$$

with coordinates in S and zero from some index onward. Equality in X means coordinatewise equality, and sums are defined coordinatewise: if $x = (a_0, a_1, a_2, \ldots)$ and $y = (b_0, b_1, b_2, \ldots)$, then

$$x + y = (a_0 + b_0, a_1 + b_1, a_2 + b_2, \ldots).$$

For $a \in S$ and $x = (a_0, a_1, a_2, \ldots) \in X$ we define a sort of 'scalar multiple'

$$ax = (aa_0, aa_1, aa_2, \ldots);$$

the axioms (1)–(9) of Definition 1.3.1 (for a vector space) are verified, except that S is not assumed to be a field. In particular, the zero element of X is the sequence $(0, 0, 0, \ldots)$ with all entries zero, and the negative of $(a_0, a_1, a_2, \ldots)$ is $(-a_0, -a_1, -a_2, \ldots)$. {The technical term is that X is a unital *left* S-*module*[1]; here 'left' refers to the fact that elements of S are written on the left of elements of X when forming scalar multiples, and 'unital' refers to the property $1x = x$.} Products in X are defined by

$$xy = (a_0b_0, a_0b_1 + a_1b_0, a_0b_2 + a_1b_1 + a_2b_0, \ldots)$$

(the ∞-ple terminates in zeros, hence belongs to X); this makes X a ring with unity element

$$u = (1, 0, 0, \ldots).$$

[1] S. MacLane and G. Birkhoff, *Algebra*, 3rd. edn, Chelsea, 1988, p. 160.

Defining $t = (0, 1, 0, 0, \ldots)$, we see that

$$t^2 = (0, 0, 1, 0, \ldots), \quad t^3 = (0, 0, 0, 1, 0, \ldots)$$

and so on. Write $u = t^0$ to round out the definition of powers, and denote this element simply 1. For an arbitrary element $x = (a_0, a_1, a_2, \ldots)$, we have

$$\begin{aligned} x &= a_0(1, 0, \ldots) + a_1(0, 1, 0, \ldots) + a_2(0, 0, 1, 0, \ldots) + \ldots \\ &= a_0 1 + a_1 t + a_2 t^2 + \ldots . \end{aligned}$$

The preferred notation for X is S[t], and its elements are called **polynomials** with coefficients in S. When S is noncommutative, so is S[t]; nevertheless, t commutes with every polynomial, thus the elements t^n ($n \in \mathrm{N}$) belong to the *center* of the polynomial ring S[t].

Polynomial rings S[t] are invented so that ring elements can be 'substituted' for t. Suppose T is a ring with unity element 1, containing S as a subring (with same unity element), and let z be an element of the center of T; then there exists a unique ring homomorphism $S[t] \to T$ for which $1 \mapsto 1$ and $t \mapsto z$, explicitly,

$$a_0 1 + a_1 t + \ldots + a_m t^m \mapsto a_0 + a_1 z + \ldots + a_m z^m .$$

The proof is tedious but elementary and straightforward.

Let's get back down to earth: how shall we make sense of the equation $M_n(R)[t] = M_n(R[t])$, R any ring with unity? (For our application, we need only consider $R = F$.) Let $S = M_n(R)$, $T = M_n(R[t])$; our objective is to 'identify' S[t] with T. View R as a subring of R[t] in the natural way; then $M_n(R)$ is a subring of $M_n(R[t]) = T$. The matrix

$$tI = \operatorname{diag}(t, \ldots, t)$$

belongs to the center of $M_n(R[t])$. It follows from the preceding discussion, with $z = tI$, that there exists a unique ring homomorphism $\Phi: M_n(R)[t] \to M_n(R[t])$ such that $\Phi(I1) = I$ and $\Phi(t) = tI$; explicitly,

$$\begin{aligned} \Phi(A_0 1 + A_1 t + \ldots + A_m t^m) &= A_0 + A_1(tI) + \ldots + A_m(tI)^m \\ &= A_0 + (tI)A_1 + \ldots + (tI)^m A_m \\ &= A_0 + tA_1 + \ldots + t^m A_m . \end{aligned}$$

{In the last expression on the right, we are regarding $M_n(R[t])$ as a left R[t]-module in the natural way: if $P = (p_{ij}) \in M_n(R[t])$ and $q \in R[t]$, then $qP = (qp_{ij})$.} It is straightforward to check that Φ is an isomorphism of rings (see the discussion surrounding the formula (*) in the proof of 11.1.3).

12

Unitary spaces

A unitary space is, so to speak, a 'complex Euclidean space', that is, a finite-dimensional complex vector space with a suitable (complex-valued) 'inner product'. At the outset, the theory of complex inner product spaces resembles superficially its real-valued cousin; echoes of the results proved for real spaces in Chapters 5 and 9 are everywhere apparent in the present chapter. However, the deeper study of linear mappings in the complex case takes a markedly different turn from the real case[1], due mainly to the properties of complex conjugation in the complex field **C** and the fact that **C** is algebraically closed (8.3.11).

The conjugate of a complex number c will also be denoted by c^*; thus, if $c = a + bi$ (a and b real numbers, $i^2 = -1$), then $c^* = \bar{c} = a - bi$.

12.1 Complex inner product spaces, unitary spaces

12.1.1 Definition A **complex inner product space** (briefly, complex IP-space) is a complex vector space E equipped with a function $\mathrm{E} \times \mathrm{E} \rightarrow \mathbf{C}$ that assigns, to each ordered pair (x, y) of vectors of E, a complex number called the **inner**

[1] Compare, for example, the simplicity of the structure of 'normal' linear mappings in the complex case (Theorem 12.7.4) with the analogous result in Euclidean spaces (§9.4, Exercise 7).

product of x and y (in that order), denoted $(x|y)$, having the following properties: for $x, y, z \in \mathrm{E}$ and $c \in \mathbf{C}$,

(1) $x \neq \theta \Rightarrow (x|x) > 0$ (strict positivity)

(2) $(x|y)^* = (y|x)$ (conjugate symmetry)

(3) $(x + y|z) = (x|z) + (y|z)$

(4) $(cx|z) = c(x|z)$.

A finite–dimensional complex inner product space is called a **unitary space**. Convention: $\{\theta\}$ is admitted as a 0-dimensional unitary space, with $(\theta|\theta) = 0$.

Conditions (3), (4) of Definition 12.1.1 say that for each vector $z \in \mathrm{E}$, the function $x \mapsto (x|z)$ is a linear form on E.

12.1.2 Example $\mathbf{C}^{\mathrm{n}}$ is an n-dimensional unitary space for the **canonical inner product**

$$(x|y) = \sum_{i=1}^{n} a_{\mathrm{i}} b_{\mathrm{i}}^*,$$

where $x = (a_1, \ldots, a_{\mathrm{n}})$, $y = (b_1, \ldots, b_{\mathrm{n}})$. {It would be nice to avoid using i as an index in contexts involving complex numbers, but the price of abandoning notations such as a_{ij} for matrix entries is too high.}

12.1.3 Example Let T be a nonempty set, $\mathcal{F}(\mathrm{T}, \mathbf{C})$ the vector space of all functions $x{:}\mathrm{T} \rightarrow \mathbf{C}$ (with the pointwise linear operations), and E the set of all $x \in \mathcal{F}(\mathrm{T}, \mathbf{C})$ whose 'support' $\{t \in \mathrm{T}{:}\ \ x(t) \neq 0\}$ is finite. Then E is a complex IP-space for the inner product

$$(x|y) = \sum_{t \in \mathrm{T}} x(t) y(t)^*$$

(a finite sum, even if T is infinite). When $\mathrm{T} = \{1, \ldots, n\}$ we recapture the unitary space $\mathbf{C}^{\mathrm{n}}$; when T is infinite, E is infinite–dimensional (hence is not a unitary space).

12.1.4 Example $\mathrm{M}_{\mathrm{n}}(\mathbf{C})$ is an n^2-dimensional unitary space for the inner product

$$(A|B) = \mathrm{tr}\,(AB^*),$$

where $B^* = (b_{\mathrm{ij}}^*)'$ is the **conjugate-transpose** of $B = (b_{\mathrm{ij}})$, that is, $B^* = (c_{\mathrm{ij}})$ with $c_{\mathrm{ij}} = b_{\mathrm{ji}}^*$. Explicitly,

$$(A|B) = \sum_{i,j=1}^{n} a_{\mathrm{ij}} b_{\mathrm{ij}}^*, \quad (A|A) = \sum_{i,j=1}^{n} |a_{\mathrm{ij}}|^2,$$

thus $\mathrm{M}_{\mathrm{n}}(\mathbf{C})$ can be identified with the unitary space $\mathbf{C}^{n^2}$.

12.1.5 Example Let $\mathrm{E} = \mathcal{C}([a, b], \mathbf{C})$ be the vector space of all continuous functions $x{:}[a, b] \rightarrow \mathbf{C}$, with the pointwise linear operations, and define inner products by the formula

$$(x|y) = \int_a^b x(t) y(t)^* dt;$$

E is an infinite-dimensional complex IP-space (assuming $a < b$), hence is not a unitary space.

12.1.6 Theorem *In a complex* IP-*space*,

(5) $(x|y + z) = (x|y) + (x|z)$,

(6) $(x|cy) = c^*(x|y)$,

(7) $(x|\theta) = (\theta|y) = 0$.

Proof. Similar to that of 5.1.4. For example,

$$(x|cy) = (cy|x)^* = (c(y|x))^* = c^*(y|x)^* = c^*(x|y). \blacksquare$$

12.1.7 Definition Let E be a complex inner product space. For a vector $x \in$ E, the **norm** of x is the number $\|x\|$ defined by the formula $\|x\| = (x|x)^{1/2}$. If $\|x\| = 1$, x is called a **unit vector**. For vectors $x, y \in$ E, the number $\|x - y\|$ is called the **distance** from x to y.

12.1.8 Theorem *In a complex* IP-*space*,

(8) $\|\theta\| = 0$; $\|x\| > 0$ *when* $x \neq \theta$;

(9) $\|cx\| = |c| \cdot \|x\|$,

(10) (*Polarization identity*)

$$(x|y) = \tfrac{1}{4}\{\|x + y\|^2 - \|x - y\|^2 + i\|x + iy\|^2 - i\|x - iy\|^2\},$$

(11) (*Parallelogram law*)

$$\|x + y\|^2 + \|x - y\|^2 = 2\|x\|^2 + 2\|y\|^2,$$

(12) $|(x|y)| \leq \|x\| \, \|y\|$ (*Cauchy–Schwarz inequality*),

(13) $\|x + y\| \leq \|x\| + \|y\|$ (*Triangle inequality*).

Proof. (8) Obvious.

(9) $\|cx\|^2 = (cx|cx) = cc^*(x|x) = |c|^2\|x\|^2$.

(10) By (3)–(5),

(a) $$\|x + y\|^2 = \|x\|^2 + \|y\|^2 + (x|y) + (y|x).$$

Replace y by $-y$ in (a) and multiply by -1:

(b) $$-\|x - y\|^2 = -\|x\|^2 - \|y\|^2 + (x|y) + (y|x).$$

Replace y by iy in (a) and multiply by i:

(c) $$i\|x + iy\|^2 = i\|x\|^2 + i\|y\|^2 + (x|y) - (y|x).$$

Replace y by $-y$ in (c) and multiply by -1:

(d) $$-i\|x - iy\|^2 = -i\|x\|^2 - i\|y\|^2 + (x|y) - (y|x).$$

Addition of (a)–(d) yields $4(x|y)$ on the right side.

(11) Subtract (b) from (a).

(12) The inequality is trivial if $(x|y) = 0$.

Assuming $(x|y) \neq 0$, x and y are nonzero vectors. Let $u = \|x\|^{-1}x$, $v = \|y\|^{-1}y$; then u and v are unit vectors and it will suffice to show that $|(u|v)| \leq 1$. We can suppose that $(u|v)$ is real; for, $c = |(u|v)|/(u|v)$ is a complex number of absolute value 1 such that $|(u|v)| = c(u|v) = (cu|v)$ and $\|cu\| = \|v\| = 1$. The polarization formula then reduces to

$$(u|v) = \tfrac{1}{4}\{\|u + v\|^2 - \|u - v\|^2\},$$

whence

$$\begin{aligned} |(u|v)| &\leq \tfrac{1}{4}\{\|u + v\|^2 + \|u - v\|^2\} \\ &= \tfrac{1}{4}\{2\|u\|^2 + 2\|v\|^2\} = \tfrac{1}{4}\{2 + 2\} = 1 \end{aligned}$$

by the triangle inequality in $\mathbf{R}$ and the parallelogram law.

(13) Since

$$(x|y) + (y|x) = 2\,\mathrm{Re}\,(x|y) \leq 2|(x|y)| \leq 2\|x\|\,\|y\|$$

by the Cauchy–Schwarz inequality, it follows from (a) that

$$\|x + y\|^2 \leq \|x\|^2 + \|y\|^2 + 2\|x\|\,\|y\| = (\|x\| + \|y\|)^2. \blacksquare$$

12.1.9 Corollary *In a complex* IP-*space*, $|(x|y)| = \|x\|\,\|y\| \;\Leftrightarrow\;$ *x, y are linearly dependent.*

Proof. $\Rightarrow$: Same as in 5.1.8.

$\Leftarrow$: With notations as in 5.1.8, we can suppose that $(u|v) = 1$ (as in the above proof of the Cauchy–Schwarz inequality, replace u by cu for a suitable complex number of absolute value 1); then also $(v|u) = 1$ and the proof that $u - v = \theta$ continues as in 5.1.8. ■

12.1.10 Definition For each vector y in a complex IP-space E, we write y' for the linear form on E defined by $y'(x) = (x|y)$. Thus $y' \in \mathrm{E}'$ (the dual space of E).

12.1.11 Lemma *If* E *is a complex* IP-*space, then the mapping* $\mathrm{E} \to \mathrm{E}'$ *defined by* $y \mapsto y'$ *is injective and conjugate-linear*: $(x + y)' = x' + y'$, $(cx)' = c^*x'$.

Proof. The only change from 5.2.2 is the computation

$$(cy)'(x) = (x|cy) = c^*(x|y) = c^*y'(x) = (c^*y')(x). \blacksquare$$

12.1.12 Corollary *If* T *is a nonempty set*, E *is a complex* IP-*space, and* $f\colon \mathrm{T} \to \mathrm{E}$, $g\colon \mathrm{T} \to \mathrm{E}$ *are functions such that* $(f(t)|x) = (g(t)|x)$ *for all* $t \in \mathrm{T}$ *and* $x \in \mathrm{E}$, *then* $f = g$.

Proof. Same as 5.2.3. ■

In contrast with the real case (5.2.4):

12.1.13 Theorem *If* E *is a complex* IP-*space and* $S, T \in \mathcal{L}(\mathrm{E})$ *are linear mappings such that* $(Sx|x) = (Tx|x)$ *for all* $x \in \mathrm{E}$, *then* $S = T$.

Proof. For any linear mapping $R \in \mathcal{L}(\mathrm{E})$, the expression $(Rx|y)$ is a linear function of x and a conjugate–linear function of y; the same argument used in proving the polarization identity shows that

$$4(Rx|y) = (R(x+y)|x+y) - (R(x-y)|x-y) \\ + i(R(x+iy)|x+iy) - i(R(x-iy)|x-iy).$$

It follows that if $(Rx|x) = 0$ for all x, then $(Rx|y) = 0$ for all x and y, consequently $R = 0$ (cf. 12.1.12).

For the theorem at hand, we need only apply the preceding remark to $R = S - T$. ■

12.1.14 Definition With notations as in 12.1.11, we write $J_{\mathrm{E}}:\mathrm{E} \to \mathrm{E}'$ for the injective conjugate-linear mapping (called the *canonical mapping* of E into E′) defined by $J_{\mathrm{E}}y = y'$.

12.1.15 Theorem (Self-duality of unitary spaces) *If* E *is a unitary space, then the conjugate-linear mapping* $J_{\mathrm{E}}:\mathrm{E} \to \mathrm{E}'$ *is bijective.*

Proof (essentially the same as Theorem 5.2.5). The mapping $\mathrm{E} \to \mathrm{E}'$ in question preserves linear independence (because it is injective and conjugate-linear) and its range is a linear subspace of E′; the image of a basis of E must therefore be a basis of E′ (3.5.13). ■

▶ ***Exercises***

1. (i) If $\mathrm{E}_1, \ldots, \mathrm{E}_n$ are complex inner product spaces and $\mathrm{E} = \mathrm{E}_1 \times \ldots \times \mathrm{E}_n$ is the product vector space (1.3.11), then the formula

$$(x|y) = (x_1|y_1) + \ldots + (x_n|y_n),$$

for $x = (x_1, \ldots, x_n)$, $y = (y_1, \ldots, y_n)$ in E, defines an inner product on E.

(ii) If, moreover, the E_k are unitary spaces, then so is E.

2. For nonzero vectors x, y in a complex inner product space, $\|x+y\| = \|x\| + \|y\|$ if and only if $y = cx$ with $c > 0$.

{Hint: If $\|x+y\| = \|x\| + \|y\|$, inspection of the proof of the triangle inequality in 12.1.8 shows that $\mathrm{Re}\,(x|y) = |(x|y)| = \|x\|\,\|y\|$. It follows that $(x|y) = \mathrm{Re}\,(x|y) > 0$ and that x, y are linearly dependent.}

12.2 Orthogonality

12.2.1 Definition Vectors x, y in a complex inner product space are said to be **orthogonal**, written $x \perp y$, if $(x|y) = 0$.

Note that $(x|y) = y'(x)$, thus $x \perp y \Leftrightarrow y'(x) = 0 \Leftrightarrow x \in \mathrm{Ker}\, y'$. Since $(y|x) = (x|y)^*$, it is clear that $x \perp y \Leftrightarrow y \perp x$. Also, $x \perp x \Leftrightarrow x = \theta$; and if $x \perp y_i$ for $i = 1, \ldots, n$ then $x \perp y$ for every linear combination $y = c_1y_1 + \ldots + c_ny_n$.

12.2.2 Theorem (Pythagorean theorem) *In a complex* IP-*space,* $x \perp y \Rightarrow \|x+y\|^2 = \|x\|^2 + \|y\|^2$.

Proof. Immediate from formula (a) in the proof of 12.1.8. ■

Note that in a complex IP-space,

$$\begin{aligned}\|x+y\|^2 &= \|x\|^2 + (x|y) + (y|x) + \|y\|^2\\ &= \|x\|^2 + \|y\|^2 + (x|y) + (x|y)^*\\ &= \|x\|^2 + \|y\|^2 + 2\,\mathrm{Re}\,(x|y),\end{aligned}$$

thus the implication of 12.2.2 is not reversible (as it was for real IP-spaces in 5.1.10).

12.2.3 Definition Subsets A and B of a complex IP-space E are said to be **orthogonal**, written $A \perp B$, if $x \perp y$ for all $x \in A$ and $y \in B$. If $\{x\} \perp B$ we also write $x \perp B$.

The set of all vectors $x \in E$ such that $x \perp A$ is denoted $A^\perp$ (verbalized 'A perp') and is called the (orthogonal) **annihilator** of A in E; the annihilator of $A^\perp$ is denoted $A^{\perp\perp}$, thus $A^{\perp\perp} = (A^\perp)^\perp$.

Convention: $\varnothing^\perp = E$.

12.2.4 Theorem *For subsets* A, B *of a complex* IP-*space* E,

(1) $A^\perp$ *is a linear subspace of* E,

(2) $A \subset A^{\perp\perp}$,

(3) $A \subset B \Rightarrow B^\perp \subset A^\perp$,

(4) $(A^{\perp\perp})^\perp = (A^\perp)^{\perp\perp} = A^\perp$.

(5) $A \cap A^\perp \subset \{\theta\}$.

Proof. Same as 5.1.14. ■

12.2.5 Theorem. (Projection theorem) *If* M *is a finite-dimensional linear subspace of a complex* IP-*space* E, *then* $E = M \oplus M^\perp$.

Proof. Same as 5.2.7 (for the notation, see 9.1.1). ■

12.2.6 Corollary *If* E *is a unitary space and* M *is a linear subspace of* E, *then* $\dim M^\perp = \dim E - \dim M$.

Proof. Same as 5.2.8. ■

12.2.7 Corollary *If* M *is a linear subspace of a unitary space, then* $M = M^{\perp\perp}$.

Proof. Same as 5.2.11. ■

12.2.8 Definition With notations as in 12.2.6, $M^\perp$ is called the **orthogonal complement** of M in E.

12.2.9 Corollary *If* A *is a subset of a unitary space, then the linear span of* A *is* $A^{\perp\perp}$, *that is,* $[A] = A^{\perp\perp}$.

Proof. Same as 5.2.12. ■

Exercises

1. Let E be a unitary space, M a linear subspace of E. Define a linear mapping $P:E \to E$ as follows: for every $x \in E$, write $x = y + z$ with (unique) $y \in M$ and $z \in M^{\perp}$ (Theorem 12.2.5) and define $Px = y$; thus, Px is the unique vector in M such that $x - Px \in M^{\perp}$. Prove:

   (i) P is linear, that is, $P \in \mathcal{L}(E)$;

   (ii) $P^2 = P$;

   (iii) $\operatorname{Im} P = M$ and $\operatorname{Ker} P = M^{\perp}$;

   (iv) $(Px|y) = (x|Py)$ for all $x, y \in E$.

   Writing $P = P_M$ to indicate the dependence of P on M (P_M is called the **projection** with range M), we have

   (v) $P_{M^{\perp}} = I - P_M$.

2. Let E be a unitary space, M and N linear subspaces of E. With notations as in Exercise 1,

   (i) $M \subset N \Leftrightarrow P_M = P_N P_M$;

   (ii) $M \perp N \Leftrightarrow P_M P_N = 0$.

3. If M and N are linear subspaces of a unitary space E, then $(M + N)^{\perp} = M^{\perp} \cap N^{\perp}$ and $(M \cap N)^{\perp} = M^{\perp} + N^{\perp}$.

12.3 Orthonormal bases, isomorphism

12.3.1 Definition A list of vectors $x_1, \ldots, x_n$ in a complex inner product space is said to be **orthogonal** if $x_i \perp x_j$ for all $i \neq j$; the list is said to be **orthonormal** if it is orthogonal and if every x_i is a unit vector, in other words $(x_i|x_j) = \delta_{ij}$ for all i and j.

12.3.2 Theorem *If $x_1, \ldots, x_n$ is an orthogonal list of nonzero vectors in a complex inner product space, then the x_i are linearly independent.*

Proof. Same as 5.1.12. ■

12.3.3 Theorem *Every unitary space $\neq \{\theta\}$ has an orthonormal basis.*

Proof. Let E be a unitary space of dimension n. The proof is by induction on n. If $n = 1$ there is nothing to prove (by convention, any unit vector will serve as an 'orthonormal basis of length 1'). Let $n \geqslant 2$ and assume that all's well for $n - 1$.

Let u_n be any unit vector, $M = Cu_n$ the 1-dimensional subspace spanned by u_n. Then $M^{\perp}$ is a unitary space of dimension $n - 1$ (12.2.6); by the induction hypothesis, $M^{\perp}$ has an orthonormal basis $u_1, \ldots, u_{n-1}$, whence the desired orthonormal basis $u_1, \ldots, u_{n-1}, u_n$ of E. ■

12.3.4 Corollary *If E and F are unitary spaces of the same dimension, then there exists a linear bijection $T:E \to F$ such that $(Tx|Ty) = (x|y)$ for all x, y in E.*

Proof. Say $\dim \mathrm{E} = \dim \mathrm{F} = n$. Let $u_1, \ldots, u_n$ and $v_1, \ldots, v_n$ be orthonormal bases of E and F, respectively, and let $T:\mathrm{E} \to \mathrm{F}$ be the bijective linear mapping such that $Tu_i = v_i$ for all i (cf. 3.8.1, 3.7.4). If $x, y \in \mathrm{E}$, say

$$x = \sum_{i=1}^{n} a_i u_i, \quad y = \sum_{i=1}^{n} b_i u_i,$$

then

$$(Tx|Ty) = \left(\sum_{i=1}^{n} a_i v_i \,\middle|\, \sum_{i=1}^{n} b_i v_i\right)$$

$$= \sum_{i,j} a_i b_j^* (v_i|v_j)$$

$$= \sum_{i,j} a_i b_j^* \delta_{ij} = (x|y). \blacksquare$$

12.3.5 Definition Complex inner product spaces E, F are said to be **isomorphic**, written $\mathrm{E} \cong \mathrm{F}$, if there exists a linear bijection $T:\mathrm{E} \to \mathrm{F}$ that preserves inner products, that is, $(Tx|Ty) = (x|y)$ for all x, y in E; such a mapping T is called an **isomorphism** of complex inner product spaces.

12.3.6 Corollary *If* E *is a unitary space of dimension* n, *then* $\mathrm{E} \cong \mathrm{C}^n$ *as complex inner product spaces.*

Proof. Immediate from 12.3.4. ■

12.3.7 Definition Let E be a complex IP-space. A linear mapping $T:\mathrm{E} \to \mathrm{E}$ is called a **unitary** mapping on E if it is an isomorphism of complex IP-spaces in the sense of Definition 12.3.5. (So to speak, T is an **automorphism** of E as a complex IP-space.)

12.3.8 Theorem *If* E *is a unitary space and* $T \in \mathcal{L}(\mathrm{E})$, *then* T *is unitary* $\Leftrightarrow \|Tx\| = \|x\|$ *for all vectors* x.

Proof. $\Rightarrow$: Put $y = x$ in 12.3.5.

$\Leftarrow$: Clearly $\mathrm{Ker}\, T = \{\theta\}$ so T is injective, hence bijective (3.7.4); the proof that $(Tx|Ty) = (x|y)$ is formally the same as in '(b) $\Rightarrow$ (c)' of 5.4.2, except that the polarization identity has four terms instead of two (12.1.8). ■

▶ **Exercises**

1. If $u_1, \ldots, u_k$ are orthonormal vectors in a unitary space E of dimension $n > k$, then E has an orthonormal basis $u_1, \ldots, u_k, u_{k+1}, \ldots, u_n$.

2. (*Gram-Schmidt process*) If $x_1, \ldots, x_n$ are linearly independent vectors in a complex inner product space, then there exist orthonormal vectors $u_1, \ldots, u_n$ such that $[\{x_1, \ldots, x_k\}] = [\{u_1, \ldots, u_k\}]$ for $k = 1, \ldots, n$.
{Hint: Cf. §5.2, Exercise 12.}

3. Let $\mathrm{E} = \mathcal{C}([0, 1], \mathrm{C})$ with inner product as in 12.1.5 and let M be the 3-dimensional linear subspace spanned by the functions $1, t, t^2$. Apply the Gram–Schmidt process (Exercise 2) to find an orthonormal basis of M.

4. With E as in Exercise 3, let N be the 3-dimensional linear subspace spanned by the functions $x_n(t) = \sin 2\pi nt$ $(n = 1, 2, 3)$. Apply the Gram–Schmidt process to find an orthonormal basis of N.

5. With notations as in Exercise 4, write $y_n(t) = \cos 2\pi nt$. Find an orthonormal basis for the 4-dimensional linear subspace $[\{y_1, x_1, x_2, x_3\}]$.

6. Let $\mathrm{E} = \mathscr{C}([0, 2\pi], \mathbf{C})$ with inner product as in 12.1.5 and write $z_n(t) = e^{int} = \cos nt + i \sin nt$ $(n \in \mathbf{Z},\ 0 \leqslant t \leqslant 2\pi)$. Find an orthonormal basis for the 4-dimensional linear subspace $[\{z_0, z_1, z_2, z_3\}]$.

12.4 Adjoint of a linear mapping

If $T{:}\mathrm{E} \to \mathrm{F}$ is a linear mapping between unitary spaces, the self-duality of unitary spaces (12.1.15) suggests a way of defining a linear mapping $\mathrm{F} \to \mathrm{E}$ 'like' the transpose of T (but acting in the given spaces rather than their duals). Let $J_{\mathrm{E}}{:}\mathrm{E} \to \mathrm{E}'$ be the conjugate–linear bijection $J_{\mathrm{E}}x = x'$ of 12.1.15, and similarly for $J_{\mathrm{F}}{:}\mathrm{F} \to \mathrm{F}'$.

12.4.1 Definition With the preceding notations, the **adjoint** of $T{:}\mathrm{E} \to \mathrm{F}$ is the mapping $T^*{:}\mathrm{F} \to \mathrm{E}$ defined by the formula $T^* = J_{\mathrm{E}}^{-1} T' J_{\mathrm{F}}$.

Since T^* is the composite of a conjugate–linear mapping, followed by a linear mapping, followed by another conjugate-linear mapping, T^* is *linear*. The diagram

$$\begin{array}{ccc} & T^* & \\ \mathrm{E} & \longleftarrow & \mathrm{F} \\ J_{\mathrm{E}} \downarrow & & \downarrow J_{\mathrm{F}} \\ \mathrm{E}' & \longleftarrow & \mathrm{F}' \\ & T' & \end{array}$$

is 'commutative' in the sense that $J_{\mathrm{E}} T^* = T' J_{\mathrm{F}}$.

12.4.2 Theorem *Let* E *and* F *be unitary spaces,* $T{:}\mathrm{E} \to \mathrm{F}$ *a linear mapping,* $T^*{:}\mathrm{F} \to \mathrm{E}$ *the adjoint of* T (12.4.1). *Then*:

(1) T^* *is a linear mapping,*

(2) $(Tx|y) = (x|T^*y)$ *for all* $x \in \mathrm{E}$, $y \in \mathrm{F}$.

(3) *If* $S{:}\mathrm{F} \to \mathrm{E}$ *is a mapping such that* $(Tx|y) = (x|Sy)$ *for all* $x \in \mathrm{E}$, $y \in \mathrm{F}$, *then necessarily* $S = T^*$.

Proof.

(1) Already noted above.

(2) Adopt the notations of 12.4.1. For all $x \in \mathrm{E}$ and $y \in \mathrm{F}$,

$$\begin{aligned} (Tx|y) &= y'(Tx) = (T'y')(x) \\ &= (T'J_{\mathrm{F}}y)(x) = (J_{\mathrm{E}}T^*y)(x) \\ &= (T^*y)'(x) = (x|T^*y). \end{aligned}$$

(3) For all x and y, $(x|Sy) = (Tx|y) = (x|T^*y)$, therefore $S = T^*$ by Corollary 12.1.12. ■

12.4.3 Theorem *If E and F are unitary spaces, then the mapping $\mathscr{L}(\mathrm{E}, \mathrm{F}) \to \mathscr{L}(\mathrm{F}, \mathrm{E})$ defined by $T \mapsto T^*$ is conjugate-linear and bijective (in a sense, 'self-inverse'):*

(4) $(S + T)^* = S^* + T^*$, $(cT)^* = c^*T^*$

(5) $(T^*)^* = T$

for $S, T \in \mathscr{L}(\mathrm{E}, \mathrm{F})$ and $c \in \mathbf{C}$. Moreover,

(6) $T^*T = 0 \Rightarrow T = 0$, *and* $TT^* = 0 \Rightarrow T = 0$.

Proof.
(4) $(cT)^* = J_{\mathrm{E}}^{-1}(cT)'J_{\mathrm{F}} = J_{\mathrm{E}}^{-1}(cT')J_{\mathrm{F}} = c^*J_{\mathrm{E}}^{-1}T'J_{\mathrm{F}}$ (because J_{E}^{-1} is conjugate–linear) $= c^*T^*$; the other formula of (4) follows similarly from $(S + T)' = S' + T'$.

(5) For all $y \in \mathrm{F}$ and $x \in \mathrm{E}$,

$$(T^*y|x) = (x|T^*y)^* = (Tx|y)^* = (y|Tx),$$

therefore $T = (T^*)^*$ by (3) of 12.4.2.

(6) $\|Tx\|^2 = (Tx|Tx) = (T^*Tx|x)$, so if $T^*T = 0$ then $Tx = \theta$ for all x, thus $T = 0$. The second implication follows from the first, since $TT^* = (T^*)^*T^*$. ■

12.4.4 Theorem *If E, F, G are unitary spaces and $T{:}\mathrm{E} \to \mathrm{F}$, $S{:}\mathrm{F} \to \mathrm{G}$ are linear mappings, then $(ST)^* = T^*S^*$.*

$$\mathrm{E} \xrightarrow{T} \mathrm{F} \xrightarrow{S} \mathrm{G}$$
$$\mathrm{E} \xleftarrow[T^*]{} \mathrm{F} \xleftarrow[S^*]{} \mathrm{G}$$

Proof. For all $x \in \mathrm{E}$ and $z \in \mathrm{G}$,

$$((ST)x|z) = (S(Tx)|z) = (Tx|S^*z) = (x|T^*(S^*z)) = (x|(T^*S^*)z),$$

therefore $T^*S^* = (ST)^*$ by (3) of 12.4.2. ■

12.4.5 Corollary *If E, F are unitary spaces and $T{:}\mathrm{E} \to \mathrm{F}$ is a linear bijection, then $T^*{:}\mathrm{F} \to \mathrm{E}$ is also bijective and $(T^*)^{-1} = (T^{-1})^*$.*

Proof. Same as 5.3.4. ■

12.4.6 Theorem *If E is a unitary space and $T \in \mathscr{L}(\mathrm{E})$, the following conditions are equivalent:*

(a) *T is unitary in the sense of* 12.3.7;

(b) *T is bijective and $T^{-1} = T^*$;*

(c) $T^*T = I$;

(c′) $TT^* = I$.

Proof.
(b) $\Leftrightarrow$ (c) $\Leftrightarrow$ (c′) by finite-dimensionality (3.7.5). In particular, all of the conditions (a)–(c′) imply that T is bijective.

(a) $\Leftrightarrow$ (c): Since $(T^*Tx|y) = (Tx|Ty)$, condition (a) says that $(T^*Tx|y) = (Ix|y)$ for all vectors x and y, and this is equivalent to condition (c) by 12.1.12 (or 12.1.13). ■

It follows that T is unitary if and only if T^* is unitary.

12.4.7 Theorem *Let* E *and* F *be unitary spaces,* $\dim \mathrm{E} = n$, $\dim \mathrm{F} = m$, *and let* $T{:}\mathrm{E} \to \mathrm{F}$ *be a linear mapping. Choose* ***orthonormal*** *bases*

$$e_1, \ldots, e_n \quad \textit{and} \quad f_1, \ldots, f_m$$

of E *and* F, *respectively, and let* A *be the matrix of* T *relative to these bases.*

Then, the matrix of T^* *relative to the bases* $f_1, \ldots, f_m$ *and* $e_1, \ldots, e_n$ *is the conjugate-transpose* A^* *of* A.

Concisely: relative to a pair of orthonormal bases, mat $T^* =$ (mat T)*.

Proof. Let $A = (a_{ij})$ and $B = (b_{ji})$ be the indicated matrices of T and T^*, respectively; thus

$$Te_j = \sum_{i=1}^{m} a_{ij} f_i \qquad (j = 1, \ldots, n),$$

$$T^*f_i = \sum_{j=1}^{n} b_{ji} e_j \qquad (i = 1, \ldots, m).$$

Then

$$b_{ji} = (T^*f_i|e_j) = (f_i|Te_j) = (Te_j|f_i)^* = a_{ij}^*$$

for all i and j, in other words, $B = A^*$. ■

12.4.8 Corollary *Let* E *be a unitary space,* $T \in \mathcal{L}(\mathrm{E})$, *and* A *the matrix of* T *relative to an orthonormal basis of* E. *The following conditions are equivalent:*

(a) T *is unitary* (cf. 12.3.7),

(b) $A^*A = I$,

(b′) $AA^* = I$.

Proof. (a) $\Leftrightarrow$ (b): Relative to the orthonormal basis in question,

$$\operatorname{mat}(T^*T) = (\operatorname{mat} T^*)(\operatorname{mat} T) = (\operatorname{mat} T)^*(\operatorname{mat} T) = A^*A,$$

thus

$$T^*T = I \quad \Leftrightarrow \quad \operatorname{mat}(T^*T) = \operatorname{mat} I \quad \Leftrightarrow \quad A^*A = I.$$

In view of Theorem 12.4.6, this means that T is unitary if and only if $A^*A = I$.

Similarly, (a) $\Leftrightarrow$ (b′). {Alternatively, (b) $\Leftrightarrow$ (b′) by the theory of matrix invertibility (4.10.3).} ■

12.4.9 Definition A matrix $A \in \mathrm{M}_n(\mathbf{C})$ is said to be **unitary** if $A^*A = I$.

An equivalent condition is that A is invertible and $A^{-1} = A^*$. The unitary matrices form a group under multiplication.

12.4.10 Theorem *Let* E *be an n-dimensional unitary space,* $C = (c_{ij}) \in \mathrm{M}_n(\mathbf{C})$ *and* $x_1, \ldots, x_n$ *an orthonormal basis of* E. *Define vectors* $y_1, \ldots, y_n$ *by the formula*

$$y_j = \sum_{i=1}^{n} c_{ij}x_i \quad (j = 1, \ldots, n).$$

Then, C *is a unitary matrix if and only if* $y_1, \ldots, y_n$ *is an orthonormal basis of* E.

Proof. Let $T \in \mathcal{L}(\mathrm{E})$ be the linear mapping such that $Tx_j = y_j$ for all j. Then C is the matrix of T relative to the basis $x_1, \ldots, x_n$; in view of 12.4.8, the problem is to prove the equivalence

$$T \text{ unitary} \Leftrightarrow y_1, \ldots, y_n \text{ orthonormal}.$$

$\Rightarrow$: Since T preserves inner products, $(y_i|y_j) = (Tx_i|Tx_j) = (x_i|x_j) = \delta_{ij}$.

$\Leftarrow$: By hypothesis,

$$(Tx_i|Tx_j) = (y_i|y_j) = \delta_{ij} = (x_i|x_j)$$

for all i and j. It follows that for each i, $(Tx_i|Ty) = (x_i|y)$ for every vector y (because y is a linear combination of the x_j); the same reasoning shows that for each vector y, $(Tx|Ty) = (x|y)$ for all x, thus T preserves inner products. In particular, T preserves norms, hence is injective, hence bijective (cf. 3.7.4), therefore unitary (12.3.7). ∎

The next big target (attained in §12.7): If E is a unitary space, and $T \in \mathcal{L}(\mathrm{E})$ is a linear mapping such that $T^*T = TT^*$ (such mappings are called *normal*) then there exists an orthonormal basis of E for which the matrix of T is diagonal.

▶ ***Exercises***

1. Let M be a linear subspace of a unitary space E and let $P = P_{\mathrm{M}}$ be the projection with range M (§12.2, Exercise 1). Then $P^* = P = P^2$.

2. Let E be a unitary space, $T \in \mathcal{L}(\mathrm{E})$ a linear mapping such that $T^2 = T$ and $T^* = T$, and let $\mathrm{M} = T(\mathrm{E})$. Prove:

(i) $\mathrm{M} = \{x \in \mathrm{E}: Tx = x\} = \mathrm{Ker}(I - T)$;

(ii) $\mathrm{M}^\perp = \mathrm{Ker}\, T$;

(iii) T is the projection with range M, that is, $T = P_{\mathrm{M}}$ in the sense of §12.2, Exercise 1.

For this reason, linear mappings $T \in \mathcal{L}(\mathrm{E})$ such that $T^* = T = T^2$ are called *projections*.

3. For every linear mapping $T: \mathrm{E} \to \mathrm{F}$ between unitary spaces, (i) T and T^* have the same rank, and (ii) T and T^*T have the same kernel and the same rank.

4. Give a proof of Theorem 12.4.4 based on $(ST)' = T'S'$.

12.5 Invariant and reducing subspaces

The concepts and arguments of §9.1 for Euclidean spaces carry over easily to unitary spaces.

12.5.1 Theorem *Let* E *and* F *be unitary spaces,* $T:\mathrm{E}\to\mathrm{F}$ *a linear mapping,* M *a linear subspace of* E, *and* N *a linear subspace of* F. *Then*

$$T(\mathrm{M})\subset\mathrm{N} \;\Leftrightarrow\; T^*(\mathrm{N}^\perp)\subset\mathrm{M}^\perp.$$

Proof. $\Rightarrow$: Formally the same as 5.3.6.

$\Leftarrow$: If $T^*(\mathrm{N}^\perp)\subset\mathrm{M}^\perp$ then, by the implication already proved, $T^{**}(\mathrm{M}^{\perp\perp})\subset\mathrm{N}^{\perp\perp}$, in other words, $T(\mathrm{M})\subset\mathrm{N}$ (12.4.3, 12.2.7). ■

12.5.2 Definition Let E be a unitary space, $T\in\mathcal{L}(\mathrm{E})$, and M a linear subspace of E. If $T(\mathrm{M})\subset\mathrm{M}$ then M is said to be **invariant** under T (9.1.3). If both M and $\mathrm{M}^\perp$ are invariant under T, then M is said to **reduce** T (or to be *reducing* for T); equivalently (12.5.1), M is invariant under both T and T^*.

12.5.3 Theorem *If* E *is a unitary space,* $T\in\mathcal{L}(\mathrm{E})$, *and* M *is a linear subspace of* E *that reduces T, then* $(T|\mathrm{M})^*=T^*|\mathrm{M}$ *and* $(T|\mathrm{M}^\perp)^*=T^*|\mathrm{M}^\perp$.

Proof. Same as for (2) of 9.1.5. ■

▶ *Exercises*

1. Let E and F be unitary spaces, $T:\mathrm{E}\to\mathrm{F}$ a linear mapping, A a subset of E, and B a subset of F. Prove:

(i) $(T(\mathrm{A}))^\perp=(T^*)^{-1}(\mathrm{A}^\perp)$;

(i′) $(T^*(\mathrm{B}))^\perp=T^{-1}(\mathrm{B}^\perp)$;

(ii) $(T(\mathrm{E}))^\perp=\operatorname{Ker}T^*$, therefore $\mathrm{F}=T(\mathrm{E})\oplus\operatorname{Ker}T^*$;

(ii′) $(T^*(\mathrm{F}))^\perp=\operatorname{Ker}T$, therefore $\mathrm{E}=T^*(\mathrm{F})\oplus\operatorname{Ker}T$;

(iii) $T(\mathrm{A})\subset\mathrm{B}\Rightarrow T^*(\mathrm{B}^\perp)\subset\mathrm{A}^\perp$.

2. If M and N are invariant (or reducing) subspaces for T, then so are $\mathrm{M}+\mathrm{N}$ and $\mathrm{M}\cap\mathrm{N}$.

3. Let V be a vector space over F, $S, T\in\mathcal{L}(\mathrm{V})$ and $c\in\mathrm{F}$. Let $\mathrm{M}=\{x\in\mathrm{V}: Tx=cx\}=\operatorname{Ker}(T-cI)$. Prove: If $ST=TS$ then M is invariant under S.

4. Let E be a unitary space, $T\in\mathcal{L}(\mathrm{E})$, M a linear subspace of E, and $P=P_{\mathrm{M}}$ the projection with range M (cf. §12.2, Exercise 1). Then:

(i) M is invariant under $T\Leftrightarrow PTP=TP$;

(ii) M reduces $T\Leftrightarrow TP=PT$.

5. If E is a unitary space and $T\in\mathcal{L}(\mathrm{E})$ is a linear mapping such that $TP=PT$ for all projections P (cf. §12.4, Exercise 2), then T is a scalar multiple of the identity. In fact, it suffices that $TP = PTP$ for all projections P with 1-dimensional range.

{Hint: By Exercise 4, every nonzero vector in E is an eigenvector for T.}

6. (i) Let $\mathcal{S}$ be a set of linear mappings in the unitary space E, such that $ST = TS$ for all S, T in $\mathcal{S}$.

Prove: There exists an orthonormal basis of E such that, relative to this basis, the matrix of every $S \in \mathcal{S}$ is triangular (with zeros below the main diagonal).

(ii) If $\mathcal{M}$ is a set of $n \times n$ complex matrices such that $AB = BA$ for all $A, B \in \mathcal{M}$, then there exists a unitary matrix C such that C^*AC is triangular (with zeros below the main diagonal) for every $A \in \mathcal{M}$.

{Hint: (i) Say $\dim \mathrm{E} = n$. By §8.3, Exercise 7, there exists a unit vector $x_n \in \mathrm{E}$ that is an eigenvector for every S^* $(S \in \mathcal{S})$; let $\mathrm{M} = \mathbf{C}x_n$. The $(n-1)$-dimensional unitary space $\mathrm{M}^{\perp}$ is invariant under every $S \in \mathcal{S}$ (Theorem 12.5.1); induction, anyone?}

12.6 Special linear mappings and matrices

Notations fixed for the section: E is a unitary space and $T \in \mathcal{L}(\mathrm{E})$.

Let A be the matrix of T relative to some orthonormal basis of E. According to 12.4.8, T is unitary $(T^*T = I)$ if and only if A is a unitary matrix $(A^*A = I)$; for such a linear mapping, $T^*T = TT^*$ (both sides are equal to I).

12.6.1 Definition We say that T is **normal** if $T^*T = TT^*$, **self-adjoint** if $T^* = T$, and **skew-adjoint** if $T^* = -T$.

If T is unitary (self-adjoint, skew-adjoint) then it is normal; T is skew-adjoint if and only if iT is self-adjoint. For every linear mapping T, the mappings T^*T and $T + T^*$ are self-adjoint, and $T - T^*$ is skew-adjoint.

12.6.2 Definition A matrix $A \in \mathrm{M}_n(\mathbf{C})$ is said to be **normal** if $A^*A = AA^*$, **Hermitian** if $A^* = A$, and **skew-Hermitian** if $A^* = -A$.

12.6.3 Theorem *Let* E *be a unitary space,* $T \in \mathcal{L}(\mathrm{E})$, A *the matrix of* T *relative to some orthonormal basis of* E. *Then*:

(i) T *is normal* $\Leftrightarrow$ A *is normal*;

(ii) T *is unitary* $\Leftrightarrow$ A *is unitary*;

(iii) T *is self-adjoint* $\Leftrightarrow$ A *is Hermitian*;

(iv) T *is skew-adjoint* $\Leftrightarrow$ A *is skew-Hermitian*.

Proof. The equivalence (ii) is proved in 12.4.8, and the others are proved similarly. ■

12.6.4 Remark Let $A \in M_n(C)$ and let $C \in M_n(C)$ be a unitary matrix. If A is normal (Hermitian, skew-Hermitian, unitary) then so is $C^{-1}AC = C^*AC$.

{**Proof.** This can be verified by direct computation. Alternatively, let $T \in \mathcal{L}(C^n)$ be the linear mapping whose matrix relative to the canonical orthonormal basis of C^n is A. If $y_1, \ldots, y_n$ is the basis of C^n for which C is the 'change of basis matrix', then $y_1, \ldots, y_n$ is an orthonormal basis (12.4.10). Since $C^{-1}AC$ is the matrix of T relative to the basis $y_1, \ldots, y_n$, it follows that if A is normal (self-adjoint, etc.), then so is T (12.6.3), therefore so is $C^{-1}AC$ (12.6.3 again).}

12.6.5 Theorem *Let E be a unitary space, $T \in \mathcal{L}(E)$.*

(1) *There exist unique self-adjoint linear mappings R, $S \in \mathcal{L}(E)$ such that $T = R + iS$.*

(2) *T is normal $\Leftrightarrow RS = SR$.*

Explicitly,

$$R = \frac{1}{2}(T + T^*), \quad S = \frac{1}{2i}(T - T^*).$$

Proof.

(1) Uniqueness: If R and S are self-adjoint linear mappings such that $T = R + iS$, then $T^* = R - iS$, so $T + T^* = 2R$ and $T - T^* = 2iS$; thus R and S are given by the stated formulas.

Existence: The linear mappings defined by these formulas meet the requirements for R and S.

(2) With notations as in (1),

$$T^*T = R^2 + S^2 + i(RS - SR),$$
$$TT^* = R^2 + S^2 - i(RS - SR);$$

thus $T^*T - TT^* = 2i(RS - SR)$ and the assertion of (2) is clear. ■

12.6.6 Theorem *Let E be a unitary space, $T \in \mathcal{L}(E)$. Then T is self-adjoint $\Leftrightarrow (Tx|x)$ is real for all vectors x.*

Proof. For all vectors x,

$$\begin{aligned}((T^* - T)x|x) &= (T^*x|x) - (Tx|x) \\ &= (x|Tx) - (Tx|x) \\ &= (Tx|x)^* - (Tx|x);\end{aligned}$$

in view of 12.1.13,

$$T^* - T = 0 \quad \Leftrightarrow \quad (Tx|x)^* = (Tx|x) \quad \text{for all } x. \ ■$$

▶ Exercises

1. For a linear mapping $T \in \mathcal{L}(E)$, the following conditions are equivalent:

(a) T is normal;

(b) if M is an invariant subspace for T, then M reduces T.

{Hint: (a) $\Rightarrow$ (b): Cf. 9.4.5. Note that the analogue of 9.4.2 for unitary spaces is true without restriction on R (12.1.13). (b) $\Rightarrow$ (a): Consider the 1-dimensional subspace spanned by an eigenvector.}

2. If $A \in M_n(C)$ is normal and triangular, then it is diagonal. {Hint: Corollary 9.4.6.}

3. If $A \in M_n(C)$ is normal (unitary, hermitian, skew-hermitian) then so is its transpose A'.

4. A linear mapping $T:E \to F$ between unitary spaces is called an *isometry* (or is said to be *isometric*) if $\|Tx\| = \|x\|$ for all $x \in E$. An equivalent condition is that $T^*T = I$ (the identity mapping on E).

5. Let M be a linear subspace of a unitary space E. A linear mapping $W \in \mathcal{L}(E)$ such that $\|Wy\| = \|y\|$ for all $y \in M$ and $Wz = \theta$ for all $z \in M^\perp$ is called a *partial isometry* with *support* M.

(i) A linear mapping $W \in \mathcal{L}(E)$ is a partial isometry with support M if and only if $W^*W = P_M$, where P_M is the projection with range M (§12.2, Exercise 1).

(ii) If $W \in \mathcal{L}(E)$ and $WW^*W = W$, then $P = W^*W$ is a projection and W is a partial isometry with support $P(E)$.

6. A linear mapping $T \in \mathcal{L}(E)$ is normal if and only if $T^* = UT$ for some unitary $U \in \mathcal{L}(E)$. {Hint: Cf. §9.4, Exercise 4.}

7. If $P, Q \in \mathcal{L}(E)$ are projections (§12.4, Exercise 2), then the following conditions are equivalent: (a) PQ is a projection; (b) PQ is self-adjoint; (c) $PQ = QP$. When this is the case, $\operatorname{Im} PQ = (\operatorname{Im} P) \cap (\operatorname{Im} Q)$. (See also §12.8, Exercise 3.)

12.7 Normal linear mappings, Spectral Theorem

For Euclidean spaces, the 'Spectral Theorem' is a structure theorem for *self-adjoint* linear mappings (Theorem 9.3.4); for unitary spaces, it is a structure theorem for *normal* linear mappings. The proof of the Spectral Theorem given in this section depends on knowing that the characteristic polynomial of the mapping has a complex root, thus the proof rests on determinant theory and the Fundamental theorem of algebra.[1] The first step is to characterize normality in terms of the norm:

12.7.1 Lemma *The following conditions on $T \in \mathcal{L}(E)$ are equivalent*:

(a) *T is normal*;

[1] A proof that depends only on elementary properties of sequences of real numbers is given in the next section, but Lemmas 12.7.1–12.7.3 are needed for both proofs. The use of determinant theory is convenient (especially for computation) but not indispensable (Exercise 13).

(b) $\|Tx\| = \|T^*x\|$ *for all* $x \in \mathrm{E}$.

Proof. After squaring, condition (b) says that $(T^*Tx|x) = (TT^*x|x)$ for all $x \in \mathrm{E}$, which is equivalent to $T^*T = TT^*$ (12.1.13). {Cf. 9.4.3.}. ■

In particular, if T is normal then $\operatorname{Ker} T = \operatorname{Ker} T^*$.

12.7.2 Lemma *If* $T \in \mathcal{L}(\mathrm{E})$ *is normal and* M *is a linear subspace of* E *that reduces* T, *then* $T|\mathrm{M}$ *and* $T|\mathrm{M}^\perp$ *are normal.*

Proof. Immediate from 12.5.3. ■

12.7.3 Lemma *Let* $T \in \mathcal{L}(\mathrm{E})$ *be normal,* $x \in \mathrm{E}$, *and* $c, d \in \mathbf{C}$. *Then:*

(1) $T - cI$ *is normal;*
(2) $Tx = cx \Leftrightarrow T^*x = c^*x$;
(3) $\operatorname{Ker}(T - cI)$ *reduces* T.
(4) *If* $c \neq d$ *then* $\operatorname{Ker}(T - cI) \perp \operatorname{Ker}(T - dI)$.

Proof.
(1) Since T commutes with T^*, it follows that $(T - cI)^* = T^* - c^*I$ commutes with $T - cI$.
(2) By (1) and 12.7.1, $\|(T - cI)x\| = 0 \Leftrightarrow \|(T - cI)^*x\| = 0$.
(3) Let S be either T or T^*; in either case, S commutes with T. If $Tx = cx$ then

$$T(Sx) = (TS)x = (ST)x = S(Tx) = S(cx) = c(Sx);$$

thus $x \in \operatorname{Ker}(T - cI) \Rightarrow Sx \in \operatorname{Ker}(T - cI)$, that is, $\operatorname{Ker}(T - cI)$ is invariant under S.
(4) If $Tx = cx$ and $Ty = dy$ then, citing (2),

$$d(y|x) = (dy|x) = (Ty|x) = (y|T^*x) = (y|c^*x) = c(y|x),$$

thus $(d - c)(y|x) = 0$; since $c \neq d$, necessarily $(y|x) = 0$. ■

12.7.4 Theorem. (Spectral Theorem) *If* E *is a unitary space and* $T \in \mathcal{L}(\mathrm{E})$ *is normal, then* E *has an orthonormal basis consisting of eigenvectors of* T.

Proof. The proof is by complete induction on $n = \dim \mathrm{E}$. If $n = 1$ then T is a scalar multiple of the identity and any unit vector serves as orthonormal basis.

Let $n \geq 2$ and assume that all's well for dimensions $< n$. Since $\mathbf{C}$ is algebraically closed (cf. 8.3.11), the characteristic polynomial of T has a complex root c. By 8.3.9, c is an eigenvalue of T, so the linear subspace $\mathrm{M} = \operatorname{Ker}(T - cI)$ of E is nonzero; by 12.7.3, M reduces T. If $\mathrm{M} = \mathrm{E}$ then $T = cI$ and *every* orthonormal basis of E consists of eigenvectors of T. If $\mathrm{M} \neq \mathrm{E}$ then $\mathrm{M}^\perp \neq \{\theta\}$ (12.2.7), $T|\mathrm{M}^\perp$ is normal (12.7.2) and $\dim \mathrm{M}^\perp < \dim \mathrm{E}$ (12.2.6). By the induction hypothesis, $\mathrm{M}^\perp$ has an orthonormal basis consisting of eigenvectors of $T|M^{\perp}$ (i.e. of T); combining them with any orthonormal basis of M, we obtain an orthonormal basis of E consisting of eigenvectors of T. ■

12.7.5 Corollary *If $A \in M_n(C)$ is normal then there exists a unitary matrix U such that U^*AU is diagonal.*

Proof. Let E be an n-dimensional unitary space, choose an orthonormal basis $x_1, \ldots, x_n$ of E, and let $T \in \mathscr{L}(E)$ be the linear mapping whose matrix is A relative to this basis. By 12.6.3, T is normal, so by the Spectral Theorem there is an orthonormal basis $y_1, \ldots, y_n$ of E consisting of eigenvectors of T. The matrix B of T relative to the basis $y_1, \ldots, y_n$ is diagonal. Express the y's as linear combinations of the x's and let U be the matrix of coefficients; by 12.4.10, U is a unitary matrix, and $U^{-1}AU = B$ by 4.10.5. ■

12.7.6 Definition Matrices $A, B \in M_n(C)$ are said to be **unitarily equivalent** if there exists a unitary matrix U such that $U^*AU = B$.

12.7.7 Corollary *A matrix $A \in M_n(C)$ is normal if and only if it is unitarily equivalent to a diagonal matrix.*

Proof. Suppose $B = U^*AU$ (U unitary). If A is normal, then U can be chosen so that B is diagonal (12.7.5). Conversely, if B is diagonal then it is obviously normal, therefore so is A (12.6.4). ■

12.7.8 Corollary *Let $A \in M_n(C)$ be normal and let U be a unitary matrix such that $U^*AU = \operatorname{diag}(c_1, c_2, \ldots, c_n)$. Then:*

(1) $A^* = A \Leftrightarrow$ *the c_i are real;*

(2) $A^*A = I \Leftrightarrow |c_i| = 1$ *for all i;*

(3) $A^* = -A \Leftrightarrow$ *the c_i are pure-imaginary.*

Proof. A condition on the left side is valid for A if and only if it is valid for U^*AU (12.6.4), and the equivalences are obvious when A is already diagonal. ■

12.7.9 Corollary *With notations as in 12.7.8, $p_A = (t - c_1)(t - c_2) \ldots (t - c_n)$, thus the c_i are the characteristic roots of A.*

Proof. The characteristic polynomial of A is equal to the characteristic polynomial of the similar matrix U^*AU (8.3.4), for which the stated formula is obviously correct. ■

12.7.10 Corollary *Let E be a unitary space, $T \in \mathscr{L}(E)$* **normal**, *and let $c_1, \ldots, c_r$ be the distinct eigenvalues of T. Then:*

(1) $T^* = T \Leftrightarrow$ *the c_i are real;*

(2) $T^*T = I \Leftrightarrow |c_i| = 1$ *for all* i;
(3) $T^* = -T \Leftrightarrow$ *the* c_i *are pure-imaginary*.

Proof. By the Spectral Theorem, E has an orthonormal basis relative to which the matrix A of T is diagonal; since the diagonal entries of A are eigenvalues of T, the corollary is immediate from 12.7.8 and 12.6.3. ■

12.7.11 Definition A linear mapping $T \in \mathcal{L}(\mathrm{E})$ is said to be **positive**, written $T \geqslant 0$, if $(Tx|x) \geqslant 0$ for all vectors $x \in \mathrm{E}$.

12.7.12 Theorem *For a linear mapping* $T \in \mathcal{L}(\mathrm{E})$, *the following conditions are equivalent*:

(a) $T \geqslant 0$;
(b) $T^* = T$ *and the eigenvalues of* T *are* $\geqslant 0$;
(c) $T = R^2$ *with* $R \in \mathcal{L}(\mathrm{E})$, $R \geqslant 0$;
(d) $T = S^*S$ *for some* $S \in \mathcal{L}(\mathrm{E})$.

Proof.
(a) ⇒ (b): If $T \geqslant 0$ then $T^* = T$ by 12.6.6, and if $Tu = ru$, u a unit vector, then $r = r(u|u) = (Tu|u) \geqslant 0$.
(b) ⇒ (c): Let $x_1, \ldots, x_n$ be an orthonormal basis of E with $Tx_i = c_i x_i$ (12.7.4). By assumption, $c_i \geqslant 0$ for all i. Let $d_i = \sqrt{c_i}$ and let $R \in \mathcal{L}(\mathrm{E})$ be the linear mapping such that $Rx_i = d_i x_i$ for all i. If $x \in \mathrm{E}$, say $x = a_1x_1 + \ldots + a_nx_n$, then

$$(Rx|x) = (d_1a_1x_1 + \ldots + d_na_nx_n|a_1x_1 + \ldots + a_nx_n)$$
$$= d_1|a_1|^2 + \ldots + d_n|a_n|^2 \geqslant 0,$$

thus $R \geqslant 0$; and $(T - R^2)x_i = 0$ for all i, so $T = R^2$.
(c) ⇒ (d): Trivial.
(d) ⇒ (a): $(Tx|x) = (S^*Sx|x) = (Sx|Sx) \geqslant 0$. ■

12.7.13 Remark If $T \geqslant 0$ then the positive linear mapping R of 12.7.12, (c) is unique; it is called the **positive square root** of T, written $R = \sqrt{T}$ (or $T^{1/2}$).

{**Proof.** With notations as in the proof of (b) ⇒ (c) of 12.7.12, we can suppose that $c_1, \ldots, c_r$ are the distinct eigenvalues of T (and that $c_{r+1}, \ldots, c_n$ are clones of the c_i with $i \leqslant r$). For $k = 1, \ldots, r$, let

$$\mathrm{M}_k = \operatorname{Ker}(T - c_kI);$$

the M_k are pairwise orthogonal by (4) of 12.7.3, and $\mathrm{M}_1 + \ldots + \mathrm{M}_r = \mathrm{E}$ by the Spectral Theorem. Note that

$$\mathrm{M}_k = \operatorname{Ker}(R - d_kI)$$

as well; for, $d_1, \ldots, d_r$ are the distinct eigenvalues of R, so if $\mathrm{N}_k = \operatorname{Ker}(R - d_kI)$ then the N_k are pairwise orthogonal with sum E, and obviously $\mathrm{N}_k \subset \mathrm{M}_k$, whence equality.

Suppose $S \in \mathcal{L}(E)$ and $ST = TS$; let us show that $RS = SR$. The M_k are invariant under S; for, if $x \in M_k$ then

$$T(Sx) = S(Tx) = S(c_k x) = c_k(Sx),$$

whence $Sx \in M_k$. (It follows that the M_k *reduce* S; for example, $M_1^\perp = M_2 + \ldots + M_r$ is invariant under S, therefore M_1 is invariant under S^*.) Thus if $x \in M_k$ we have $Sx \in M_k = \operatorname{Ker}(R - d_k I)$, so

$$R(Sx) = d_k(Sx) = S(d_k x) = S(Rx),$$

whence $RS = SR$.

Suppose now that $S \in \mathcal{L}(E)$ with $S \geqslant 0$ and $T = S^2$; we are to show that $S = R$. Obviously $ST = TS$, so $SR = RS$ by the preceding paragraph; it follows that

$$0 = T - T = R^2 - S^2 = (R + S)(R - S),$$

whence $(R - S)(R + S)(R - S) = 0$. Thus

$$0 = (R - S)^*R(R - S) + (R - S)^*S(R - S).$$

Since $R \geqslant 0$ and $S \geqslant 0$, it follows easily (cf. 12.1.13) that

$$(R - S)^*R(R - S) = (R - S)^*S(R - S) = 0,$$

that is,

$$(R - S)R(R - S) = (R - S)S(R - S) = 0,$$

whence $(R - S)^3 = 0$. Then

$$0 = (R - S)^4 = (R - S)^{2*}(R - S)^2,$$

so $0 = (R - S)^2 = (R - S)^*(R - S)$, and finally $R - S = 0$.}

▶ Exercises

1. Let $T \in \mathcal{L}(E)$ be normal, let $c_1, \ldots, c_r$ be the distinct characteristic roots (= eigenvalues) of T, and let $M_k = \operatorname{Ker}(T - c_k I)$ for $k = 1, \ldots, r$; the M_k are reducing subspaces for T and are pairwise orthogonal (12.7.3). Give an alternate proof of the Spectral Theorem by showing that $M_1 + \ldots + M_r = E$.

{Hint: Let $M = M_1 + \ldots + M_r$, assume to the contrary that $M \neq E$, and contemplate an eigenvector of $T|M^\perp$.}

2. (i) Let $\mathcal{M}$ be a set of $n \times n$ normal matrices such that $AB = BA$ for all $A, B \in \mathcal{M}$. Prove: There exists a unitary matrix $C \in M_n(\mathbf{C})$ such that C^*AC is diagonal for every $A \in \mathcal{M}$.

(ii) Let $\mathcal{S}$ be a set of normal linear mappings in the unitary space E, such that $ST = TS$ for all $S, T \in \mathcal{S}$. Prove: There exists an orthonormal basis of E consisting of vectors that are eigenvectors for every $S \in \mathcal{S}$.

{Hint: (i) §12.5, Exercise 6, (ii) and §12.6, Exercise 2.}

3. Let A and B be $n \times n$ complex matrices.

(i) If A is normal and $AB = BA$, then $AB^* = B^*A$.

(ii) If A and B are both normal and $AB = BA$, then AB and $A + B$ are normal.

(iii) Exhibit matrices A, B (with $n = 2$) such that $AB = BA$ but $AB^* \neq B^*A$.

{Hints: (i), (ii) Cf. §9.4, Exercise 8.

(iii) Consider $A = B$ with A nonnormal.}

4. If $x_1, \dots, x_n$ is any finite list of vectors in a complex inner product space and if $a_{ij} = (x_j|x_i)$ for all i and j, then the matrix $A = (a_{ij})$ is unitarily equivalent to a matrix $\operatorname{diag}(r_1, \dots, r_n)$ with the r_i real and ≥ 0; in particular, $\det A \geq 0$.
{Hint: Note that $A^* = A$; cf. the hint for §9.3, Exercise 1.}

5. Let E be a unitary space, $T \in \mathcal{L}(\mathrm{E})$.

(i) If $T = UR$ with U, $R \in \mathcal{L}(\mathrm{E})$, U unitary and $R \geq 0$, then $R = (T^*T)^{1/2}$.

(ii) If $R = (T^*T)^{1/2}$ then $\|Tx\| = \|Rx\|$ for all $x \in \mathrm{E}$.

(iii) There exists a factorization $T = UR$ with U unitary and $R \geq 0$.

(iv) If T is bijective then the factorization $T = UR$ of (iii) is unique; it is called the *polar decomposition* of T. Explicitly, $R = (T^*T)^{1/2}$ and $U = TR^{-1}$.

{Hint: (iii) Define $U_1{:}R(\mathrm{E}) \to T(\mathrm{E})$ by $U_1(Rx) = Tx$ for all $x \in \mathrm{E}$; U_1 is well-defined and is an isomorphism of unitary spaces. Infer that $R(\mathrm{E})^\perp$ and $T(\mathrm{E})^\perp$ have the same dimension; let $U_2{:}R(\mathrm{E})^\perp \to T(\mathrm{E})^\perp$ be any isomorphism of unitary spaces. The mapping $y + z \mapsto U_1y + U_2z$ ($y \in R(\mathrm{E})$, $z \in R(\mathrm{E})^\perp$) is the desired unitary mapping U.}

6. Let E be a unitary space, $T \in \mathcal{L}(\mathrm{E})$, $R = (T^*T)^{1/2}$.

(i) There exists a unique partial isometry W with support $T^*(\mathrm{E})$ (§12.6, Exercise 5) such that $T = WR$.

(ii) The range of W is $T(\mathrm{E})$; thus W maps $T^*(\mathrm{E})$ isometrically onto $T(\mathrm{E})$ and is 0 on the orthogonal complement of $T^*(\mathrm{E})$.

The factorization $T = WR$ is called the *canonical factorization* of T; when T is bijective, this is the same as the polar decomposition of Exercise 5.

7. If $R, S, T \in \mathcal{L}(\mathrm{E})$ with $R \geq 0$ and $S \geq 0$, and if c is a real number ≥ 0, then $R + S \geq 0$, $cR \geq 0$ and $T^*RT \geq 0$. If, moreover, $RS = SR$, then $RS \geq 0$ and $(RS)^{\frac{1}{2}} = R^{\frac{1}{2}}S^{\frac{1}{2}} = S^{\frac{1}{2}}R^{\frac{1}{2}}$.

8. Let T be a linear mapping in a unitary space and write $T = UR$ with U unitary and $R \geq 0$ (Exercise 5). Then T is normal if and only if $UR = RU$.
{Hint: If T is normal then $R^2 = (URU^*)^2$.}

9. If $T \in \mathcal{L}(\mathrm{E})$ is normal and $R = (T^*T)^{\frac{1}{2}}$, then $T(\mathrm{E}) = T^*(\mathrm{E}) = R(\mathrm{E})$.
{Hint: $\operatorname{Ker} T^* = \operatorname{Ker} T = \operatorname{Ker} R$.}

10. Let $T = WR$ be the canonical factorization of $T \in \mathcal{L}(\mathrm{E})$ (Exercise 6). Then T is normal if and only if $WR = RW$.

{Hint: If T is normal then the linear subspace $R(\mathrm{E})$ reduces W (Exercise 9); let $P = W^*W$, $U = W + (I - P)$ and look at Exercise 8. On the other hand, if $WR = RW$ then $R(\mathrm{E})$ reduces W and $W|R(\mathrm{E})$ is a unitary mapping in $R(\mathrm{E})$.}

11. A matrix $A \in \mathrm{M}_n(\mathbf{C})$ is said to be *positive*, written $A \geqslant 0$, if $A^* = A$ and the characteristic roots of A (real, by Corollary 12.7.8) are $\geqslant 0$.

If E is an n-dimensional unitary space, $T \in \mathcal{L}(\mathrm{E})$ and A is the matrix of T relative to an orthonormal basis of E, show that $A \geqslant 0$ if and only if $T \geqslant 0$. State and prove the matricial analogues of the foregoing results on positive linear mappings (starting with Theorem 12.7.12).

12. Let E be a unitary space, $T \in \mathcal{L}(\mathrm{E})$.

(i) T is normal if and only if $T^* \in \{T\}''$ (see §8.2, Exercise 4 for the notation).

(ii) T is normal if and only if $T^* = p(T)$ for some polynomial $p \in \mathbf{C}[t]$.

(iii) Infer an alternate proof of Exercise 3, (i), (ii).

{Hint: (i), (ii) If T is normal, then it is diagonalizable in the sense of §8.2, Exercise 5. By (iii) of that exercise, $T^* \in \{T\}''$; cite (vi) of the same exercise.}

13. The appearance of determinant theory (via the characteristic polynomial) in the proof of the Spectral Theorem (12.7.4) is avoidable. {Hint: §8.3, Exercise 8.}

14. If E is a unitary space and $T \in \mathcal{L}(\mathrm{E})$ is normal, then $\mathrm{E} = T(\mathrm{E}) \oplus \operatorname{Ker} T$. {Hint: §12.5, Exercise 1.}

12.8 The Spectral Theorem: another way

In this section the Spectral Theorem for a normal linear mapping (12.7.4) will be given an 'analytic' proof that avoids the Fundamental theorem of algebra (as well as determinants); the key ideas occur in the earlier discussion of self-adjoint linear mappings in Euclidean spaces (9.3.4) and the effective substitute for the Fundamental theorem of algebra is the Weierstrass–Bolzano theorem.

Notations fixed for the section: E is a unitary space, $\mathrm{E} \neq \{\theta\}$ (to keep things interesting), and $T \in \mathcal{L}(\mathrm{E})$. The three lemmas on normality in the preceding section (12.7.1–12.7.3) will be needed here as well.

Every complex vector space—for example, every unitary space—can be regarded as a real vector space (by restricting scalar multiplication to real scalars).

12.8.1 Lemma *The mapping $W : \mathbf{R}^{2n} \to \mathbf{C}^n$ defined by*

$$W(a_1, b_1, a_2, b_2, \ldots, a_n, b_n) = (a_1 + ib_1, \ldots, a_n + ib_n)$$

is a real-linear bijection such that $\|Wx\| = \|x\|$ *for all* $x \in \mathbf{R}^{2n}$, *where the norms are derived from the canonical inner products in the Euclidean space* $\mathbf{R}^{2n}$ *and the unitary space* $\mathbf{C}^n$.

Proof. $|a_k + ib_k|^2 = |a_k|^2 + |b_k|^2$. ■

12.8.2 Lemma *For every linear mapping* $T \in \mathcal{L}(\mathrm{E})$, *there exists a constant* $K \geq 0$ *such that* $\|Tx\| \leq K\|x\|$ *for all* $x \in \mathrm{E}$.

Proof. Same as 9.2.1. ■

12.8.3 Lemma *If* $T \in \mathcal{L}(\mathrm{E})$ *then, for vectors* x_k, x *in* E,

$$\|x_k - x\| \to 0 \Rightarrow |(Tx_k|x_k) - (Tx|x)| \to 0.$$

Proof. Formally the same as in 9.2.8 (with $\mathbf{R}$ replaced by $\mathbf{C}$) and 9.2.10. ■

Recall that when T is self-adjoint, $(Tx|x)$ is real for all vectors x (12.6.6).

12.8.4 Lemma *If* $T^* = T$, *then there exists a unit vector* v *such that* $(Tu|u) \leq (Tv|v)$ *for all unit vectors* u, *consequently* $(Tx|x) \leq \|x\|^2(Tv|v)$ *for all vectors* x.

Proof. We can suppose $\mathrm{E} = \mathbf{C}^n$ (12.3.6). With notations as in 12.8.1, W transforms the unit vectors of the Euclidean space $\mathbf{R}^{2n}$ into the unit vectors of $\mathbf{C}^n$. It follows from the Weierstrass–Bolzano theorem in $\mathbf{R}$ that for every sequence (y_k) of unit vectors in $\mathbf{R}^{2n}$, there exist a unit vector y and a subsequence (y_{k_i}) of (y_k) such that $\|y_{k_i} - y\| \to 0$ (proof of 9.2.11); it follows that the analogous remark holds for a sequence of unit vectors in $\mathbf{C}^n$ (apply W^{-1}, etc.).

With K as in 12.8.2, we have $|(Tu|u)| \leq \|Tu\|\,\|u\| \leq K$ for all unit vectors u in E; if

$$M = \sup\{(Tu|u): \ \|u\| = 1\}$$

then, by the same argument as for 9.2.11 (making use of the remark in the preceding paragraph), there exists a unit vector v such that $M = (Tv|v)$.

Now let $x \in \mathrm{E}$ be arbitrary. If $x = \theta$, the inequality $(Tx|x) \leq \|x\|^2(Tv|v)$ reduces to $0 \leq 0$; if $x \neq \theta$, the inequality follows from $(Tu|u) \leq (Tv|v)$, where $u = \|x\|^{-1}x$. ■

12.8.5 Lemma *If* $T^* = T$ *and if* $x, y \in \mathrm{E}$ *are vectors such that* $(Tx|y)$ *is real, then*

$$(Tx|y) = \tfrac{1}{4}\{(T(x+y)|x+y) - (T(x-y)|x-y)\}.$$

Proof. As in the proof of 9.3.1, the strategy is to imitate the proof of the polarization identity (12.1.8, in the present context). Defining $[x, y] = (Tx|y)$ for arbitrary vectors x and y, we see that $[x, y]$ is linear in x and conjugate–linear in y. Since $T^* = T$, we know from 12.6.6 that for all vectors z, $[z, z] = (Tz|z)$ is real (but not necessarily ≥ 0, so we decline to take its square root). A proof analogous to that of the polarization identity yields

$$4[x, y] = [x + y, x + y] - [x - y, x - y] + i[x + iy, x + iy] - i[x - iy, x - iy].$$

If x and y are vectors such that $[x, y] = (Tx|y)$ is a real number then, since $[z, z]$ is real for all vectors z, the above equation simplifies to

$$4[x, y] = [x + y, x + y] - [x - y, x - y],$$

that is,

$$4(Tx|y) = (T(x + y)|x + y) - (T(x - y)|x - y),$$

as we wished to show. ■

12.8.6 Lemma *With notations as in 12.8.4, if also $(Tx|x) \geqslant 0$ for all vectors x, then $|(Tx|y)| \leqslant (Tv|v)$ for all unit vectors x and y.*

Proof. Write $M = (Tv|v)$. Suppose first that $(Tx|y)$ is real. Then, by 12.8.5 and 12.8.4,

$$|(Tx|y)| \leqslant \tfrac{1}{4}\{\|x + y\|^2 M + \|x - y\|^2 M\}$$
$$= \frac{M}{4}\{2\|x\|^2 + 2\|y\|^2\} = M,$$

as desired.

If $(Tx|y)$ is not real, let $c = |(Tx|y)|/(Tx|y)$. Then $|c| = 1$ and $|(Tx|y)| = c(Tx|y) = (T(cx)|y)$, where cx and y are unit vectors, so $|(Tx|y)| \leqslant M$ by the preceding case. ■

12.8.7 Lemma *With notations and assumptions as in 12.8.6, $Tv = rv$ for a suitable real number r.*

Proof. If $Tv = \theta$ we are done ($r = 0$ works); assume $Tv \neq \theta$. By 12.8.6, $|(Tv|y)| \leqslant (Tv|v)$ for all unit vectors y; applying this to $y = \|Tv\|^{-1}Tv$, we have $\|Tv\| \leqslant (Tv|v)$. On the other hand

$$(Tv|v) = |(Tv|v)| \leqslant \|Tv\|\,\|v\| = \|Tv\|,$$

whence equality. But then $|(Tv|v)| = \|Tv\| = \|Tv\|\,\|v\|$, so Tv is proportional to v, say $Tv = rv$ (12.1.9), and $r = r(v|v) = (Tv|v)$ is real (in fact, $\geqslant 0$). ■

12.8.8 Lemma *If $T^* = T$ then T has an eigenvector (with associated eigenvalue a real number).*

Proof. Let $m = \inf\{(Tu|u)\colon \|u\| = 1\}$. Then $(Tx|x) \geqslant m\|x\|^2$ for all vectors x, so $((T - mI)x|x) \geqslant 0$ for all vectors x. By 12.8.7, $T - mI$ has an eigenvector v, say $(T - mI)v = rv$; then $Tv = (m + r)v$ and we are done. ■

12.8.9 Lemma *If $T \in \mathcal{L}(\mathrm{E})$ is normal, then T has an eigenvector.*

Proof. Write $T = R + iS$ with R and S self-adjoint (12.6.5); since T is normal, we know that $RS = SR$.

By Lemma 12.8.8, R has an eigenvector u, say $Ru = ru$. Then the linear subspace $\mathrm{M} = \mathrm{Ker}(R - rI)$ is nonzero and $Rx = rx$ for all $x \in \mathrm{M}$. If $x \in \mathrm{M}$ then

$$R(Sx) = (RS)x = (SR)x = S(Rx) = S(rx) = r(Sx),$$

so $Sx \in \mathrm{M}$. Thus M is invariant under $S = S^*$, so M reduces S and $S|\mathrm{M}$ is self-adjoint (12.5.3); by 12.8.8, $S|\mathrm{M}$ has an eigenvector $v \in \mathrm{M}$, say $Sv = sv$. But also $Rv = rv$ (because $v \in \mathrm{M}$), so $Tv = Rv + iSv = (r + is)v$. ■

12.8.10 Theorem (Spectral Theorem) *If* E *is a unitary space and* $T \in \mathscr{L}(\mathrm{E})$ *is normal, then* E *has an orthonormal basis consisting of eigenvectors of* T.

Proof. The proof is by induction on $n = \dim \mathrm{E}$. If $n = 1$ then T is a scalar multiple of the identity and any unit vector serves as an orthonormal basis. Suppose $n \geqslant 2$ and assume that all's well for dimensions $<n$.

By 12.8.9, T has an eigenvalue $c \in \mathbf{C}$; then $\mathrm{M} = \mathrm{Ker}(\mathrm{T} - cI)$ is a nonzero linear subspace of E that reduces T (12.7.3). If $\mathrm{M} = \mathrm{E}$ then $T = cI$ and any orthonormal basis of E will do. If $\mathrm{M} \neq \mathrm{E}$ then $\mathrm{M}^\perp \neq \{\theta\}$ (12.2.7), $T|\mathrm{M}^\perp$ is normal (12.7.2), and $\dim \mathrm{M}^\perp < \dim \mathrm{E}$ (12.2.6), so by the induction hypothesis $\mathrm{M}^\perp$ has an orthonormal basis consisting of eigenvectors for $T|\mathrm{M}^\perp$ (in other words, for T); combining these vectors with any orthonormal basis of M, we get an orthonormal basis of E consisting of eigenvectors for T. ■

▶ ***Exercises***

1. If $T \in \mathscr{L}(\mathrm{E})$ and $(Tx|x) \geqslant 0$ for all vectors $x \in \mathrm{E}$, prove that $|(Tx|y|^2 \leqslant (Tx|x)(Ty|y)$ for all vectors x and y. {Hint: With $[x, y]$ defined as in the proof of Lemma 12.8.5, the problem is to prove the analogue of the Cauchy–Schwarz inequality.}

2. Reprove the assertions in §12.7, Exercise 2 (about commuting sets of normals) without appealing to the fact that the complex field is algebraically closed.
{Hint: First prove assertion (ii), along the lines of the argument suggested in §8.3, Exercise 7; use Lemma 12.8.9 and §12.6, Exercise 1 to get the ball rolling.}

3. (i) If $T \in \mathscr{L}(\mathrm{E})$, the **norm** of T is defined to be the number

$$\|T\| = \sup\{\|Tx\|\colon\ \|x\| \leqslant 1\}$$

(cf. 12.8.2, 9.2.7). Prove that $\|T\| \leqslant 1$ if and only if $I - T^*T \geqslant 0$ in the sense of Definition 12.7.11.

(ii) If $P \in \mathscr{L}(\mathrm{E})$ satisfies $P^2 = P$ and $\|P\| \leqslant 1$, then $P^* = P$ (thus P is a projection in the sense of §12.4, Exercise 2).

(iii) If P and Q are projections such that $PQ = (PQ)^2$, then PQ is a projection (cf. §12.6, Exercise 7).

{Hint: (ii) It suffices to show that $P = PP^*$. If $R = P - PP^*$ then $RR^* = -P(I - P^*P)P^*$; infer from (i) that $(RR^*x|x) \leqslant 0$ for every vector x. (iii) $\|PQ\| \leqslant 1$.}

13

Tensor products

V and W are **finite-dimensional** vector spaces over a field F, fixed for the rest of the chapter (to be definite, say $m = \dim V$, $n = \dim W$). There is also a theory of tensor products for infinite-dimensional spaces[1], but the assumption of finite-dimensionality greatly expedites the development of the theory; even with this restriction, the results obtained are adequate for valuable applications to the theory of representation of groups by matrices.

To set the flavor of the enterprise: addition is a way of combining matrices (of the same size) so that $\operatorname{tr}(A + B) = \operatorname{tr} A + \operatorname{tr} B$; multiplication is a way of combining matrices (of appropriate sizes) so that $\operatorname{tr}(AB) = \operatorname{tr}(BA)$; the *tensor product* is a way of combining matrices so that $\operatorname{tr}(A \otimes B) = (\operatorname{tr} A)(\operatorname{tr} B)$. (Of course we will have to define what $A \otimes B$ means!) All three formulas are of capital importance in the theory of matricial representations of groups; this is not the place to go into such applications[2], but it's reassuring to know they exist before immersing ourselves in tensor products.

13.1 Tensor product $V \otimes W$ of vector spaces

13.1.1 Definition We define $V \otimes W$ to be the vector space $\mathcal{L}(V', W)$, where V' is the dual space of V. It's as simple as that:

$$V \otimes W = \mathcal{L}(V', W).$$

We call $V \otimes W$ the **tensor product** of V and W (in that order). For the

[1] Cf. S. MacLane and G. Birkhoff, *Algebra*, Chelsea, 1988, Ch. IX, §8.

[2] Cf. D. Gorenstein, *Finite groups*, Harper & Row, 1968, p. 116; M. Hall, Jr., *The theory of groups*, Macmillan, 1959, p. 277; W. Ledermann, *Introduction to group characters*, Cambridge University Press, 1977, p. 29.

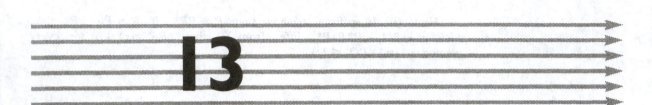

moment, $V \otimes W$ is offered as nothing more than an alternative notation for $\mathcal{L}(V', W)$, but in fact it is the formal properties of the notation that are the subject of this chapter.

13.1.2 Theorem $\dim(V \otimes W) = (\dim V)(\dim W)$.

Proof. $\dim \mathcal{L}(V', W) = (\dim V')(\dim W) = (\dim V)(\dim W)$ by 3.8.2 and 3.9.2. ∎

The notation $V \otimes W$ is motivated by the fact that it is possible to associate, with each pair of vectors $x \in V$ and $y \in W$, a linear mapping $x \otimes y: V' \to W$, in such a way that *every* linear mapping $T: V' \to W$ is a finite sum of such special linear mappings $x \otimes y$. The details are as follows.

13.1.3 Definition For each $x \in V$ and $y \in W$, define $T: V' \to W$ by the formula $Tf = f(x)y$ $(f \in V')$. We write $T = x \otimes y$; thus

$$(x \otimes y)(f) = f(x)y \quad (f \in V').$$

It is clear that $x \otimes y$ is linear, that is $x \otimes y \in \mathcal{L}(V', W) = V \otimes W$.

The reason one is led to such mappings will be clear from the proof of Theorem 13.1.5. Eventually, when sufficiently many properties of the notation $x \otimes y$ have been developed, we can stop thinking about the elements of $V \otimes W$ as linear mappings and treat $V \otimes W$ as an abstract vector space with certain characteristic mapping properties.

13.1.4 Theorem *The mapping* $V \times W \to V \otimes W$ *defined by* $(x, y) \mapsto x \otimes y$ *is bilinear, that is,*

$$(x_1 + x_2) \otimes y = x_1 \otimes y + x_2 \otimes y,$$
$$x \otimes (y_1 + y_2) = x \otimes y_1 + x \otimes y_2,$$
$$(cx) \otimes y = c(x \otimes y) = x \otimes (cy).$$

Proof. For example,

$$[(cx) \otimes y](f) = f(cx)y = cf(x)y = c(x \otimes y)(f) = [c(x \otimes y)](f)$$

for all $f \in V'$, therefore $(cx) \otimes y = c(x \otimes y)$. ∎

13.1.5 Theorem *Every* $T \in \mathcal{L}(V', W) = V \otimes W$ *is a finite sum of elements of the form* $x \otimes y$.

Proof. Let $y_1, \dots, y_n$ be a basis of W. We will show that there exist vectors $u_1, \dots, u_n \in V$ such that $T = u_1 \otimes y_1 + \dots + u_n \otimes y_n$.

Let $g_1, \dots, g_n$ be the basis of W' dual to $y_1, \dots, y_n$ (3.9.4). If $y \in W$, say

$$y = \sum_{k=1}^{n} c_k y_k,$$

then $g_k(y) = c_k$ for $k = 1, \dots, n$, thus

$$y = \sum_{k=1}^{n} g_k(y) y_k.$$

In particular for $y = Tf$ $(f \in V')$, we have

$$Tf = \sum_{k=1}^{n} g_k(Tf)y_k = \sum_{k=1}^{n} (T'g_k)(f)y_k,$$

where T' is the transpose of T (3.9.7). But $T'g_k \in (V')' = V''$, so by the principle of duality (3.9.17) there exists a vector $u_k \in V$ such that $T'g_k = u_k''$. Then, for all $f \in V'$,

$$(T'g_k)(f)y_k = u_k''(f)y_k = f(u_k)y_k = (u_k \otimes y_k)(f),$$

therefore

$$Tf = \sum_{k=1}^{n} (u_k \otimes y_k)(f);$$

thus $T = \sum u_k \otimes y_k$, as we wished to show. ■

13.1.6 Corollary *If $x_1, \ldots, x_m$ is a basis of V and $y_1, \ldots, y_n$ is a basis of W, then the mn elements $x_j \otimes y_k$ are a basis of $V \otimes W$.*

Proof. Since $\dim(V \otimes W) = mn$ (13.1.2), we need only show that the $x_j \otimes y_k$ generate $V \otimes W$ (3.5.13). Indeed, every $x \otimes y$ is a linear combination of the $x_j \otimes y_k$ by the bilinearity of the operation $\otimes$ (13.1.4), and every element of $V \otimes W$ is a finite sum of elements $x \otimes y$ (13.1.5). ■

13.1.7 Theorem *If*

$$\sum_{j=1}^{r} x_j \otimes y_j = 0$$

and $x_1, \ldots, x_r$ are linearly independent, then $y_1 = y_2 = \ldots = y_r = \theta$.

Proof. Choose $f_1, \ldots, f_r \in V'$ so that $f_i(x_j) = \delta_{ij}$ for $i, j = 1, \ldots, r$ (expand $x_1, \ldots, x_r$ to a basis of V and let $f_1, \ldots, f_r$ be the first r elements of the dual basis of V'). For $i = 1, \ldots, r$,

$$\theta = \sum_{j=1}^{r} (x_j \otimes y_j)(f_i) = \sum_{j=1}^{r} f_i(x_j)y_j = y_i. \ ■$$

Tensor products provide a way of transforming bilinear mappings into linear ones:

13.1.8 Theorem *If U is a vector space over F and $\beta: V \times W \to U$ is a bilinear mapping* (cf. 13.1.4), *then there exists a unique linear mapping $T_\beta: V \otimes W \to U$ such that*

$$T_\beta(x \otimes y) = \beta(x, y)$$

for all $x \in V$ and $y \in W$.

Proof. Uniqueness is clear from 13.1.5.

Existence: Choose bases $x_1, \ldots, x_m$ of V and $y_1, \ldots, y_n$ of W. Given $z \in V \otimes W$, express z (uniquely) as a linear combination of the $x_j \otimes y_k$ (13.1.6), say

$$z = \sum_{j,k} c_{jk}(x_j \otimes y_k),$$

and define

$$T_\beta z = \sum_{j,k} c_{jk}\beta(x_j, y_k).$$

Since each coefficient c_{jk} is a linear function of z, it is clear that the mapping $T_\beta{:}\mathrm{V} \otimes \mathrm{W} \to \mathrm{U}$ is linear. Finally, if $x \in \mathrm{V}$ and $y \in \mathrm{W}$, say

$$x = \sum_{j=1}^{m} a_j x_j, \quad y = \sum_{k=1}^{n} b_k y_k,$$

then

$$x \otimes y = \sum_{j,k} a_j b_k (x_j \otimes y_k),$$

therefore

$$T_\beta(x \otimes y) = \sum_{j,k} a_j b_k \beta(x_j, y_k) = \beta(x, y)$$

by the definition of T_β and the bilinearity of β. ■

13.1.9 Definition An (abstract) **tensor product** of V and W (as contrasted with the 'concrete' one defined in 13.1.1) is a pair $(\mathrm{P}, \odot)$, where P is a vector space over F and $(x, y) \mapsto x \odot y$ is a bilinear mapping $\mathrm{V} \times \mathrm{W} \to \mathrm{P}$, having the following property: *for every bilinear mapping* $\beta{:}\mathrm{V} \times \mathrm{W} \to \mathrm{U}$ (into a vector space U over F), *there exists a unique linear mapping* $S_\beta{:}\mathrm{P} \to \mathrm{U}$ *such that* $S_\beta(x \odot y) = \beta(x, y)$ *for all* $x \in \mathrm{V}$ *and* $y \in \mathrm{W}$.

The preceding results show that $(\mathrm{V} \otimes \mathrm{W}, \otimes)$ is a tensor product in the sense of 13.1.9; that is, a tensor product (in the sense of 13.1.9) *exists*. The next theorem will show that, in the appropriate sense, such a tensor product is *unique*; we separate out part of the argument as a lemma:

13.1.10 Lemma *If* $(\mathrm{P}, \odot)$ *is a tensor product of* V *and* W *and if* $S{:}\mathrm{P} \to \mathrm{P}$ *is a linear mapping such that* $S(x \odot y) = x \odot y$ *for all* $x \in \mathrm{V}$ *and* $y \in \mathrm{W}$, *then* $S = I$.

Proof. For the bilinear mapping $\beta{:}\mathrm{V} \times \mathrm{W} \to \mathrm{P}$ defined by $\beta(x, y) = x \odot y$, S and I are linear mappings $\mathrm{P} \to \mathrm{P}$ such that $S(x \odot y) = x \odot y = I(x \odot y)$ for all x and y, so $S = I$ by uniqueness in 13.1.9. ■

13.1.11 Theorem *If* $(\mathrm{P}, \odot)$ *is a tensor product of* V *and* W *in the sense of* 13.1.9, *then there exists a vector space isomorphism* $S{:}\mathrm{P} \to \mathrm{V} \otimes \mathrm{W}$ *such that*

$$S(x \odot y) = x \otimes y$$

for all $x \in \mathrm{V}$ *and* $y \in \mathrm{W}$.

Proof. If $\beta{:}\mathrm{V} \times \mathrm{W} \to \mathrm{V} \otimes \mathrm{W}$ is the bilinear mapping $\beta(x, y) = x \otimes y$, then there exists a linear mapping $S{:}\mathrm{P} \to \mathrm{V} \otimes \mathrm{W}$ such that $S(x \odot y) = \beta(x, y) = x \otimes y$ for all x and y (13.1.9). On the other hand, if $\gamma{:}\mathrm{V} \times \mathrm{W} \to \mathrm{P}$ is the bilinear mapping $\gamma(x, y) = x \odot y$, then there exists a linear mapping

$T{:}\mathrm{V}\otimes\mathrm{W}\to\mathrm{P}$ such that $T(x\otimes y)=\gamma(x,y)=x\odot y$ for all x and y (13.1.8). Then $TS{:}\mathrm{P}\to\mathrm{P}$ is a linear mapping such that

$$(TS)(x\odot y)=T(S(x\odot y))=T(x\otimes y)=x\odot y$$

for all x and y, so $TS=I$ by the lemma. Similarly $ST=I$, so S is an isomorphism. ∎

13.1.12 Corollary *If* $(\mathrm{P},\odot)$ *is a tensor product of* V *and* W *(in the sense of* 13.1.9*), then every element of* P *is a finite sum of elements* $x\odot y$.

Proof. Immediate from the theorem and 13.1.5. ∎

Theorem 13.1.11 allows us to forget that $\mathrm{V}\otimes\mathrm{W}$ started life as a space of linear mappings, provided we use only the properties of 13.1.9 and their consequences. Thus it is appropriate to back up and restate 13.1.6 in abstract form:

13.1.13 Theorem *If* $(\mathrm{P},\odot)$ *is a tensor product of* V *and* W *(in the sense of* 13.1.9*),* $x_1,\ldots,x_{\mathrm{m}}$ *is a basis of* V*, and* $y_1,\ldots,y_{\mathrm{n}}$ *is a basis of* W*, then the vectors* $x_{\mathrm{j}}\odot y_{\mathrm{k}}$ *are a basis of* P.

Proof. Immediate from 13.1.6 and 13.1.11. ∎

The 'abstract' form of 13.1.7 follows similarly from 13.1.7 and 13.1.11. There is also a converse of 13.1.13:

13.1.14 Theorem *Let* P *be a vector space over* F*,* $(x,y)\mapsto x\odot y$ *a bilinear mapping* $\mathrm{V}\times\mathrm{W}\to\mathrm{P}$*. If there exist bases* $x_1,\ldots,x_{\mathrm{m}}$ *of* V *and* $y_1,\ldots,y_{\mathrm{n}}$ *of* W *such that the vectors* $x_{\mathrm{j}}\odot y_{\mathrm{k}}$ *are a basis of* P*, then* $(\mathrm{P},\odot)$ *is a tensor product of* V *and* W.

Proof. Formally the same as the proof of 13.1.8. ∎

▶ ***Exercises***

1. Deduce 13.1.6 from 13.1.7 by arguing that the elements $x_{\mathrm{j}}\otimes y_{\mathrm{k}}$ are linearly independent.

2. If x and y are vectors such that $x\otimes y=\theta$, then either $x=\theta$ or $y=\theta$.

3. Let U, V, W be vector spaces over F, with V and W finite-dimensional.

(i) There is an isomorphism $\mathcal{B}(\mathrm{V}\times\mathrm{W},\mathrm{U})\to\mathcal{L}(\mathrm{V}\otimes\mathrm{W},\mathrm{U})$, where $\mathcal{B}(\mathrm{V}\times\mathrm{W},\mathrm{U})$ is the space of all bilinear mappings $\mathrm{V}\times\mathrm{W}\to\mathrm{U}$ (with the pointwise linear operations). {Hint: Theorem 13.1.8.}

(ii) $\mathcal{B}(\mathrm{V}\times\mathrm{W},\mathrm{F})\cong(\mathrm{V}\otimes\mathrm{W})'$, where $\mathcal{B}(\mathrm{V}\times\mathrm{W},\mathrm{F})$ is the space of all bilinear forms $\mathrm{V}\times\mathrm{W}\to\mathrm{F}$ (cf. 7.1.1).

(iii) $\mathrm{V}\otimes\mathrm{W}\cong\mathcal{B}(\mathrm{V}\times\mathrm{W},\mathrm{F})'$ (the dual space of the space of bilinear forms).

(iv) The theory of tensor products can be based on *defining* $\mathrm{V}\otimes\mathrm{W}$ to be the dual space of $\mathcal{B}(\mathrm{V}\times\mathrm{W},\mathrm{F})$, with associated mapping $(x,y)\mapsto x\otimes y$, where $x\otimes y$ is defined by the formula $(x\otimes y)(\beta)=\beta(x,y)$ for all $\beta\in\mathcal{B}(\mathrm{V}\times\mathrm{W},\mathrm{F})$.

4. If P is a vector space over F, $(x, y) \mapsto x \odot y$ is a bilinear mapping $V \times W \to P$, and $T:V \otimes W \to P$ is a linear bijection such that $T(x \otimes y) = x \odot y$ for all $x \in V$ and $y \in W$, then $(P, \odot)$ is a tensor product of V and W.

{Hint: If $\beta:V \times W \to U$ is bilinear and $T_\beta:V \otimes W \to U$ is the unique linear mapping such that $T_\beta(x \otimes y) = \beta(x, y)$ for all $x \in V$, $y \in W$, consider $T_\beta T^{-1}:P \to U$.}

5. Suppose P is a vector space over F and $(x, y) \mapsto x \odot y$ is a bilinear mapping $V \times W \to P$ such that (1) every element of P is a finite sum of elements $x \odot y$, and (2) if $x_1, \ldots, x_m$ is a basis of V and $w_1, \ldots, w_m$ are vectors in W such that $\sum_i x_i \odot w_i = \theta$, then necessarily $w_1 = \ldots = w_m = \theta$. (In a sense, $x_1, \ldots, x_m$ are 'linearly independent over W' with $\odot$ playing the role of the 'scalar multiple' operation.) Then $(P, \odot)$ is a tensor product of V and W.

{Hint: Let $S:V \otimes W \to P$ be the unique linear mapping such that $S(x \otimes y) = x \odot y$ for all x and y. It is clear from (1) that S is surjective. It follows from (2) that S is injective; for, if $x_1, \ldots, x_m$ and $y_1, \ldots, y_n$ are bases of V and W and if $u = \sum a_{ij} x_i \otimes y_j$ is an element of $V \otimes W$ such that $Su = \theta$, then $\theta = \sum_i x_i \odot (\sum_j a_{ij} y_j)$, therefore $a_{ij} = 0$ for all i and j (by (2) and the independence of $y_1, \ldots, y_n$), whence $u = 0$. Now cite Exercise 4.}

6. Prove that $(V \otimes W)' = V' \otimes W'$. More precisely (with 13.1.9 in mind), let $P = (V \otimes W)'$ and consider the bilinear mapping $V' \times W' \to P$ defined by $(f, g) \mapsto f \odot g$, where $f \odot g$ is the unique linear form on $V \otimes W$ such that $(f \odot g)(x \otimes y) = f(x)g(y)$ for all $x \in V$ and $y \in W$.

{Hint: For the existence of such $f \odot g \in (V \otimes W)' = \mathcal{L}(V \otimes W, F)$, consider the bilinear form $V \times W \to F$ defined by $(x, y) \mapsto f(x)g(y)$. Let $T:V' \otimes W' \to P$ be the unique linear mapping such that $T(f \otimes g) = f \odot g$ for all $f \in V'$ and $g \in W'$. By Exercise 4, it suffices to show that T is bijective; on grounds of dimensionality, it is enough to show that T is injective. Suppose $Tu = 0$. Let $x_1, \ldots, x_m$ be a basis of V and $f_1, \ldots, f_m$ the dual basis of V', write $u = \sum f_i \otimes g_i$ for suitable $g_1, \ldots, g_m \in W'$, and calculate $0 = (Tu)(x_i \otimes y)$ for arbitrary $y \in W$.}

7. Given linear subspaces $M \subset V$ and $N \subset W$, $M \otimes N$ is a linear subspace of $V \otimes W$ in the following sense: if P is the linear subspace of $V \otimes W$ generated by the vectors $x \otimes y$ with $x \in M$ and $y \in N$ (in other words, P is the set of all finite sums of such vectors), and if $\beta:M \times N \to P$ is the bilinear mapping $\beta(x, y) = x \otimes y$ $(x \in M, y \in N)$, then the pair (P, β) is a tensor product of M and N. {Hint: Theorem 13.1.14.}

8. Let M, M_1, M_2 be linear subspaces of V, and N, N_1, N_2 linear subspaces of W; regard $M \otimes N$, $M_i \otimes N_i$, etc., as linear subspaces of $V \otimes W$ (Exercise 7). The following progression of statements[3] culminates in an important dimension formula.

[3] Inspired by W. H. Greub, *Multilinear algebra*, Springer, 1967, Ch. 1, §4.

(i) $(V \otimes N_1) \cap (V \otimes N_2) = V \otimes (N_1 \cap N_2)$. In particular, if $N_1 \cap N_2 = \{\theta\}$ then $(V \otimes N_1) \cap (V \otimes N_2) = \{\theta\}$.

(ii) If $W = N_1 \oplus N_2$ then $V \otimes W = (V \otimes N_1) \oplus (V \otimes N_2)$.

(iii) $(M \otimes W) \cap (V \otimes N) = M \otimes N$.

(iv) Prove that

$$\dim(M \otimes W + V \otimes N) = (\dim M)(\dim W) + (\dim V)(\dim N) - (\dim M)(\dim N).$$

{Hints:

(i) The inclusion $\supset$ is obvious. Suppose $u \in (V \otimes N_1) \cap (V \otimes N_2)$. Let $x_1, \ldots, x_m$ be a basis of V, write $u = \sum x_i \otimes y_i = \sum x_i \otimes z_i$ with $y_i \in N_1$ and $z_i \in N_2$, and apply Theorem 13.1.7 to the relation $\sum x_i \otimes (y_i - z_i) = \theta$.

(ii) It is obvious that $V \otimes W = V \otimes N_1 + V \otimes N_2$; cf. (i).

(iii) Let N^* be a linear subspace of W such that $W = N \oplus N^*$. By (ii), $M \otimes W = (M \otimes N) \oplus (M \otimes N^*)$; noting that $V \otimes N \supset M \otimes N$, apply the modular law (§1.6, Exercise 5) to calculate $(V \otimes N) \cap (M \otimes W)$.

(iv) Cf. Theorem 13.1.2 and §3.6, Exercise 2.}

9. With notations as in Exercise 8, $(M_1 \otimes N_1) \cap (M_2 \otimes N_2) = (M_1 \cap M_2) \otimes (N_1 \cap N_2)$.

10. If E and F are Euclidean spaces, there exists a unique inner product on $E \otimes F$ such that $(x \otimes y | u \otimes v) = (x|u)(y|v)$ for all $x, u \in E$ and $y, v \in F$.

{Hint: Let $x_1, \ldots, x_m$ be an orthonormal basis of E. By Theorem 13.1.7, every $a \in E \otimes F$ has a unique representation $a = \sum x_i \otimes y_i$ with $y_1, \ldots, y_m \in F$; if also $b = \sum x_i \otimes z_i$ with $z_1, \ldots, z_m \in F$, define $(a|b) = \sum (y_i|z_i)$.}

11. With notations as in Exercise 10, if $x_1, \ldots, x_m$ is an orthonormal basis of E and $y_1, \ldots, y_n$ is an orthonormal basis of F, then the vectors $x_i \otimes y_j$ form an orthonormal basis of $E \otimes F$.

12. The analogues of Exercises 10 and 11 are valid for unitary spaces. {Hint: With notations as in the hint for Exercise 10, observe that $(a|b)$ is a conjugate linear function of b; otherwise, the argument is identical.}

13.2 Tensor product $S \otimes T$ of linear mappings

We assume henceforth that $(V \otimes W, \otimes)$ is *any* tensor product of V and W (not necessarily the one defined in 13.1.1; cf. 13.1.11).

13.2.1 Theorem *If $S \in \mathscr{L}(V)$ and $T \in \mathscr{L}(W)$, there exists a unique linear mapping $R \in \mathscr{L}(V \otimes W)$ such that*

$$R(x \otimes y) = Sx \otimes Ty$$

for all $x \in \mathrm{V}$ *and* $y \in \mathrm{W}$.

Proof. Let $\beta: \mathrm{V} \times \mathrm{W} \to \mathrm{V} \otimes \mathrm{W}$ be the bilinear mapping defined by $\beta(x, y) = Sx \otimes Ty$ and let $R: \mathrm{V} \otimes \mathrm{W} \to \mathrm{V} \otimes \mathrm{W}$ be the unique linear mapping provided by 13.1.9. ∎

13.2.2 Definition With notations as in 13.2.1, we write $R = S \otimes T$ and call it the **tensor product** of S and T. (The terminology will be justified shortly.) Thus

$$(S \otimes T)(x \otimes y) = Sx \otimes Ty$$

for all $x \in \mathrm{V}$ and $y \in \mathrm{W}$.

13.2.3 Theorem *With notations as in* 13.2.2, *the mapping*

$$\mathscr{L}(\mathrm{V}) \times \mathscr{L}(\mathrm{W}) \to \mathscr{L}(\mathrm{V} \otimes \mathrm{W})$$

defined by $(S, T) \mapsto S \otimes T$ *is bilinear*:

$$(S_1 + S_2) \otimes T = S_1 \otimes T + S_2 \otimes T,$$
$$S \otimes (T_1 + T_2) = S \otimes T_1 + S \otimes T_2,$$
$$(cS) \otimes T = c(S \otimes T) = S \otimes (cT).$$

Moreover, $I \otimes I = I$ *(where the same letter* I *stands for the appropriate three identity mappings).*

Proof. For example,

$$\begin{aligned}[S \otimes (T_1 + T_2)](x \otimes y) &= Sx \otimes (T_1 + T_2)y \\ &= Sx \otimes (T_1 y + T_2 y) \\ &= Sx \otimes T_1 y + Sx \otimes T_2 y \\ &= (S \otimes T_1)(x \otimes y) + (S \otimes T_2)(x \otimes y) \\ &= (S \otimes T_1 + S \otimes T_2)(x \otimes y)\end{aligned}$$

for all $x \in \mathrm{V}$ and $y \in \mathrm{W}$, therefore $S \otimes (T_1 + T_2) = S \otimes T_1 + S \otimes T_2$ (because the $x \otimes y$ generate $\mathrm{V} \otimes \mathrm{W}$). ∎

13.2.4 Theorem *With notations as in* 13.2.2,

$$(S_1 \otimes T_1)(S_2 \otimes T_2) = S_1 S_2 \otimes T_1 T_2$$

for all $S_1, S_2 \in \mathscr{L}(\mathrm{V})$ *and* $T_1, T_2 \in \mathscr{L}(\mathrm{W})$.

Proof. For all $x \in \mathrm{V}$ and $y \in \mathrm{W}$.

$$\begin{aligned}[(S_1 \otimes T_1)(S_2 \otimes T_2)](x \otimes y) &= (S_1 \otimes T_1)(S_2 x \otimes T_2 y) \\ &= (S_1 S_2 x) \otimes (T_1 T_2 y) \\ &= (S_1 S_2 \otimes T_1 T_2)(x \otimes y),\end{aligned}$$

whence the asserted equality. ∎

13.2.5 Corollary *Suppose that the spaces* V *and* W *are nonzero and let* $S \in \mathcal{L}(\mathrm{V})$, $T \in \mathcal{L}(\mathrm{W})$. *Then*

$$S \otimes T \textit{ is bijective} \Leftrightarrow S \textit{ and } T \textit{ are bijective},$$

in which case $(S \otimes T)^{-1} = S^{-1} \otimes T^{-1}$.

Proof. $\Leftarrow$: By Theorems 13.2.4 and 13.2.3,

$$(S \otimes T)(S^{-1} \otimes T^{-1}) = SS^{-1} \otimes TT^{-1} = I \otimes I = I,$$

and similarly for the product in the reverse order.

$\Rightarrow$: By hypothesis, $S \otimes T$ is injective; it will suffice to show that S and T are injective (3.7.4). If $x \in \mathrm{V}$ and $y \in \mathrm{W}$ are nonzero, then $x \otimes y \neq \theta$ (13.1.7), therefore $Sx \otimes Ty = (S \otimes T)(x \otimes y)$ is nonzero by the injectivity of $S \otimes T$, consequently $Sx \neq \theta$ and $Ty \neq \theta$. {Note that if, for example, $\mathrm{W} = \{\theta\}$, then $\mathrm{V} \otimes \mathrm{W} = \{\theta\}$, therefore $S \otimes T = S \otimes 0 = 0$ is trivially bijective, regardless of whether or not S is bijective.} ■

13.2.6 Theorem *With notations as in* 13.2.2, $(\mathcal{L}(\mathrm{V} \otimes \mathrm{W}), \otimes)$ *is a tensor product of* $\mathcal{L}(\mathrm{V})$ *and* $\mathcal{L}(\mathrm{W})$; *suggestively,*

$$\mathcal{L}(\mathrm{V} \otimes \mathrm{W}) = \mathcal{L}(\mathrm{V}) \otimes \mathcal{L}(\mathrm{W}).$$

Proof. Let $x_1, \ldots, x_m$ be a basis of V, and $y_1, \ldots, y_n$ a basis of W. For i, $j = 1, \ldots, m$ let $S_{ij} \in \mathcal{L}(\mathrm{V})$ be the linear mapping such that

$$S_{ij}x_r = \begin{cases} x_i & \text{if } r = j \\ \theta & \text{if } r \neq j \end{cases}$$

(cf. 3.8.1); thus $S_{ij}x_r = \delta_{jr}x_i$ for $r = 1, \ldots, m$. Similarly, for h, $k = 1, \ldots, n$, let $T_{hk} \in \mathcal{L}(\mathrm{W})$ be the linear mapping such that $T_{hk}y_s = \delta_{ks}y_h$ for $s = 1, \ldots, n$. It is easy to see that the m^2 mappings S_{ij} are a basis of $\mathcal{L}(\mathrm{V})$. {Sketch of proof: By 3.8.2 and 3.5.13, it suffices to show that they are linearly independent; apply a vanishing linear combination of the S_{ij} to a vector x_r.} Similarly, the n^2 mappings T_{hk} are a basis of $\mathcal{L}(\mathrm{W})$.

By 13.1.14, it will suffice to show that the m^2n^2 mappings $S_{ij} \otimes T_{hk}$ are a basis of $\mathcal{L}(\mathrm{V} \otimes \mathrm{W})$; by 13.1.2, $\mathcal{L}(\mathrm{V} \otimes \mathrm{W})$ has dimension $(mn)^2$, so we need only show that the mappings are linearly independent. Suppose

$$\sum_{i,j,h,k} c_{ijhk}(S_{ij} \otimes T_{hk}) = 0.$$

Applying both sides to a vector $x_r \otimes y_s$ ($r = 1, \ldots, m$; $s = 1, \ldots, n$), we have

$$\sum_{i,j,h,k} c_{ijhk}(S_{ij}x_r) \otimes (T_{hk}y_s) = \theta,$$

thus

$$\sum_{i,j,h,k} c_{ijhk}\delta_{jr}\delta_{ks}(x_i \otimes y_h) = \theta,$$

that is,

$$\sum_{i,h} c_{irhs}(x_i \otimes y_h) = \theta;$$

since the $x_i \otimes y_h$ are linearly independent (13.1.13), it follows that $c_{irhs} = 0$ for all i and h (and for all r and s), as we wished to show. ■

▶ Exercises

1. With $V \otimes W = \mathcal{L}(V', W)$ as in 13.1.1, let $S \in \mathcal{L}(V)$, $T \in \mathcal{L}(W)$. Let $R: V \otimes W \to V \otimes W$ be the mapping defined by $X \mapsto TXS'$:

$$\begin{array}{ccc} V' & \xrightarrow{\ X\ } & W \\ S' \uparrow & & \downarrow T \\ V' & & W \end{array}$$

(i) Show that $R = S \otimes T$. {Hint: Calculate $[T(x \otimes y)S']f$ for $x \in V$, $y \in W$, $f \in V'$.}

(ii) Deduce the formulas of this section from (i). {For example, the proof of the formula in 13.2.4 will entail $(S_1S_2)' = S_2'S_1'$.}

2. If $T_1: V_1 \to W_1$ and $T_2: V_2 \to W_2$ are linear mappings (where V_i, W_i are finite-dimensional vector spaces over a field F), show that there exists a unique linear mapping (denoted) $T_1 \otimes T_2: V_1 \otimes V_2 \to W_1 \otimes W_2$ such that $(T_1 \otimes T_2)(x_1 \otimes x_2) = T_1x_1 \otimes T_2x_2$ for all vectors $x_1 \in V_1$ and $x_2 \in V_2$. {Hint: Consider the mapping $(x_1, x_2) \mapsto T_1x_1 \otimes T_2x_2$.}

3. If $S \in \mathcal{L}(V)$ and $T \in \mathcal{L}(W)$, then $S(V) \otimes T(W)$ can be identified with the linear subspace $(S \otimes T)(V \otimes W)$ of $V \otimes W$ (cf. §13.1, Exercise 7), whence a formula for the rank of $S \otimes T$: $\rho(S \otimes T) = \rho(S)\rho(T)$.

4. If $S \in \mathcal{L}(V)$ and $T \in \mathcal{L}(W)$, prove that $\operatorname{Ker}(S \otimes T) = (\operatorname{Ker} S) \otimes W + V \otimes (\operatorname{Ker} T)$.

{Hint: Both sides are linear subspaces of $V \otimes W$ (§13.1, Exercise 7) and the inclusion $\supset$ is obvious, so it suffices to show that both sides have the same dimension; cf. Exercise 3 and §13.1, Exercise 8 (iv).}

5. If E and F are Euclidean spaces and $E \otimes F$ is the Euclidean space defined in §13.1, Exercise 10, then $(S \otimes T)^* = S^* \otimes T^*$ for all $S \in \mathcal{L}(E)$ and $T \in \mathcal{L}(F)$.

6. If c is an eigenvalue of $S \in \mathcal{L}(V)$ and d is an eigenvalue of $T \in \mathcal{L}(W)$, then cd is an eigenvalue of $S \otimes T$.

{Hint: $S \otimes T - cdI = S \otimes (T - dI) + (S - cI) \otimes dI$, where the I's stand for the appropriate identity mappings. See also §13.3, Exercise 5.}

13.3 Matrices of tensor products

Let $S \in \mathscr{L}(V)$, choose a basis $x_1, \ldots, x_m$ of V, and let $A = (a_{ij})$ be the matrix of S relative to this basis:

$$Sx_j = \sum_{i=1}^{m} a_{ij} x_i.$$

Similarly, let $T \in \mathscr{L}(W)$ and let $B = (b_{hk})$ be the matrix of T relative to a basis $y_1, \ldots, y_n$ of W:

$$Ty_k = \sum_{h=1}^{n} b_{hk} y_h.$$

We know that the $x_j \otimes y_k$ are a basis of $V \otimes W$ (Theorem 13.1.13). Question: What is the matrix of $S \otimes T$ relative to this basis?

The question can't be answered until we specify an **ordering** of the basis vectors $x_j \otimes y_k$. Once this is settled in a reasonable way, the answer to the question will be the $mn \times mn$ block matrix $A \otimes B$ defined by the formula

$$A \otimes B = \begin{pmatrix} a_{11}B & a_{12}B & \cdots & a_{1m}B \\ a_{21}B & a_{22}B & \cdots & a_{2m}B \\ \vdots & \vdots & & \vdots \\ a_{m1}B & a_{m2}B & \cdots & a_{mm}B \end{pmatrix}$$

where, for example, $a_{11}B = (a_{11}b_{hk})$ is the $n \times n$ minor in the northwest corner.

The ordering of the $x_j \otimes y_k$ that accomplishes this is the one effected by the 'lexicographic order' on the index pairs: $(j, k) < (j', k')$ means that either (1) $j < j'$, or (2) $j = j'$ and $k < k'$. Thus, we let $j = 1$ and list the $x_1 \otimes y_k$ in order of increasing k, then let $j = 2$ and list the $x_2 \otimes y_k$ in order of increasing k, and so on:

$$x_1 \otimes y_1, \ldots, x_1 \otimes y_n; x_2 \otimes y_1, \ldots, x_2 \otimes y_n; \ldots; x_m \otimes y_1, \ldots, x_m \otimes y_n.$$

13.3.1 Theorem *The matrix of $S \otimes T$ relative to the preceding basis of $V \otimes W$ is the matrix $A \otimes B$ defined above.*

Proof. Let C be the matrix of $S \otimes T$ relative to the basis in question. We have

$$\begin{aligned}(S \otimes T)(x_j \otimes y_k) &= Sx_j \otimes Ty_k \\ &= \left(\sum_{i=1}^{m} a_{ij} x_i\right) \otimes \left(\sum_{h=1}^{n} b_{hk} y_h\right) \\ &= \sum_{(i,h)=(1,1)}^{(m,n)} a_{ij} b_{hk} (x_i \otimes y_h),\end{aligned}$$

where, in the latter sum, (i, h) increases from $(1, 1)$ to (m, n) in the sense of the ordering described above. Thus, the (j, k)'th column of C is

$$\begin{pmatrix} a_{1j}b_{1k} \\ a_{1j}b_{2k} \\ \vdots \\ a_{1j}b_{nk} \\ a_{2j}b_{1k} \\ a_{2j}b_{2k} \\ \vdots \\ a_{2j}b_{nk} \\ \vdots \\ a_{mj}b_{1k} \\ a_{mj}b_{2k} \\ \vdots \\ a_{mj}b_{nk} \end{pmatrix}$$

This needs to be organized a little more succinctly! Writing β_k for the k'th column of B ($k = 1, \ldots, n$), the (j, k)'th column of C can be written

$$\begin{pmatrix} a_{1j}\beta_k \\ a_{2j}\beta_k \\ \vdots \\ a_{mj}\beta_k \end{pmatrix}.$$

For fixed j, as k runs from 1 to n—so that (j, k) increases from $(j, 1)$ to (j, n)—the columns β_k describe the matrix

$$(\beta_1 \ldots \beta_n) = B,$$

and the entries of C in question describe the block matrix

$$\begin{pmatrix} a_{1j}B \\ a_{2j}B \\ \vdots \\ a_{mj}B \end{pmatrix}$$

(an $mn \times n$ matrix). Now letting j increase from 1 to m, the preceding block matrices trace out the matrix $A \otimes B$ described earlier. ■

13.3.2 Definition We call $A \otimes B$ the **tensor product** (or *Kronecker product*) of the matrices A and B.

The message of 13.3.1: relative to the indicated bases, the matrix of the tensor product is the tensor product of the matrices.

There is some arbitrariness in this definition; had we chosen to order the index pairs (j, k) the *other* natural way (with the second entry taking precedence over the first), the matrix of $S \otimes T$ would have come out otherwise (can you guess its form?).

To facilitate the study of such matrices $A \otimes B$, we must develop some algebraic properties of the notation:

13.3.3 Theorem *The mapping* $M_m(F) \times M_n(F) \to M_{mn}(F)$ *defined by* $(A, B) \mapsto A \otimes B$ *is bilinear:*

(1) $(A_1 + A_2) \otimes B = A_1 \otimes B + A_2 \otimes B$,
(2) $A \otimes (B_1 + B_2) = A \otimes B_1 + A \otimes B_2$,
(3) $(cA) \otimes B = c(A \otimes B) = A \otimes (cB)$.

Moreover,

(4) $I \otimes I = I$

(*where* I *stands for the identity matrices of the appropriate sizes*), *and*

(5) $(A \otimes B)(C \otimes D) = AC \otimes BD$

for all $A, C \in M_m(F)$ *and* $B, D \in M_n(F)$.

Proof. For example, let us prove (5). Choose bases $x_1, \ldots, x_m$ of V and $y_1, \ldots, y_n$ of W. Let $S, P \in \mathcal{L}(V)$ be the linear mappings with matrices A, C relative to the basis $x_1, \ldots, x_m$; then SP has matrix AC relative to this basis. Similarly, let $T, Q \in \mathcal{L}(W)$ be the linear mappings with matrices B, D relative to the basis $y_1, \ldots, y_n$; then TQ has matrix BD relative to this basis.

We know from 13.3.1 that, relative to the basis $x_j \otimes y_k$ ordered in the indicated way,

$$\begin{aligned} S \otimes T &\quad \text{has matrix} \quad A \otimes B, \\ P \otimes Q &\quad \text{has matrix} \quad C \otimes D, \\ SP \otimes TQ &\quad \text{has matrix} \quad AC \otimes BD; \end{aligned}$$

since $(S \otimes T)(P \otimes Q) = SP \otimes TQ$ by Theorem 13.2.4, we conclude that $(A \otimes B)(C \otimes D) = AC \otimes BD$.

The other assertions of the theorem follow similarly from 13.2.3. ■

13.3.4 Corollary $A \otimes B$ *is invertible if and only if* A *and* B *are both invertible, in which case* $(A \otimes B)^{-1} = A^{-1} \otimes B^{-1}$.

Proof. After choosing bases of V and W and introducing linear mappings S and T whose matrices relative to these bases are A and B, respectively, the corollary follows easily from 13.2.5 and 13.3.1. ■

13.3.5 Corollary *If* A_1, A_2 *are similar and* B_1, B_2 *are similar, then* $A_1 \otimes B_1$, $A_2 \otimes B_2$ *are similar.*

Proof. This follows easily from 13.3.4 and (5) of 13.3.3. ■

The payoff:

13.3.6 Theorem *If* $A \in M_m(F)$ *and* $B \in M_n(F)$, *then*

(6) $\operatorname{tr}(A \otimes B) = (\operatorname{tr} A)(\operatorname{tr} B)$,
(7) $|A \otimes B| = |A|^n |B|^m$.

Proof. Formula (6) follows immediately from the definition of $A \otimes B$: the diagonal of $A \otimes B$ is obtained by stringing together the diagonals of $a_{11}B$, $a_{22}B, \ldots, a_{mm}B$, thus

$$\begin{aligned}\operatorname{tr}(A \otimes B) &= a_{11}(\operatorname{tr} B) + a_{22}(\operatorname{tr} B) + \ldots + a_{mm}(\operatorname{tr} B)\\ &= (a_{11} + a_{22} + \ldots + a_{mm})(\operatorname{tr} B)\end{aligned}$$

as claimed.

To prove formula (7), we can assume that the characteristic polynomials of A and B factor completely over F. {Let $\mathrm{K} \supset \mathrm{F}$ be an overfield in which the characteristic polynomials of A and B factor completely into linear factors (Appendix B.5.3); the determinants to be calculated are the same, whether we regard the matrix entries as being in F or in K.} It then follows that A and B are both similar to triangular matrices (cf. the proof of 8.3.12), say with 0's below the main diagonal; in view of 13.3.5 and the invariance of determinants under similarity (8.1.4), we can suppose that A and B are already triangular. It is then clear that $A \otimes B$ is triangular,

$$A \otimes B = \begin{pmatrix} a_{11}B & a_{12}B & \cdots & a_{1m}B \\ & a_{22}B & \cdots & a_{2m}B \\ & & \ddots & \vdots \\ \huge 0 & & & a_{mm}B \end{pmatrix};$$

since the determinant of a triangular matrix is the product of the diagonal elements, we have

$$\begin{aligned}|A \otimes B| &= (a_{11})^{n}|B| \cdot (a_{22})^{n}|B| \ldots (a_{mm})^{n}|B|\\ &= (a_{11}a_{22} \ldots a_{mm})^{n}|B|^{m}\\ &= |A|^{n}|B|^{m}. \blacksquare\end{aligned}$$

The terminology and notation of 13.3.2 are justified by the following theorem (which could have been proved immediately after 13.3.3):

13.3.7 Theorem $(\mathrm{M}_{mn}(\mathrm{F}), \otimes)$ *is a tensor product of* $\mathrm{M}_m(\mathrm{F})$ *and* $\mathrm{M}_n(\mathrm{F})$, *where* $\otimes$ *is the bilinear mapping* $(A, B) \mapsto A \otimes B$ *of* 13.3.3.

Proof. Let's verify the criterion of Theorem 13.1.14. Let $E_{ij} \in \mathrm{M}_m(\mathrm{F})$ be the matrix with 1 in the (i, j) position and 0's elsewhere; the m^2 matrices E_{ij} (called the *matrix units* of order m) are a basis of $\mathrm{M}_m(\mathrm{F})$. Similarly, let $F_{hk} \in \mathrm{M}_n(\mathrm{F})$ be the n^2 matrix units of order n. The m^2n^2 matrices $E_{ij} \otimes F_{hk}$ are matrix units of order mn and are pairwise distinct (a change in any one of i, j, h, k will change the location of the entry 1), hence they are the totality of matrix units of order mn. In particular, the $E_{ij} \otimes F_{hk}$ are a basis of $\mathrm{M}_{mn}(\mathrm{F})$, as we wished to show. ■

▶ Exercises

1. If I is the $n \times n$ identity matrix, then $\operatorname{tr} I = n$. When F is a field of 'characteristic 0' (for example, **Q**, **R** or **C**), deduce the formula for the dimension of $\mathrm{V} \otimes \mathrm{W}$ (13.1.2) from formula (6) of Theorem 13.3.6.

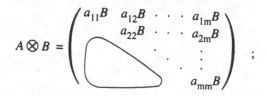

2. $(A \otimes B)' = A' \otimes B'$.

3. If A and B are real matrices of even order, then the product of the characteristic roots of $A \otimes B$ is $\geqslant 0$.

4. Let $A, C \in \mathrm{M_m(F)}$ and $B, D \in \mathrm{M_n(F)}$. By Corollary 13.3.5, if A, C are similar and B, D are similar, then $A \otimes B, C \otimes D$ are similar. Prove that if K is an overfield of F in which the characteristic polynomials of A and B factor completely (cf. Appendix B.5.3), then the characteristic roots of $A \otimes B$ are the elements of K of the form ab, where a is a characteristic root of A and b is a characteristic root of B.

{Hint: As in the proof Theorem 13.3.6, A and B are similar over K to triangular matrices C and D (cf. §8.3, Exercise 2).}

5. If F is algebraically closed, V is a finite-dimensional vector space over F, and $S, T \in \mathcal{L}(\mathrm{V})$, then

$$\sigma(S \otimes T) = \{cd\colon c \in \sigma(S) \text{ and } d \in \sigma(T)\}.$$

{Hint: Here σ stands for spectrum (§8.3, Exercise 9). Cf. Exercise 4.}

A

Foundations

A.1 A dab of logic

A.1.1 A **proposition** is a statement that is either true or false, but not both. For example, if P and Q are the statements

P: '2 < 3'

Q: '3 is an even integer'

then P is a true proposition and Q is a false proposition; the statement

R: 'small is beautiful'

is not a proposition. It is psychologically useful to 'say it with numbers': if a proposition P is true, its 'truth-value' is said to be 1; if P is false, its truth-value is said to be 0. We assume the usual arithmetic of 0 and 1 ($0+1=1$, $1-1=0$, $0\cdot 1=0$, $0<1$, etc.) For convenience, we write $|P|$ for the truth-value of a proposition P. In the above examples, $|P|=1$ and $|Q|=0$.

A.1.2 If P is a proposition, the proposition 'not P' is called the **negation** of P and is denoted P′ (or ~P); if P is true then P′ is false, and if P is false then P′ is true. Thus

$$|P'| = 1 - |P|.$$

For the example Q of A.1.1,

Q′: not '3 is an even integer';

this is expressed more gracefully as '3 is not an even integer'.

A.1.3 If P and Q are propositions, then

$$P \,\&\, Q$$

(read 'P and Q'; also written $P \wedge Q$) is the proposition obtained by stating P and Q (in one breath); it is true when both P and Q are true, and is false otherwise. Thus, $|P \,\&\, Q|$ is 1 if $|P| = |Q| = 1$, and it is 0 in the other three cases ($|P| = 1$, $|Q| = 0$; $|P| = 0$, $|Q| = 1$; $|P| = |Q| = 0$). This is expressed by the formula

$$|P \,\&\, Q| = |P| \cdot |Q|.$$

For the examples of A.1.1, the proposition

P & Q: '2 < 3 and 3 is an even integer'

is false (because Q is false), whereas P & Q′ is true.

For every proposition P, the proposition P & P′ is false (**Law of contradiction**).

A.1.4 If P and Q are propositions, then

$$P \text{ or } Q$$

(also written $P \vee Q$) is the proposition stating P or Q (in one breath); it is true if either P is true or Q is true (or both), and it is false otherwise. We have the formulas

$$\begin{aligned}|P \text{ or } Q| &= |P| + |Q| - |P| \cdot |Q| \\ &= |P| + |Q|(1 - |P|) \\ &= |Q| + |P|(1 - |Q|)\end{aligned}$$

(if either $|P| = 1$ or $|Q| = 1$, we get 1; otherwise we get 0). For the examples of A.1.1, the proposition

P or Q: '2 < 3 or 3 is an even integer'

is true (because P is true).

For every proposition P, the proposition P or P′ is true (**Law of the excluded middle**).

A.1.5 Let P and Q be propositions. The expression

$$P \Rightarrow Q$$

means that **if** P is true, **then** Q is true; it is read 'P implies Q' (shorthand for 'P true implies Q true'). For example, if P and Q are the propositions

P: '*x* is an integer'

Q: '*x* is a rational number'

then $P \Rightarrow Q$ (because every integer is a rational number). The meaning of $P \Rightarrow Q$ is that the truth of P assures the truth of Q, in other words $|P| = 1$

assures that $|Q| = 1$; a condition that expresses this numerically is the inequality

$$|P| \leqslant |Q|$$

{Another such condition is $|P|(1 - |Q|) = 0$.} Note that if $|P| = 0$ then the inequality is surely true (whatever the value of $|Q|$); thus, if P is a false proposition, then $P \Rightarrow Q$ for *any* proposition Q. (This is not very exciting; it is called *vacuous implication*.) For example, if

P: 'the integer 9 is even'
Q: '$\sqrt{2}$ is a rational number'

then $P \Rightarrow Q$ (not interesting). Another example: if

P: 'the integer 9 is odd'
Q: 'the integer 10 is even'

then $P \Rightarrow Q$ (also not interesting). Another example: if

P: 'x is a real number $\geqslant 0$'
Q: '$x = y^2$ for some real number y'

then $P \Rightarrow Q$; this is interesting, because the demonstration that a nonnegative real number has a real square root uses deep properties of the real number system.

Such is the grist for the mathematician's mill: the discovery of implications $P \Rightarrow Q$ whose demonstrations are interesting. When we demonstrate that $P \Rightarrow Q$, we say that we have proved a **theorem**, whose **hypothesis** is P and whose **conclusion** is Q. Clearly there are interesting theorems and there are boring theorems. In the above examples, the interesting ones involve propositions that contain variables (such as x) that make the propositions sometimes true and sometimes false.

A.1.6 Let P and Q be propositions. The implication $P \Rightarrow Q$ is expressed by the inequality $|P| \leqslant |Q|$. The inequality is false if $|P| = 1$ and $|Q| = 0$, and it is true in the remaining three cases. We can regard $P \Rightarrow Q$ itself as a proposition, whose truth value is given by

$$|P \Rightarrow Q| = \begin{cases} 0 & \text{when } |P| = 1 \text{ and } |Q| = 0 \\ 1 & \text{in the remaining three cases.} \end{cases}$$

Thus, $P \Rightarrow Q$ is false when P is true and Q is false (that is, when P is true yet Q is not); and $P \Rightarrow Q$ is true in the remaining cases (the 'vacuous' cases where P is false, and the case where P is true and so is Q).

When $P \Rightarrow Q$ is false, we write $P \not\Rightarrow Q$ (P does *not* imply Q). For example, if

P: 'x is a real number'
Q: '$x = y^2$ for some real number y'

then $P \not\Rightarrow Q$ (the choice $x = -1$ makes P true but Q false).

A.1.7 If both $P \Rightarrow Q$ and $Q \Rightarrow P$ then the propositions P and Q are said to be **equivalent** and one writes

$$P \Leftrightarrow Q.$$

The notation $P \equiv Q$ is also used to indicate equivalence. One calls $Q \Rightarrow P$ the **converse** of $P \Rightarrow Q$. The implication $Q \Rightarrow P$ says that P is true **if** Q is true; $P \Rightarrow Q$ says that P is true **only if** Q is true as well. Thus, the equivalence $P \Leftrightarrow Q$ says that P is true **if and only if** Q is true; the 'if' part is $Q \Rightarrow P$, the 'only if' part is $P \Rightarrow Q$.

Another style of expressing $P \Leftrightarrow Q$: For P to be true, it is **necessary and sufficient** that Q be true. The 'necessity' part is $P \Rightarrow Q$, the 'sufficiency' part is $Q \Rightarrow P$.

For example, if

P: 'x is a real number $\geqslant 0$'

Q: '$x = y^2$ for some real number y'

then $P \Leftrightarrow Q$ (a real number is nonnegative if and only if it is the square of a real number); the proof of $Q \Rightarrow P$ is easy, the proof of its converse is hard. If R is the proposition

R: 'x is a positive integer'

then $R \Rightarrow Q$ but the converse is false ($Q \not\Rightarrow R$), thus R and Q are *not* equivalent, expressed by writing $R \not\Leftrightarrow Q$.

An equivalence $P \Leftrightarrow Q$ means that both $|P| \leqslant |Q|$ and $|Q| \leqslant |P|$, in other words

$$|P| = |Q|;$$

that is, P and Q have the same truth-values (both 1 or both 0). To prove that a proposition is true, it suffices to prove that an equivalent proposition is true.

A.1.8 If $P \Rightarrow Q$ then $Q' \Rightarrow P'$. {The first implication says that $|P| \leqslant |Q|$, the second that $1 - |Q| \leqslant 1 - |P|$; they say the same thing.} In fact, $P \Rightarrow Q$ and $Q' \Rightarrow P'$ are equivalent propositions, so we may write

$$(P \Rightarrow Q) \Leftrightarrow (Q' \Rightarrow P').$$

It is perhaps easier on the eye to write

$$(P \Rightarrow Q) \equiv (Q' \Rightarrow P').$$

In proving a theorem $P \Rightarrow Q$, it is sometimes more congenial to prove $Q' \Rightarrow P'$ (called the **contrapositive** form of $P \Rightarrow Q$).

A.1.9 As noted in A.1.7, an equivalence $P \Leftrightarrow Q$ means that P and Q are either both true or both false; thus,

$$(P \Leftrightarrow Q) \equiv ((P \;\&\; Q) \text{ or } (P' \;\&\; Q')).$$

There is a similar 'formula' for implication:

$$(P \Rightarrow Q) \equiv (Q \text{ or } P').$$

{For example, suppose that either Q or P′ is true. If P is true, then P′ is false; but either Q or P′ is true, so it must be that Q is true. We have shown that $P \Rightarrow Q$. That's half the battle.}

A.1.10 $(P')' \equiv P$ for every proposition P. {For, $|(P')'| = 1 - |P'| = 1 - (1 - |P|) = |P|$.}

A.1.11 When is $P \,\&\, Q$ false? Precisely when one of P, Q is false, that is, when one of P′, Q′ is true. This suggests the equivalence

$$(P \,\&\, Q)' \equiv P' \text{ or } Q'.$$

{Indeed, $|(P \,\&\, Q)'| = 1 - |P \,\&\, Q| = 1 - |P| \cdot |Q|$, and

$$\begin{aligned} |P' \text{ or } Q'| &= |P'| + |Q'| - |P'| \cdot |Q'| \\ &= (1 - |P|) + (1 - |Q|) - (1 - |P|)(1 - |Q|), \end{aligned}$$

which simplifies to $1 - |P| \cdot |Q|$.} Similarly,

$$(P \text{ or } Q)' \equiv P' \,\&\, Q'.$$

These formulas are known as **De Morgan's laws**. For example, $(P \,\&\, P')' \equiv P \text{ or } P'$ (cf. A.1.3, A.1.4).

A.1.12 (*Law of syllogism*) If $P \Rightarrow Q$ and $Q \Rightarrow R$, then $P \Rightarrow R$. {For, if $|P| \leq |Q|$ and $|Q| \leq |R|$ then $|P| \leq |R|$.}

A.1.13 Here are some common formats for proving theorems:

Theorem *If* P *then* Q.

Proof (direct). Assuming P to be true, one demonstates that Q is also true. ■[1]

Theorem *If* P *then* Q.

Proof (contrapositive form). Supposing Q false, one demonstrates that P is false. (In other words, one gives a direct proof of 'If Q′ then P′.') ■

Theorem *If* P *then* Q.

Proof (by contradiction). Assuming P to be true, we suppose 'to the contrary' that Q is false, that is, that Q′ is true. On the basis of the truth of P and Q′, we demonstrate that there is a proposition R such that both R and R′ are true (the cunning is in finding R). At this point, we say 'contradiction!' and declare the proof to be over. {Why? We have demonstrated that $(P \,\&\, Q') \Rightarrow (R \,\&\, R')$, that is, *if* $P \,\&\, Q'$ is true then $R \,\&\, R'$ is true. But we know $R \,\&\, R'$ to be false (A.1.3); therefore $P \,\&\, Q'$ can't be true. But P *is* true (by hypothesis); so Q′ must be false, therefore Q is true, as we wished to show.}. ■

[1] 'End of proof' symbol.

Theorem *If P then Q.*

Proof (mortgaged). One proves that $P \Rightarrow R$ and one cites Qwerty's theorem that $R \Rightarrow Q$, whence $P \Rightarrow Q$ by the law of syllogism (A.1.12). {Our proof is mortgaged to the correctness of Qwerty's proof. The reader who hasn't already verified that Qwerty's theorem is true will have to do so before our proof can be considered to be complete.} ■

Theorem P *if and only if* Q.

Proof (double implication). One demonstrates that $P \Rightarrow Q$ and that $Q \Rightarrow P$ (using, say, one of the above methods of proof). ■

In a different vein:

Theorem *For every positive integer n, the proposition* P_n *is true.*

Proof (by induction). One demonstrates that P_1 is true. *Assuming* P_k to be true (this is called the 'induction hypothesis') one demonstates that P_{k+1} must be true; that is, one demonstrates

$$P_k \Rightarrow P_{k+1}$$

(this is called the 'induction step' of the proof). At this point one declares the proof to be complete 'by induction'. {Why? If P_n were false for some n, there would be a smallest m for which P_m is false. Then $m \geqslant 2$ (because P_1 is certified to be true) and $k = m - 1$ is an embarrassment: P_k is true (by the minimality of m) yet $P_{k+1} = P_m$ is false, and we are at odds with the demonstated $P_k \Rightarrow P_{k+1}$.} ■

A.2 Set notations

A.2.1 The notation $x \in A$ (read x **belongs** to A) means that x is an element of the set A. If x is *not* an element of A, we write $x \notin A$.

A.2.2 If A and B are sets, the notation $A \subset B$ (read A is **contained** in B, or A is a **subset** of B) means that every element of A is an element of B, that is, $x \in A \Rightarrow x \in B$; the notation $B \supset A$ means the same thing (read B **contains** A, or B is a **superset** of A).

A.2.3 For sets A and B, $A = B$ means that both $A \subset B$ and $B \subset A$, that is, $x \in A \Leftrightarrow x \in B$. The negation of $A = B$ is written $A \neq B$; this means that either A is not contained in B, or B is not contained in A.

A.2.4 If $A \subset B$ and $A \neq B$, then A is said to be a **proper** subset of B.

A.2.5 The set with *no* elements is denoted $\varnothing$ and is called the **empty set**. A set with only one element, say a, is denoted $\{a\}$ and is called a **singleton.** The relation $x \in A$ means that $A \supset \{x\}$; one also says that A contains the element x (though this is a slight misuse of the terminology). The set whose only elements are a and b is denoted $\{a, b\}$; $a = b$ is permitted, in which case $\{a, b\} = \{a\}$. The notation $\{a, b, c, \ldots\}$ stands for the set whose elements are $a, b, c, \ldots$.

A.2.6 Let X be a set. Suppose that for each $x \in X$ a proposition $P(x)$ is given (which may be true or false). By $\{x \in X\colon\ P(x)\}$ we mean the set of all elements x of X for which the proposition $P(x)$ is *true*. For example, if $X = \{2, 3, 4\}$ then

$$\{x \in X\colon\ x \text{ is even}\} = \{2, 4\}.$$

One can write simply $\{x\colon\ P(x)\}$ if there is a tacit understanding as to the set X to which the elements x belong. For example, if we are talking about integers, then

$$\{x\colon -3 < x \leq 1\}$$

is the set $\{-2, -1, 0, 1\}$ with four elements; if we are talking about real numbers, it represents an interval on the real line (see A.2.10).

A.2.7 Let X be a set, A and B subsets of X. The **union** of A and B is the set

$$A \cup B = \{x \in X\colon\ x \in A \quad \text{or} \quad x \in B\};$$

the **intersection** of A and B is the set

$$A \cap B = \{x \in X\colon\ x \in A \;\&\; x \in B\};$$

the (set-theoretic) **difference** 'A minus B' is the set

$$A - B = \{x \in X\colon\ x \in A \;\&\; x \notin B\};$$

the **complement** of A in X is the set

$$X - A = \{x \in X\colon\ x \notin A\}$$

(also denoted $\complement A$, or A', when there is a tacit understanding as to the set X).

There is even a notation for the set of *all* subsets of X, namely $\mathcal{P}(X)$; thus, $A \in \mathcal{P}(X)$ means that $A \subset X$. For example, if $X = \{a, b\}$ then $\mathcal{P}(X) = \{\varnothing, \{a\}, \{b\}, X\}$. If X is a finite set with n elements, then $\mathcal{P}(X)$ has 2^n elements; for this reason, $\mathcal{P}(X)$ is called the **power set** of X (for any set X)

A.2.8 If X and Y are sets, the set of all ordered pairs (x, y) with $x \in X$ and $y \in Y$ is denoted $X \times Y$ and is called the **cartesian product** of X and Y (in that order); thus

$$X \times Y = \{(x, y)\colon\ x \in X, y \in Y\}.$$

In this set, $(x, y) = (x', y')$ means that $x = x'$ and $y = y'$. Similarly, the cartesian product of a finite list of sets $X_1, \ldots, X_n$ is denoted $X_1 \times \ldots \times X_n$ and consists of all the ordered n-ples $(x_1, \ldots, x_n)$ with $x_i \in X_i$ for $i = 1, \ldots, n$.

A.2.9 Some specific sets for which there is some concensus on a notation:

$\mathbf{N} = \{0, 1, 2, 3, \ldots\}$, the set of all *natural numbers* (or *nonnegative integers*);
$\mathbf{Z} = \{0, \pm 1, \pm 2, \pm 3, \ldots\}$, the set of all *integers*;
$\mathbf{Q} = \{m/n\colon \ m \in \mathbf{Z},\ n \in \mathbf{Z},\ n \neq 0\}$, the set of all *rational numbers*;
$\mathbf{R} =$ the set of all *real numbers*;
$\mathbf{C} = \{a + bi\colon \ a, b \in \mathbf{R}\}$, the set of all *complex numbers*.

No concensus at all on the following notation (but it's suggestive):

$\mathbf{P} = \{1, 2, 3, \ldots\}$, the set of all *positive integers*

(also denoted $\mathbf{N}^* = \mathbf{N} - \{0\}$). One has the inclusions

$$\mathbf{P} \subset \mathbf{N} \subset \mathbf{Z} \subset \mathbf{Q} \subset \mathbf{R} \subset \mathbf{C}$$

(all of them proper).

A.2.10 The intervals in $\mathbf{R}$ of finite length, with endpoints a, b in $\mathbf{R}$ $(a \leqslant b)$ are the subsets

$$\begin{aligned} (a, b) &= \{x\colon a < x < b\}, \\ [a, b) &= \{x\colon a \leqslant x < b\}, \\ (a, b] &= \{x\colon a < x \leqslant b\}, \\ [a, b] &= \{x\colon a \leqslant x \leqslant b\}; \end{aligned}$$

the first of these is called an **open interval**, the last a **closed interval**, and the ones in between 'semi-closed' (or 'semi-open') intervals. For all $a \in \mathbf{R}$, $[a, a] = \{a\}$ (called a 'degenerate' closed interval), and $(a, a) = [a, a) = (a, a] = \varnothing$. The intervals of infinite length are the subsets

$$\begin{aligned} (a, +\infty) &= \{x\colon x > a\} \\ [a, +\infty) &= \{x\colon x \geqslant a\} \\ (-\infty, a) &= \{x\colon x < a\} \\ (-\infty, a] &= \{x\colon x \leqslant a\} \end{aligned}$$

and $(-\infty, +\infty) = \mathbf{R}$.

A.2.11 Some useful shorthand:

$\ni$ such that
$\forall$ for all (for every)
$\exists$ there exists
$\exists!$ there exists a unique (one and only one).

For example,

$$(\forall y \in \mathbf{R}) \qquad y^2 \geqslant 0$$

says that 'for every real number y, $y^2 \geq 0$'. Also, for a real number x,

$$x \geqslant 0 \Leftrightarrow (\exists\ y \in \mathbf{R} \ni x = y^2);$$

this is, in symbolic shorthand, the equivalence mentioned in A.1.7. Finally,

$$(\forall\, x \in \mathbf{R},\, x \geqslant 0)\ \exists!\ y \in \mathbf{R} \ni y \geqslant 0\ \&\ y^2 = x;$$

in words, every nonnegative real number has a unique nonnegative square root.

A.2.12 A book written in symbolic shorthand would be ugly indeed. The symbols are tolerable in print if they are surrounded by prose, and they are useful because they abbreviate concepts whose verbal form is relatively long and logically structured. On the other hand, one must squint and concentrate to make sense of a line of symbols such as those in A.2.11; the symbols are useful not because they save thought (they don't) but because they help organize it. They also save time in lecturing at a blackboard (acceptable there because the lecturer is also verbalizing what is going on) and in taking notes; the practice of transcribing such notes into readable form is a useful exercise.

A.3 Functions

A.3.1 A **function** f:X→Y assigns, to each element x of the set X, a corresponding element $f(x)$ of the set Y (cf. Fig. 24).

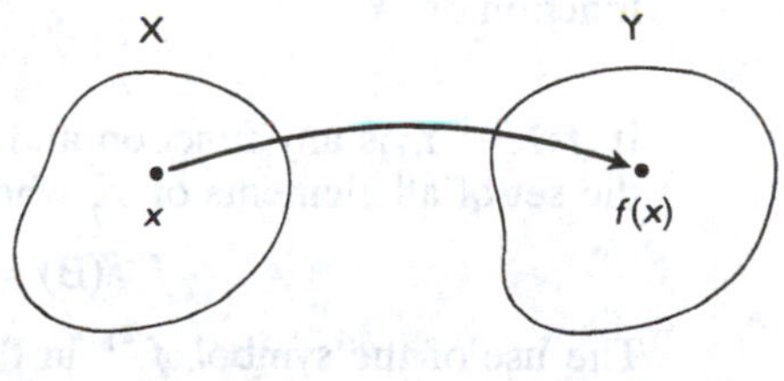

Fig. 24

The term **mapping** (also 'transformation', or 'operator') is synonymous with 'function'. One calls X the **initial set** of f, Y the **final set** of f; X is also called the **domain** of f. If $x \in$ X, the element $f(x)$ of Y is called the **value** of f at x, or the **image** of x under f. For a subset A of X, f(A) denotes the set of images under f of the elements of A:

$$f(\mathrm{A}) = \{f(x): x \in \mathrm{A}\}.$$

In particular, f(X) is the set of *all* values of f; it is called the **range** (or **image**) of f and is denoted Range f (or Im f). The notation

$$\mathrm{Im}\, f = \{f(x): x \in \mathrm{X}\}$$

has the advantage of brevity. {The natural abbreviation for 'range' is too likely to suggest 'rank' (Definition 3.6.3).}

A.3.2 The range of f:X→Y is a subset of Y; if Im f = Y then f is said to be **surjective** (or 'onto'). A function f:X→Y is said to be **injective** (or 'one-one') if it sends distinct elements of X to distinct elements of Y:

$$x, x' \in X, x \neq x' \Rightarrow f(x) \neq f(x'),$$

or, in contrapositive form,

$$f(x) = f(x') \Rightarrow x = x'.$$

If a function is both injective and surjective, it is said to be **bijective** (or 'one-one onto').

An injective function is called an **injection**; a surjective function, a **surjection**; and a bijective function, a **bijection.**

A.3.3 A bijection $f{:}X \to Y$ admits an **inverse function** $f^{-1}{:}Y \to X$. The mission of f^{-1} is to undo everything that f does. Thus, if $f(x) = y$ then $f^{-1}(y) = x$. Consequency $f^{-1}(f(x)) = x$ for all $x \in X$ and $f(f^{-1}(y)) = y$ for all $y \in Y$.

A.3.4 Two functions $f{:}X \to Y$ and $g{:}X \to Y$ are regarded as **equal**, written $f = g$, if $f(x) = g(x)$ for all $x \in X$. The set of all functions $f{:}X \to Y$ is denoted $\mathcal{F}(X, Y)$; when $X = Y$, this is abbreviated to $\mathcal{F}(X)$.

A.3.5 If $f{:}X \to Y$ and $g{:}Y \to Z$, then the **composite** function $g \circ f{:}X \to Z$ is defined by $(g \circ f)(x) = g(f(x))$ for all $x \in X$. Thus if $f{:}X \to Y$ is bijective, then $f^{-1} \circ f = \text{id}_X$ and $f \circ f^{-1} = \text{id}_Y$, where $\text{id}_X{:}X \to X$ is the **identity function** on X (that is, $\text{id}_X(x) = x$ for all $x \in X$) and id_Y is the identity function on Y.

A.3.6 If $f{:}X \to Y$ is any function and if $B \subset Y$, the **inverse image** of B under f is the set of all elements of X whose image is in B; it is denote $f^{-1}(B)$, thus

$$f^{-1}(B) = \{x \in X{:}\ \ f(x) \in B\}.$$

The use of the symbol f^{-1} in this connection is not meant to imply that f is bijective (i.e., that an inverse function $f^{-1}{:}Y \to X$ exists), but when f is bijective the two interpretations of $f^{-1}(B)$ agree: the inverse image of B under f is equal to the 'direct' image of B under f^{-1} (basically, because $f(x) = y$ if and only if $f^{-1}(y) = x$).

A.3.7 A function $f{:}X \to Y$ sends each $x \in X$ to $f(x) \in Y$; this is sometimes indicated by writing $x \mapsto f(x)$ $(x \in X)$, particularly useful when a function is being defined by a formula. {For example, the function $x \mapsto x^2 + 5$ $(x \in \mathbf{R})$ has for its graph a parabola.) If there is no reason to pin down a letter for the function itself, we may speak noncommittally of a function $X \to Y$. {For example, the arctangent function is a bijection $\mathbf{R} \to (-\pi/2, \pi/2)$.}

A.3.8 Let $f{:}X \to Y$ and let A be a subset of X. The function $A \to Y$ defined by $x \mapsto f(x)$ for $x \in A$ is called the **restriction** of f to A and is denoted $f|A$; thus $f|A{:}A \to Y$, $(f|A)(x) = f(x)$ for all $x \in A$. The restriction of the identity mapping id_X to the subset A of X is denoted $i_A{:}A \to X$ and is called the **insertion mapping** of A into X; note that $f|A = f \circ i_A$.

A.4 The axioms for a field

A.4.1 Definition A **field** is a nonempty set F with two operations, called **addition** and **multiplication**.

The operation of addition assigns, to each pair of elements a, b of F, an element $a + b$ of F, called the **sum** of a and b, subject to the following axioms:

(1) $(a + b) + c = a + (b + c)$ for all a, b, c in F;

(2) there exists an element 0 in F such that $a + 0 = a = 0 + a$ for all a in F (this element is unique and is called the **zero element** of F);

(3) for each element a of F there exists an element b of F such that $a + b = 0 = b + a$ (the element b us uniquely determined by a and is called the **negative** of a, written $-a$);

(4) $a + b = b + a$ for all a,b in F.

The operation of multiplication assigns, to each pair of elements a, b of F, an element ab of F, called the **product** of a and b, subject to the following axioms:

(5) $a(b + c) = ab + ac$ and $(a + b)c = ac + bc$ for all a, b, c in F;

(6) $(ab)c = a(bc)$ for all a, b, c in F;

(7) there exists a nonzero element 1 in F such that $a1 = a = 1a$ for all a in F;

(8) for each nonzero element a of F, there exists an element b of F such that $ab = 1 = ba$ (the element b is also nonzero; it is uniquely determined by a and is called the **inverse** or **reciprocal** of a, written a^{-1} or $1/a$);

(9) $ab = ba$ for all a, b in F.

A.4.2 Examples F = **Q**, **R** or **C**, with the usual operations (see A.2.9). The world's smallest field: F = $\{0, 1\}$, with $1 + 1 = 0$ the only surprise.

A.4.3 Axiom (1) alone says that we have a set with an associative operation (also called a **semigroup**). As each axiom is added, the algebraic structure gets richer; there is a name for each of the 'intermediate' structures:

(1)–(2): F is a **monoid** for addition;

(1)–(3): F is a **group** for addition;

(1)–(4): F is an **abelian** (or 'commutative') group;

(1)–(5): F is a **ring** for the two operations;

(1)–(6): F is an **associative** ring;

(1)–(7): F is an associative ring with **unity element**;

(1)–(8): F is a **division ring**.

Thus, a field is a division ring for which multiplication is commutative.

B

Integral domains, factorization theory

B.1 The field of fractions of an integral domain

An **integral domain** is a commutative ring with unity in which there are 'no divisors of 0':

$$ab = 0 \quad \Rightarrow \quad a = 0 \text{ or } b = 0.$$

Another way of expressing the latter condition:

$$a \neq 0 \;\&\; ab = ac \quad \Rightarrow \quad b = c$$

(*law of cancellation*).

B.1.1 Theorem *If* R *is an integral domain, there exists a field* F *such that* (1) R *is a subring of* F, *and* (2) *every element of* F *is of the form* ab^{-1} *with* $a,b \in$ R *and* $b \neq 0$.

Proof. Let G be the set of all ordered pairs (a, b) with $a,b \in$ R and $b \neq 0$; thus

$$\text{G} = \text{R} \times (\text{R} - \{0\}).$$

{The underlying strategy, stated informally: think of (a, b) as a 'fraction' a/b, allowing that different ordered pairs may represent the same 'fraction'.} Introduce operations of addition and mulitiplication in G by the formulas

$$(a, b) + (a', b') = (ab' + a'b, bb'),$$
$$(a, b)(a', b') = (aa', bb');$$

since $bb' \neq 0$, G is closed under these operations. It is straightforward to check that both operations in G are commutative and associative.

If R has at least two nonzero elements (i.e., R is not the field of two elements) then G is not a ring for the above operations; for example, if b is a nonzero element of R other than 1, then

$$(0, b)\{(0, 1) + (0, 1)\} = (0, b)(0, 1) = (0, b)$$

whereas

$$(0, b)(0, 1) + (0, b)(0, 1) = (0, b) + (0, b) = (0, b^2),$$

thus the distributive law fails. {The underlying cause: $(0, 1)$ and $(0, b)$ are different pairs, but they ought to represent the same fraction (zero).}

The remedy is an equivalence relation: declare $(a, b) \equiv (c, d)$ if $ad = bc$. Write $[a, b]$ for the equivalence class of (a, b), and $F = G/\equiv$ for the quotient set; thus $(a, b) \mapsto [a, b]$ is the quotient mapping $G \to F$. It is straightforward to check that if $(a, b) \equiv (c, d)$ and $(a', b') \equiv (c', d')$, then

$$(a, b) + (a', b') \equiv (c, d) + (c', d'),$$
$$(a, b)(a', b') \equiv (c, d)(c', d');$$

it follows that sums and products in F may be defined by the formulas

$$[a, b] + [a', b'] = [ab' + a'b, bb'],$$
$$[a, b][a', b'] = [aa', bb'].$$

With these operations, F is a commutative ring with unity element $[1, 1]$; for example, the proof of the distributive law follows from the computations

$$(a, b)\{(a', b') + (a'', b'')\} = (aa'b'' + aa''b', bb'b''),$$
$$(a, b)(a', b') + (a, b)(a'', b'') = (aa'bb'' + aa''bb', bb'bb'')$$
$$= (aa'b'' + aa''b', bb'b'')(b, b),$$

and the fact that $[b, b] = [1, 1]$.

The zero element of F is $[0, 1]$. If $[a, b]$ is nonzero—in other words $a \neq 0$—then $[a, b][b, a] = [1, 1]$ shows that $[a, b]$ is invertible with $[a, b]^{-1} = [b, a]$. This proves that F is a field. Moreover, for every element $[a, b]$ of F,

$$[a, b] = [a, 1][1, b] = [a, 1][b, 1]^{-1}.$$

Define a mapping $\theta: R \to F$ by the formula $\theta(a) = [a, 1]$; one checks easily that θ is a monomorphism of rings, and

$$[a, b] = \theta(a)\theta(b)^{-1}$$

for every element $[a, b]$ of F. All that remains is to 'identify' R with its image in F (throw away R and regard its image in F as its rightful heir). An alternative to identification is the following strategy.

Let X be a set such that (1) R and X are disjoint, and (2) there exists a bijection $\sigma : F - \theta(R) \to X$. Informally, X is a 'set-theoretic copy' of $F - \theta(R)$ that is disjoint from R. For example, let

$$X = (F - \theta(R)) \times \{1\},$$

where $\{1\}$ is the set whose only element is 1; the mapping $[a, b] \mapsto ([a, b], 1)$ is a bijection $F - \theta(R) \to X$, and

$$c \neq ([a, b], 1) \quad (\forall c \in R)$$

shows that R and X have no elements in common. Let $Y = R \cup X$ (the terms of the union are disjoint). On the other hand, $F = \theta(R) \cup (F - \theta(R))$ is also a disjoint union.

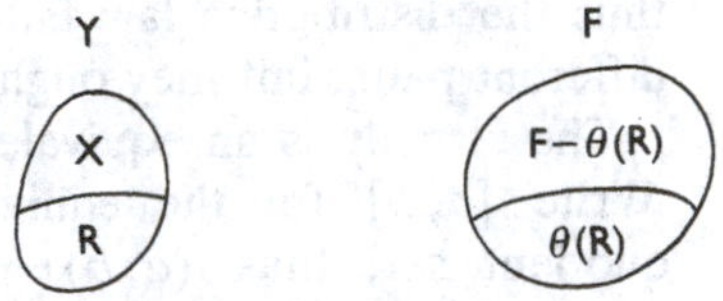

Let $\varphi : Y \to F$ be the bijection that agrees with θ on R and with σ^{-1} on X. There is a unique ring structure on Y that makes φ an isomorphism of rings (hence makes Y a field): for elements y, y' of Y, define

$$y + y' = \varphi^{-1}(\varphi(y) + \varphi(y'))$$
$$yy' = \varphi^{-1}(\varphi(y)\varphi(y')).$$

Because of the way φ is defined on R, the 'new operations' on R coincide with its original operations — in other words, R is a subring of the field Y. The field Y has the properties (1), (2) required in the statement of the theorem. ■

B.1.2 Definition With notations as in B.1.1, F is called the **field of fractions** of the integral domain R.

{The reason for the definite article 'the' is that F is unique up to isomorphism in the following sense: if F′ is another field with the properties (1), (2), then there exists a ring isomorphism $F \to F'$ that leaves fixed the elements of R, explicitly, writing $a \mapsto a$ for the insertion mapping $R \to F$, and $a \mapsto a'$ for the insertion mapping $R \to F'$, the mapping $ab^{-1} \mapsto a'(b')^{-1}$ is an isomorphism with the desired properties.}

B.1.3 Examples The field of fractions of **Z** is **Q**. The field of fractions of a polynomial ring $K[t]$ (K a field) is the field $K(t)$ of rational forms p/q, where $p, q \in K[t]$ and $q \neq 0$. A field K is an integral domain; it is its own field of fractions.

B.2 Divisibility in an integral domain

Let R be an integral domain.

B.2.1 Definition Let $a, b \in \mathrm{R}$. We say that a **divides** b, written $a|b$, if b is a multiple of a in the sense that $b = ra$ for some $r \in \mathrm{R}$.

Writing $(a) = \mathrm{R}a = \{ra: r \in \mathrm{R}\}$ for the ideal of R consisting of all multiples of a (called the **principal ideal generated** by a), we have

$$a|b \quad \Leftrightarrow \quad (a) \supset (b).$$

Some elementary properties of divisibility:

B.2.2 Theorem

(i) $a|a \quad (\forall a \in \mathrm{R})$;

(ii) $a|b \;\&\; b|c \Rightarrow a|c$;

(iii) $a|0 \quad (\forall a \in \mathrm{R})$;

(iv) $0|a \Leftrightarrow a = 0$.

B.2.3 Definition If $u|1$ (that is, $uv = 1$ for some $v \in \mathrm{R}$) then u is called a **unit** of R.

The units of R are the invertible elements of R; they form a group under multiplication, denoted $\mathrm{U_R}$ and called the **group of units** of R.

B.2.4 Examples If **Z** is the ring of integers, then $\mathrm{U_Z} = \{1, -1\}$. The ring $\mathrm{F}[t]$ of polynomials in an indeterminate t, with coefficients in a field F, is an integral domain with $\mathrm{U_{F[t]}} = \mathrm{F} - \{0\}$. In the ring $\mathbf{Z}_n$ of integers modulo n, the units are the cosets $k + (n)$ with k relatively prime to n (i.e. the only common divisors of k and n are ± 1).

B.2.5 Definition Elements $a, b \in \mathrm{R}$ are said to be **associate**, written $a \sim b$, if $a|b$ and $b|a$, in other words, $(a) = (b)$.

The relation $\sim$ is an equivalence relation, and $a \sim b \Leftrightarrow a = bu$ for some unit u; the equivalence class of $a \in \mathrm{R}$ for $\sim$ is the set $a\mathrm{U_R} = \{au: u \in \mathrm{U_R}\}$. The class of 0 is $\{0\}$.

B.2.6 Examples In the ring of integers **Z**, the associates of $a \in \mathbf{Z}$ are $\pm a$; in a polynomial domain $\mathrm{F}[t]$, the associates of p are the scalar multiples cp with $c \neq 0$.

B.2.7 Theorem *Let $a \in \mathrm{R}$, $a \neq 0$. The following conditions on a are equivalent:*

(1) $a = bc \Rightarrow a \sim b$ *or* $a \sim c$;

(2) $a = bc \Rightarrow$ *either* b *or* c *is a unit of* R.

Proof. (1) $\Rightarrow$ (2): Let $a = bc$. By (1), $a \sim b$ or $a \sim c$. If $a \sim b$, say $a = bu$ with u a unit, then $bc = a = bu$, so $c = u$ by cancellation; similarly, $a \sim c$ implies that b is a unit.

(2) $\Rightarrow$ (1): Obvious from the remarks following B.2.5. ■

B.2.8 Definition An element $a \in \mathrm{R}$ is said to be **irreducible** if (1) $a \neq 0$, (2) a is not a unit, and (3) the only divisors of a are units and associates (cf. B.2.7).

An element $a \in \mathrm{R}$ is said to be a **prime** if (1) $a \neq 0$, (2) a is not a unit, and (3′) $a|bc \Rightarrow a|b$ or $a|c$. If a is prime and $a|b_1b_2 \ldots b_n$ then $a|b_i$ for some i (by an easy induction argument).

B.2.9 Theorem *In any integral domain* R, *prime* $\Rightarrow$ *irreducible.*

Proof. Let $a \in \mathrm{R}$ be a prime (in particular, $a \neq 0$ and a is not a unit) and suppose $a = bc$; of course $b|a$ and $c|a$. Since a divides $a = bc$ and a is prime, either $a|b$ or $a|c$, hence either $a \sim b$ or $a \sim c$. ■

B.2.10 Example The converse of B.2.9 is in general false. For example, let $m > 1$ be a square-free positive integer and let $\mathrm{R} = \mathbf{Z} + i\mathbf{Z}\sqrt{m}$ be the set of all complex numbers of the form

$$x = a + ib\sqrt{m} \qquad (a, b \in \mathbf{Z});$$

for such an x, define

$$\mathrm{N}(x) = x\bar{x} = a^2 + mb^2,$$

called the **norm** of x. Since $\mathrm{N}(x)$ is an integer, $\mathrm{N}(xy) = \mathrm{N}(x)\mathrm{N}(y)$ and $\mathrm{N}(1) = 1$, it is clear that

$$x \text{ is a unit of R} \quad \Leftrightarrow \quad \mathrm{N}(x) = 1 \quad \Leftrightarrow \quad x = \pm 1.$$

If p is a prime in the ring of integers $\mathbf{Z}$ and if p is not irreducible in R, then $p = a^2 + mb^2$ for suitable $a, b \in \mathbf{Z}$ and therefore $b \neq 0$ and $p \geq m$. {For, if $p = xy$ with x and y not units, then $p^2 = \mathrm{N}(p) = \mathrm{N}(x)\mathrm{N}(y)$ shows that $\mathrm{N}(x) = \mathrm{N}(y) = p$.} It follows that if p is a prime in $\mathbf{Z}$ with $p < m$, then p is irreducible in R.

If, for example, $m = 3$ and $p = 2$, then 2 is irreducible in $\mathrm{R} = \mathbf{Z} + i\mathbf{Z}\sqrt{3}$. However,

$$2 \cdot 2 = 4 = (1 + i\sqrt{3})(1 - i\sqrt{3});$$

writing $x = 1 + i\sqrt{3}$, $y = 1 - i\sqrt{3}$, we have $2|xy$ but it is obvious that 2 divides neither x nor y. Thus, in the integral domain $\mathbf{Z} + i\mathbf{Z}\sqrt{3}$, 2 is irreducible but is not a prime.

B.2.11 Remark In the integral domain $\mathrm{R} = \mathbf{Z} + i\mathbf{Z}\sqrt{m}$ of the preceding example, every nonzero nonunit x is a finite product of irreducibles. {Sketch of proof: If x is irreducible, we are done. Otherwise $x = yz$ with y and z nonunits. Then $\mathrm{N}(x) = \mathrm{N}(y)\mathrm{N}(z)$ with $\mathrm{N}(y) > 1$, $\mathrm{N}(z) > 1$, so $\mathrm{N}(x) > \mathrm{N}(y)$ and $\mathrm{N}(x) > \mathrm{N}(z)$. This can't go on forever.} If, for example, $m = 3$ and $x = 4$, the factorizations

$$4 = 2 \cdot 2 = (1 + i\sqrt{3})(1 - i\sqrt{3})$$

show that the irreducible factors in one factorization need not be associates of those in another factorizarion.

The superiority of primes over irreducibles is demonstrated by the following theorem:

B.2.12 Theorem *In an integral domain, if*

$$p_1 \cdots p_m = q_1 \cdots q_n,$$

where the p_i *and* q_j *are primes, then* $m = n$ *and, after a rearrangement of the* q*'s,* $p_i \sim q_i$ *for all* i.

Proof. The proof is by induction on n. If $n = 1$ then $m = 1$ by the irreducibility of q_1.

Let $n \geq 2$ and assume that all's well for $n - 1$. Since $p_m | q_1 \dots q_n$, p_m must divide one of the q's; rearranging the q's we can suppose that $p_m | q_n$. Since q_n is prime, $p_m \sim q_n$; let u be a unit with $q_n = up_m$. Then

$$p_1 \dots p_{m-1} p_m = q_1 \dots q_{n-1}(up_m).$$

Cancelling p_m, we have $p_1 \dots p_{m-1} = q_1 \dots q_{n-1}u$; by the induction hypothesis, $m - 1 = n - 1$ and, after a further rearrangement of the q's,

$$p_1 \sim q_1, \dots, p_{m-2} \sim q_{m-2}, \qquad p_{m-1} \sim q_{m-1}u$$

(hence also $p_{m-1} \sim q_{m-1}$). ■

B.3 Principal ideal rings

B.3.1 Definition A **principal ideal ring** (briefly, PIR) is an integral domain R in which every ideal I is principal, that is, of the form $I = (a)$ for some $a \in R$.

Note that the generator a of I is unique up to associate (B.2.5).

For the rest of the section, R *denotes a principal ideal ring.*

B.3.2 Examples The ring of integers **Z** is a PIR (every nonzero ideal I is generated by the smallest positive element of I). The polynomial ring $F[t]$ (F a field) is a PIR (every nonzero I is generated by the unique monic polynomial in I of minimal degree). For the details, see B.4.2 and B.4.3 in the next section.

The polynomial ring $\mathbf{Z}[t]$ is *not* a principal ideal ring: the set I of polynomials with even constant term is an ideal that is not principal. {Since $2 \in I$, the only candidates for generator of I are ± 2; but $t + 2 \in I$ and is not a multiple of 2.}

B.3.3 Theorem *Let* $a, b \in R$. *Then:*

(1) *There exists elements* $d, m \in R$ *(unique up to associate) such that*

$$(a) + (b) = (d) \quad \textit{and} \quad (a) \cap (b) = (m).$$

(2) *For* $c \in R$,

$$c|a \;\&\; c|b \;\Leftrightarrow\; c|d,$$
$$a|c \;\&\; b|c \;\Leftrightarrow\; m|c.$$

Proof. (1) is immediate from the definition of a PIR. For $c \in R$,

$$(c) \supset (a) \;\&\; (c) \supset (b) \;\Leftrightarrow\; (c) \supset (a) + (b)$$

and

$$(a) \supset (c) \;\&\; (b) \supset (c) \;\Leftrightarrow\; (a) \cap (b) \supset (c). \;■$$

B.3.4 Definition With notations as in B.3.3, we call d a **greatest common divisor** (GCD) of a and b, written

$$d = (a, b),$$

and m a **least common multiple** (LCM) of a and b, written

$$m = [a, b].$$

The GCD and LCM are only unique up to associate; to say that $d = (a, b)$ means nothing more nor less than $(d) = (a) + (b)$, and $m = [a, b]$ means that $(m) = (a) \cap (b)$.

B.3.5 Remark For $a, b, d \in \mathrm{R}$, the following conditions are equivalent:

(i) $d = (a, b)$;

(ii) $d|a$, $d|b$ and $d = ra + sb$ for some $r, s \in \mathrm{R}$.

{Proof: Condition (ii) says that $(d) \supset (a)$, $(d \supset (b)$ and $(d) \subset (a) + (b)$.}

B.3.6 Definition Elements $a, b \in \mathrm{R}$ are said to be **relatively prime** if $(a, b) = 1$, in other words, if $(a) + (b) = (1) = \mathrm{R}$; equivalently, $1 = ra + sb$ for suitable $r, s \in \mathrm{R}$.

B.3.7 Remarks Suppose $d = (a, b)$ and $d \neq 0$ (in other words, not both a and b are 0). Write $a = a_1 d$, $b = b_1 d$ (d is a common divisor of a and b). For suitable $r, s \in \mathrm{R}$ (cf. B.3.5)

$$d = ra + sb = (ra_1 + sb_1)d,$$

so $1 = ra_1 + sb_1$; thus a_1, b_1 are relatively prime (B.3.6).

B.3.8 Theorem *If $a|bc$ and $(a, b) = 1$, then $a|c$.*

Proof. Say $bc = ax$ and $ra + sb = 1$; then

$$c = rac + sbc = rac + sax = a(rc + sx),$$

so $a|c$. ■

B.3.9 Theorem *If $a|c$, $b|c$ and $(a, b) = 1$, then $ab|c$.*

Proof. Say $c = ax$, $c = by$ and $ra + sb = 1$; then

$$c = rac + sbc = ra(by) + sb(ax) = ab(ry + sx),$$

so $ab|c$. ■

B.3.10 Lemma *If $a, b \in \mathrm{R}$ and a is irreducible, then either $a|b$ or $(a, b) = 1$.*

Proof. Let $d = (a, b)$; since $d|a$ and a is irreducible, either $d \sim a$ or d is a unit, thus either $(a) = (d) \supset (b)$ or $(d) = (1)$. ■

B.3.11 Theorem *In a principal ideal ring R, prime $\Leftrightarrow$ irreducible.*

Proof. In view of B.2.9, we need only show that if $a \in \mathrm{R}$ is irreducible then it is prime. Assuming $a|bc$, we have to show that $a|b$ or $a|c$. If a does not divide b, then $(a, b) = 1$ by B.3.10, therefore $a|c$ by B.3.8. ■

B.3.12 Lemma *In a principal ideal ring* R, *if* $I_1, I_2, I_3, \ldots$ *is a sequence of ideals such that*

$$I_1 \subset I_2 \subset I_3 \subset \ldots,$$

then the sequence is constant from some index onward.

Proof Let $I = \cup I_k$ be the union of the I_k, that is,

$$I = \{x \in R: \quad x \in I_k \quad \text{for some } k\};$$

since the sequence (I_k) is increasing, I is an ideal of R. Say $I = (a)$. Then $a \in I_n$ for some n; for all $k \geq n$, $I = (a) \subset I_n \subset I_k \subset I$, so $I_k = I$. ■

B.3.13 Theorem *In a principal ideal ring* R, *every nonzero nonunit is a finite product of irreducibles (= primes, by* B.3.11).

Proof. We can suppose that R is not a field (in a field there aren't any nonzero nonunits, so the assertion is vacuously true). Let D be the set of all elements of R that have a factorization of the desired sort:

$$D = \{x \in R: x \text{ is a product of irreducibles}\}.$$

At the moment, we don't know that D has any elements (i.e., that R has any irreducibles), but in any case it is clear that

$$x, y \in D \Rightarrow xy \in D;$$

stated contrapositively.

$$(*) \qquad xy \notin D \Rightarrow x \notin D \text{ or } y \notin D.$$

Let $x \in R$ be a nonzero nonunit; the claim is that $x \in D$. Assume to the contrary that $x \notin D$. In particular, x is not irreducible, so there is a factorization $x = x_1y_1$ such that neither x_1 nor y_1 is an associate of x (B.2.8, B.2.7). Then $(x) \subset (x_1)$ (because $x_1|x$) but $(x) \neq (x_1)$ (because x, x_1 are not associate). In other words, $(x) \subset (x_1)$ properly. Similarly $(x) \subset (y_1)$ properly.

Since $x \notin D$, it follows from $(*)$ that one of x_1, y_1 must fail to belong to D. Suppose, for example, that $x_1 \notin D$. Repeating the preceding argument indefinitely, we construct a strictly increasing sequence of ideals $(x) \subset (x_1) \subset (x_2) \subset (x_3) \ldots$, in contradiction with the lemma. ■

B.3.14 Corollary *In a principal ideal ring, every nonzero nonunit* a *can be written as a product of primes,* $a = p_1 \ldots p_n$; *if* $a = q_1 \ldots q_m$ *is another factorization into primes then* $m = n$ *and, after a rearrangement of the q's,* $p_i \sim q_i$ *for all* i.

Proof. Immediate from B.3.13, B.3.11 and B.2.12. ■

B.3.15 If, in B.3.14, the p's are arranged so that $p_1, \ldots, p_r$ are pairwise 'distinct' (i.e., not associate) and the p_i for $i > r$ are associate to them, then a can be written

$$a = up_1^{\alpha_1} \dots p_r^{\alpha_r}$$

with u a unit and the α_i positive integers; this is called the *prime-power factorization* of a. {In general the unit u can't be 'absorbed' into the p's; for example, in the ring **Z**, consider $a = -4$.}

B.4 Euclidean integral domains

B.4.1 Definition An integral domain R is said to be **Euclidean** if there exists a function $\delta: \mathrm{R} - \{0\} \to \mathbf{N}$, assigning to each nonzero element a of R a nonnegative integer $\delta(a)$, in such a way that the *division algorithm* holds:

$$a, b \in \mathrm{R}, b \neq 0 \Rightarrow \exists q, r \in \mathrm{R} \ni a = bq + r \text{ with } r = 0 \text{ or } \delta(r) < \delta(b).$$

Such a function δ is called a 'rank function'; we abbreviate 'Euclidean integral domain' to EID.

The 'quotient' q and 'remainder' r of the division algorithm are in general not uniquely determined by a and b.

B.4.2 Examples $\mathrm{R} = \mathbf{Z}$ with $\delta(n) = |n|$ for nonzero integers n; $\mathrm{R} = \mathrm{F}[t]$, F a field, with $\delta(p) = \deg p$ for nonzero polynomials p.

B.4.3 Theorem EID ⇒ PIR.

Proof. The claim is that every Euclidean integral domain R is a principal ideal ring. Let I be a nonzero ideal of R. The set $\mathrm{S} = \{\delta(a): a \in \mathrm{I} - \{0\}\}$ is a nonempty subset of **N**; choose $b \in \mathrm{I} - \{0\}$ with $\delta(b)$ minimal. Of course $(b) \subset \mathrm{I}$; we assert that $(b) = \mathrm{I}$. Assuming $a \in \mathrm{I}$, we have to show that a is a multiple of b. Write $a = bq + r$ as in the division algorithm. Then $r = a - bq \in \mathrm{I}$, so we must have $r = 0$; for, the alternative $r \neq 0$ would imply $\delta(r) < \delta(b)$, contrary to the minimality of $\delta(b)$. Thus $a = bq \in (b)$. ∎

It follows that in a Euclidean domain, every pair of elements has a GCD; the division algorithm is the basis of an algorithm for calculating the GCD (B.4.5 below).

B.4.4 Lemma *In a commutative ring* R *with unity, if* $a = bq + r$ *then* $(a) + (b) = (b) + (r)$. *Thus if* R *is a principal ideal ring, then* $(a, b) = (b, r)$.

Proof. Since (b) is contained in both sides, the crux of the matter is to show that $(a) \subset (b) + (r)$ and $(r) \subset (a) + (b)$; these inclusions are obvious from $a = qb + r$ and $r = a + (-q)b$. The second assertion of the lemma is immediate from Definition B.3.4. ∎

B.4.5 (Euclid's algorithm) Let R be a Euclidean integral domain, with rank function δ, and let $a, b \in \mathrm{R}$; Euclid's algorithm is a technique for calculating the GCD (a, b) of a and b. We can suppose $a \neq 0$, $b \neq 0$ (if $a = 0$ or $b = 0$, who needs an algorithm?).

The first step is to 'divide b into a', that is, write $a = q_1 b + r_1$ with $r_1 = 0$ or $\delta(r_1) < \delta(b)$. By the lemma, $(a, b) = (b, r_1)$. If $r_1 = 0$ the procedure comes to an end: $b|a$ and $(a, b) = (b, 0) = b$.

If $r_1 \neq 0$, divide it into b (the preceding 'divisor'): write $b = q_2 r_1 + r_2$ with $r_2 = 0$ or $\delta(r_2) < \delta(r_1)$. By the lemma, $(b, r_1) = (r_1, r_2)$, in other words, $(a, b) = (r_1, r_2)$. If $r_2 = 0$ the procedure ends: $(a, b) = (r_1, 0) = r_1$.

If $r_2 \neq 0$, divide it into r_1 (the preceding 'divisor'): write $r_1 = q_3 r_2 + r_3$ with $r_3 = 0$ or $\delta(r_3) < \delta(r_2)$. By the lemma, $(r_1, r_2) = (r_2, r_3)$, in other words, $(a, b) = (r_2, r_3)$. If $r_3 = 0$ the procedure ends: $(a, b) = (r_2, 0) = r_2$.

Since $\delta(b) > \delta(r_1) > \delta(r_2) > \delta(r_3) > \ldots$, this procedure must end in a finite number of steps, say n; then $r_n = 0$ and $(a, b) = (r_{n-1}, 0) = r_{n-1}$, thus (a, b) is *the last nonzero remainder* in the procedure.

A fringe benefit of Euclid's algorithm is an explicit formula expressing (a, b) as a linear combination of a and b. If the procedure stops at the first step, then $(a, b) = 0a + 1b$; if it stops at the second step, then $(a, b) = r_1 = 1a + (-q_1)b$. Suppose the procedure continues further. At step $i > 2$ one has $r_i = r_{i-2} - q_i r_{i-1}$, which expresses r_i in terms of earlier remainders; starting at $(a, b) = r_{n-1} = r_{n-3} - q_{n-1} r_{n-2}$ and working backwards through these formulas, (a, b) is eventually expressed as a linear combination of a and b.

To illustrate the method in $\mathbf{Z}$, let $a = 58$, $b = 12$. Keeping track of a, b and remainders by means of overbars, we have

$$\overline{58} = 4 \cdot \overline{12} + \overline{10}, \quad \overline{12} = 1 \cdot \overline{10} + \bar{2}, \quad \overline{10} = 5 \cdot \bar{2} + 0,$$

therefore

$$\bar{2} = 1 \cdot \overline{12} + (-1)\overline{10} = 1 \cdot \overline{12} + (-1)\{\overline{58} + (-4)\overline{12}\} = 5 \cdot \overline{12} + (-1)\overline{58}$$

thus $(12, 58) = 5 \cdot 12 + (-1) \cdot 58$.

B.5 Factorization in overfields

If F is a field and $p \in F[t]$ is a nonconstant polynomial, then there exists an overfield $K \supset F$ such that, in the polynomial ring $K[t]$, p factors completely into linear factors. The details are as follows.

B.5.1 Lemma *If R is a principal ideal ring and $a \in R$ is a prime, then the quotient ring $R/(a)$ is a field.*

Proof. Let $u \in R/(a)$, say $u = x + (a)$. Let (a, x) be the GCD of a and x (B.3.4). Since a is prime, either $(a, x) = 1$ or $(a, x) = a$ (cf. B.3.11, B.3.10). If $u \neq 0$ then a does not divide x, so $(a, x) = a$ is ruled out and we must have $(a, x) = 1$; then $1 = ra + sx$ for suitable $r, s \in R$, and passage to quotients modulo the ideal (a) shows that u is invertible (with inverse equal to $s + (a)$). ■

B.5.2 Lemma *If $p \in F[t]$ is nonconstant then there exists an overfield $K \supset F$ in which p has a root.*

Proof. Since $F[t]$ is a principal ideal ring (B.4.2, B.4.3), p has a prime (= irreducible, by B.3.11) divisor q in $F[t]$. Let $(q) = F[t]q$ be the principal ideal of $F[t]$ generated by q and let $K = F[t]/(q)$ be the quotient ring. By the preceding lemma, K is a field.

Write $f^\# = f + (q)$ for $f \in F[t]$; thus $f \mapsto f^\#$ is the quotient mapping $F[t] \to K$. In particular, $q^\# = 0^\#$ is the zero element of K, and since $p \in (q)$ we have also $p^\# = 0^\#$; we are beginning to smell 'a root of p in K', but there are some notations to wade through before things are crystal clear.

Regard F as a subring of $F[t]$ (identify $a \in F$ with the constant polynomial $a1$). The restriction of the quotient mapping $f \mapsto f^\#$ to F is a monomorphism of rings $F \to K$; write $F^\# = \{a^\# : a \in F\}$ for the image of F, which is a subfield of K that is isomorphic to F.

Say $p = a_0 1 + a_1 t + \ldots + a_m t^m$. Applying the quotient homomorphism, we have

$$(*) \qquad 0^\# = a_0{}^\# 1^\# + a_1{}^\# t^\# + \ldots + a_m{}^\# (t^\#)^m,$$

thus $t^\#$ is a root of a polynomial with coefficients in the subfield $F^\#$ of K. There remains one bit of formalism to pin down.

The isomorphism $a \mapsto a^\#$ of F with $F^\#$ extends to an isomorphism $F[t] \to F^\#[u]$ of the polynomial rings for which $t \mapsto u$; under this isomorphism $F[t] \to F^\#[u]$, the polynomial

$$f = b_0 1 + b_1 t + \ldots + b_r t^r \in F[t]$$

corresponds to the polynomial (let us call it $\tilde{f}$)

$$\tilde{f} = b_0{}^\# 1 + b_1{}^\# u + \ldots + b_r{}^\# u^r \in F^\#[u].$$

In particular,

$$\tilde{p} = a_0{}^\# 1 + a_1{}^\# u + \ldots + a_m{}^\# u^m.$$

We are now ready to interpret the equation $(*)$: K is an overfield of $F^\#$, $t^\# \in K$ and $\tilde{p}(t^\#) = 0^\#$. Thus, identifying F with $F^\#$, and p with $\tilde{p}$ (via the isomorphisms described above), we have constructed an overfield K of F in which p has a root $t^\#$. ■

B.5.3 Theorem *If $p \in F[t]$ is a nonconstant polynomial with coefficients in a field F, then there exists an overfield $K \supset F$ such that p is a product of linear factors in the polynomial ring $K[t]$.*

Proof. The proof is by induction on $n = \deg p$. If $n = 1$ there is nothing to prove. Let $n \geq 2$ and assume that all is well for polynomials of degree $n - 1$. By the preceding lemma, there exists an overfield $G \supset F$ of F in which p has a root c. By the 'factor theorem', $p = (t - c)q$, where q has degree $n - 1$. {Apply the division algorithm in the Euclidean domain $G[t]$: $p = (t - c)q + r$, where $r = 0$ or $\deg r < 1$; substituting c for t, conclude that $r = 0$ is the only possibility.} By the induction hypothesis, there is an overfield $K \supset G$ in which q factors completely into linear factors. ■

With the aid of the preceding theorem, Euclid's algorithm yields a simple test for repeated roots:

B.5.4 Theorem *If $p \in \mathrm{F}[t]$ is nonconstant, then p has a repeated root in some overfield $\mathrm{K} \supset \mathrm{F} \Leftrightarrow (p, p') \neq 1$.*

Proof. Here (p, p') is the greatest common divisor (GCD) of p and the 'derivative' polynomial p' (formed by the usual rules); the GCD (unique up to a nonzero constant factor) is easily calculated using Euclid's algorithm (B.4.5). Since (p, p') is the generator of the ideal $(p) + (p')$, it is nonzero; thus $(p, p') = 1$ means that (p, p') is a nonzero constant, and $(p, p') \neq 1$ means that (p, p') is a nonconstant polynomial.

Note that if $f, g \in \mathrm{F}[t]$ and K is an overfield of F, then (f, g) is the same whether calculated in $\mathrm{F}[t]$ or in $\mathrm{K}[t]$; for, Euclid's algorithm leads to the same GCD whether we view the computation as taking place in $\mathrm{F}[t]$ or in $\mathrm{K}[t]$.

$\Rightarrow$: Suppose p has a repeated root c in some overfield K of F. Then $p = (t - c)^2 q$ for a suitable polynomial $q \in \mathrm{K}[t]$, so

$$p' = (t - c)^2 q' + 2(t - c)q = (t - c)q_1$$

with $q_1 \in \mathrm{K}[t]$. Thus $t - c$ divides both p and p' in $\mathrm{K}[t]$, so $(p, p') \neq 1$ in $\mathrm{K}[t]$; since p and p' both belong to $\mathrm{F}[t]$, it follows that $(p, p') \neq 1$ in $\mathrm{F}[t]$.

$\Leftarrow$: Let $f = (p, p')$ and suppose f is nonconstant. By the preceding theorem (or its lemma), there exists an overfield K of F in which f has a root c; since f is a common divisor of p and p', c is a root of both p and p'.

Write $p = (t - c)^m p_1$, where $p_1 \in \mathrm{K}[t]$ and $p_1(c) \neq 0$ (thus $m \geq 1$ is the 'multiplicity' of c in p); to complete the proof, we need only show that $m \geq 2$. Assuming to the contrary that $m = 1$, we have

$$p' = (t - c)p_1' + p_1,$$

so $p'(c) = p_1(c) \neq 0$; this contradicts the fact that c is a root of p'. ■

C

Weierstrass–Bolzano theorem

The Weierstrass–Bolzano theorem says that *every bounded sequence of real numbers has a convergent subsequence*. The theorem is generally presented early in a course in real analysis (it is just a step away from the 'completeness axiom'[1] for the field of real numbers). Since the theorem plays such an important role in the Spectral Theorem (cf. 9.2.11, 12.8.4), its proof is reviewed here for the reader's convenience.

Lemma 1 *Every bounded monotone (increasing or decreasing) sequence of real numbers is convergent.*

Proof. Suppose, for example, that (a_n) is bounded and increasing. Among all upper bounds for the sequence there is a smallest, say a: (1) $a_n \leqslant a$ for all n, and (2) if $a_n \leqslant b$ for all n then $a \leqslant b$. Expressing (2) in contrapositive form, if $b < a$ then $b < a_n$ for at least one index n.

We assert that (a_n) converges to a. Given any $\varepsilon > 0$, choose an index N such that $a - \varepsilon < a_N$. Since the sequence is increasing and is bounded above by a, $a_N \leqslant a_n \leqslant a$ for all $n \geqslant N$, thus $a - \varepsilon < a_n < a + \varepsilon$ for all $n \geqslant N$ and the assertion is proved. ■

Lemma 2 *Every sequence of real numbers has a monotone subsequence.*

Proof. Let (a_n) be any sequence of real numbers. Call an index n a *peak point* for the sequence if $a_n \geqslant a_i$ for all $i \geqslant n$. If there are infinitely many peak points $n_1 < n_2 < n_3 < \ldots$ then the subsequence (a_{n_k}) of (a_n) is decreasing. The alternative is that, from some index N onward, there are no peak points; assuming this to be the case, we construct an increasing subsequence as follows. Let $n_1 > N$. Since n_1 is not a peak point, there is an index $n_2 > n_1$ such that $a_{n_2} > a_{n_1}$; since n_2 is not a peak point, there is an index $n_3 > n_2$ such that $a_{n_3} > a_{n_2}$, etc. ■

[1] K. A. Ross, *Elementary analysis: The theory of calculus*, Springer, 1980, p. 17.

Theorem (Weierstrass–Bolzano) *Every bounded sequence of real numbers has a convergent subsequence.*

Proof.[2] Select a monotone subsequence (Lemma 2) and cite Lemma 1. ■

[2] This elegant proof is taken from W. Maak, *An introduction to modern calculus*, Holt, 1963, p. 30.

Index of notations

Notation	Reference	Meaning
E^{n}	5.1.2	Euclidean n-space
$\|x\|$	5.1.5, 12.1.7	norm of a vector x
$x \perp y$	5.1.11, 12.2.1	orthogonal vectors
$\mathrm{A} \perp \mathrm{B}$	5.1.13, 12.2.3	orthogonal subsets
$\mathrm{J}_{\mathrm{E}}: \mathrm{E} \to \mathrm{E}'$	5.2.6, 12.1.15	canonical mapping of an IP-space E
$\mathrm{M}^{\perp}$	5.2.7	orthogonal complement of M
T^*	5.3.2, 12.4.1	adjoint of a linear mapping T
$\|A\|$, det A	6.1.3, 6.3.5	determinant of A (2×2, 3×3)
$x \times y$	6.2.2	cross product of vectors in $\mathbf{R}^3$
p_A	6.4.1, 8.3.1	characteristic polynomial of A
$\mathcal{A}(\mathrm{V})$	7.1.3	space of alternate forms
$\mathbf{S}_{\mathrm{n}}$	7.1.5	symmetric group of degree n
sgn σ	7.1.9	sign of a permutation σ
$\mathbf{A}_{\mathrm{n}}$	7.1.13	alternating group
$\|T\|$, det T	7.2.1	determinant of a linear mapping T
$\|A\|$, det A	7.3.1	determinant of a square matrix A
tr A	7.3.7	trace of a matrix A
adj A	7.4.3	adjoint matrix of A
$\rho(A)$, $\nu(A)$	§7.4, Exercise 6	rank and nullity of a matrix A
$A \sim B$	8.1.1	similarity of matrices
$\{T\}'$	§8.2, Exercise 4	commutant of a linear mapping T
$\{T\}''$	§8.2, Exercise 4	bicommutant of a linear mapping T
p_T	8.3.7	characteristic polynomial of T
$\sigma(T)$	§8.3, Exercise 9	spectrum of a linear mapping T
$\mathrm{M} \oplus \mathrm{N}$	9.1.1	direct sum of subspaces M, N
P_{M}	§9.1, Exercise 7	projection with range M
$\|T\|$	9.2.5	norm of a linear mapping T
$\mathcal{S}'$	§9.4, Exercise 12	commutant of a set of linear maps
$\mathcal{S}''$	§9.4, Exercise 12	bicommutant of a set of linear maps
U_{R}	10.1.3	group of units of the ring R
A equiv B	10.2.2	equivalence of matrices
$\mathrm{D}_{\mathrm{r}}(A)$	10.4.1	determinantal divisors of A
m_T	§11.1, Exercise 1	minimal polynomial of $T \in \mathcal{L}(\mathrm{V})$
C_p	11.2.1	companion matrix of p
m_A	11.3.4	minimal polynomial of a matrix A
$\mathrm{J}_{\mathrm{m}}(c)$	11.4.5	Jordan block matrix of order m
$T \geq \mathrm{O}$	12.7.11	positive linear mapping
$\sqrt{T}$	12.7.13	square root of a positive mapping
$A \geq \mathrm{O}$	§12.7, Exercise 11	positive matrix
$\mathrm{V} \otimes \mathrm{W}$	13.1.1	tensor product of vector spaces
$x \otimes y$	13.1.3	tensor product of vectors
$S \otimes T$	13.2.2	tensor product of linear mappings
$A \otimes B$	13.3.2	Kronecker product of matrices

Index

Errata and Comments

Errata

p. xiii, ℓ. −5. For "301" read "307".

p. 53, ℓ. 19, (1) of proof.

"…by assumption, A is nonempty…" (suppress comma after the A)

p. 73, ℓ. −4.

"…subspaces…" (restore missing 's')

p. 77, Exer. 15.

"The following…" (restore missing 'n')

p. 136, Exer. 10, (ii).

Read: "(ii) $(A \cup B)^{\perp} = A^{\perp} \cap B^{\perp} \subset (A+B)^{\perp}$, with equality when $A \cap B \neq \varnothing$."

I owe this correction, and the clever observation about equality, to R. Ramabadran (Ahmedabad, India).

The inclusion can be proper if $A \cap B = \varnothing$. For example, in the Euclidean plane $E = \mathbf{R}^2$, let $A = \{e_1\}$, $B = \{e_2\}$. Then $A^{\perp} \cap B^{\perp} = \{\theta\}$ but $\{e_1 + e_2\}^{\perp}$ is 1-dimensional.

Proof of $(A+B)^{\perp} \subset A^{\perp} \cap B^{\perp}$ when $A \cap B \neq \varnothing$: Suppose $c \in A \cap B$. Let $x \in (A+B)^{\perp}$. Then $x \perp c + c$, whence $x \perp c$. If $a \in A$, write $a = (a+c) - c$; since $x \perp a + c$ and $x \perp c$, also $x \perp a$; thus $x \in A^{\perp}$. Similarly $x \in B^{\perp}$.

p. 272, ℓ. 20.

Read 11.1.8 (instead of 11.1.18).

p. 332, ℓ. 13.

"…one demonstrates…" (restore missing 'r')

p. 332, ℓ. 21.

"…demonstrated…" (restore missing 'r')

Comments

p. 96, Def. 4.2.2.

Without fidelity to the historical order of events, let us savor the power of a well-chosen notation and terminology. Given a linear mapping T on a finite-dimensional vector space, its effect on any vector is completely determined by its effect on the vectors of a given basis; the organization of the resulting doubly indexed coefficients into rows and columns of a square matrix (4.2.1) is a felicitous move that leads to powerful tools for the study of T.

Matrices and linear mappings appear to be coming from different environments: systems of linear equations (coefficient matrices) and calculus (differentiation, integration), respectively. Each comes with a concept of multiplication, arising from substitution-of-variables and composition, respectively.

By a guided miracle, the two multiplications are in harmony (4.3.3) and the same is true of the linear operations (4.2.7); thus, with respect to any chosen basis of an n-dimensional space over a field F, the correspondence $T \mapsto \mathrm{M}_T$ (in the notation of 4.2.7) is an isomorphism of the algebras(*) $\mathscr{L}(\mathrm{V})$ and $\mathrm{M}_n(\mathrm{F})$ (§2.3, §4.4). Consequently a property of the matrix M_T proved in the context of the algebra $\mathrm{M}_n(\mathrm{F})$ is reflected in a property of T in the context of the algebra $\mathscr{L}(\mathrm{V})$, which in turn is reflected in a property of the matrix of T relative to any other basis—in other words, of any matrix similar to M_T (8.1.3).

The study of the solution of n linear equations in n unknowns (§4.8) leads readily (cf., Ch. 6) to the concept of determinant of a square matrix, a predictor of invertibility (7.2.6), and providing a formula for calculating the inverse when it exists (7.4.5) and building blocks for canonical forms (10.4.1, 11.1.1, 11.2.4). In particular, the Fundamental Theorem of Similarity (11.1.7) provides an algorithm for deciding when two matrices represent the same linear mapping relative to two different bases (11.1.8).

p. 199, Def. 7.2.1.

The treatment of determinants in Chapter 7 via multilinear algebra may be found in the "Elements" of N. Bourbaki (Book II, *Algebra I*, Ch. III, §8). I learned of it in a course on linear algebra taught by Irving Kaplansky at the University of Chicago in the Fall of 1950, on which Part II depends heavily, including, in particular, the theory of similarity exposed in Chapter 11, a Chicago legacy passed down in the books of Leonard E. Dickson (*Modern algebraic theories*, B. H. Sanborn & Co., Chicago, 1926, Ch. V) and A. Adrian Albert (*Modern higher algebra*, The University of Chicago Press, Chicago, 1937, Ch. IV).

p. 205, Th. 7.4.2.

The formula for the expansion of a determinant by cofactors can be proved by showing that the right side of the formula has the formal properties that characterize the determinant function (7.3.9). A proof is written out in the author's article posted at the University of Texas archival web site *http://www.ma.utexas.edu/mp_arc* as item 11–136 (the 136th item in the folder for 2011).

p. 223, Exer. 5, (i).

Hint: p. 84, Exer. 7. This leads to a determinant-free proof that every linear mapping on a finite-dimensional vector space over an algebraically closed field has an eigenvalue (p. 224, Exer. 8).

The existence of such a proof has served as the basis of an interesting exposition of linear algebra by S. Axler ("Down with determinants", *Amer. Math. Monthly*, **102** (1995), 139–154; *Linear algebra done right*, Springer–Verlag, 1996, 2nd. edn. 1997). Particularly effective is the use of the concept of 'generalized eigenspace' (cf. W. H. Greub, *Linear algebra*, Springer–Verlag, 3rd. edn. 1967, p. 375*ff*).

S.K.B. (August 27, 2013)

(*) An *algebra* is a vector space equipped with a 'multiplication' $(x, y) \mapsto xy$ that is *associative* ($x(yz) = (xy)z$ for all x, y, z), *distributive* ($x(y+z) = xy + xz$ and $(x+y)z = xz + yz$ for all x, y, z)), and *compatible* with multiplication by a scalar, in the sense that $c(xy) = (cx)y = x(cy)$ for all scalars c and all vectors x, y.

A CATALOG OF SELECTED
DOVER BOOKS
IN SCIENCE AND MATHEMATICS

Astronomy

CHARIOTS FOR APOLLO: The NASA History of Manned Lunar Spacecraft to 1969, Courtney G. Brooks, James M. Grimwood, and Loyd S. Swenson, Jr. This illustrated history by a trio of experts is the definitive reference on the Apollo spacecraft and lunar modules. It traces the vehicles' design, development, and operation in space. More than 100 photographs and illustrations. 576pp. 6 3/4 x 9 1/4. 0-486-46756-2

EXPLORING THE MOON THROUGH BINOCULARS AND SMALL TELESCOPES, Ernest H. Cherrington, Jr. Informative, profusely illustrated guide to locating and identifying craters, rills, seas, mountains, other lunar features. Newly revised and updated with special section of new photos. Over 100 photos and diagrams. 240pp. 8 1/4 x 11. 0-486-24491-1

WHERE NO MAN HAS GONE BEFORE: A History of NASA's Apollo Lunar Expeditions, William David Compton. Introduction by Paul Dickson. This official NASA history traces behind-the-scenes conflicts and cooperation between scientists and engineers. The first half concerns preparations for the Moon landings, and the second half documents the flights that followed Apollo 11. 1989 edition. 432pp. 7 x 10. 0-486-47888-2

APOLLO EXPEDITIONS TO THE MOON: The NASA History, Edited by Edgar M. Cortright. Official NASA publication marks the 40th anniversary of the first lunar landing and features essays by project participants recalling engineering and administrative challenges. Accessible, jargon-free accounts, highlighted by numerous illustrations. 336pp. 8 3/8 x 10 7/8. 0-486-47175-6

ON MARS: Exploration of the Red Planet, 1958-1978--The NASA History, Edward Clinton Ezell and Linda Neuman Ezell. NASA's official history chronicles the start of our explorations of our planetary neighbor. It recounts cooperation among government, industry, and academia, and it features dozens of photos from Viking cameras. 560pp. 6 3/4 x 9 1/4. 0-486-46757-0

ARISTARCHUS OF SAMOS: The Ancient Copernicus, Sir Thomas Heath. Heath's history of astronomy ranges from Homer and Hesiod to Aristarchus and includes quotes from numerous thinkers, compilers, and scholasticists from Thales and Anaximander through Pythagoras, Plato, Aristotle, and Heraclides. 34 figures. 448pp. 5 3/8 x 8 1/2. 0-486-43886-4

AN INTRODUCTION TO CELESTIAL MECHANICS, Forest Ray Moulton. Classic text still unsurpassed in presentation of fundamental principles. Covers rectilinear motion, central forces, problems of two and three bodies, much more. Includes over 200 problems, some with answers. 437pp. 5 3/8 x 8 1/2. 0-486-64687-4

BEYOND THE ATMOSPHERE: Early Years of Space Science, Homer E. Newell. This exciting survey is the work of a top NASA administrator who chronicles technological advances, the relationship of space science to general science, and the space program's social, political, and economic contexts. 528pp. 6 3/4 x 9 1/4. 0-486-47464-X

STAR LORE: Myths, Legends, and Facts, William Tyler Olcott. Captivating retellings of the origins and histories of ancient star groups include Pegasus, Ursa Major, Pleiades, signs of the zodiac, and other constellations. "Classic." – *Sky & Telescope.* 58 illustrations. 544pp. 5 3/8 x 8 1/2. 0-486-43581-4

A COMPLETE MANUAL OF AMATEUR ASTRONOMY: Tools and Techniques for Astronomical Observations, P. Clay Sherrod with Thomas L. Koed. Concise, highly readable book discusses the selection, set-up, and maintenance of a telescope; amateur studies of the sun; lunar topography and occultations; and more. 124 figures. 26 halftones. 37 tables. 335pp. 6 1/2 x 9 1/4. 0-486-42820-6

Chemistry

MOLECULAR COLLISION THEORY, M. S. Child. This high-level monograph offers an analytical treatment of classical scattering by a central force, quantum scattering by a central force, elastic scattering phase shifts, and semi-classical elastic scattering. 1974 edition. 310pp. 5 3/8 x 8 1/2. 0-486-69437-2

HANDBOOK OF COMPUTATIONAL QUANTUM CHEMISTRY, David B. Cook. This comprehensive text provides upper-level undergraduates and graduate students with an accessible introduction to the implementation of quantum ideas in molecular modeling, exploring practical applications alongside theoretical explanations. 1998 edition. 832pp. 5 3/8 x 8 1/2. 0-486-44307-8

RADIOACTIVE SUBSTANCES, Marie Curie. The celebrated scientist's thesis, which directly preceded her 1903 Nobel Prize, discusses establishing atomic character of radioactivity; extraction from pitchblende of polonium and radium; isolation of pure radium chloride; more. 96pp. 5 3/8 x 8 1/2. 0-486-42550-9

CHEMICAL MAGIC, Leonard A. Ford. Classic guide provides intriguing entertainment while elucidating sound scientific principles, with more than 100 unusual stunts: cold fire, dust explosions, a nylon rope trick, a disappearing beaker, much more. 128pp. 5 3/8 x 8 1/2. 0-486-67628-5

ALCHEMY, E. J. Holmyard. Classic study by noted authority covers 2,000 years of alchemical history: religious, mystical overtones; apparatus; signs, symbols, and secret terms; advent of scientific method, much more. Illustrated. 320pp. 5 3/8 x 8 1/2. 0-486-26298-7

CHEMICAL KINETICS AND REACTION DYNAMICS, Paul L. Houston. This text teaches the principles underlying modern chemical kinetics in a clear, direct fashion, using several examples to enhance basic understanding. Solutions to selected problems. 2001 edition. 352pp. 8 3/8 x 11. 0-486-45334-0

PROBLEMS AND SOLUTIONS IN QUANTUM CHEMISTRY AND PHYSICS, Charles S. Johnson and Lee G. Pedersen. Unusually varied problems, with detailed solutions, cover of quantum mechanics, wave mechanics, angular momentum, molecular spectroscopy, scattering theory, more. 280 problems, plus 139 supplementary exercises. 430pp. 6 1/2 x 9 1/4. 0-486-65236-X

ELEMENTS OF CHEMISTRY, Antoine Lavoisier. Monumental classic by the founder of modern chemistry features first explicit statement of law of conservation of matter in chemical change, and more. Facsimile reprint of original (1790) Kerr translation. 539pp. 5 3/8 x 8 1/2. 0-486-64624-6

MAGNETISM AND TRANSITION METAL COMPLEXES, F. E. Mabbs and D. J. Machin. A detailed view of the calculation methods involved in the magnetic properties of transition metal complexes, this volume offers sufficient background for original work in the field. 1973 edition. 240pp. 5 3/8 x 8 1/2. 0-486-46284-6

GENERAL CHEMISTRY, Linus Pauling. Revised third edition of classic first-year text by Nobel laureate. Atomic and molecular structure, quantum mechanics, statistical mechanics, thermodynamics correlated with descriptive chemistry. Problems. 992pp. 5 3/8 x 8 1/2. 0-486-65622-5

ELECTROLYTE SOLUTIONS: Second Revised Edition, R. A. Robinson and R. H. Stokes. Classic text deals primarily with measurement, interpretation of conductance, chemical potential, and diffusion in electrolyte solutions. Detailed theoretical interpretations, plus extensive tables of thermodynamic and transport properties. 1970 edition. 590pp. 5 3/8 x 8 1/2. 0-486-42225-9

Engineering

FUNDAMENTALS OF ASTRODYNAMICS, Roger R. Bate, Donald D. Mueller, and Jerry E. White. Teaching text developed by U.S. Air Force Academy develops the basic two-body and n-body equations of motion; orbit determination; classical orbital elements, coordinate transformations; differential correction; more. 1971 edition. 455pp. 5 3/8 x 8 1/2. 0-486-60061-0

INTRODUCTION TO CONTINUUM MECHANICS FOR ENGINEERS: Revised Edition, Ray M. Bowen. This self-contained text introduces classical continuum models within a modern framework. Its numerous exercises illustrate the governing principles, linearizations, and other approximations that constitute classical continuum models. 2007 edition. 320pp. 6 1/8 x 9 1/4. 0-486-47460-7

ENGINEERING MECHANICS FOR STRUCTURES, Louis L. Bucciarelli. This text explores the mechanics of solids and statics as well as the strength of materials and elasticity theory. Its many design exercises encourage creative initiative and systems thinking. 2009 edition. 320pp. 6 1/8 x 9 1/4. 0-486-46855-0

FEEDBACK CONTROL THEORY, John C. Doyle, Bruce A. Francis and Allen R. Tannenbaum. This excellent introduction to feedback control system design offers a theoretical approach that captures the essential issues and can be applied to a wide range of practical problems. 1992 edition. 224pp. 6 1/2 x 9 1/4. 0-486-46933-6

THE FORCES OF MATTER, Michael Faraday. These lectures by a famous inventor offer an easy-to-understand introduction to the interactions of the universe's physical forces. Six essays explore gravitation, cohesion, chemical affinity, heat, magnetism, and electricity. 1993 edition. 96pp. 5 3/8 x 8 1/2. 0-486-47482-8

DYNAMICS, Lawrence E. Goodman and William H. Warner. Beginning engineering text introduces calculus of vectors, particle motion, dynamics of particle systems and plane rigid bodies, technical applications in plane motions, and more. Exercises and answers in every chapter. 619pp. 5 3/8 x 8 1/2. 0-486-42006-X

ADAPTIVE FILTERING PREDICTION AND CONTROL, Graham C. Goodwin and Kwai Sang Sin. This unified survey focuses on linear discrete-time systems and explores natural extensions to nonlinear systems. It emphasizes discrete-time systems, summarizing theoretical and practical aspects of a large class of adaptive algorithms. 1984 edition. 560pp. 6 1/2 x 9 1/4. 0-486-46932-8

INDUCTANCE CALCULATIONS, Frederick W. Grover. This authoritative reference enables the design of virtually every type of inductor. It features a single simple formula for each type of inductor, together with tables containing essential numerical factors. 1946 edition. 304pp. 5 3/8 x 8 1/2. 0-486-47440-2

THERMODYNAMICS: Foundations and Applications, Elias P. Gyftopoulos and Gian Paolo Beretta. Designed by two MIT professors, this authoritative text discusses basic concepts and applications in detail, emphasizing generality, definitions, and logical consistency. More than 300 solved problems cover realistic energy systems and processes. 800pp. 6 1/8 x 9 1/4. 0-486-43932-1

THE FINITE ELEMENT METHOD: Linear Static and Dynamic Finite Element Analysis, Thomas J. R. Hughes. Text for students without in-depth mathematical training, this text includes a comprehensive presentation and analysis of algorithms of time-dependent phenomena plus beam, plate, and shell theories. Solution guide available upon request. 672pp. 6 1/2 x 9 1/4. 0-486-41181-8

Mathematics–Bestsellers

HANDBOOK OF MATHEMATICAL FUNCTIONS: with Formulas, Graphs, and Mathematical Tables, Edited by Milton Abramowitz and Irene A. Stegun. A classic resource for working with special functions, standard trig, and exponential logarithmic definitions and extensions, it features 29 sets of tables, some to as high as 20 places. 1046pp. 8 x 10 1/2. 0-486-61272-4

ABSTRACT AND CONCRETE CATEGORIES: The Joy of Cats, Jiri Adamek, Horst Herrlich, and George E. Strecker. This up-to-date introductory treatment employs category theory to explore the theory of structures. Its unique approach stresses concrete categories and presents a systematic view of factorization structures. Numerous examples. 1990 edition, updated 2004. 528pp. 6 1/8 x 9 1/4. 0-486-46934-4

MATHEMATICS: Its Content, Methods and Meaning, A. D. Aleksandrov, A. N. Kolmogorov, and M. A. Lavrent'ev. Major survey offers comprehensive, coherent discussions of analytic geometry, algebra, differential equations, calculus of variations, functions of a complex variable, prime numbers, linear and non-Euclidean geometry, topology, functional analysis, more. 1963 edition. 1120pp. 5 3/8 x 8 1/2. 0-486-40916-3

INTRODUCTION TO VECTORS AND TENSORS: Second Edition–Two Volumes Bound as One, Ray M. Bowen and C.-C. Wang. Convenient single-volume compilation of two texts offers both introduction and in-depth survey. Geared toward engineering and science students rather than mathematicians, it focuses on physics and engineering applications. 1976 edition. 560pp. 6 1/2 x 9 1/4. 0-486-46914-X

AN INTRODUCTION TO ORTHOGONAL POLYNOMIALS, Theodore S. Chihara. Concise introduction covers general elementary theory, including the representation theorem and distribution functions, continued fractions and chain sequences, the recurrence formula, special functions, and some specific systems. 1978 edition. 272pp. 5 3/8 x 8 1/2. 0-486-47929-3

ADVANCED MATHEMATICS FOR ENGINEERS AND SCIENTISTS, Paul DuChateau. This primary text and supplemental reference focuses on linear algebra, calculus, and ordinary differential equations. Additional topics include partial differential equations and approximation methods. Includes solved problems. 1992 edition. 400pp. 7 1/2 x 9 1/4. 0-486-47930-7

PARTIAL DIFFERENTIAL EQUATIONS FOR SCIENTISTS AND ENGINEERS, Stanley J. Farlow. Practical text shows how to formulate and solve partial differential equations. Coverage of diffusion-type problems, hyperbolic-type problems, elliptic-type problems, numerical and approximate methods. Solution guide available upon request. 1982 edition. 414pp. 6 1/8 x 9 1/4. 0-486-67620-X

VARIATIONAL PRINCIPLES AND FREE-BOUNDARY PROBLEMS, Avner Friedman. Advanced graduate-level text examines variational methods in partial differential equations and illustrates their applications to free-boundary problems. Features detailed statements of standard theory of elliptic and parabolic operators. 1982 edition. 720pp. 6 1/8 x 9 1/4. 0-486-47853-X

LINEAR ANALYSIS AND REPRESENTATION THEORY, Steven A. Gaal. Unified treatment covers topics from the theory of operators and operator algebras on Hilbert spaces; integration and representation theory for topological groups; and the theory of Lie algebras, Lie groups, and transform groups. 1973 edition. 704pp. 6 1/8 x 9 1/4. 0-486-47851-3

A SURVEY OF INDUSTRIAL MATHEMATICS, Charles R. MacCluer. Students learn how to solve problems they'll encounter in their professional lives with this concise single-volume treatment. It employs MATLAB and other strategies to explore typical industrial problems. 2000 edition. 384pp. 5 3/8 x 8 1/2. 0-486-47702-9

NUMBER SYSTEMS AND THE FOUNDATIONS OF ANALYSIS, Elliott Mendelson. Geared toward undergraduate and beginning graduate students, this study explores natural numbers, integers, rational numbers, real numbers, and complex numbers. Numerous exercises and appendixes supplement the text. 1973 edition. 368pp. 5 3/8 x 8 1/2. 0-486-45792-3

A FIRST LOOK AT NUMERICAL FUNCTIONAL ANALYSIS, W. W. Sawyer. Text by renowned educator shows how problems in numerical analysis lead to concepts of functional analysis. Topics include Banach and Hilbert spaces, contraction mappings, convergence, differentiation and integration, and Euclidean space. 1978 edition. 208pp. 5 3/8 x 8 1/2. 0-486-47882-3

FRACTALS, CHAOS, POWER LAWS: Minutes from an Infinite Paradise, Manfred Schroeder. A fascinating exploration of the connections between chaos theory, physics, biology, and mathematics, this book abounds in award-winning computer graphics, optical illusions, and games that clarify memorable insights into self-similarity. 1992 edition. 448pp. 6 1/8 x 9 1/4. 0-486-47204-3

SET THEORY AND THE CONTINUUM PROBLEM, Raymond M. Smullyan and Melvin Fitting. A lucid, elegant, and complete survey of set theory, this three-part treatment explores axiomatic set theory, the consistency of the continuum hypothesis, and forcing and independence results. 1996 edition. 336pp. 6 x 9. 0-486-47484-4

DYNAMICAL SYSTEMS, Shlomo Sternberg. A pioneer in the field of dynamical systems discusses one-dimensional dynamics, differential equations, random walks, iterated function systems, symbolic dynamics, and Markov chains. Supplementary materials include PowerPoint slides and MATLAB exercises. 2010 edition. 272pp. 6 1/8 x 9 1/4. 0-486-47705-3

ORDINARY DIFFERENTIAL EQUATIONS, Morris Tenenbaum and Harry Pollard. Skillfully organized introductory text examines origin of differential equations, then defines basic terms and outlines general solution of a differential equation. Explores integrating factors; dilution and accretion problems; Laplace Transforms; Newton's Interpolation Formulas, more. 818pp. 5 3/8 x 8 1/2. 0-486-64940-7

MATROID THEORY, D. J. A. Welsh. Text by a noted expert describes standard examples and investigation results, using elementary proofs to develop basic matroid properties before advancing to a more sophisticated treatment. Includes numerous exercises. 1976 edition. 448pp. 5 3/8 x 8 1/2. 0-486-47439-9

THE CONCEPT OF A RIEMANN SURFACE, Hermann Weyl. This classic on the general history of functions combines function theory and geometry, forming the basis of the modern approach to analysis, geometry, and topology. 1955 edition. 208pp. 5 3/8 x 8 1/2. 0-486-47004-0

THE LAPLACE TRANSFORM, David Vernon Widder. This volume focuses on the Laplace and Stieltjes transforms, offering a highly theoretical treatment. Topics include fundamental formulas, the moment problem, monotonic functions, and Tauberian theorems. 1941 edition. 416pp. 5 3/8 x 8 1/2. 0-486-47755-X

Mathematics–Logic and Problem Solving

Mathematics–Algebra and Calculus

VECTOR CALCULUS, Peter Baxandall and Hans Liebeck. This introductory text offers a rigorous, comprehensive treatment. Classical theorems of vector calculus are amply illustrated with figures, worked examples, physical applications, and exercises with hints and answers. 1986 edition. 560pp. 5 3/8 x 8 1/2. 0-486-46620-5

ADVANCED CALCULUS: An Introduction to Classical Analysis, Louis Brand. A course in analysis that focuses on the functions of a real variable, this text introduces the basic concepts in their simplest setting and illustrates its teachings with numerous examples, theorems, and proofs. 1955 edition. 592pp. 5 3/8 x 8 1/2. 0-486-44548-8

ADVANCED CALCULUS, Avner Friedman. Intended for students who have already completed a one-year course in elementary calculus, this two-part treatment advances from functions of one variable to those of several variables. Solutions. 1971 edition. 432pp. 5 3/8 x 8 1/2. 0-486-45795-8

METHODS OF MATHEMATICS APPLIED TO CALCULUS, PROBABILITY, AND STATISTICS, Richard W. Hamming. This 4-part treatment begins with algebra and analytic geometry and proceeds to an exploration of the calculus of algebraic functions and transcendental functions and applications. 1985 edition. Includes 310 figures and 18 tables. 880pp. 6 1/2 x 9 1/4. 0-486-43945-3

BASIC ALGEBRA I: Second Edition, Nathan Jacobson. A classic text and standard reference for a generation, this volume covers all undergraduate algebra topics, including groups, rings, modules, Galois theory, polynomials, linear algebra, and associative algebra. 1985 edition. 528pp. 6 1/8 x 9 1/4. 0-486-47189-6

BASIC ALGEBRA II: Second Edition, Nathan Jacobson. This classic text and standard reference comprises all subjects of a first-year graduate-level course, including in-depth coverage of groups and polynomials and extensive use of categories and functors. 1989 edition. 704pp. 6 1/8 x 9 1/4. 0-486-47187-X

CALCULUS: An Intuitive and Physical Approach (Second Edition), Morris Kline. Application-oriented introduction relates the subject as closely as possible to science with explorations of the derivative; differentiation and integration of the powers of x; theorems on differentiation, antidifferentiation; the chain rule; trigonometric functions; more. Examples. 1967 edition. 960pp. 6 1/2 x 9 1/4. 0-486-40453-6

ABSTRACT ALGEBRA AND SOLUTION BY RADICALS, John E. Maxfield and Margaret W. Maxfield. Accessible advanced undergraduate-level text starts with groups, rings, fields, and polynomials and advances to Galois theory, radicals and roots of unity, and solution by radicals. Numerous examples, illustrations, exercises, appendixes. 1971 edition. 224pp. 6 1/8 x 9 1/4. 0-486-47723-1

AN INTRODUCTION TO THE THEORY OF LINEAR SPACES, Georgi E. Shilov. Translated by Richard A. Silverman. Introductory treatment offers a clear exposition of algebra, geometry, and analysis as parts of an integrated whole rather than separate subjects. Numerous examples illustrate many different fields, and problems include hints or answers. 1961 edition. 320pp. 5 3/8 x 8 1/2. 0-486-63070-6

LINEAR ALGEBRA, Georgi E. Shilov. Covers determinants, linear spaces, systems of linear equations, linear functions of a vector argument, coordinate transformations, the canonical form of the matrix of a linear operator, bilinear and quadratic forms, and more. 387pp. 5 3/8 x 8 1/2. 0-486-63518-X